MODERN PERSPECTIVES IN ENERGY
Series Editors: David J. Rose, Richard K. Lester, and John Andelin

CARBON
NITROGEN
SULFUR

Human Interference in Grand Biospheric Cycles

VACLAV SMIL

The University of Manitoba
Winnipeg, Manitoba, Canada

The effects of human interference in the grand biospheric cycles of carbon, nitrogen, and sulfur—evidenced by concerns about global climatic change, nitrates in the water and food, and acid rain—have come to international attention inside and outside the scientific community.

Here, Vaclav Smil brings to bear more than twenty years of experience in researching these cycles, presenting a broad overview of the enormous advances made in recent years in understanding both the intricacies of the cycles themselves and the effects of human interfernce in them. Smil assess the situation as it now stands and offers a critical perspective of what might be done—and what should not be done—in the future.

A signal achievement in the field, this book will be essential reading for those involved in environmental engineering, chemical engineering, pollution research, ecology, and related fields of inquiry.

CARBON NITROGEN SULFUR

HUMAN INTERFERENCE IN GRAND BIOSPHERIC CYCLES

MODERN PERSPECTIVES IN ENERGY

Series Editors: David J. Rose, Richard K. Lester, and John Andelin

ENERGY : The Conservation Revolution
John H. Gibbons and William U. Chandler

STRUCTURAL MATERIALS IN NUCLEAR POWER SYSTEMS
J. T. Adrian Roberts

BIOMASS ENERGIES: Resources, Links, Constraints
Vaclav Smil

CARBON–NITROGEN–SULFUR: Human Interference in Grand Biospheric Cycles
Vaclav Smil

LEARNING ABOUT ENERGY
David J. Rose

CARBON
NITROGEN
SULFUR

HUMAN INTERFERENCE IN GRAND BIOSPHERIC CYCLES

Vaclav Smil

The University of Manitoba
Winnipeg, Manitoba, Canada

PLENUM PRESS • NEW YORK AND LONDON

Library of Congress Cataloging in Publication Data

Smil, Vaclav.
Carbon-nitrogen-sulfur.

(Modern perspectives in energy)
Bibliography: p.
Includes index.
1. Energy development—Environmental aspects. 2. Carbon cycle (Biogeochemistry) —Environmental aspects. 3. Nitrogen cycle—Environmental aspects. 4. Sulphur—Environmental aspects. 5. Man—Influence on nature. I. Title. II. Series.
TD195.E49S65 1985 363.7'38 85-25758
ISBN 978-1-4684-8841-8 ISBN 978-1-4684-8839-5 (eBook)
DOI 10.1007/978-1-4684-8839-5

Softcover reprint of the hardcover 1st edition 1985
A Division of Plenum Publishing Corporation
233 Spring Street, New York, N.Y. 10013

SETTING TRAPS FOR TIGERS

When the hunter sets traps only for rabbits,
Tigers and dragons are left uncaught.

—Li Bo, *To his three friends*
(Shigeyoshi Obata translation)

Li Bo's ancient admonition has been heeded in this book, which tries to trap some large and unruly creatures between its covers: rapidly evolving understanding of human interferences in three global biogeochemical cycles whose functioning is critical for the survival of the biosphere. Much has been written recently on these topics, almost always in multiauthored edited volumes, an obvious solution to tackle the huge scope and intricacy inherent in these studies.

No single author can have very good understanding of all far-flung outposts of this huge realm, ranging from water wells (with nitrates and risk of infant methemoglobinemia) to bronchioles (with their specific sensitivity to inhaled sulfates), from complex soil chemistry (where microorganisms fix, nitrify, denitrify volatilize, and bind nitrogen) to even more involved atmospheric reactions (where photochemistry of the free hydroxyl radical controls oxidation and removal rates for numerous trace gases).

Still, a single scientist should not be intimidated from setting the difficult trap: inevitable shortcomings accompanying one man's effort may well be balanced by a much more unified approach than even the best editing can bring to a collection of specialized contributions, and the absence of many particularistic in-depth treatments may be an advantage where a systematic elucidation of principal components and their links is a major goal.

Substantial previous experience with more than one small segment of a single cycle is, however, a must in undertaking the task and, fortunately, I have had many opportunities to study critical parts of all three cycles. Indeed, the origins of this book go back many years. My thesis at the Faculty of Natural Sciences of Carolinum University in Prague in 1965 dealt with environmental factors in location of large coal-fired power plants with a particular application to the North Bohemian Brown Coal Mining Region, one of Europe's largest concentrations of coal mining, mine-mouth thermal electricity generation, and chemical industry. This large industrial aggregation and the location of the area in a relatively deep intermountain valley lying across rather than along the direction of the prevailing winds create severe air pollution problems.

And so when the notion of acid deposition came to the forefront of scientific attention, first in Scandinavia in the early 1970s, a few years later in North America, I considered myself an old hand: when I started to study the environment of the North Bohemian region in 1963, the ecosystemic changes and health effects resulting from extremely high concentrations and deposition of sulfurous and nitrogenous air pollutants and particulate matter could not be ignored. When I returned to the area in 1966 to work there for nearly three years as a consultant in energy and environmental affairs, I came to realize the difficulties of efficiently controlling the problem.

Hiking on the crest of the Ore Mountains overlooking the valley, I saw much destruction and degradation of coniferous plantings—but I was also repeatedly surprised by the contrast of the withering tops and stunted dried-out growth of spruces and firs with the magnificent beech trees and the healthy understory of shrubs and wild flowers. I recall this impressive lesson of ecosystemic vulnerability and resistance every time I read sweeping generalizations about the environmental effects of acid deposition.

At the same time, in the second half of the 1960s, I was introduced by a friend, an engineer working in analytical chemistry and biochemistry, to some of the mysteries of enzymes; this led me to nitrogenase, one of the most incredible substances on this planet, and to an interest in various aspects of the nitrogen cycle, which was further strengthened by my later work on the energy cost of crop production, involving inevitable comparisons between natural nitrogen fixation and Haber–Bosch ammonia synthesis.

And after I came to Penn State in 1969, computers brought me to the carbon cycle: I had never worked with them before and was quickly captivated by the simulation opportunities they offered. My continuing interest in complex energy and environment systems made the carbon cycle a natural choice for modeling. My constructs assembled in the early 1970s were poor-looking creatures compared to the more recent work in the field of carbon cycle modeling, but they taught me much about the ubiquitous quantification gaps and approximations and made me appreciate that unending uncertainties are the dominant features of all such efforts.

Later during the 1970s I did something else, writing above all on long-range energy prospects, returning to the energy and economy of China and energy in the poor world and criticizing the prospects of renewable energies. Meanwhile, general developments caught up with my old interests and went on to surpass my never fully abandoned but slipping grasp of the three cycles. Symbolized by nitrates in drinking water, acid rain, and a warmer planet earth, the three cycles became not only subjects of extensive research by many individual scientists but also a focus of national and international scientific and government bureaucracies and welcome discoveries in the media's never-ending search for new bad news.

And so this book is many things to me: a return to some of my research beginnings, a catch-up with a decade I spent elsewhere, an opportunity to put some old knowledge to new use, a challenging learning experience, and undoubtedly the most difficult project I have undertaken so far. Its aims are clear: to survey the enormous advances made, mostly during the 1970s, in the elucidation of the intricate workings of the three vital cycles, to appraise where we stand, and to look critically at what we might be doing and what we should avoid in the future.

Thanks to the work of hundreds of fellow researchers, I have been able to gather a large amount of evidence and to consider a stimulating variety of opinions. Omissions and emphases are mine, as are the final interpretations and conclusions. Perservering readers will be best able to judge if some tigers were caught. For snaring dragons I never held much hope.

Vaclav Smil

Winnipeg

ACKNOWLEDGMENTS

First and foremost, I am indebted to more than one thousand of my colleagues, whose work made it possible to undertake this interdisciplinary inquiry into the three cycles. Then to the Social Sciences and Humanities Research Council of Canada for giving me time off from my lecturing in order to prepare most of the manuscript. Finally, my thanks to those who typed, retyped, and copied, and who drew and reproduced the graphics; alphabetically they are Mrs. H. Chambers, Mrs. B. Drebert, Mrs. E. Gaetz, Mrs. M. Halmarson, Mrs. J. Michalski, and Mr. E. Pachanuk.

CONTENTS

1 ENERGY AND THE ENVIRONMENT

We as a species, aided by the industries at our command, have now significantly altered some of the major chemical cycles of the planet. We have increased the carbon cycle by 20 percent, the nitrogen cycle by 50 percent, and the sulphur cycle by over 100 percent. As our numbers and our use of fossil fuels increase, those perturbations will grow likewise. What are the most likely consequences?

—James Lovelock
Gaia (1979)

Any retrospective look at the fundamental changes of industrial civilization in the seventh decade of the 20th century will have to note the sudden emergence of two critical concerns as energy supplies and the state of the environment were added to perennial worries ranging from economic growth and unemployment to crime and the arms race. And the longer the hindsight, I believe, the greater will be the amazement that it took so long to elevate these essential concerns to their rightfully prominent places in scientific inquiry and in the making of public policy.

Accumulated scientific knowledge pointed to such an assessment for many decades before the 1970s but rapid rates of new oil and gas discoveries and relative cheapness and continuing decline in real energy prices on the one hand, and the tolerance of deteriorating environment in the context of otherwise rising living standards and environmental degradation limited to cities and heavily industrialized regions on the other acted as powerful moderators of any acute concerns. Only when new hydrocarbon discoveries had to be made with much greater effort and expense, when OPEC's sudden crude oil price rise put an end to a generation of falling energy prices, when the unprecedented mass affluence of industrialized nations spilled over into greater attention to intangibles, and when various forms of environmental pollution started to affect not only densely settled urban areas but some of the most remote sites of this planet—only then the concerns about energy and the environment rose to their long overdue prominence.

There are countless interfaces and links between the provision of energies to modern civilization and the preservation or degradation of environment and this book will focus on the three interferences which matter most—because they involve the three largest fluxes of pollutants generated by energy conversion, because the compounds of the three elements involved move through soils and waters and atmosphere and are thus distributed on large regional, continental, and even global scales and because these anthropogenic interferences affect the biosphere's three most vulnerable grand biogeochemical cycles.

As these foundation-laying statements need more justification the rest of this introductory chapter will be devoted to their more detailed explication before moving on to detailed discussions of individual cycles and human interferences in their working.

1.1. FOSSIL-FUELED CIVILIZATION

Writing on matters of energy and society in the mid-1980s is easier than two decades ago—and much easier than in the mid-1970s. A generation ago, energy systems analysis was virtually nonexistent, writings on energy were almost exclusively confined to technical journals, and discussions of engineering problems dominated the field. Ten years ago, one may have been forgiven for believing that everyone was an energy expert and one could have gathered with ease scores of learned opinions such as predictions of the whole world trembling before OPEC in perpetuity or the planet running out of all fossil fuels within less time than Mozart took to write his piano concerti.

Compared to the 1970s, energy is of marginal interest in the mid-1980s but so much so the better. With most of the hyperbole and repetitive misinformation gone, the work—of which there is no shortage—can go on in a more sensible atmosphere. And, of course, those nearly 10 years of glowing interest *did* bring much excellent and innovative writing on complex energy problems, a foundation making further steps so much easier.

Hence, I shall assume that it will be mostly the survivors of the energy decade committed to serious study of energetics who will pick up this book—or the novices who cannot be today so easily impressed by the promises of speedy deliverance by OTECs, central solar stations, and fuel alcohols. If so, then the brevity of this section will be appreciated: in fact, its message will need only the next few paragraphs.

Industrial civilization requires huge quantities of power which must flow incessantly in direct ways to deliver heat, electricity, and speedy locomotion and indirectly in countless products synthesized, smelted, machined, and assembled with considerable fossil fuel energy subsidies. And in this civilization, even the production of food, impossible without the renewing flux of solar radiation, is critically dependent on fossil fuels which not only power the field, crop processing, and water pumping machinery but, above all, provide both feedstock and process energy for synthesis of nitrogenous fertilizers without whose extensive application it would be impossible to feed the world's current population.

The intrinsically gradual nature of the energy substitution process makes it impossible for any new energy source to capture an appreciable market share without decades of gradual climb, and this means that the current dominance of fossil fuels—even when adjustments are made for traditional uses of biomass energies, coal, oil, and gas comprise no less than 80% of the global primary energy supply—shall continue for *at least* the next half a century. Their shares will undoubtedly shift but regardless of these particulars—will coal really be resurgent, will lower use and some grand offshore discoveries extend oil's leading role, will natural gas surpass oil's share soon?—there is no doubt that in the third and fourth decades of the next century we shall still derive *most* of our primary energy needs from fossil fuels with all attendant environmental effects and complications.

That exploratory, extractive, transportation, conversion, transmission, storage, and disposal activities forming energy supply systems of modern societies have numerous environmental implications is self-evident. All occupy space, more often than not good, flat, accessible land which used to be in crops or in pasture, some by just sitting on it (power plants, transforming stations, refineries), some by leaving it largely intact but preempting other uses (pipelines, transmission lines), some by degrading it and changing it even beyond restoration (surface mines, deep mines). Some use large volumes of water and discharge it either much warmer (thermal power plants) or considerably polluted (coal cleaning plants, refineries).

But it is the combustion of fossil fuels which has the greatest environmental impact. Particulate matter mobilized by combustion is only a small fraction of natural emissions but gaseous releases have grown to notable shares. CO_2 from fossil combustion represents rapid increments much in excess of both the additional photosynthetic capability and absorption by the ocean; hence, the gas accumulates in the atmosphere and although one cannot as yet point to any definite sign of resulting climatic effects, the potential for complex changes is undeniable. SO_2 does not reside in the atmosphere long enough to become a global pollutant but stays sufficiently long to be oxidized into sulfates, transported over large regions, even across continental distances, and contribute to higher levels of acid deposition. Oxidation of nitrogen oxides will produce similar effects and the human interference with nitrogen's flows is even greater owing to fertilization of crops.

The fact that carbon, nitrogen, and sulfur are undoubtedly the three elements mobilized in the largest quantities by the managed energy conversions ranging from producing metals to growing crops and from transporting goods to cooking meals would be sufficient to give their flows special attention. But are not these interferences even more noteworthy because they affect three of the planet's key biogeochemical cycles? To answer this question one has to first evaluate the relative importance of these cycles.

1.2. RANKING THE CYCLES

Unexceptional departure in a search for critical cycles may be to state that those which would qualify must be the most vital flows enabling the continuation of life on the earth. In the foreword to the 13th SCOPE report the Chairman of the Project on Biogeochemical Cycles writes that "the carbon cycle is the most important and most complex cycle of all, as it is the pacemaker for the other cycles which in turn codetermine flow rates in the carbon cycle" (La Rivière, 1979).

Clearly, the second part of the statement undermines the first one and a few pages later one can read a better appraisal: "The biosphere, as it exists today, has evolved in a complex interplay between carbon and many other elements; primarily hydrogen, oxygen, the basic nutrient elements, nitrogen, phosphorus, sulphur, and some metals in minor quantities that are fundamental to the development of life. It is for this reason that the carbon cycle cannot be dealt with independently of the cycles of other elements involved in the biogeochemical system'' (Bolin et al., 1979).

Similarly, most participants in the 1976 symposium on the cycling of minerals in agricultural systems were reluctant to rank the importance of N, P, and K, the three top plant nutrients: ``They are all of equal importance, all three can be crop production limiting and they share this property with Ca, the trace elements and water'' (Frissel, 1977a). Nor is there much difference in terms of their theoretical availability. Nitrogen is plentiful in the atmosphere, while the earth's crust contains very large resources of potassium and the nutrient can be easily thought of as inexhaustible over a time span of even a much enduring civilization (say 10^4 years). At the present price level, phosphates would ``run out'' first but, again, there are large mineable resources of phosporous-containing rocks which will shift into the category of economic reserves once the prices go up or the extraction technologies make the task easier.

Clearly, no ranking according to importance makes sense as there can be no meaningful ordering of elements, conditions, and processes where life is concerned: departure of one of numerous critical variables beyond the limits of normal tolerance wrecks the whole structure. But one may rank the elements of the planetary biogeochemistry according to their cycling intensity, either in terms of their turnover rates or in terms of their mass flows. Again, such rankings are not very meaningful: the figures may be intriguing, often thought-provoking, but the realizations that sulfur compounds will stay airborne for only hours or days while it will take carbon compounds hundreds of years to mix from the wind-churned upper layer of the ocean to the deep strata are inevitable consequences of scavenging caused by recurrent precipitation on the one hand and slow exchange in stratified ocean waters on the other.

The only sensible ranking of biogeochemical cycles is thus according to their vulnerability, their proneness to human interference resulting in undesirable alterations and destabilization. Undesirable, I hasten to add, from the human point of view because it is hard to find noncatastrophic biogeochemical changes which would not benefit some biota somewhere on the earth.

As this planet is dominated by water, the two most abundant elements of the biosphere are, inevitably, hydrogen and oxygen. And it is not only the earth's milieu which is largely aqueous: water dominates the composition of all living creatures, accounting for more than four-fifths of lush grasses and leaves, more than two-thirds of animal and human bodies. Even when the biomass is dried in an oven, the moisture-free matter is still predominantly hydrogen and oxygen in a proportion closely resembling water.

The water cycle is thus easily the best known and undoubtedly the most studied of all grand biospheric circulations. The reservoirs are all well known—97.6% of the total in the oceans, some 1.8% locked in polar ice caps and glaciers, 0.6% in lakes, underground aquifers, and streams, and a mere 1/1000th of a percent in the atmosphere. The flows are obvious—evaporation, precipitation, runoff—and readily quantifiable and the whole cycle must be in global balance. The only known "sink" is the loss of hydrogen atoms to space, a marginal loss replaced by juvenile water from volcanoes.

Human interference with the cycle is ubiquitous but the effects are just in the form of delays or accelerations of local or regional balances caused by construction of reservoirs, by irrigation, urbanization, and by changes in land cover and land use. There are no net additions or withdrawals from the global flows, no anthropogenic releases altering the fundamental features of the cycles. While the local and also regional characteristics of evaporation, precipitation, runoff, surface and underground storages may be altered appreciably, the global water volume and the rapidity of its cycling are immune to human interference.

If hydrogen is no candidate for studies of human interference in the principal biogeochemical flows, oxygen's criticality in animal and human metabolism may appear to make it a more vulnerable element whose cycling should be worthy of closer attention. But it turns out that atmospheric oxygen is not easy to interfere with: fossil fuel combustion is, of course, the most extensive anthropogenic method of oxygen removal but simple calculations show an extremely comfortable margin of safety. Complete oxidation of each atom of carbon requires two atoms of oxygen and hence at the current rates of fossil fuel combustion, about 13 billion t of oxygen is sequestered in CO_2 and another 8 billion or so is removed by biomass burning and grassland and forest fires.

Since the atmosphere contains about 9.6×10^{14} t of oxygen, annual burning of all fossil fuels and of all biomass removes merely 0.002% of atmospheric oxygen, a totally insignificant amount. Even should we do the impossible and burn all of the currently known deposits of fossil fuels containing some 4 trillion t of carbon, we would reduce the atmospheric oxygen by merely 1.1%, a depletion of no significance whatsoever for the life on this planet.

With neither hydrogen nor oxygen qualifying as vital elements whose cycling might be significantly influenced by human actions, we have to turn to carbon, the single largest elemental constituent of living matter (about 45% of dry weight of planetary biomass) and to a group of elements which are present in the biosphere in much lower quantities than hydrogen, oxygen, and carbon (the differences are huge, at least two or three orders of magnitude!) but which are well known to be essential plant macro- and micronutrients or to rather rare biospheric constituents introduced into the environment in urban wastes and by industrial and farming practices.

But the choices among this group narrow down rapidly as most of these elements do not really cycle: insoluble minerals stay put unless mechanically moved by water or wind erosion, soluble elements flow just one way, from continents to oceans, with temporary interruptions en route. To interfere with such flows on a global scale would not be easy. While local or regional inputs or alterations may be worrisome and outright dangerous—e.g., dumping mercury into rivers and coastal waters—they would not lead to global or continental changes unless vast amounts of anthropogenic pollution were dumped continuously over large areas of the earth.

But heavy metals such as mercury and lead as well as numerous lighter elements, including important plant micronutrients such as zinc, boron, or molybdenum, are not needed by modern civilization in prodigious amounts (e.g., the annual global output of lead is just 3 million t, that of molybdenum less than 100,000 t), are not dumped indiscriminately over large portions of inhabited continents or along most coastlines, and their escapes to the environment can be reduced to negligible rates by economic recycling and proper applications.

Phosporus is a vital nonsubstitutable ingredient in any terrestrial life, a principal plant macronutrient, a constituent of nucleic acids, a key presence in adenosine triphosphate, the carrier of energy in plants and in animal and human muscles.

However, elemental phosphorus and its compounds are not, as just relayed, atmospheric pollutants affecting ecosystems on a continental scale, application of phosphatic fertilizers leads to no releases of compounds with potentially harmful effects on human health or on stratospheric composition, and the element is in no way implicated in any long-range, irreversible, or difficult-to-correct degradation of the environment.

The only undesirable invironmental consequences arise from direct discharges of phosphorous compounds into stationary or slow-moving waters. Concentrations of iron and aluminum, the two major elements immobilizing phosphorus, are, not surprisingly, much lower in waters than in soil, and phosphate ions can thus be much more readily used by aquatic plants than by terrestrial species. Phytoplankton and algae respond first with vigorous growth and serious eutrophication problems may ensue. But these problems do not appear on a continental or even a regional scale, though they may affect some rather large lakes receiving extensively industrial wastes, sewage, and phosphorus-containing detergents.

So the focus must be on those elements which are introduced into the biosphere by essential and ubiquitous human actions in large quantities and which are doubly mobile, i.e., both water soluble as well as airborne. Only three elements are in this choice class—carbon, nitrogen, and sulfur—and all three must be recycled together along with water to sustain life and "it may not be an accident that all three are more reduced in the biosphere than they are in the external world. Be that as it may, they all seem to belong to the biosphere, which is otherwise mainly water" (Deevey, 1970).

And they are, locked in countless compounds, everywhere—as carbonates, nitrates, and sulfates in the earth's crust and dissolved in both fresh and ocean waters; in every plant, with C:N:S ratios typically 100:1:0.1, and with carbon alone accounting for nearly half of the living mass; and in the atmosphere as carbon dioxide, carbon monoxide, methane, free nitrogen, ammonia, nitrogen oxides, nitrates, sulfur dioxide, hydrogen sulfide, and sulfates to name just the principal constituents of airborne segments of their respective cycles.

Clearly, focus on carbon, nitrogen, and sulfur is inevitable not only because of the magnitude of their flows but even more owing to the double mobility of the elements and hence the unusual susceptibility of their cycles to human interference. This book shall thus look at natural reservoirs and flows of these three elements and, in much greater detail, at their anthropogenic releases and their subsequent climatic, ecosystemic, health, economic, and social effects. But before entering on the lengthy journeys through the complexities of the three cylces, I must outline how their study and attention to human interferences in their functioning gathered interest and evolved in the recent past to elevate them among the most prominent and richly controversial topics of modern scientific inquiry.

Gaining the Attention

In both North America and Europe the onset of widespread interest in environmental quality, ecology, and the fate of complex ecosystems can be dated to the early 1960s. Many actions served as important catalysts and ingredients of this multifaceted process, ranging from publication of new books (Rachel Carson's *Silent Spring*) and republication of older ones (Aldo Leopold's *Sand County Almanac*) to local resistance to dubious new engineering projects (from power plants to multilane highways) and legislative efforts to prepare new environmental protection laws (from forbidding dumping of hazardous wastes to preserving visibility in national parks).

A brief recounting of major scientific advances in this challenging interdisciplinary effort must start with the International Geophysical Year (IGY; 1957–1958) when for the first time concentrations and fluxes of a wide variety of atmospheric and oceanic compounds were measured in a systematic manner in locations spanning the planet from the Arctic to the Antarctic. And the program did not really end in 1958: it rather laid foundations for subsequent continuation and considerable expansion of continuous monitoring of compounds such as carbon dioxide or ozone.

Out of these endeavors grew the Geophysical Monitoring for Climate Change (GMCC), a program run by the National Oceanic and Atmospheric Administration since 1972 and monitoring CO_2, ozone, stratospheric water vapor, halocarbons, N_2O, and surface aerosols.

What the IGY, its follow-ups, and GMCC have accomplished in terms of achieving a much better understanding of the biosphere's air and water envelopes the International Biological Program (1964–1974) has done for increasing knowledge of the planet's principal ecosystems as the work of thousands of researchers in far-flung locations yielded detailed information on primary productivity of forests, grasslands, wetlands, arid and Arctic biomes, as well as on the heterotrophic organisms and decomposers.

All these endeavors have brought invaluable contributions to a deeper understanding of various segments of biogeochemical cycles and together with rapid advances in disciplines ranging from geochemistry to remote sensing and from computer modeling to oceanography set the stage for publication of detailed reviews and analyses focusing specifically on biogeochemical cycles. Whole series of such publications originated from the initiative of the Scientific Committee on Problems of the Environment (SCOPE) which was formed by the International Council of Scientific Unions in 1967.

Study of biogeochemical cycles was given a prominent place among SCOPE's activities and of the 17 reports published to date, four focus on global cycling: No. 7 on nitrogen, phosphorus, and sulfur (Svensson and Söderlund, 1976), Nos. 13 and 16 on carbon (Bolin et al., 1979; Bolin, 1981), and No. 17 a broad overview of carbon, and nitrogen, sulfur cycles (Likens, 1981). Other major volumes of environmental biogeochemical studies published since the mid-1970s are those edited by Nriagu (1976), Stumm (1977), Trudinger et al. (1980), Hutzinger (1981), and Hallberg (1983).

Specific literature on the three cycles to which this book is devoted is much more extensive than general biogeochemical treatises and I will review it chronologically in separate sections tracing the recent rapid rise of these studies to scientific, and even public, prominence. Interestingly, this vault to professional and public media attention occurred almost simultaneously for all three cycles during the late 1960s and early 1970s, although strict chronology established by frequency of publications, conferences, and controversies shows that in North America, nitrogen came first, sulfur next, and carbon last.

NITROGEN IN THE ENVIRONMENT Chronologically, anthropogenic interference with the nitrogen cycle was the first one to receive broad public attention following Barry Commoner's lectures and writings about the leaching of nitrogenous fertilizers in the late 1960s and early 1970s (Commoner, 1968, 1971, 1975; Kohl et al., 1971). His conclusions were that owing to the rapid introductions of inorganic forms of nitrogen in oxidized forms (by combustion but above all by fertilizers), we were confronted with "a number of problems which are large in their magnitude, difficult in their complexity, and grave in their import for the nation. In sum, we have in the United States, thrown the nitrogen cycle seriously out of balance. . . . The present stress on the nitrogen cycle has already produced environmental hazards and carries the risk of equally serious medical hazards. Clearly, corrective measures are urgent" (Commoner, 1975).

Commoner's perception of the future interference was, if possible, even gloomier because he argued that the difficulties with technical steps needed to correct the situation would be "enormously magnified by the fact that the practices which have produced the imbalance are now deeply embedded in the nation's economy." Consequently, he believed that "these difficulties will eventually require the limitation of the current high rate of use of inorganic nitrogen fertilizer. I hardly need emphasize the explosive consequences of imposing such a restriction on the current farm economy."

Moreover, he felt that the state of overcapacity in the nitrogen fertilizer industry would lead to a search for "new ways to sell its products—for example, large scale use of nitrogen to fertilize timber crops. If this is done our present environmental problems will become worse." The gravity he ascribed to the whole problem can be best judged by the fact that he expected "great public debate in connection with the artificial intrusion of nitrogen into the biosphere'' and likened the importance of choices to those faced during the radioactive fallout controversy which led to the Nuclear Test Ban Treaty in 1963. And ''the people as a whole'' were called by Commoner to make the grave choices.

If all these warnings and exhortations had proved correct, then today we would be in the midst of a severe environmental crisis with considerable health hazards—especially as the use of inorganic nitrogenous fertilizers has increased and some of these applications have indeed gone to forest fertilization. But there is no such crisis upon us, no great public debates on the nitrogen cycle enliven our legislatures, scientific or news magazines. Special panels of the National Academy of Sciences studied virtually every aspect of nitrates in the environment (NAS, 1972, 1978a) as well as the effects of nitrogen oxides (NAS, 1977a) but their conclusions were decidedly uncatastrophic. And just 5 years after Commoner's cited warnings were printed, the United States Environmental Protection Agency's appraisal of the concerns of a new decade (USEPA, 1980a) did not devote even a single paragraph to problems judged so recently to be ''large in magnitude'' and ''grave in their import for the nation.''

But new nitrogen-related concerns arose in the mid-1970s owing to the rapid rise of hydrocarbon prices in 1973–1974: prospects for continuing fast growth of synthetic fertilizer production based on the availability of reliable and inexpensive natural gas and oil supplies became uncertain, and as a result much attention was focused not only on energy cost of nitrogen fertilizer production but also on the fundamental processes of nitrogen fixation in natural and farming ecosystems and on the possibilities of human manipulation of biofixation of nitrogen.

And yet another nitrogen-related concern emerged to gain both public and scientific prominence during the 1970s: the use of nitrite in curing meats, as the compound—widely added to processed meats, and also to some fish products and cheeses—was identified as a potential carcinogen, and lively debates developed regarding the acceptable levels of the additive, its metabolic fate, and possible consequences of its elimination or substitution. NAS publications treat these developments in great detail (NAS, 1978a, 1981b) and Section 3.4.3.3 will review the current state of our knowledge.

Finally, all these concerns with provision, availability, and environmental and health effects of nitrogenous compounds led to greatly intensified interest in all facets of the element's intricate cycling through the soils and its exchange between the soils, plants, waters, and atmosphere. This broad, interdisciplinary interest is best exemplified by a series of major international meetings between 1975 and 1979.

The first of these conferences, in Copenhagen in August 1975, was a thorough examination of nitrogen as a water pollutant (Harremoës, 1975). In December of that year a meeting in Örsundsbro (Sweden) dealt not only with nitrogen but also with general problems of phosphorus and sulfur cycles (Svensson and Söderlund, 1976). Another international meeting held in Sweden, in late August and early September 1976 as Nobel Symposium No. 38, focused solely on nitrogen as ''an essential life factor and a growing environmental hazard'' (Bolin and Arrhenius, 1977).

The first detailed examination of mineral nutrient cycling in agro-ecosystems was conducted in Amsterdam in the late spring of 1976, with nitrogen taking, naturally, the most prominent place (Frissel, 1977a). The broad topic of nitrogen in the environment discussed at Lake Arrowhead, California, in February 1977 resulted in publication of a two-volume set of papers (Nielsen and MacDonald, 1978). Finally, in Österfornebo (Sweden) in September 1979, terrestrial nitrogen cycles were analyzed from many interdisciplinary viewpoints (Clark and Rosswall, 1981). Among single-author books, Aldrich's (1980) thorough and balanced treatment of nitrogen use and its consequences in modern American farming is outstanding.

Complex storages, flows, and transformations of soil nitrogen have always been of high interest to soil scientists, and two American Society of Agronomy reviews edited by Bartholomew and Clark (1965) and its truly monumental update edited by Stevenson (1982a) are certainly the best comprehensive sources. And, of course, thousands of new research papers in scores of scientific journals, ranging from such traditional studies as yield response of rice cultivars to nitrogen fertilization or NO_x combustion controls to fashionable topics such as nitrate vegetables or nitrites in saliva, have appeared during some 15 years of unprecedented interest in nitrogen in its many guises and roles.

ACID RAIN Although human-induced distortions in the sulfur cycle started to gather major public attention at about the same time as the news of potentially dangerous human interference in nitrogen flows, concern with acid rain has in a short time outperformed—in general awareness and in concentration of research and legislative interests—any of the possible worries arising from lack or excess of nitrogen in the environment. As the specters of hopelessly nitrate-polluted waters and worsening nitrogen oxide air pollution were gradually receding, the dangers of fishless lakes, acidified soils, and crippled forests came to be portrayed in more pressing terms.

The origins of this concern, however, are surprisingly old. Various scientists of the late 17th and 18th centuries wrote about the acrid properties of smoke and the presence of sulfur in rain and dew, but the foundations of systematic acid rain studies were laid by the Englishman R. A. Smith in the middle of the 19th century. In an 1852 paper, Smith noted corrosion of metals and color fading of fabrics in urban air owing to sulfuric acid or acid sulfate and two decades later, in a pioneering book, he actually used the term ''acid rain'' and outlined key components of the phenomenon, ranging from coal combustion to long-distance airborne transport (Smith, 1872).

In the following decades, infrequent contributions were made in various European countries but it was not until the 1950s and 1960s when the suddenly expanding sciences of atmospheric chemistry, biogeochemistry, and ecology furnished enough information to begin formation of a modern synthesis of the acid deposition problem. Detailed historical reviews of this multifaceted progress can be found, together with scores of appropriate references, in Gorham (1981) and Cowling (1982a, b) and only major highlights will be mentioned here.

Swedish contributions were essential: in 1948 Hans Egnér set up the first large-scale precipitation chemistry network and Erik Eriksson expanded it in the early 1950s to cover much of western Europe; this continuing record is the best available base for secular comparisons of ionic composition and acidity of precipitation. At the same time, Eville Gorham started to publish, first in England and later in Canada, his many findings about acid precipitation and degradation of soils and aquatic ecosystems. Between 1955 and 1965 he described just about every major facet of the problem and must be considered the eminent pioneer of acid deposition studies in North America.

But, curiously, Gorham's writings did not usher the intensive interest in acid rain; they did not travel beyond scientific journals and it was the Swedish publicity which brought the problem into public focus. Svante Odén, basing his claims on results from Scandinavian surface water chemistry measurements that he started in the early 1960s and on the European Air Chemistry Network data, was the first to suggest, in Stockholm's leading daily *Dagens Nyheter* in October 1967 and in 1968 in an Ecology Committee Bulletin, that long-range transport of acid air pollutants is responsible for degradation of aquatic ecosystems, leaching of toxic metals from soils to waters, and impaired forest growth.

Then the Swedish government took over and as its major contribution to the 1972 United Nations Stockholm Conference on the Human Environment it offered, under Bolin's leadership, a detailed interdisciplinary report on sulfur in air and precipitation (Royal Ministry for Foreign Affairs, 1971). This report was the real beginning of a massive scientific effort readily supported by governmental grants in Scandinavia and North America. In the following years hundreds of specialized papers ranging from topics in synoptic meteorology to heavy metal toxicity were published in dozens of different journals and the evolving knowledge of acid deposition was brought into more manageable confines by a series of interdisciplinary research programs, conferences, and edited volumes which became the most often cited sources of information.

Chronologically, the first was the joint project of three Norwegian organizations designed to elucidate the effects of acid precipitation on forest and fish whose final report was published in 1980 (Overrein et al., 1980) but which started in 1972; this project sponsored the first large international conference on acidification in Telemark in June 1976 (Braekke, 1976) and another meeting at Sandefjord in March 1980 (Drabløs and Tollan, 1980). The Organization for Economic Cooperation and Development supported a continentwide study of the long-range transport of air pollutants in Europe in the years 1973–1975 (OECD, 1977) and by that time acid rain started to become a fashionable research topic in North America.

The first international symposium on acid precipitation and forest ecosystems was organized by the United States Forest Service (Dochinger and Seliga, 1976), the National Research Council of Canada published a massive multiauthor volume on sulfur in the environment (NRC, 1977), and sulfur in the environment was the topic of two volumes edited by Nriagu (1978a, b). North American participation was strong at the International Symposium on sulfur in the atmosphere in Dubrovnik in September 1977 (Husar et al., 1978) and at the NATO Conference on ecological effects of acid precipitation in Toronto in May 1978 (Hutchinson and Havas, 1980).

Both the United States and Canadian governments entered the scene: they set up monitoring networks [the Canadian Network for Sampling Precipitation (CANSAP) in 1976, the National Atmospheric Deposition Program (NADP) in 1978], established a Bilateral Research Consultation Group in 1978, and signed a Memorandum of Intent to conclude a bilateral agreement on controls of transboundary air pollution. Money gates opened—especially with the announcement of a decade-long program of research on causes and consequences of acid deposition by President Carter in 1979—and intricate government bureaucracies were set up. The Interagency Task Force on Acid Precipitation (ITFAP) includes the NOAA, DOA, EPA, CEQ, DOC, DOE, DOL, DOS, DMHS, NASA, NSF, and TVA and reading documents of its effort is a punishing task of acronymic acrobacy. Canada has its own version of the same.

To return to key references marking the progress of our understanding, the Electric Power Research Institute (Palo Alto, Calif.) and the Central Electricity Generating Board sponsored a symposium on ecological efforts in 1978 (EPRI, 1979) and the Oak Ridge National Laboratory organized a broader meeting on environmental and health effects in 1979 (Shriner et al., 1980). Yet another multi-author volume on biotic effects was edited by D'Itri (1982).

Finally, the top scientific bureaucracies of the United States and Canada entered with their assessments, the NAS first more generally on sulfur oxides (NAS, 1978b), then in 1981 (NAS, 1981a) as part of an appraisal of atmosphere–biosphere interaction, and finally in much greater detail in 1983 (NAS, 1983a); the NRC issued its evaluation in 1981 (NRC, 1981).

One of the most informative recent publications is the book commissioned by the Swedish Ministry of Agriculture (1982) as that country's contribution to the 1982 Stockholm Conference on the Acidification of the Environment; its comparison with the 1971 Swedish report indicates best the advances achieved during the 1970s and confirmation or shifts in the appraisal of causes, effects, and outlook.

And, of course, after filling up countless pages of environmental journals and providing newspaper editors with welcome opportunities to set headlines about killing rain and death from the sky, acid deposition even made it to the cover of *Time* in October 1982. And so in the 10 years since Stockholm 1972 and the Swedish revelations to the world, acid rain grew to occupy, with genetic engineering, machine intelligence, and other such perennials, the prestigious sphere of scientific problems in public possession—with all attendant perils of misinterpretations and political influence. But interest in acid rain now has a fierce competitor—the planetary doom from CO_2-induced warming, another problem with a long scientific history and fresh rise to the limelight.

WARMING THE EARTH The first awareness of carbon dioxide's effect on atmospheric temperature is about as old as the first systematic attempts to study sulfurous air pollution: Tyndall's 1861 paper was the first clear description of what came to be known, not quite precisely, as the "greenhouse effect" (Tyndall, 1861) and just before the end of the century Arrhenius did the first calculations of global temperature rise following doubling of CO_2 concentrations (Arrhenius, 1896): his result, an 8°C rise, is not that much different from the current computerized efforts (see Section 2.4.1.1).

Chamberlin (1897) followed shortly afterwards with his comprehensive theory of CO_2-driven climatic change, but little interest was focused on the problem until the late 1930s. Callendar (1938) brought it back in a major paper followed by many similar writings published over the next 23 years. Among the American researchers, Plass (1956) was the first resurrector of the concern, followed soon by Revelle and Suess (1957).

The IGY (1957–1958) brought the first reliable measurements of background concentrations of CO_2 in remote, clean places, and continuation of this monitoring soon showed a steady rise of average global CO_2 levels. Clear proof of a CO_2 rise led to an era of assorted quantitative models constructed usually with a single goal: determination of the mean global temperature increase with doubling of CO_2 levels. Soon the implications of temperature increases predicted by a variety of new climatic models started to gain wider attention in popularizing publications and among policymakers.

In 1965 the CO_2 question was included in the report of the Environmental Pollution Panel of the President's Science Advisory Committee, thus achieving for the first time an acknowledged status in a government-sponsored scientific agenda (Revelle et al., 1965). Five years later it figured prominently in the Report of the Study of Critical Environmental Problems (SCEP, 1970) although a year later in the Report of the Study of Man's Impact on Climate (SMIC, 1971) it was given a very ordinary mention as just one of numerous anthropogenic interferences.

But it was the first rapid rise of crude oil prices in 1973–1974 which must be credited, among so many other effects, with making atmospheric CO_2 enhancement one of the top scientific problems of the decade as judged by the number of publications in general research journals, numerous books and conferences, broad public appeal, and extensive intervention of scientific and government bureaucracies. A sudden massive increase in energy studies was an obvious stimulant for CO_2 research and the latter half of the 1970s and the early 1980s literally overflow with related publications.

In 1980 the Oak Ridge National Laboratory published a partially annotated bibliography of 1000 citations on carbon cycles and climate (Olson et al., 1980) and just a year later the laboratory's newly created Carbon Dioxide Information Center had a computerized bibliographic file of over 2100 indexed references (Allison and Talmage, 1982).

The most comprehensive treatments among this flood of often inevitably repetitive publications have been the volumes edited by Woodwell and Pecan (1973) and Andersen and Malahoff (1977); the proceedings of a World Meteorological Organization workship (WMO, 1977) and of another workshop sponsored by the WMO, the United Nations Environment Program, and the SCOPE at the International Institute of Applied System Analysis at Laxenburg, Austria, in February 1978 edited by Williams (1978a); the 13th SCOPE report a year later (Bolin et al., 1979); the three assessments by special panels of the NAS (NAS, 1979c, 1982a, 1983b); and a summary volume edited by Clark (1982). Repeated entry of the NAS is a clear sign of the importance with which the CO_2 problem came to be seen by a leading scientific bureaucracy, and the formation of the United States Department of Energy institutionalized the governmental attention and enhanced the opportunities for additional research funding.

With the heightened attention given to general problems of global climatic changes (an interest which rose in the wake of the Sahelian drought of the early 1970s and has been sustained mainly by the concerns about future food production capacity), the CO_2 question has also ranked high in publications and research priorities of a growing number of workers trying to interpret the past trends and forecast the future shifts. Here the most useful summaries are the volumes by the Geophysics Study Committee of the National Research Council (NAS, 1977c), the proceedings of the World Climate Conference (WMO, 1979), and the volumes edited by Bach et al. (1979, 1981).

Later discussion will show that there is no shortage of uncertainties surrounding the effects of CO_2 on global climate and on the biosphere but, in contrast to studies on nitrogen and sulfur, a revisionist wave has barely started to roll over the unusually uniform view of the most likely impacts of CO_2 in the future. Perhaps the most obvious way to explain this difference is to point to the richness of the experimental and field evidence of so many facets of the nitrogen and sulfur cycles, evidence so complex, ambiguous, and often even contradictory that it does not allow simple conclusions and easy committee consensus.

On the other hand, the principal worry about long-term CO_2 effects—planetary warming—rests *solely* on theoretical modeling exercises now made so much easier with computers while *no* empirical evidence exists for warming attributable to CO_2 for the simple reason that there has been *no* clearcut proof of long-term warming in general: the mean annual surface air temperature over the Northern Hemisphere shows, ever since the beginning of reliable instrumental measurements some 100 years ago, just inconclusive fluctuations.

So with CO_2 we have a problem much unlike that with nitrogen compounds in well water, in cabbage leaves, or in urban smog or that with acid deposition in already acid lakes or on sensitive plant leaves. The former is the mold of complex computer models, of conjectures relating to the next or even to the 22nd century; the latter arise from easy-to-see outright damage or easily measurable gradual deterioration we wish to stop.

And besides the disparate time scales the spatial dimension is different, too. Rise of atmospheric CO_2 is a truly global environmental problem: there is less of it in the Southern Hemisphere but this owes to the natural lag caused by difficulties of transequatorial mixing compared to intrahemispheric diffusion. In contrast, in spite of the heavy human interference with anthropogenic fluxes accounting for major shares of total global circulation, both nitrogen and sulfur pollution remain often merely local problems, and though the latter has spread its effects on a large regional and even a continental scale in Europe and North America, it is not decidedly global in terms of worrisome concentrations and future effects.

But otherwise I shall try to treat the three cycles in very similar ways. In the case of carbon I will not concentrate just on possible future CO_2 temperature effects; I will look first in some detail at the element's storage and fluxes in the biosphere, then at its anthropogenic outputs and at overall global cycling and atmospheric concentration increases.

2 CARBON

In no other circumstances have I ever seen vegetation so vigorous as in this kind of air, which is immediately fatal to animal life . . . . This observation led me to conclude that plants, instead of affecting the air in the same manner with animal respiration, reverse the effects of breathing, and tend to keep the atmosphere sweet and wholesome, when it is become noxious in consequence of animals either living and breathing, or dying and putrefying in it.

—JOSEPH PRIESTLEY
Experiments and Observations on Different Kinds of Air (1790)

Bioplasm is mainly a dispersion of carbon compounds in water.

—A. E. NEEDHAM
The Uniqueness of Biological Materials (1965)

Life on this planet can be seen simply as a seemingly endless variation of carbon compounds. Biomass is basically a dispersion of carbonaceous matter in water—upon dehydration, about 45% of it, a share much higher than that of any other element, is carbon. Even in the biosphere taken as a whole, with its enormous stores of ocean water, carbon accounts for one-quarter of all atoms and shares second place with oxygen (hydrogen atoms are, obviously, most abundant).

Sixth in Mendeleev's periodical table, a dozen times as heavy as hydrogen and three-quarters as massive as oxygen, carbon is a dense solid (3.51 g/cm^3) at all temperatures encountered in the biosphere (its melting point is 3570°C) and, unlike oxygen, nitrogen, or sulfur, a rather rare ingredient in the earth's crust (a mere 0.032%). It is carbon's place in the central group of the periodic system that gives the element its ambivalent affinity in forming compounds with both electropositive and electronegative elements via a covalent bond as well as its ready joining with many different radicals.

And yet another property of the element fits its ambivalent image. Its covalent bonds are not formed easily without catalytic activation, but once in place they are distinguished as much for their stability as for their lability, combining, as Needham (1965) put it, momentum with inertia. A true *yin–yang* element, carbon is a metastable entity creating the living world around us—but this creation is not in complete equilibrium with the environment, ready to react and to change again.

The physical diversity of carbon and its compounds is also notable. In its elemental form it spans the opposites of hard, transparent diamond and soft, black graphite; among its common compounds at temperatures normally encountered in the biosphere, CO_2 and CH_4 are gases, ethyl alcohol, acetone, light paraffins are liquids, and higher paraffins are solids. Similarly wide-ranging are the solubilities of these compounds in water, from extremely high to virtually nonexistent.

The diversities of carbon have helped to create much more than our living environment. The past ages of photosynthesis passed on huge stores of fossil carbon in coals, oils, and gases which became the energetical foundations of modern industrial civilization. Their combustion has released increasingly large volumes of carbon dioxide and this process and its possible environmental consequences will be the main focus of this chapter. But carbon dioxide is the final stable product of complete oxidation—abiotic or biotic—of any carbon compound and it is, of course, the only form in which plants assimilate the element for their growth. So the first logical step before appraising human interference in carbon transfers is to look at biotic carbon reduction and oxidation, that is, at photosynthesis and respiration, whose reversibility was so well noted by Priestley (1790) almost two centuries ago. This will be followed by quantifications of primary productivity in different ecosystems, global storages of biomass and their changes owing to deforestation, crop cultivation, and fuelwood combustion.

2.1. CARBON AND BIOTA

All complex life on this planet is based on continuous exchange of carbon between the atmosphere and the biosphere as myriads of autotrophs—photosynthesizing plants ranging from simple, tiny, and ancient blue-green algae to giant and highly diversified temperate conifers—reduce, powered by solar radiation, atmospheric CO_2 to use carbon as a basic building block of complex organic compounds and as these compounds are recycled, partially by the respiration of plants and mostly through decomposition mediated largely by microorganisms, CO_2 reenters the air.

This biospheric breath has four distinct frequencies and amplitudes: rhythmic inhalations and exhalations are easily noted as day and night and summer and winter contrasts by measuring ambient CO_2 concentrations. The third fluctuation is much longer and irregular, requiring a time scale of civilizations and deducible from historical records and archeological evidence: the eras when Greenland was green as was the heartland of the Sahara, after which mighty changes obliterated these thriving communities. And the fourth one demands a planetary yardstick of grand geological eras.

Even when leaving the oldest documented autotrophs aside, eukaryotic photosynthesizers started to colonize the earth at least 1 billion years ago so that the higher life forms have been around for roughly one-fifth of the planet's age and one does not have to subscribe to Lovelock's (1979) Gaia theory to acknowledge the immensely important role of plants in shaping the climate and hence also governing the fates of the hydrosphere and lithosphere. And as carbon is, in mass terms, the cornerstone of this ever-changing balance governing the survival of life, any major interference with its critical exchange must be viewed with caution. And to understand its implications, one must first appreciate at least the basic characteristics, requirements, and limitations of the process creating new matter and storing it in the greenness of forests, grasslands, and crops.

2.1.1. Primary Productivity

General features of the photosynthetic process, as well as the magnitudes of biomass storages in different plant formations, have been appreciated for many decades but our knowledge of the actual functioning of multistep cycles performing CO_2 fixation dates only from the late 1950s (Bassham and Calvin, 1957) and our quantification of primary productivities, autotrophic and heterotrophic respiration, and standing phytomass in different ecosystems reached a satisfactory level only during the last decade.

Major uncertainties remain, none more important than the extrapolation of detailed and reliable findings on productivities and storages from a few scores of studied sites to ecosystems covering billions of hectares, a jump so obviously error-prone. Moreover, as will be shown later (Section 2.1.2.1), often we are not even reasonably certain about the area by which to multiply! Consequently, local appraisals may be very accurate but global aggregates continue to be a matter of disagreements. But before reviewing the global totals, I will introduce essential quantitative information about the photosynthetic process.

Photosynthesis and Productivity What is perhaps most striking about the light-harvesting reaction which sustains all life on this planet is that we have both imperfect knowledge of its complex processes as well as limited understanding of its interactions with other aspects of plant growth. Radmer and Kok (1977) illustrate this best when they write that "domesticated photosynthesis has made clever use of nature's storeroom, but has not gone beyond it; lacking other inputs, we have manipulated symptoms without having access to the underlying causes."

Still, we know enough to appreciate the limitations. First, there is the unbreachable barrier of the maximum conversion efficiency given by the energetics of the photochemical reactions—27.5% of the photosynthetically active radiation (for detailed calculations see Bassham, 1977; Good and Bell, 1980). As the active wavelengths between 400 and 700 nm account for 43% of overall flux, maximum efficiencies for the full solar spectrum reaching the ground are no more than 12%. But the actual performance is always considerably below this level as two additional limitations intervene. Plant surfaces are obviously not perfect absorbers so there is a reflective loss of at least 10%, and respiration during the night in all higher plants and also photorespiration during the day in so-called C_3 species, consumes a minimum of 10% of daily production.

Consequently, ideal plants would conserve no more than 10% of incident solar radiation in new phytomass. A perfect leaf facing full sunlight at a 90° angle would thus convert about 250 mg CO_2/hr per dm^2 and a hard-to-set-up perfect field completely covered with such leaves would, on a bright summer day, produce about 1.5 t of biomass per ha. A year-round crop such as sugarcane or algal cultures could thus yield maximum theoretical harvests of 550 t/ha—but we have yet to come across such performers. Peak daily efficiencies of the best known producers are half or one-third of the calculated value and typical annual yields of crops or natural vegetation are easily less than 1/50th of that impressive total.

Reasons for this huge difference are manifold. Ironically, the bright sun postulated in the above example may not make much difference as the photosynthetic rates of most plants level off at just one-tenth to one-fifth of bright sunlight. No less fundamentally, both of the naturally occurring compounds providing irreplaceable carbon and hydrogen for the synthesis, CO_2 and H_2O, are almost always present in concentrations too low to support the best possible conversion rate and, moreover, acquisition of the former by the plant is always paid for by the losses of the latter.

Much will be written later in this chapter about CO_2 levels and the rates of photosynthesis and so here just a note on the high CO_2–H_2O exchange rate. Even under the best imaginable conditions, no plant can absorb CO_2 without losing more than 100 times as much water (Good and Bell, 1980)—and actual exchanges are much more unfavorable, between 450 and 600 moles of water transpired per mole of CO_2 fixed in the most efficient autotrophs such as tropical grasses or corn and up to 1400 moles per mole for other plants. Quite clearly, availability of water places a limit on photosynthetic rates even when the plants grow in very humid environments!

Lack of nutrients is another well-appreciated limitation of which much more will be mentioned in chapters on nitrogen (Section 3.2.4) and sulfur (Section 4.4.3.3). Nor are the structural necessities a negligible drain on photosynthetic gains. Large mass of nonphotosynthetic tissues in higher plants, especially in trees, represents a high respiratory burden, a price to pay for better access to light, water, and nutrients. Inevitable shading of the leaves limits the maximum leaf surface area per unit of land and as a result this measure is on the average only twice as large in tropical rain forests than in semiarid savannas!

Details on these and other limitations of carbon fixation may be found in many excellent writings of the past two decades. My selective list includes Bassham and Calvin (1957), Zelitch (1971), Moss and Musgrave (1971), Barber (1976–1979), Radmer and Kok (1977), and Good and Bell (1980). Most of this work is, naturally, in the realm of biochemistry whose focus is largely microscale, whereas ecosystem energetics studies employ a much larger scale and their standardized measures will be briefly defined here.

Gross primary productivity (GPP) is the most embracing measure, a sum of all newly photosynthesized matter given as either grams per square meter per year or tonnes per hectare per year; when diminished by autotrophic respiration (R_A), it becomes net primary productivity (NPP), a value used more often than any other in ecological productivity accounts. NPP does not, of course, represent net increments of new phytomass, as heterotrophic respiration (R_H), ranging from microorganismic decomposition to mammalian herbivory, may reduce it to usually very low net ecosystem productivities (NEP).

Consequently, in a fertilized and irrigated field planted to an annual crop, NPP may not be much lower than GPP as the cultural practices minimize R_A (which is relatively low to begin with as there are no accumulated nonphotosynthesizing tissues), and when the crop is sprayed by pesticides (minimizing R_H) NEP may be rather close to NPP (and, hence, to GPP). On the other hand, in a climax forest with its huge mass of respiring tissues (high R_A) and complex webs of heterotrophs rapidly recycling the newly fixes phytomass, GPP may be very large, NPP less than a quarter of it, and NEP can easily approach or equal zero. For the study of carbon fixation as a part of the element's global cycling, knowledge of primary productivities is essential to establish biosphere–atmosphere fluxes. Of no lesser importance is knowledge of the mass of the fixed carbon reservoir, the total living stores of the planet, and hence the so-called standing phytomass values will be discussed together with productivities in the following account of global photosynthesis.

2.1.2. Global Storages and Fluxes

Recent ecosystemic studies have shown great ranges of primary productivities and phytomass storages among the world's major ecosystems as well as considerable variability within individual vegetation units. Still, for estimates of global storages and fluxes, average values must be chosen—but not without being accompanied by the most likely ranges of uncertainty. And while getting the information on areas covered by various forests, grasslands, and other biomes might seem, in comparison with quantification of productivity and phytomass, a straightforward task of perusing a variety of statistical sources, the coming section will show again the prevalence of major uncertainties. With all of these caveats in mind, I shall then proceed to present the best estimates of global fluxes and storages—those of other researchers and my own.

2.1.2.1. Ecosystem Areas and Changes Even truly representative averages of a particular ecosystem's productivities and standing phytomass would be of little use in determining global fluxes and storages if the second term needed to determine the product—areas of vegetation units—is very inaccurate. Unfortunately, such is the case with the most important biome, as our knowledge of areas of tropical rain and seasonal forests is uncomfortably tentative, an uncertainty largely responsible for major differences in available global productivity and phytomass summaries. Substantial discrepancies in the area values exist also for other forest ecosystems, as well as for grasslands (only farmlands are known with satisfactory, though far from perfect, accuracy), but these differences do not add up to such errors as over- or undercounts of tropical forests.

Readily available international statistics are of little help as globally no more than half of all forestland has been appropriately surveyed and many of these surveys are greatly outdated. For example, in the tropics, Indonesia has never had a national forest inventory; for Thailand the latest survey figure goes back to 1961–1966; and the repeatedly given Indian total derives from 1954 surveys and may exaggerate the actual area by as much as 66% (FAO, 1976; Venkatasubramanian and Bowonder, 1980). Use of LANDSAT imagery eases the task of large-scale inventories, but limits of the technology do not allow accurate interpretation without very large amount of ground truth.

Considerable possibilities of substantial errors are best illustrated by comparing the available estimates of forested areas published during the past three decades in the forestry and ecological literature. Forests will be, naturally, the first ones to consider. FAO's world forest inventories listed the global areas at 38.37 million km^2 in 1955, 44.05 million km^2 in 1958, and 41.26 million km^2 in 1963 (FAO, 1955, 1960, 1963). An update of the 1963 inventory in 1971 put the total at 37.12 million km^2, virtually identical with the 1955 value.

Persson (1974) tried to improve the FAO estimates by distinguishing between closed forests with canopies covering at least 20% of the ground when seen from above and open woodlands with scattered tree presence giving ground cover of just 5–19%. In this way he arrived at 28 million km^2 of closed forests and a few years later he adjusted this total to 26.57 million km^2 so that with woodlands at 15.78 million km^2 the total forested land came to 42.35 million km^2 (Persson, 1978). Close to Persson's total is Windhorst's (1974) calculation of 24.926 million km^2 of closed forests in the early 1970s.

An estimate by Bazilevich et al. (1971) for pre-agricultural era forests at least some 5000 years ago—66 million km^2—is of purely theoretical note while the values used by American ecologists in global productivity and phytomass calculation are of the highest interest as they served most often as the bases for carbon storage calculations. Olson's (1970) total is 47.2 million km^2 (Olson, 1970), his latest revision being 32.499 million km^2 in what he labels "mostly closed forests" and 24.26 million km^2 in "mostly open woodlands or mixtures" (Olson, 1982). Whittaker and Likens's (1975) figure referring to the year 1950 is 48.5 million km^2 plus another 8.5 million km^2 of "woodland and shrubland." Profusion of ill-defined or undefined and hence incomparable categories is no small problem in all these comparisons!

Even when limiting the comparisons to the values published since 1970, one finds the total for global forests anywhere between 37 and 57 million km^2 with closed forests between 25 and 32.5 million km^2. In the first case there is a 54% difference between the extremes, and in the second, 30%, clearly a very unsatisfactory situation.

In view of the major, and inescapable, uncertainties in areal estimates of the richest and the most productive plant formations, it makes little sense to look for any detailed disaggregated estimates for grasslands or semideserts which produce much less and store very little. Consequently, I will use only very approximate values for the world's biomes (Table 2-1), a choice which makes even more sense

TABLE 2-1
APPROXIMATE AREAS OF MAJOR ECOSYSTEMS AROUND 1980[a]

Ecosystem	Total area (10^6 km^2)
Forests	35
Tropical moist forests (evergreen and seasonal)	10
Temperate forests (evergreen and deciduous)	10
Boreal forests	15
Woodland and shrubland	10
Grasslands	30
Tropical grassland (semiarid and humid)	20
Temperate grasslands	10
Cultivated land	15
Settlements, transportation	5
Tundras	10
Deserts and semideserts	20
Glaciers	15
Wetlands	5

[a]Sources: author's estimates based mainly on Windhorst (1974), Whittaker and Likens (1975), Persson (1974, 1978), Myers (1980), and Olson (1982).

when the tenuousness of many productivity and storage averages is taken into account.

The values listed in Table 2-1 are those of a moment sometime at the beginning of this decade; they change with the continuous losses of arable land (to suburban housing and highways in the rich countries, to new village housing, industrial plants, roads, and irrigation canals in the densely inhabited lowlands of the poor world), with the deforestation in tropical and subtropical regions which is one of the major causes of another widespread land cover change, desertification, and with the conversion of grasslands to new grain fields. I will try to establish the most plausible global totals of these shifts.

In terms of phytomass and primary productivity losses, the severest impoverishment is, of course, that owing to tropical deforestation. Reports on the disappearance of moist tropical forests became very frequent during the late 1970s but attempts at systematic evaluation of annual global losses are much less frequent. Recently the most quoted assessment has been that by Myers (1980). His estimates put the areas annually affected as follows: shifting cultivation at some 200,000 km^2 (primarily in Southeast Asia), legal logging up to 87,000 km^2, cutting of fuelwood about 25,000 km^2, and clearing for ranching at least 20,000 km^2.

These figures cannot be added as shifting cultivators most often follow loggers or do much of their clearing in secondary growth and as areas affected by industrial logging or fuelwood cutting are not completely cleared of vegetation. Myers's (1980) best estimate is thus that some 245,000 km^2 of primary tropical rain forest is converted annually, an area equal to that of the United Kingdom or West Germany.

Sommer (1976) put the total annual loss in the mid-1970s at about 2.1% of the remaining rain forest area of 9.53 million km^2, but as only less than half of this clearing takes place in natural forests, his annual conversion value is as little as 50,000 and no more than 100,000 km^2. He found a further support for this value by extrapolating the best available national totals to the whole ecosystem, a procedure yeilding an annual conversion of 112,000 km^2.

Finally, Lanly and Gillis (1980) worked with the most recent forest maps and LANDSAT images for Brazil, Mexico, Peru, and Bolivia to estimate annual conversion of closed moist Latin American forests at 41,300 km^2, or 0.6% of the remaining total. Extended to the global area of moist tropical forests which Lanly and Clement (1979) put at 11.32 million km^2, this rate would amount to about 70,000 km^2/year.

I believe that for estimates of global carbon storage changes, use of somewhat conservative values is in order. Myers (1980) may well be correct in his estimate of the areal total affected by various conversion processes—in fact, that aggregate may be even higher—but to treat this whole area as a complete loss of previously rich standing phytomass would be a mistake. I will assume that annually 10,000 km^2 is lost completely to new settlements, roads, reservoirs, and soil erosion, that 100,000 km^2 is turned into woodlands and grasslands (storing no more than 25 t/ha), and that an equivalent of another 100,000 km^2 is degraded so that it loses at least one-third of its standing phytomass.

Changes in other ecosystems have not recently been so drastic—and even where large areas of grasslands, or woodlands in subtropical or temperate regions, have turned into fields or desertified, loss of the phytomass was relatively much smaller than in tropical deforestation. For agriculturally induced vegetation changes, we can turn to a comprehensive study by Prentice and Coiner (1980) who used the USDA series of global cereal production for the years 1950–1975 to chart the changes in harvested grain areas. Obviously, such data do not reflect all farmland expansion but they undoubtedly cover most of the plant cover shifts caused by farming. Moreover, by disaggregating their data into 11 latitudinal categories, the authors made it possible to separate the extratropical changes from the tropical vegetation shifts.

While in tropical regions displacement of forests by fields lowers phytomass storage, in arid and semiarid areas poor native vegetation has been converted to higher phytomass crops and in many regions in temperate latitudes abandonment of low-productivity fields has led to shrub and forest regrowth and hence to higher phytomass storages. The net result of these three regional trends is the poleward shift of phytomass but the overall effect on the global sum of living matter appears to be negligible.

Prentice and Coiner (1980) calculated the extratropical gains of arable land at roughly 70 million ha during the 26 years of the studied period or roughly 25,000 km^2/year. But as most of this displaced vegetation is a variety of grasslands, the net phytomass loss is only about half of the original living mass, or an equivalent of losing some 12,000 km^2 of good grassland each year.

Desertification advances became a focus of much international research activity during the early 1970s, an effort culminating in a United Nations conference in Nairobi in 1977 (UNO, 1977). According to the conference's conclusions, areas undergoing severe desertification now extend over 30 million km^2, or 23% of the earth's ice-free land. Present losses to desertification amount to about 32,000 km^2 of rangelands, 25,000 km^2 of rain-fed and 1250 km^2 of irrigated farmland for a total of roughly 60,000 km^2. The maximum phytomass storage loss could be calculated by assuming the drop of standing phytomass from grassland mean (20 t/ha) to desert–semidesert average (5 t/ha) in the case of rangelands and halving of the storage for cultivated land loss (from 10 to 5 t/ha).

Estimates of vegetation losses owing to new settlement, industrial, and transportation links construction are very uncertain: they are as high as 80,000 km^2 annually (Wolman, 1983) and as low as 10,000 km^2 (Brown, 1978). I will assume that, in addition to the already noted 10,000 km^2 loss in the tropics, 40,000 km^2 is taken over annually by new development and that roughly two-thirds of it is from cultivated land or grasslands and one-third from the forests. Erosion-caused vegetation losses are no less uncertain. I will use Wolman's (1983) 30,000 km^2 and assume, again, about two-thirds of the decline occurs in farmlands and grasslands, one-third in the forests. For all remaining qualitative soil declines—waterlogging, salinization, alkalinization, and toxification—I will assume a loss of 10,000 km^2/year but here the total phytomass decrease will be only modest. This account opens the way for assemblies of global productivity and phytomass storage estimates and for calculations of annual phytomass declines owing to various anthropogenic interventions.

2.1.2.2. Productivity and Storage Estimates Although scientific interest in plant productivity became quite keen and widespread during the 19th century, it was not until the 1880s that Ebermayer used his Bavarian forest and crop productivity calculations to extrapolate the global annual CO_2 consumption by plants (Lieth, 1975). His result—24.5 billion t of carbon—was too low and it took several decades before this estimate was raised. Bazilevich et al. (1971) assembled an interesting review of primary production estimates published between 1919 and 1970 which shows the earliest estimates all just between 30 and 50 billion t/year. Deevey (1970) was the first to raise the value drastically, to 182 billion t. Afterwards the totals again fluctuated, mostly between 30 and 50 billion t and as low as 15 billion t, but by the end of the 1960s, sums in excess of 100 billion t were firmly accepted: Whittaker and Woodwell's (1969) 109 billion t, Bazilevich and colleagues' (1971) 172.5 billion t. The 1970s did not bring any radically different values.

As for the standing phytomass, all estimates up to the late 1960s fell within a very narrow range between 2.3 and 2.5 trillion t (Bazilevich et al., 1971). Only when actual ecosystem areas, rather than potential vegetation cover, started to be used in global phytomass calculations did the total begin to tumble.

Multiplication of approximate ecosystem areas (Table 2-1) by average net primary productivity and standing phytomass estimates in the first two columns of Table 2-2 yields the total production and storage values in the last two columns of this table. How do these figures compare with other global estimates published since the late 1960s? Surprisingly, one finds that such comparisons are very limited owing to the historical nature of the often-quoted Soviet and American estimates.

Detailed calculations were done by Bazilevich et al. (1971) for the reconstructed plant cover, i.e., for the preagricultural era before crop cultivation, clearing of forests for fields, settlements, and pastures altered the natural climax vegetation. Obviously, such estimates must be on the high side: net primary production was put at 171.5 billion t/year and the terrestrial standing phytomass at 2.4 trillion t.

In the review volume on primary productivity of the biosphere arising from the International Biological Program studies and edited by Lieth and Whittaker (1975), Lieth (1975) and Whittaker and Likens (1975) offer only slightly different productivity values—100.2 billion t in the first case, 117.5 billion t in the second—but both of these estimates are keyed to vegetation areas existing around 1950. For the same period, Whittaker and Likens (1975) also estimated dry matter phytomass at 1.837 trillion t.

So among the recent values, only estimates by Atjay et al. (1979) and by Olson (1970, 1982) refer to current productivities and storages. Atjay et al. put the total

TABLE 2-2
AN ESTIMATE OF AVERAGE NET PRIMARY PRODUCTIVITY AND PHYTOMASS SHORTAGE OF MAJOR ECOSYSTEMS AND OF TOTAL GLOBAL PRODUCTION AND PHYTOMASS STORAGE[a]

Ecosystem	Average NPP (t/ha)	Average phytomass (t/ha)	Total production (10^9 t)	Total storage (10^9 t)
Tropical moist forests	20	300	20	300
Temperate forests	10	250	10	250
Boreal forests	10	200	15	300
Woodland and shrubland	10	75	10	75
Tropical grasslands	10	20	20	40
Temperate grasslands	10	20	10	20
Cultivated land	10	10	15	15
Settlements, transportation	5	5	3	3
Tundras	1	5	1	5
Deserts and semideserts	1	5	2	10
Wetlands	15	75	8	40
Total	—	—	114	1058

[a]All values are in dry terms.

net primary production of terrestrial autotrophs at 60 billion t C or 133 billion t of dry phytomass and the total living phytomass at 560 billion t C or 1.244 trillion t of phytomass—but these estimates were obviously much influenced by Olson's work.

In his first set of global estimates, Olson's (1970) values for NPP and standing phytomass were, respectively, 120 billion t/year and 1.247 trillion t of dry matter phytomass. In his latest calculations summarizing detailed 0.5° × 0.5°-cell mapping of global vegetation, Olson (1982) offers a value for carbon storage totaling 560 ± 100 billion t for live vegetation, equivalent to 1.02–1.46 (mean of 1.24) trillion t of plant mass.

Comparison of Olson's painstakingly derived plant carbon total with my simple estimate using highly rounded area values and no less approximate storage means is most encouraging: my value, 1.06 trillion t, is virtually identical with Olson's lower value (1.02 trillion t) and just 15% below his mean. This difference would have been even smaller if Olson had incorporated the latest IBP work on grasslands which leads to choice of lower averages than those in previous general use.

And, moreover, Olson himself concludes that "further downward refinement seems more likely than the higher numbers sometimes cited and used in flux calculations." Consequently, it is justifiable to state that so far the most detailed global mapping of vegetation and the subsequent use of this computerized base for carbon storage calculations and my simple exercise are in excellent agreement.

Unless most of the storage averages are in great error (a very unlikely possibility) and until we get very reliable global land cover maps prepared on the basis of computer-processed satellite data, we should be content with the appealingly round total of 1 trillion t of phytomass (450 million t of carbon) stored in the aerial parts and roots of terrestrial plants. Similarly, it appears reasonably certain that the current rate of terrestrial net primary production is not much above 100 billion t of dry phytomass per year.

Unlike the estimates of terrestrial photosynthesis, those of primary production in the oceans have shown a declining trend, from values mostly over 100 billion t before 1960 (in fact, in the 1930s and 1940s it was thought that most of the earth's phytomass resided in the ocean) to totals between 30 and 60 billion t during the 1960s (Bazilevich et al., 1971). However, a carefully established range is that of Koblents-Mishke et al. (1970) at 60–72 billion t, and De Vooys (1979) evaluated the methods used to measure primary plankton productivity and concluded that the global total should be put at 43.5 billion t C, 95% of the annual primary production of all oceanic ecosystems (this would imply total production of 101 billion t of dry phytomass). Coastal macrophyta, salt marshes, and estuaries contribute 3.2% and coral reefs, highly productive but spatially not very extensive, a mere 0.7% of the global total.

As for the standing marine phytomass, it is insignificant in comparison with the terrestrial stores. Koblents-Mishke et al. put the total at 170 million t, while other recent estimates are as high as 3.9 billion t, in any case only a negligible fraction (no more than 10^{-1}, more likely only 10^{-3}) of all planetary phytomass.

Consequently, the ocean phytomass can be safely ignored in stating a planetary total of 1 trillion t of dry mass stored in living autotrophs but it increases the global NPP rate by at least one-half to two-thirds to a total of around 160 billion t annually. This total means that without constant replacement by respiration and decay, photosynthesis would deplete all atmospheric CO_2 in less than 5 years.

The obvious questions to ask—in view of the previously described (Section 2.1.2.1) changes of ecosystem areas—are, how much are these totals changing, what is the magnitude, and even more fundamentally, what is the direction of current carbon fluxes owing to shifts of plant formation? I have assembled all the previously described shifts in the areas of major ecosystems, together with typical phytomass declines, in Table 2-3: their combined effect would amount to an annual loss of roughly 5.5 billion t of plant mass (some 2.5 billion t C), an equivalent of about half a percent of all standing terrestrial phytomass. As I have not been conservative either in the selection of estimates for areas annually affected by devegetating and deteriorative processes or in the choice of average phytomass decline value, I must consider this total as nearly a maximum estimate.

Clearly, the global phytomass loss hinges on the estimates of tropical deforestation and degradation which in my calculations account for 85% of the total

TABLE 2-3
ESTIMATES OF GLOBAL ANNUAL DECLINE OF PHYTOMASS AROUND 1980[a]

	Areas affected (10^3 ha)	Standing phytomass decline (t/ha)	Total phytomass loss (10^6 t)
Tropical deforestation			4545
Development	1,000	300 → 5	295
Destruction	10,000	300 → 75	2250
Degradation	10,000	300 → 100	2000
Desertification			65
Grasslands	3,200	20 → 5	50
Fields	2,800	10 → 5	15
Development			410
Grasslands, fields	2,500	20 → 5	40
Forests	1,500	250 → 5	370
Erosion			275
Grasslands, fields	2,000	20 → 5	30
Forests	1,000	250 → 5	245
Farmland extension	2,500	20 → 10	25
All other losses	1,000	20 → 5	15
Total	37,500		5335

[a]All mass values are rounded to the nearest five.

and which must dominate any other similar accounts. Instead of the calculated 4.5 billion t, the annual loss may not surpass 3 billion t, but it is certainly not lower than 2.5 billion t. Hence, the likely range is about 3–5.5 billion t—but this loss must be adjusted for gains owing to natural tree regrowth and afforestation, above all in temperate regions.

Abandonment of less productive farmland in North America and in most European countries has led to considerable displacement of fields by woodland and shrubland as well to major increases in forested areas. Comparisons of available historical land use–land cover statistics for many European nations, the United States, and Canada indicate forest and woodland gains of almost 50 million ha during the past three decades. Assuming fairly evenly spread gains, the average age of this new growth would be just 15 years so that the additional phytomass storage should not be put at more than 150 t/ha and the annual gains would be no more than 250 million t of phytomass. If woody regrowth and afforestation elsewhere around the world would double this total, then about half a billion t of new phytomass would have to be balanced against losses of 3–5.5 billion t for a net loss of 2.5–5 billion t annually.

Comparison of this range with Olson's (1982) detailed estimate of current plant carbon shifts is quite interesting. Olson's total—derived in a very different way and including releases of organic carbon from burning during new land clearing, from rapid decay of new debris, and those owing to ecosystem changes and higher storages resulting from the regrowth including afforestation on nonwooded land and woody regrowth on abandoned land—comes to a net overall shift of nonfossil carbon to the atmosphere or water of 0.8 billion t, an equivalent of 1.8 billion t of phytomass.

But Olson (1982) also warns that the uncertainty band around the sum of carbon releases owing to ecosystem category changes is at least 1–2 billion t C/year and uncertainty ranges of up to 1 billion t are also inherent in burning and regrowth estimates so that the overall net carbon outflow from the plants may be as high as 4 billion t, or there may be an actual storage gain of 2 billion t, although he believes that there is most likely a loss fitting somewhere between 0.5 and 2.0 billion t C/year. As my range is 1.1–2.2 billion t of net terrestrial carbon loss, the two estimates are virtually identical, and until much better evidence concerning tropical deforestation and other vegetation losses comes along we should count the current anthropogenic interference with terrestrial ecosystems a net source of carbon at least on the order of 1 billion t/year.

But anthropogenic influences on phytomass carbon are not limited to widespread destruction and alteration of plant cover. As much of the forest clearing is done by burning and as burning is also a normal cultural practice in maintaining productivity of managed grasslands, these two practices, as well as crop residue burning on the fields and household and industrial combustion of fuelwood, straws, and dung are major contributors of carbon in the earth's atmosphere.

2.1.2.3. Biomass Combustion In spite of rapid global increases in consumption of oils, gases, and electricity since the end of World War II, most people have not left the era of biomass fuels. For hundreds of millions of peasants in Asia, Africa, and Latin America who live by subsistence farming, plant fuels—stem wood, branches, roots, and leaves of forest, grove, backyard, or roadside trees or shrubs, as well straws, stalks, and vines of field crops and, where the fuel shortages are greatest, dried animal dung—are either the sole or the principal source of energy not only for cooking and heating but also for food drying and smoking, commercial bread baking and alcohol brewing, brick and tile manufacture, and, as charcoal, for blacksmithing, metal smelting, and in restaurants.

To hope for any reliable accounting of these fuel uses is, obviously, impossible: although in many countries a surprisingly large share of fuelwood and charcoal is bought from large-scale traders or in local markets, most forest fuels and virtually all crop residues are gathered by the families themselves and used, as needed or as available, without any records. Even careful local surveys may be of little help, as extrapolations to regional, and more so to national, averages are often quite uncertain.

Global accounting of traditional biomass fuel consumption in the poor countries is thus an exercise in approximations and guesstimates and I have recently reviewed its state and outcomes in some detail (Smil, 1983), so here only some essential numbers will be presented. As with most global statistics, data gathered by a United Nations organization are used most commonly to quote the fuelwood consumption: the Forestry Division of the Food and Agriculture Organization prepares each year national estimates for all of the world's countries and these figures have recently toatled about 1.3 billion m^3. In contrast, Openshaw (1978) used a variety of actual fuelwood consumption surveys to put the mid-1970s total at almost 2.5 billion m^3 as a "conservative" estimate. Annual per capita totals in surveys cited by Openshaw (1978) are between 1 and 2.2 m^3 while Arnold's (1979) examples range just between 0.17 and 1.53 m^3.

Taking 1 m^3 per capita as the global average and assuming that each of the 3.2 billion people in the poor nations burns this volume of wood (or, in cities, its equivalent in charcoal) puts the global total at just over 3 billion m^3. I shall assume that the aggregate is now no less than 2.75 billion m^3 of roundwood equivalent or, using slightly lower than the standard conversion factor of 0.65, about 1.75 billion t of wood averaging about 14.5 MJ/kg. Global fuelwood consumption throughout the poor world would thus equal about 25,000 PJ, an equivalent of some 600 million t of oil.

Fuelwood consumption in the rich Northern Hemisphere nations, although again on the rise since the mid-1970s, adds only less than 150 million t to the poor world's use, a mass smaller than the uncertaintyin estimating the larger total. Global combustion of 1.9 billion t of wood would release, assuming complete oxidation and 45% carbon content, some 850 million t of carbon annually.

In deforested regions of the poor world (but also in the still fuelwood-rich localities during the months after crop harvests), the principal sources of biomass energy for household cooking are various crop residues, above all cereal straws. Estimates of their global combustion are certainly more accurate than those for fuelwood: the total mass of harvested crop residues can be calculated with satisfactory accuracy and although the competitive uses (above all as animal feed and bedding) account for widely varying shares in different countries, the most likely range is relatively small.

For China, the world's largest consumer of crop residues for fuel, there is now a countrywide study of rural energy showing that one-half of 450 million t harvested in recent years was burned (Smil, 1984b). In India, where the demand for feed is relatively higher, only about one-fifth of all straw is burned (Revelle, 1976). In other nations for which good estimates are available, the shares range from just 10% for rice straw in Bangladesh to three-quarters of all residues in Egypt (el Din et al., 1980).

The average for the poor world's crop residue combustion thus cannot be set higher than 50% and lower than 25%, with the most likely share at about 40%. Applied to the annual crop residue harvest of 1.3 billion t (Smil, 1983), this translates to some 500 million t (325–650 million t), containing about 7300 PJ and releasing in complete combustion 225 million t of carbon. Finally, combustion of about 100 million t of dried animal dung—mainly in India, Pakistan, Bangladesh, Chinese interior, Sahelian Africa, and Andean America—adds nearly 50 million t of carbon for a grand total of just over 1.1 billion t.

Approximate as all these values are, they are much closer to elusive reality than estimates of carbon releases by combustion of large volumes of biomass during forest wildfires, fires set to clear the vegetation in shifting cultivation, and also by grassland fires in pastoral regions whose extent is impressively documented by nighttime satellite images (Croft, 1976). In all of these cases one cannot hope for more than a proper order of magnitude.

The number of shifting cultivators was estimated by Myers (1980) to total about 140 million people in the mid-1970s but other estimates put the total at over 200 million already in the late 1950s (Nye and Greenland, 1960). Should the current number be as large as 300 million people, or about 50 million families, and should each family clear annually just 1 ha of forest, 50 million ha of trees would be burned each year. As much of this growth will be a poorer secondary forest, a standing phytomass of no more than 20 t/ha should be used to calculate the maximum release of 450 million t of carbon.

The overall extent of tropical grassland burning is very difficult to estimate. Fire as a management tool is considered necessary in most grazed tall-grass formations where annual or semiannual burning can boost herbage production by impressive margins (Gupta and Ambasht, 1979). Assuming that half of the permanent pastures in tropical Asia, Africa, and Latin America are burned each year, the total area affected would be at least 1.2 billion ha.

For comparison, Flohn (1961) assumed that about 60% of the area with tropical seasonally dry climate (Köppen's Aw type) and 5% of the adjoining climatic regions (Af, BS, Cw) are burned: this adds up to about 1.5 billion ha but this total also includes arable land on which the unharvested crop residues are also burned. And Root's (1976) estimate, also including arable land subject to annual burning, is 1.8 billion ha.

As I will account for agricultural burning separately, I will stay with a lower value of 1.2 billion ha; with a conservative average estimate of about 4 t of aboveground phytomass per ha and with 75% burn-up, the annual carbon release from this source would come to 1.6 billion t, a total rivaling CO_2 generation from coal burning.

Field burning after harvest is common both in tropical regions to dispose of unwanted crop residues before putting in a new crop as well as in some regions of temperate latitudes where residues are not so extensively needed for protection against erosion and for bedding. Based on the available accounts of crop residue management practices (Tanaka, 1973; Staniforth, 1979) I shall assume, conservatively, that in temperate latitudes about 10% of all residues are burned, in tropical regions at least 30%. These assumptions would translate in both cases into roughly 100 million t of residue whose incomplete (75%) combustion would introduce annually about 70 million t of carbon into the atmosphere.

Finally, using the United States forest wildfire statistics as a base to approximate the global total burned annually prorates to no less than 20 million ha of forest destroyed by fire. With 30 t of standing phytomass per ha and 75% burn-up, these fires would liberate 200 million t of carbon annually.

The grand total of anthropogenic biomass combustions—fuelwood (850 million t) and crop residue and dung (270 million t) household combustion, straw and stalk field burning (70 million t), vegetation burning for shifting cultivation (330 million t) and for pasture management (1.6 billion t)—is thus about 3 billion t of carbon. Considering the cumulative uncertainties inherent in all of these estimates, the actual value may be put with reasonable confidence somewhere between 2 and 4 billion t annually.

2.2. FOSSIL FUEL COMBUSTION

Return of fossil carbon—sequestered at least thousands of years ago in fresh peats but more likely tens and hundreds of millions of years ago in coals, oils, and gases—to the atmosphere is now undoubtedly by far the largest anthropogenic contribution to carbon's biospheric flows. Releases of common air pollutants such as carbon monoxide, particulate matter, nitrogen and sulfur oxides, and hydrocarbons are globally two to three orders of magnitude smaller than the annual CO_2 flux from fossil fuel combustion.

This is a necessary consequence of the world's critical dependence on carbonaceous fuels whose total extracted mass now surpasses 7 billion t annually, i.e., almost 2 t per capita each year. The only other human activity which rivals this huge solid mass transfer is the combined extraction of all nonfuel minerals, including the excavation of construction sand and rocks. As on the average some two-thirds of the burned fossil fuels is carbon, it is obvious that annually about 5 billion t of the element enters the atmosphere from our chimneys and smokestacks. The following sections will attempt to fix this flux with the greatest possible accuracy, an effort unthinkable without first appreciating the trends in fossil fuel energy consumption and difficulties in using averages to characterize fuel composition.

2.2.1. Energy Consumption

Although the extraction of fossil fuels has evolved into generally highly mechanized activities which are also highly concentrated and closely monitored, there is still surprisingly large room for uncertainty in compiling reliable consumption statistics in other than simple output mass terms. This is not only because of the often unmonitored difference between production and consumption caused by fluctuating stocks, storage and processing losses, and nonfuel uses of fossil fuels (mainly in chemical syntheses) but above all owing to the wide variety of extracted fuels hidden by standard mass or volume output statistics. Substantial compositional differences thus make it virtually impossible to calculate very accurately the global consumption of fossil fuels in energy terms, an item of information of undoubtedly greater interest than the usual tonnage data.

And, similarly, the inevitable use of representative averages for carbon, nitrogen, or sulfur shares in individual fossil fuels, and yet other averages for their actual emission factors, does not allow determination of totals whose accuracy could be tagged with less than ±10% errors. The following sections will present a detailed discussion of carbon in fossil fuels after a brief review of availability and reliability of global coal, oil, and natural gas production statistics.

2.2.1.1. Fuel Statistics Since the first round of rapid crude oil price rises in 1973–1974, the number of available statistical sources for various forms of energy has expanded considerably. Presentation of oil data ranges from monthly production reviews in *Petroleum Economist* to the CIA's detailed output, trade, and consumption compilations in the monthly *International Energy Statistical Review*. These publications are obvious choices for the most up-to-date information on specific facts on a national level. For all nations belonging to the Organization on Economic Cooperation and Development, updated, disaggregated statistics are conveniently available in the excellent *Statistics of Energy* and as these nations account for about three-fifths of the global energy use, all that is needed to obtain an almost complete world coverage is to augment the OECD publications by data for Communist nations, above all the USSR and China, to obtain coverage of some nine-tenths of the global energy use. This can be done easily by consulting the regular reports of the Council for Mutual Economic Assistance, the Soviet statistical yearbook (since 1955), and, after a gap of two decades (1958–1978), the renewed Chinese data books.

All these sources are of unmatched importance for such calculations as nitrogen oxide and sulfur oxide emissions where different coefficients must be used for individual final uses. For calculation of CO_2 generation, however, only the total global consumption figures are necessary so it would seem that this is the easiest requirement to satisfy. Indeed, there is a convenient compilation of detailed information on production, trade, and consumption of all energy sources for all of the world's countries and territories which is generally used as the most reliable source of global energy consumption trends—*World Energy Supplies*, published annually by the United Nations as Statistical Papers, Series 7 until 1979 (the volume containing data for 1973–1978) and since then superseded by the *Yearbook of World Energy Statistics* whose first volume, published in 1980, reviewed all of the 1970s (UNO, 1980).

Another advantage of this series is its historical coverage important for calculations of long-term emission changes. The first volume of the series covers production, trade, and consumption data for the years 1946–1950 so that we now have nearly four decades of basically consistent information. For tracing the secular development of CO_2 emissions we should, of course, have coverage extending much further back but the *World Energy Supplies* have global information on just two years before 1946—1929 and 1937—in the first, and the least accessible, volume. What might be seen as the predecessor of this series, compilations of the World Power Conference, does not extend much further. This organization, now known as the World Energy Conference, published the first comprehensive review of global energy production in 1929 with most of the data referring to 1924–1927 (WPC, 1929).

And so the best available series with the longest coverage is the compilation prepared in 1956 by the staff of the United Nations Department of Economic and Social Affairs for the International Conference on the Peaceful Uses of Atomic Energy (UNO, 1956): it covers the years 1860–1953 and lists only global production totals of hard coal, lignite, crude oil, natural gas, and hydroelectricity. This series was used as the principal data base to calculate global CO_2 emissions in *Carbon Dioxide Review: 1982* and as that emission table (Rotty, 1982) is the longest one readily available in the CO_2 literature, it is bound to be widely used and quoted.

However, the data in the original table are not very reliable because for the period 1860–1913, "they have been estimated chiefly on the basis of reported production (or capacity) in the USA, and the US share in the world production at the end of the period" (UNO, 1956). This was, obviously, the easiest path to be taken by the then American dominated staff of the Department aware of the excellent fuel production series readily available in the United States Historical Statistics (for the latest edition, see U.S. Bureau of the Census, 1975)—but it is surely a source of numerous errors as the relative position of the United States fossil fuel output changed drastically between 1860 and 1913. For example, in 1913 the country mined 430 million t of coal, 2.25 times more than the United Kingdom—while in 1860 the U.K. produced more than nine times as much bituminous coal as the USA (Daniel, 1956).

For the years after 1913, when the estimates were based on data covering most producing countries, the errors are certainly smaller but the basic difficulty remains: this is an output series, not a consumption one, and hence its use inevitably overestimates any emissions generated during actual combustion. Consequently, we have no satisfactorily reliable global fossil energy consumption series until the late

1940s—and even during these past four decades the United Nations energy statistics should not be used uncritically. There are primarily two important reasons for this. The first are the errors arising from rough and often outdated coefficients used in the UNO statistics in conversion to hard coal equivalents (for more on this, see the next section), and second, the major errors introduced by estimates of Chinese energy consumption between 1958 and 1978.

2.2.1.2. COALS "There is no such thing as coal, there are only coals" goes a common comment of those knowledgeable about this heterogeneous group of solid fuels. Ultimate analysis of coals show a very wide variation of all essential constituents with carbon ranging from just 15% in the poorest lignites to 98% in the best meta-anthracites, volatile matter as low as 0% and as high as 85%, and ash content between 1 and 45% (Averitt, 1975; World Energy Conference, 1974). Overall, there is about a fourfold difference between the heat value of the worst and the best coals (8.3 versus 36.3 MJ/kg) and there is no simple way to find the total carbon content of the global coal extraction.

Still, simplifying approaches are inevitable for estimating globally applicable coverages. Hirschler (1981) simply calculated average composition of hard coal from 16 representative samples of various ranks listed in Considine (1977). His typical hard coal has 58.1% of fixed carbon and 25.8% of volatile matter (nitrogen- and sulfur-free), whose approximate composition taken as $C_3H_6O_2$ adds another 12.6% of carbon. The total of 70.7% is then reduced by 3% to account for cleaning losses and aerosol formation.

Keeling (1973a) took a different simplifying approach by relying on the United Nations coal satistics which have since 1960 listed values for both bituminous coal and lignite in terms of hard coal equivalent and assuming an average carbon content of 70% for hard coal and 63% for lignite. For the period before 1960 when lignite totals were not converted to hard coal equivalents, he multiplied by 0.44, the mean of post-1960 conversions. With bituminous coal dominant, Keeling's (1973a) average carbon share for all coals ranged from a high of 69.5% in 1929 to 68.6% in 1969, the last value being identical with Hirschler's (1981) independently derived mean for net emissions.

There are at least two nontrivial problems with both of these approaches. To begin with, owing to the great variety of coals, any averaging needs a weighted mean based on compositions of principal coals extracted in several top coal-mining countries rather than an arithmetic average of some coal samples. Moreover, such a mean could not be permanently fixed: short-term differences might be negligible but in countries extracting large amounts of progressively lower-quality coals, mean carbon content can easily slip by more than 10% in two decades.

The uninitiated researcher might think that these quality changes are accounted for by conversion coefficients used in the United Nations statistics to express all coal extraction in hard coal terms. Yet this is not the case as these approximate coefficients have now been used unchanged for over two decades, the time during which especially the quality of lignites extracted by the world's three leading producers (Germany, USSR, and Czechoslovakia) declined by more than 10%.

Especially large errors arise from inappropriate conversion of the Soviet coal production whose average energy content declined by about 17% between the mid-1950s and the early 1980s (CIA, 1980; Tsentralnoye Statisticheskoye Upravleniye, 1982). In 1981 the country produced 544.213 million t of black and 159.831 million t of brown coal and according to the official Soviet data this equals only 470.5 million t of standard coal (i.e., the mean heat value of all Soviet coal is now just 19.42 MJ/kg).

However, the UNO conversion coefficients are 0.82–0.92 for low-grade black coal and 0.5 for brown coal, so even when assuming that all black coal is of lower quality (which it is not) and when multiplying by the lowest value (0.82), the USSR's bituminous coal output would equal no less than 451.09 million t of standard fuel, that of brown coal 79.91 million t for a total of 531.0 million t, about 13% higher than the actual value. And similar corrections are necessary for the huge East German lignite output as well as the Indian bituminous coal extraction and, to a lesser extent, the hard coal production in many coal basins around the world.

Whatever the actual specific differences, simplifying estimates are thus definitely overestimating the carbon content of lignites currently used for combustion. The second important cause of overestimation arises from the differences between raw (run-of-the-mine) coal and processed fuel. Although coal processing is now common in all industrialized nations, it is still far from universal even there—and it is largely absent in the poor countries, most notably in China.

So, once again, the uninitiated researcher may assume that each tonne of coal listed in the United Nations energy statistics is identical as long as it is expressed in hard coal equivalent but in reality much of the extracted coal is not processed before combustion while the rest is prepared by washing, sorting, and grading for specific uses requiring uniform quality.

By far the greatest discrepancy concerns China, the world's third largest producer of bituminous coal: that country's coal output totals were always treated in international statistics as though referring to the international standard of washed coal at the pithead—while in reality the Chinese continue to report in raw coal terms as only 10% of their nonmetallurgical coal is washed. The average energy content of China's bituminous coal production is thus only 20.8 MJ/kg rather than the standard 29 MJ/kg, a difference of nearly 30%. Ignoring this fact alone leads, with an annual Chinese coal output now around 700 million t, to an overestimate of about 200 million t, a tonnage greater than the annual bituminous output of all but the world's three largest bituminous coal producers!

I have tried to make all of these corrections in calculating the most likely hard coal equivalent of global coal extraction. As a result, the 1981 production of roughly 3.7 billion t (2.7 billion t of bituminous coals, 1 billion t of brown coal and lignites) shrinks to just 2.6 billion t of hard coal equivalent, and the 1975 output of 3.2 billion t converts to 2.34 billion t. Compared to the conversion done with the UNO multipliers, these totals are about 10% smaller.

Consequently, I would argue that it is possible to use the UNO global coal production totals for the 1970s and early 1980s with greater accuracy by simply reducing them by 10%. For older data, different correction factors would be needed. The importance of these clearly justified corrections is obvious: there has been less carbon in coal than implied by the universally used UNO data and hence less CO_2 generated. In calculating coal-generated CO_2 I will reduce my production estimate by 5% to account for cleaning, handling, and storage losses and for a part of the carbon not oxidized during combustion, and then multiply by the carbon emission factor of 0.71.

2.2.1.3. HYDROCARBONS Crude oils are much more homogeneous fuels than coals, with ultimate analysis showing a very narrow carbon range, between 83.9 and 86.8% (King et al., 1973). Consequently, good results should be obtained simply by multiplying the amount of burned petroleum products by 0.85 to get carbon emissions with perfect combustion. Using a separate emission factor for each of the major refined products (gasoline, kerosene, various fuel oils) would not alter the total by more than a few percent.

Similarly, Keeling (1973a) uses Brame and King's (1967) average of 84% carbon, and Hirschler (1981) assumes that the total heteroelement content of crude oil is just 1% and that the remainder has an average composition of $CH_{1.8}$, with approximately 3% remaining unburnt in typical operations; these assumptions translate into an emission factor of 0.835 kgC/kg refined petroleum products.

Global crude oil output must be reduced by tonnages lost in refining operations, during ocean shipping, land transfers, tank storage, and vehicle filling, by the fractions used for petrochemical feedstocks and as lubricants on road-surfacing compounds. Refinery losses are usually negligible and there may actually be volumetric gains during the feedstock processing. But transportation, storage, and handling losses should not be neglected. As most of the world's oil trade involves tanker shipping, it is not surprising that losses from regular tank flushing and, to a much lesser degree, from accidental spills, coastal refining, and offshore drilling, add up to millions of tonnes each year. Available estimates differ but the total is almost certainly no less than 2 million t, possibly as much as 15 million t/year.

Typical specific losses owing to evaporation from storage tanks, loading and unloading of storage and transportation containers, and filling of motor vehicles are detailed in the EPA's emission factors summaries (USEPA, 1973a). Application of these averages to global stores and flows of crude oil and refined products gives annual losses of almost 10 million t. But the emission factors used apply best to United States conditions and so the global loss may easily be much higher.

For polyethylene, polypropylene, and polystyrene, the feedstocks average about 46 GJ/t of final product; for PVC, about 22 GJ/t (Berry et al., 1975). With an average of 42.7 MJ/kg crude oil, these values translate, respectively, to 1077 and 515 kg of feedstock crude per t of plastics. Using the weighted mean of the United States plastics production, the world's largest, the best representative value appears to be about 950 kg of crude per t of plastics. Applied to the global plastics production of over 50 million t at the beginning of this decade, this means that some 50 million t of crude oil was diverted as a raw material for this rapidly growing industry.

Nonenergy oil products—above all, lubricating oils, paraffins, and asphalt—are fairly accurately monitored in international statistics. Several decades ago they accounted for nearly one-fifth of the total refinery output but lately they have slipped to about 8%. To determine the recent refined fuel consumption total, I will thus reduce the annual crude oil output (including gas condensate liquids) by 1% to account for shipping, handling, and storage losses and, in turn, this total will be lowered by 10%, 2% going to petrochemicals and 8% for nonenergy uses, so that the current consumption rates for combustion would be about 89% of the total extraction.

For greater accuracy one should not remove completely these nonenergy uses of refined products from consideration as CO_2 sources. While carbon in asphalt will be buried under new paving or removed and discarded in landfills and hence very effectively prevented from oxidation, many plastics will eventually be burned and most lubricants will in time oxidize and a lag factor should be used for the entry of their carbon into the atmosphere. Keeling (1973a) estimated such turnover times but I do not think it necessary because asphalt, the single largest nonenergy use of crude oil products (about two-fifths), is virtually unaffected by oxidation, and plastics, another large diversion, are for the most part not disposed of by burning. The most likely underestimate then amounts to no more than a few percent, a negligible total amidst all other uncertainties.

Natural gas is the most homogeneous of the three principal fossil fuels: it is extracted either as a nearly pure methane or as a mixture of methane homologs dominated by CH_4 with lesser amounts of C_2H_6, C_3H_8, C_4H_{10} and with a sometimes nonnegligible presence of nitrogen and very small amounts of hydrogen, sulfur, and CO_2. Common processing before the compression in pipelines further reduces the content of minor constituents so that the actually consumed fuel is of considerable homogeneity and there is no necessity to resort to lengthy compilations of individual gas compositions.

Further confirmation of this uniformity comes from comparisons of readily available average energy contents for the world's major natural gas producers. Most notably, the difference between American and Soviet averages (the two countries produce roughly three-quarters of the global output) is a mere 5%. Consequently, one might simply assume that average natural gas is a mixture of nine-tenths of methane, with a lower heat value of 35.5 MJ/Nm^3 and specific density of 0.717 kg/Nm^3, and 8% of higher hydrocarbons expressed as ethane, with 63.9 MJ/Nm^3 and 1.34 kg/Nm^3, hence a fuel with about 37.1 MJ/Nm^3 and typical density of 0.78 kg/Nm^3.

These assumptions result in an average heating value very close to that used in the UNO statistics where 1 m^3 of natural gas is put to be an equivalent of 1.332 kg of standard coal (29 MJ $\times$ 1.332 = 38.6 MJ/m^3). As carbon is 75% of CH_4 and 80% of C_2H_6, the average carbon content of such a mixture is 75.5% and perfect combustion would release about 0.59 kg of carbon for each cubic meter of natural gas burned at 0°C. At an ambient temperature of 15°C, the emission factor would be about 5% lower, or 0.56 kg/m^3.

For comparison, Hirschler (1981) assumes that typical natural gas contains 38.9 MJ/Nm^3 and 71.25% of carbon and weighs 0.7 kg/m^3 to yield an emission factor of 0.5 kg C/m^3 natural gas, while Keeling (1973a) uses the average composition given by Brame and King (1967)—CH_4 89.6%, its homologs 5.5%, CO_2 0.8%, N_2 4.1%—to calculate a density of 0.802 kg/Nm^3 at 0°C and a carbon content of 0.565 kg/Nm^3 (0.536 kg at 15°C).

My simply derived emission factor is thus less than 5% higher than Keeling's multiplier and about 11% above Hirschler's value so it makes sense to use the arithmetic average of 0.53 as the single convertor while being aware that the uncertainty about actual emission is most likely confined within 10% around this mean. But accounting of natural gas combustion has more complications as relatively large volumes may be spent by repressuring of hydrocarbon fields and, above all, by flaring.

Flaring losses would not arise if oil and gas were not present in the same fields. In that case, gas wells could simply be capped and stored for later extraction. However, most of the time, oil and gas are found together, so that the "associated gas" accounts for the bulk of global gas output, and most oil cannot be produced without appreciable gas flows. Separation of the two fuels is not a forbidding technical problem but is not done in very many cases either because there is no distribution infrastructure and no developed market for the gas (a common situation in the hydrocarbon-rich Middle Eastern nations) or because the processing of the gas and its transportation from smaller remote fields would be uneconomical.

The extent and magnitude of natural gas flaring is best demonstrated by nighttime images acquired by the United States Air Force's Defense Meteorological Satellite Program: the brightness of the largest flaring zones in Saudi Arabia, Iran, Nigeria, and Siberia rivals the lights of the world's major cities and conurbations (Croft, 1976). Pinpointing the locations of natural gas flaring is thus relatively easier than calculating satisfactory estimates of total flared volumes. Rotty (1974) was the first to collate or derive such estimates on a global scale.

Subsequent revisions and extensions led to natural gas flaring estimates for the years 1950–1980 (Rotty, 1982) indicating that the practice wasted fuel equivalent to about 25% of the global marketed production in 1950 and to less than 15% by 1980. Undoubtedly, with higher energy prices and with gradual increases of liquefied natural gas exports, this share will continue to shrink although it is certain that some huge Middle Eastern flares will continue to light the desert skies for decades to come.

Natural gas used as feedstock in the chemical industry goes predominantly into synthesis of nitrogenous fertilizers as the donor of hydrogen for subsequent ammonia synthesis (see Section 3.2.3.2) so that all carbon is released as CO_2 and hence the feedstock natural gas can be treated as if burned for the purpose of CO_2 accounting. In the United States, unlike in Europe or Japan, natural gas is also used as the principal feedstock to ethylene, but this diversion amounts to just a few million tonnes per year, a total much smaller than the uncertainty connected with gas flaring and hence easily neglected.

2.2.2. CO_2 Generation

With both fuel statistics and appropriate emission factors discussed, critically evaluated, and carefully chosen for global applications, all is in place for calculating the current rates of CO_2 generation and for a relatively more tenuous account of the past increments. Before the final summation, however, two remaining anthropogenic gaps must be filled: combustion of solid wastes and CO_2 generation by cement production.

Accounting for carbon releases in cement making is much the easier task of the two. Limestone, the principal raw material in the process, is transformed into CO_2 (42%) and CaO (53%) (the other 5% being impurities in the feedstock rock); typical portland cement contains about 64% CaO which means that production of 1 t of cement generates about 138 kg of carbon. With the annual global cement production now at about 800 million t, some 110 million t of carbon is transferred to the atmosphere each year.

There are no global statistics on the total amount of solid wastes. Data available from many industrialized countries show values as high as 5 kg/day per capita in the United States, up to 2 kg in West Germany, about 0.5 kg in poorer European countries; however, a considerable portion of this waste is not combustible and most of it—often as much as 90–95%—is just dumped anyway. A chain of assumptions is thus needed to derive at least a proper order of magnitude range: providing that the total global generation of combustible waste is about 1 billion t, that its average heat content is 10 MJ/kg and its carbon content is about 25%, and that about half of the generated total is burned (either in incinerators or in the open) with at least 75% efficiency, annual carbon releases would be nearly 100 million t; uncertainty requires possible extreme values of at least ±50%.

Current Inventory and the Past Increments The best approximate estimates and, where data permitted, careful calculations presented in the preceding sections are assembled in Table 2-4 to give a complete summary of anthropogenic CO_2 generation. Fossil fuel combustion is by far the largest contributor with nearly 5 billion t of carbon annually, but much less appreciated is the fact that burning of biomass fuels releases about 1 billion t of carbon, or about one-fifth of the mass generated by fossil fuel consumption. An even larger contribution comes from burning for shifting cultivation and grasslands management, although these totals are certainly the least reliable of all the listed values. In sum, anthropogenic carbon combustion emissions amounted to nearly 8 billion t at the beginning of the 1980s. Before I consider the past carbon releases caused by human activities, I shall compare the entries of Table 2-4 with recent estimates of other workers.

First to be considered is fossil fuel combustion. Hirschler's (1981) detailed inventory comes to (including natural gas flaring and cement production) 5.47 billion t C for 1977. *Carbon Dioxide Review: 1982* lists for the same year 5.03 billion t C and for 1980, the last year of its carbon generation table, 5.255 billion t (Rotty, 1982). Using global fossil fuel production data for 1977 and the same

TABLE 2-4
ANTHROPOGENIC RELEASES OF CO_2 IN THE EARLY 1980s[a]

	Total production (10^9 t)	Total combustion (10^9 t)	Carbon emission factor	Total carbon generation (10^9 t)
Fossil fuel combustion	—	—	—	4.58
Coal	2.6	2.47	0.71	1.75
Liquid fuels	2.6	2.31	0.84	1.94
Natural gas (10^{12} m^3)	1.5	1.40	0.56	0.78
Natural gas flaring	—	0.21	0.56	0.11
Biomass fuel combustion		2.50	0.40	1.00
Fuelwood	1.9	1.90	0.40	0.76
Crop residues	2.3	0.50	0.40	0.20
Dried dung	—	0.10	0.40	0.04
Crop residue field burning	2.3	0.20	0.33	0.07
Grassland burning	—	4.80	0.33	1.60
Burning for shifting cultivation	—	1.00	0.33	0.33
Cement production	0.8	—	0.14	0.11
Solid waste combustion	1.0	0.5	0.19	0.09

[a]Fossil fuel combustion values are for 1981.

combustion shares and emission factors as in Table 2-4, I calculated the 1977 flux at about 4.7 billion t C.

The difference between my accounts and those in *Carbon Dioxide Review: 1982* is about 8%, a very good agreement to begin with, and the discrepancy is easily explained largely owing to my adjustments of coal consumption (mainly the lower hard coal equivalent values for the Soviet, eastern European, and Chinese coals). Hirschler's (1981) values are obviously exaggerated because of his failure to adjust for lower-quality coal extraction as well as for nonfuel uses and losses of liquid fuels.

I would thus recommend the conveniently tabulated CO_2 generation values in *Carbon Dioxide Review: 1982* as the best published emission summaries for the years after 1950 when relatively reliable global fuel statistics became available—but with the reminder that in the early period of this 30-year series reduction of the totals by about 5% and in more recent years reduction by about 8% would bring those values perhaps even closer to reality.

Estimates of carbon from fuelwood combustion are generally too low, owing to the use of FAO fuelwood statistics or outdated consumption estimates, such as those of Robinson and Robbins (1972). Consequently, Wong (1978) put the total at only 200 million t C/year and Hirschler (1981) at 320 million t, but the latter value also includes forest fire contribution. Hirschler (1981) also estimates 30 million t C from agricultural burning, the same order of magnitude as my value. I have no doubt that a relatively large amount of new information published on biomass fuels since the mid-1970s (Smil, 1983) justifies the higher carbon generation total of about 1 billion t C presented in Table 2-4.

Values for carbon generated by biomass combustion are not usually available in such a disaggregated form as presented in Table 2-4 where the total sum of all combustion and on-site burning releases is 3 billion t C. Olson (1982), using very different approaches, put the "normal" burning releases, i.e., carbon generated by grassland and forest fires, slash and burn farming, and traditional fuelwood combustion, at 3.15 billion t, an excellent agreement.

Grand rounding of the best available values to the nearest billion thus yields 5 billion t of carbon emitted from the combustion of fossil fuels and 3 billion t released from the burning of biomass. But, of course, the latter source does not differ from the former only owing to its genesis (fossil versus recent carbon) but also owing to its effects on the global carbon cycle: much of the grassland and forest burning for pasture management and shifting cultivation, as well as the field burning of crop residues and household combustion of straws, stalks, twigs, and wood has been done for millennia at a relatively steady rate and can be thought of as just an accelerated respiration.

In appraising the buildup of CO_2 in the atmosphere, one thus seeks not the total emissions generated by biomass combustion but only that emitted mass which is in excess of the "normal" photosynthesis–respiration cycle. Similarly, carbon releases resulting from deforestation, clearing of other vegetation, and decline of soil, humus, and debris carbon must be lowered by countervailing carbon storages in new plantings and in existing vegetation to arrive at net values. This is, indeed, one of the fundamental uncertainties complicating any quantification of carbon cycle: is the biosphere a net source or sink of carbon?

The ability of vegetation to influence the CO_2 content of the atmosphere in microscale as well as in global terms has been recognized for many decades: Lundegardh's (1924) classic is an excellent example as many of its conclusions can still be cited today.

But it was not until the 1970s that vegetation's role in the global carbon balance became a topic of major scientific interest—and of some quite irreconcilable disagreements as disparate estimates by Adams et al. (1977), Bolin (1977), Woodwell et al (1978), Wong (1978), Olson et al. (1978), and others followed in rapid sequence. Clark et al. (1982) tabulated 17 such estimates published during the 1970s and early 1980s and Woodwell (1983) provided a similar review.

With regard to my total of 1.1–2.2 billion t C (Section 2.1.2.2) as the net annual loss arising from the anthropogenic changes of vegetation cover and Olson's (1982) similar range of 0.5–2.0 billion t, the other available estimates have means ranging from 0 to 8 billion t and extremes going up to as much as 18–20 billion t C.

These very high values were favored by Woodwell and his colleagues in the late 1970s—the best summary is Woodwell et al. (1978)—but they are obviously untenable: by 1983 their principal author lowered the value significantly to an average of 2.6, and a maximum of 4.7 billion t C (Woodwell, 1983). Simple averaging of the 17 listed means gives 2.58 billion t C, and after the elimination of two of Woodwell's discarded values, the average declines to almost exactly 2 billion t C. A surprising degree of consensus can then be seen with regard to the magnitude of carbon releases from the biosphere as ten of the given estimates differ no more than ±50% from this mean, as also does my range.

Consequently, although the idea of the biosphere as neither a significant source nor a meaningful sink of carbon is far from extinct, the consensus today relates to the question of just how large a net carbon source are the declining biota. And although the agreement in this respect is surprisingly good considering all the uncertainties attendant in global estimates of this kind, in absolute terms the continuing impossibility of placing it confidently somewhere within the most likely range of 0.5 to 3 billion t C annually means that we cannot determine the total of excess anthropogenic carbon releases to the atmosphere more accurately than about 6.5 ± 1.5 billion t annually.

Obviously, even greater uncertainties surround any estimate of the past CO_2 generation rates. Here the middle of the 19th century is usually taken as the starting point—the industrialization of the Western world, fueled by rapid increments of coal consumption, and the large expansion of pioneer farming in North and South America, Russia, and Australia and tropical deforestation all began to accelerate after 1850. As noted in Section 2.2.1.1, the commonly used global fossil fuel consumption series starting in 1860 is not very reliable until 1913 and the CO_2 totals based on its estimates may commonly be off by 25–35%.

These errors must be kept in mind when attempting comparisons of historical CO_2 releases and assumed rise in atmospheric concentrations. During the period of the most uncertain data (1860–1914), cumulative CO_2 releases from fossil fuel combustion amounted to about 20 billion t C, an amount generated today in just 4 years, and between 1860 and 1980 the grand total is about 160 billion t (Clark, 1982). Half of this total was generated only since 1960 and by that time our global energy consumption data became fairly reliable with annual cumulative errors unlikely to surpass 10% (see Section 2.2.1.1).

Consequently, the cumulative mass of fossil fuel-derived carbon for the years 1860–1985 was almost certainly not larger than 180 billion t and definitely not smaller than 150 million t, with 165 billion t C being my best estimate. No comparable accuracy is possible in estimating the total of excess carbon liberated by pioneer farming, deforestation, and vegetation changes owing to urbanization and industrialization. One may come up with plausible global totals for increases of farmland—a good example can be seen in the study of Revelle and Munk (1977)—but disaggregation of such changes according to the original ecosystems is largely guesswork where errors become further compounded by use of very uncertain estimates of initial carbon content per hectare of preagricultural vegetation unit.

Available estimates of CO_2 releases following conversions to crop fields, again conveniently summarized in Clark et al. (1982), cover mostly the period from 1850 (or 1860) to between 1950 and 1974 and range from as little as 40 billion t (Bolin, 1977) to almost 200 billion t (Siegenthaler and Oeschger, 1978) of biospheric carbon moved into the atmosphere; in terms of annual rates, this means anywhere between 0.5 and 1.7 billion t C but the process has been, obviously, one of very substantial fluctuations. Simple averaging of all the estimates covering the period from 1860 to 1970 results in a cumulative mean of 125 billion t C and an annual average of 1.13 billion t. Using the same mean annual rate for the 1970s would raise the total to just over 135 billion t C, an equivalent of about four-fifths of carbon releases from combustion of fossil fuels.

The order of magnitude is undoubtedly right—all one has to do to check this is to assume simply that half of the world's closed forests (some 2.5 billion ha) has been replaced since 1850 by farmland, grassland, shrubland, as well as by settlements, industries, and transportation links, so that the original carbon content averaging about 125 t C/ha shrank to no more than 50 t C/ha. The result is almost 190 billion t C lost with an annual average of about 1.4 million t C, values perfectly consistent with the elaborately derived estimates cited in the preceding paragraph, and a nice proof that in uncertainty a single simple calculation does as well as complex constructs based on guesswork.

The most recent addition to these carbon loss accounts is a very detailed study of global conversions of forested land to agricultural uses prepared by Richards et al. (1983). These authors assembled a wide variety of statistics and estimates for 176 countries and territories for the periods 1860–1920 and 1920–1978, and added new estimates for conversion of land into permanent crops. Summation of national totals shows that between 1860 and 1920 some 440 million ha was converted from natural or disturbed ecosystems to farmlands, while between 1920 and 1978 the aggregate was even larger at 470 million ha.

After including the reversion of some cultivated land to woodland and scrubland, the net conversions came to, respectively, 432 and 419 million ha for the two periods and carbon releases from the biomass removed to make place for these newly cropped fields totaled nearly 39 billion t, or 330 million t C/year, with losses from oxidation of humus estimated at some 23 billion t C. This detailed account did not include losses from shifting agriculture, intensification of forestry practices, conversions to unimproved pastures and rangelands, or deforestation for settlements, industrial or transportation reasons. All of these biomass losses would easily double the 62 billion t C total and would place this latest evaluation squarely into the midrange of other available estimates of carbon losses caused by ecosystem changes.

While the values between 100 and 200 billion t C represent the most plausible range of biomass carbon loss since the middle of the last century, the lowest acceptable total should not be put at less than 70 billion t C so that the contributions of the biomass carbon would equal less than one-half of fossil fuel emissions during that period—or they would add up to a total 1.25 times larger than the combustion releases. The grand total of anthropogenic carbon generation since 1850 thus cannot be fixed closer than between 220 and 380 billion t, with 300 billion t being the most likely value I would offer when pressed for a single representative value. This total would be equivalent to almost exactly two-fifths of all carbon currently present as CO_2 in the atmosphere.

Before leaving these accounts of anthropogenic CO_2 generation, at least a brief mention must be made of a notable recent process contributing to elevated ambient CO_2 levels: drainage of wetland, organic soils. This change is more important than its annual spatial extent suggests, as it affects what were previsouly appreciable carbon sinks. Armentano (1980) offers a thorough review of this transformation and flow and I will summarize his main points.

The total global area of organic soils—histosols more than 20–40 cm deep and with at least 12–18% of carbon and histic and humic gleysols—is not known with great certainty but it could exceed 900 million ha, and all soils with large organic deposits, ranging from tropical swamps to high arctic tundras, cover about 3 billion ha, or roughly one-fifth of the continental surface. Two recent estimates of total carbon in these formations—by Ajtay et al. (1979) and Schlesinger (1983)—are 550 and 725 billion t and even these totals (at least a quarter of the world soil carbon pool) may be underestimates.

The best available information on wetland response to long-term drainage comes from the Sacramento–San Joaquin Delta in California and from the Everglades agricultural area in Florida where more than half a century of conversion to farmland brought about large subsidence and major carbon loss. Oxidation of Everglades peat has released 135 t CO_2 annually from each hectare of drained soil so that the total carbon outflow is on the order of 10 million t. Carbon losses from the entire Florida Everglades may be much heavier and the whole area, one of the world's largest organic carbon sinks, may already have been converted into a net carbon source.

Elsewhere, the large-scale drainage of Scandinavian swamps to raise wood yields and, above all, the Sudd project in the Sudan which is now under way to drain parts of the world's largest permanent swamp have started to rival North American releases and soon to be added is the drainage of the huge Pantanal swamp in Brazil. Global estimates of annual carbon releases from these localities and from other affected histosols are uncertain but may now total as much as 370 million t, or an equivalent of about 8% of the carbon emitted from fossil fuel combustion. The following sections will discuss how the rising biogenic and anthropogenic emissions translate into higher atmospheric CO_2 levels—yet another area of major uncertainties.

2.3. ATMOSPHERIC CO_2

The early atmosphere, almost certainly a secondary biogenic creation, contained abundant CO_2 (Cloud, 1980). Of course, we will never know the actual levels but Hart's (1978) computer simulations of atmospheric evolution suggest that

4.25 billion years ago, the atmospheric CO_2 content was about 1000 times the present level, and 1 billion years ago still 100 times today's level.

Evidence of these high concentrations can be seen in the huge deposits of sedimentary carbonate rocks: calcium and magnesium ions released by weathering of volcanic rocks combined with carbonate ions which formed by dissolving of abundant CO_2 in water. The total quantities of CO_2 removed in this way were staggering: when recalculated in terms of average mass per unit area over the whole earth, no less than 50 kg CO_2/cm^2 was so removed. The fate of Phanerozoic CO_2 is not known with any accuracy but there is little doubt that autotrophs emerged and spread in the epoch with much higher atmospheric CO_2 than today; the ready response of plant productivity to higher CO_2 levels in controlled experiments (see Section 2.4.2.3) is another proof of the relative CO_2 impoverishment of our current atmosphere.

Before describing these current concentrations, their measurements and global trends, I must first complete an outline of the biogeochemical carbon cycle, whose major biotic fluxes and storages have already been covered by assessment of primary production and standing phytomass. I will concentrate on only the key remaining reservoirs—above all on oceans—and on the ocean–atmosphere exchange which is critical in determining long-term averages of tropospheric CO_2.

2.3.1. Biospheric Carbon Links

As a result of great advances in modeling the global carbon cycle—a development which can perhaps be best appreciated by comparing Bolin's (1970) basic model with a voluminous collection of largely computerized complex exercises he edited a decade later (Bolin, 1981)—there is no shortage of comprehensive accounts of planetary carbon pools and fluxes as well as simulations of detailed segments of this complex cycling. Consequently, only the essentials relevant to appreciation of processes governing the atmospheric CO_2 concentrations will be given in this section.

Of the nine reservoirs in the relatively simple flow diagram of the cycle depicted in Figure 2-1, only the storages in terrestrial and oceanic phytomass were discussed earlier in this chapter (Section 2.1.2.2). Although living terrestrial phytomass provides an appreciable carbon storage—using my relatively conservative calculation, land plant storage would be equivalent to exactly two-thirds of all carbon now present in the atmosphere—dead biomass in litter and in soils represents almost certainly a much larger amount of carbon, but estimates available for this reservoir show huge disparities.

Bolin (1970) estimated terrestrial dead organic matter carbon at 700 billion t—nearly identical with atmospheric CO_2 carbon—but Keeling (1973b) put the total at 1.05 trillion t C, Duvigneaud (in Bolin et al., 1979) at 2.84 trillion t, and Bolin (1977) at 3 trillion t. Obviously, such large differences in assessing the storage will be reflected in estimates of fluxes involving the soil carbon pool. For example, Bolin (1977) calculated—assuming that peat soils, having a density of 0.2 g/cm^3

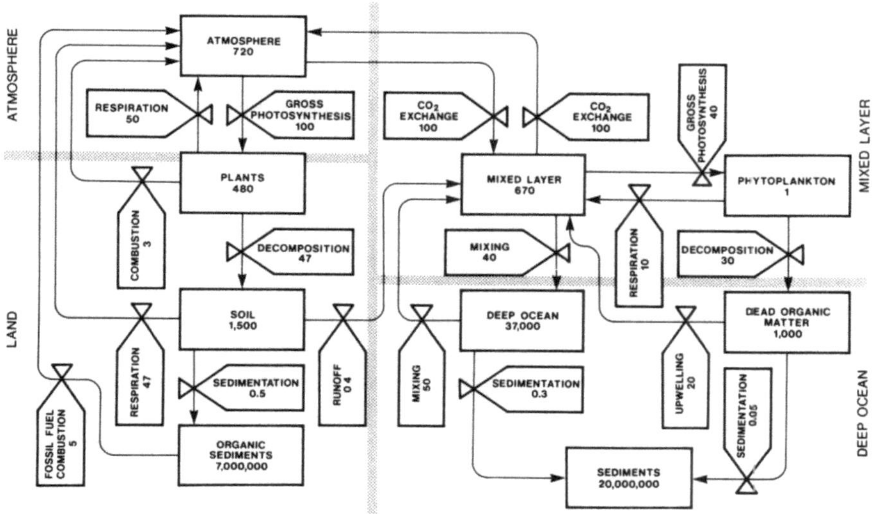

FIGURE 2-1. Global biogeochemical carbon cycle. Estimates for all storages (rectangles) and annual fluxes (valves) are in billion tonnes of carbon. Sources: Reiners (1973), Smil and Milton (1974), Baes et al. (1976), Bolin et al. (1979), Bolin (1981), Clark (1982).

and carbon content of 58%, are accumulating at an average rate of 1 mm/year over an area of 3.3 million km^2 of wetlands—that carbon is currently being sequestered in organic soils at a rate of 400 million t annually. This would be a significant sink equal to nearly 10% of annual fossil fuel CO_2 generation, but no one else joined Bolin in offering such a high estimate.

However, we are quite certain that the transport of dissolved and particulate dead organic carbon to the ocean by river flow is only a very small part of the global cycle. A variety of approximate estimates, ranging over an order of magnitude from only 100 million t to 1 billion t C/year, were published for this flux during the 1970s. The best evaluation, by Schlesinger and Melack (1981), put the global total at 370–410 million t C.

Surprisingly, the highest estimates of dead litter and soil carbon are not far from the total sequestered in fossil fuels, mostly in coals: as will be shown later in this chapter (Table 2-7), the best current estimate of ultimately recoverable resources is just over 4 billion t C, a mass insignificant in comparison with storage in carbonates. The crust contains 6.551×10^{16} t C, a mere 0.27% of all crustal rocks, with some 73% in carbonate minerals, above all in continental and oceanic sediments (Kempe, 1979). Calcium carbonate ($CaCO_3$) is the dominant biogenic compound, encountered as either calcite or aragonite; dolomite [$CaMg(CO_3)_2$] is the third most widespread carbonate mineral. Creation, transfer, and redeposition of these minerals occur within the lithosphere and hydrosphere: terrigenous and planktonic materials are first sedimented in the ocean and eventual subduction, metamorphism, or uplifting transfers these sediments to the continental crust from which weathering and erosion return the sediments to the oceans or to continental basins.

Although CO_2 is critically responsible for continental dissolution of carbonates, the gas used for this chemical destruction from soil air is recharged following the reprecipitation of the carbonate and hence this process provides no net sink for atmospheric carbon. Only the weathering of silicates consumes airborne carbon (after mineral precipitation, just one-half of the CO_2 is returned to the atmosphere) but this flux amounts to no more than 20 million t C/year, a wholly negligible total compared to biotic carbon flows.

Kempe (1979) summarized most of the available information on carbonate sedimentation rates in the ocean (net annual flux most likely no greater than 220 million t), marine organic carbon deposition (less than 50 million t/year), sedimentation in continental basins (also about 50 million t/year) as well as much more uncertain values for seafloor spreading, subduction, uplifting, and metamorphosis rates (all of them well below 100 million t/year).

Organisms play an essential role in sequestering carbon in ocean deposits: it is a long-established fact that carbonate minerals are by far the most widely used bioinorganic constituents. Lowenstam (1981) offers a fine summary showing that calcite is formed by organisms belonging to seven phylla of Protoctista and 11 phylla of Animalia; among plants, some Bryophyta and Tracheophyta are also calcite mineralizers.

A mass of 10^8 t C/year appears to be the maximum for waterborne erosion; winds raise annually a mere 1 million t C to be redeposited largely into oceans. Fascinating as the tracing of all these flows is, these transfers may easily be ignored when studying the "CO_2 problem"—as can the freshwater segment of carbon cycling—as they do not affect atmospheric CO_2 during time scales of human interest.

But other processes involving the ocean are of critical importance in understanding short-term changes of atmospheric CO_2 levels and much attention has been paid to them since the late 1960s. A comprehensive volume edited by Andersen and Malahoff (1977) and a review by Brewer (1983) are excellent introductions into these complex topics.

The ocean's huge carbon stores—more than 37 trillion t of inorganic and about 1 trillion t of organic carbon, or roughly 50 times the element's current atmospheric total—and prodigious capacity to absorb more CO_2 led to conclusions that seawater will suffice to buffer any increase in airborne CO_2. As late as the early 1960s one could find this assumption repeated in the scientific literature, but it has had to be discarded owing to the obviously layered structure of the ocean.

The topmost layer well mixed by winds and waves and warmed by daily heating is very thin (the thermocline is on the average at 75 m), and accounts for less than 2% of the total water volume—and only this limited realm is in diffusion equilibrium with the atmosphere. Currently, it contains 670 billion t C, or just 1.8%

of all inorganic carbon stores in the ocean and it exchanges annually about 100 billion t C with the atmosphere.

CO_2 uptake by the surface layer is governed by a complex buffer system: basic steps in the carbonate equilibria involve reversibilities between gaseous and aqueous CO_2; aqueous CO_2, H_2O, and H_2CO_3; and H_2CO_3, HCO_3^-, and H_2O and their ionic constituents. The primary buffering is taken care of by

$$CO_2 + H_2O + CO_3^{2-} \leftrightarrows 2HCO_3^-$$

Interaction between ocean water and increasing partial pressure of CO_2 is expressed as the ratio of fractional changes in the atmosphere and in the surface water layer; Wagener (1979) compares different algebraic definitions of this buffer factor published during the two decades between 1957 and 1977. Revelle and Suess (1957) were the first to calculate the value for oceanic average by using the formula

$$\frac{\Delta P / P^0}{\Delta \Sigma CO_2 / \Sigma CO_2^0}$$

where P is CO_2 partial pressure and "zero" values refer to preindustrial levels; their result of about 10 has since been confirmed repeatedly, with most subsequently published values ranging from 9.2 to 12.5. However, experimental data from the late 1970s suggest lower values, between 7 and 9.

The response of this huge reservoir to higher atmospheric CO_2 levels is defined above all by the transfer and exchange processes between the air and the sea and within the ocean, and by the uptake capacity of ocean water for additional CO_2—or, most concisely, by the kinetics and thermodynamics of the processes.

The kinetics has been studied most profitably via ^{14}C diffusion in the ocean with a variety of oceanic structure and mixing models, and the average global exchange of CO_2 between the atmosphere and the ocean has been determined to range from 16 to 22 moles/m^2 per year with 20 being a good mean. However, major fluctuations can result from wind conditions and the rate is most likely much higher in the stormier circumpolar areas than in the relatively quiet subtropics (Siegenthaler and Oeschger, 1978; Bolin, 1981).

Given sufficient time for mixing of CO_2-enriched surface waters with deep ocean layers—equilibration lasting no less than several hundred and up to 2000 years depending on the area and exchange model used—huge ocean carbon reservoir will determine the new equilibrium CO_2 concentrations arising from fossil fuel combustion. The neutralization reaction,

$$CO_2 + CaCO_3 + H_2O \leftrightarrows 2HCO_3^- + Ca^{2+}$$

will add bicarbonate and calcium ions to the ocean's large load of dissolved salts.

Deep oceans will thus gradually become acidified as the downward mixing progresses and their sediments will slowly dissolve.

Broecker and Takahashi (1977) estimated the amount of $CaCO_3$ available for dissolution from the calcite content of marine sediments and from the extent to which these layers are stored by benthic organisms. Their conclusion is that the available calcite is about equivalent to carbon dioxide which would be released by burning all known resources of fossil fuels. As such CO_2 release is only theoretically possible, there is no doubt about the ocean's capacity to eventually clear away virtually all anthropogenic CO_2 from the atmosphere.

Rough calculations show that burning of all estimated fossil fuel reserves and eventual reaction of each mole of the released CO_2 with 1 mole of marine $CaCO_3$ would result in atmospheric CO_2 levels only about 40% higher than at the beginning of the 19th century, i.e., before the outset of rising fossil fuel combustion (Broecker, 1973). However, the process would be, necessarily, very slow.

With linear kinetics and an average of 1 mole/m^2 per year excess dissolution from the deep ocean carbonate-bearing sediments (about half of all the ocean floor), the process would take about 1280 years, a time comparable to that needed for deep ocean mixing. In contrast, with exponential kinetics favored by Morse and Berner (1972), dissolution rates would be so rapid that all excess CO_2 would be neutralized as fast as it could reach the deep ocean. In any case, accumulation of noncalcite residues (clay minerals and opal) in the upper sediment layers would reduce further dissolution rates and hence any realistic consideration of CO_2 neutralization in deep oceans must be concerned with mixing processes, interface dissolution as well as sediment stirring.

The latter point presents a major uncertainty: the mechanical eddy diffusivity as a function of depth in the sediment remains to be determined for a variety of oceanic environments and the stirring activities of benthic organisms bear no easy generalization. On the other hand, a great deal is known about the mixing of ocean layers and this knowledge has been incorporated into a large variety of ocean exchange models whose detailed descriptions can be found in Keeling (1973b), Oeschger et al. (1975), Keeling and Bacastow (1977), Broecker and Takahashi (1977), and Bolin (1981).

Yet this proliferation of models has not brought us closer to determining with satisfactory accuracy how much carbon has been taken up by the ocean since the beginning of accelerated fossil fuel combustion and forest clearing in the middle of the last century. For example, Killough and Emanuel (1981) compared five sets of carbon depth distribution in the oceans and found that of the estimated 134 billion t of fossil carbon released between 1860 and 1975, the models estimate a net uptake range of 24–66 billion t, or anywhere between 18 and 49%.

Clearly, one of the most important research tasks in advancing our understanding of the global biogeochemical carbon cycle is to learn more about the rate at which the intermediate ocean waters receive carbon from the surface mixed layer. Meanwhile we must go along with what we feel to be the best available evidence and assumptions in assessing the past and future rates of excess CO_2 retention in the atmosphere.

2.3.2. Measurements and Concentrations

Detecting atmospheric CO_2 can be done readily and with great accuracy using a variety of nondispersive infrared gas analyzers, but setting up a long-term monitoring program with baseline stations located at several relatively clean sites representative of background concentration changes and turning out reliable and consistent measurements proved to be a challenging task which came to be successfully realized only during the 1970s with the establishment of the Geophysical Monitoring for Climatic Change program. This is an outstanding program run by the National Oceanic and Atmospheric Administration (NOAA), part of the United States Department of Commerce, whose measurements reflect the most reliable documentation not only of CO_2 increases but also of other anthropogenic atmospheric pollutants with possible climatic effects, above all ozone and halocarbons.

The program has four baseline stations—Barrow (Alaska), Mauna Loa (Hawaii), Tutuila (American Samoa), and the South Pole—and essential details will be given in the next section on the two oldest monitoring locations run by NOAA's predecessors since the late 1950s when as a part of the International Geophysical Year the first systematic CO_2 monitoring started in 1957–1958. Consequently, our oldest reliable series of CO_2 measurements are now just a quarter of a century old, a mere blink even on a civilization scale and an altogether negligible span on a geological scale. Any rigid quantitative conclusions based on such ephemeral evidence—and one not so very accurate as will shortly be seen—cannot be reached with any confidence: all they prove is haste where patience would serve better, misplaced audacity where caution is so obviously in order.

2.3.2.1. Monitoring CO_2 Levels The longest daily uninterrupted record of atmospheric CO_2 measurement is that of the Mauna Loa Observatory on the island of Hawaii. Location of this station is unparalleled in its near perfection for background monitoring of principal atmospheric constituents. The observatory, operated since 1955 by the NOAA's predecessors, sits 3400 m above sea level at 19.5°N on vast lava flows of the NNE slope of Mauna Loa, about 780 m below the summit of Hawaii's largest active volcano (Miller, 1974).

In many ways the general as well as the particular location of the observatory is ideal for CO_2 monitoring. The Hawaiian islands are quite remote from any major area of anthropogenic CO_2 generation (California, the closest continental shore, is more than 4000 km away). Persistent tropical trade winds come from the east-northeast but about three-quarters of the time they are capped by thermal inversion at an average altitude of 2 km and large-scale evening subsidence exposes the site to extremely clean air typical of the middle troposphere above the emptiness of the central Pacific.

Sadler et al. (1982) analyzed hourly values of CO_2 levels at the observatory from May 1974 to March 1981 in a search for important daily to monthly concentration fluctuations but found no evidence linking the fluctuations with large-scale variations in the atmospheric circulation. Air reaching the observatory has been so well mixed that it does not retain any evidence of specific sources or sinks, while very minor local biotic interference remains much the same year after year and intermittent volcanic outgassing can be easily eliminated from the record. Clearly, Mauna Loa is an excellent site for background CO_2 monitoring.

Precipitation below the inversion is extremely heavy (up to 7.6 m/year) but the observatory records only 500 mm annually, and as it is surrounded by seemingly endless and relatively fresh lava fields and sits about 1.5 km above the tree line, there is no vegetation at all—not even shrubs or weeds—within 15 km of the site. Remoteness and cleanliness of the site are obvious but there have been some problems and interferences. Downslope winds may bring in CO_2 from the volcano's active vents, and the trades will carry air whose concentrations were lowered as it passed distant forests and sugarcane fields; until July 21, 1967 the observatory's electricity was produced by a diesel generator causing local CO_2 contamination and since July 1969 the presence of a good road has produced automobile pollution in the daytime. Consequently, as Keeling et al. (1982) note, "one should not assume that every detail seen in the monthly averages reflects regional or global phenomena."

Numerous changes of air intakes and sampling procedures and frequent checks of calibration and required adjustments have been made over the years (detailed in Keeling et al., 1976a, 1982). The three most notable deviations of Mauna Loa measurements have been persistent diurnal variations caused by local influences, somewhat mysterious midday peaks, and the concentration dip of the 1960s.

Nocturnal spikes are explained by volcanic releases upslope from the observatory but there is no simple cause for persistent midday peaks. Arrival of downslope air enriched in CO_2 by plant respiration during the previous night appears to be a part of the surge, as does the car traffic and choice of air intakes. A leaking detector caused a long period of analyzer malfunction producing an error of up to 1 ppm below the correct value and hence all the published monthly means between May 1964 and January 1969 carry errors of that magnitude. Since that time, various procedures should have held measurement imprecision to near 0.1 ppm, but in practice, accuracy lower than 0.5 ppm for monthly averages has proved elusive.

Uncertainty of 1 ppm before 1970 and at least 0.5 ppm since then is obviously a small fraction of the total concentration but a very large part of month-to-month changes (or even larger than such variations) and of average annual rise: using differences of monthly or yearly means for precise calculations of CO_2 retention or absorption is then beset with surprisingly high uncertainty on this account alone.

The South Pole measurements are second to those at Mauna Loa in longevity but not in continuity. Sampling at the South Pole actually started earlier than on Hawaii, beginning in the spring of 1957 as a part of the International Geophysical Year, but unlike the continuous Mauna Loa monitoring, only twice-a-month discrete collections of air in glass flasks were made until 1960 and again since 1965;

continuous analysis was done only during the years 1960–1963 and no measurements at all are available for 1964. Complications, errors, and adjustments associated with the first 15 years of this long-range program are summarized in Keeling et al. (1976b).

After excluding contaminated data and performing the necessary adjustments of measured concentrations to compensate for analytical errors, adjusted monthly CO_2 index values derived from individual flask analyses show a rise from 311.23 ppm in June 1957, the month of the first sample, to 312.27 a year later to 314.86 5 years later and to 317.84 ppm after 10 years (Keeling et al., 1976b). Unfortunately, it is impossible to offer proper annual averages for the period before 1965 as the cleaned-up data set is missing at least two but usually many more monthly means (e.g., for 1960, only one monthly mean is available; for 1957 and for 1958, only two).

The other two stations of the NOAA's network of baseline observatories of geophysical monitoring for climatic change are at Barrow, Alaska, and on Tutuila Island, American Samoa. Barrow Station is only about 1 km south of North America's northernmost point and just 9 m above sea level; CO_2 measurements started there only in the spring of 1971. In American Samoa the station was set up in 1973 at Cape Matatula, at the extreme northeast of Tutuila Island; continuous CO_2 records are available from 1976.

Comparability of all these measurements—as well as other CO_2 records from Europe (where the Swedes have taken airborne readings for over two decades) and other parts of the world—is essential but is not achieved without difficulties and errors; this is so because the principal monitoring technique of atmospheric CO_2, nondispersive infrared gas analysis, is a relative measurement method requiring careful calibration with standard gases, usually mixtures of CO_2 in air or CO_2 in nitrogen.

Carrier gases have the greatest influence on readings and hence both the analyzer and the standard gases must be calibrated against an absolute system (Bischof, 1974). Such a primary standard CO_2/N_2 mixture system has existed since 1958 at the Scripps Institute of Oceanography at La Jolla, California. Still, there have been numerous complications and errors, which will be described briefly on the basis of detailed investigations carried by Bischof (1973, 1977), Pearman and Garratt (1975), and Pearman (1977).

Complications arise owing to the use of different types of analyzers—models in the most widespread use are those of the URAS line manufactured by Hartman and Braun in Frankfurt am Main and the UNOR line produced by H. Maihak in Hamburg—and different standard gases. International comparison of the major baseline CO_2 monitoring stations was done by Pearman (1977) with three CO_2/N_2 and three CO_2/air mixtures to check the precision of the intercalibration system and to fix the carrier gas error at each station. While for each of the CO_2/N_2 mixtures the mean concentrations for all stations varied by less than 0.3 ppm (and in most cases by less than 0.1 ppm), at stations using CO_2/air mixtures they measured 2–4 ppm below the all-station mean when the analyzers with parallel detector cells were used and 1.5–4 ppm above the mean where the series detectors were installed. Pearman's (1977) conclusion is that the "reported mean atmospheric CO_2 concentration differences between stations of up to 1 or 2 ppmv cannot be interpreted as evidence for large-scale horizontal gradients in the atmosphere."

As for the variation in vertical measurements, Pearman and Garratt (1975) found that for both UNOR and URAS analyzers, the indicated CO_2 concentrations relative to a CO_2/N_2 mixture depend on ambient pressure and hence all aircraft or stationary high-altitude measurements need additional correction to be comparable with near-sea-level monitoring.

And there are problems with the standard Scripps CO_2/N_2 mixture as well. The mixture was chosen in the late 1950s because it was believed that an inert gas maintained its integrity better than a gas mixture containing oxygen. However, Bischof (1977) argues that air standards, used for Stockholm devices, are more appropriate as the calibration gases should be of the same composition as the medium to be analyzed. And intercomparisons and calibrations of Stockholm standard gas tanks stored since 1963 show that these mixtures have remained stable in relation to the Scripps baseline manometric scale.

But checks and comparisons at Scripps have shown the need for important adjustments owing to errors by drift, nonlinearity, and carrier gas effect in the calibration system. So there is a new adjusted scale which requires recalculation of all past measurements. For Mauna Loa records, the necessary adjustments, except for some small changes of less than 0.2 ppm, were completed in time for *Carbon Dioxide Review: 1982,* but further refinements (above all, computations of the pressure dependency) will follow (Keeling et al., 1982). Clearly, all values that are cited in the next section are fairly unassailable when the uncertainty of ± 1 ppm is kept in mind—playing with tenths of 1 ppm is dubious, with hundredths quite useless.

2.3.2.2. GLOBAL CO_2 TRENDS Continuous long-term monitoring of CO_2 in locations remote from populated regions and vegetated areas is essential for establishing changes representative of global trends owing to natural variations of the gas—large diurnal and annual fluctuations as well as smaller irregular airborne shifts.

Diurnal fluctuations are most pronounced over vegetative surfaces with regular afternoon CO_2 minima (photosynthetic uptake) and late night maxima (plant and soil respiration). Not surprisingly, the greatest fluctuations can be measured inside the canopies and the amplitudes flatten rapidly in upward measurements although they may be easily noticeable at surprisingly long distances away from vegetated surfaces. Just two examples of these well-documented phenomena will be given, one for forests and the other for a remote site.

Woodwell et al. (1973) measured intermittently the diurnal CO_2 fluctuations at different heights above an oak–pine forest of central Long Island. At lower elevations, they commonly encountered changes of up to 100 ppm in less than 1 hr with peaks just after midnight and minima of around 4–5 ppm. Nocturnal maxima were further enhanced by temperature inversions when the differences between daylight and nighttime extreme exceeded 200 ppm, and the maxima can go to over 500 ppm. In fact, on some occasions near-ground CO_2 levels inside dense vegetation may rise to 1000 ppm or more.

Diurnal differences in Brookhaven measurements gradually diminished with altitude but were still clearly noticeable at the top of the 125-m-high meteorological tower with ranges of up to 40 ppm during the peak vegetation period. Similarly, measurements in Kenya yielded daytime CO_2 declines to 310 ppm in and around tropical rain forests and to 322 ppm over savannas with nightly buildups to more than 400 ppm; on the higher reaches of Mount Kenya, these fluctuations were reduced to just 2–6 ppm (Schnell et al., 1981).

Perhaps the most interesting documentation of the diurnal plant-induced CO_2 fluctuations comes from Hawaii, where the characteristic afternoon decrease can be measured even at Mauna Loa Observatory, 3400 m above sea level and some 1500 m above the vegetation lines on the N-NE slope of the volcano: peak CO_2 concentrations around 11 AM are clearly brought by upslope northerly winds commencing in the early morning and carrying the air above the respiring vegetation; subsequent photosynthetic activity lowers the concentration by nearly 3 ppm within 6 hr (Goldman, 1974).

In the Northern Hemisphere, there is a very obvious annual cycle with late summer–early autumn minima as plants incorporate carbon into new tissues, and late winter–late spring maxima caused mainly by decomposition of phytomass and by higher combustion of fossil fuels. The cycle's amplitude decreases with altitude although it may be small near the surface just as well owing to the influences of large ubran areas which generate huge CO_2 volumes throughout the year. A smaller amplitude in the Southern Hemisphere is explainable by its largely oceanic character (Garratt and Pearman, 1973).

Unlike annual and diurnal cycles, the third CO_2 periodicity arises irregularly from large-scale mixing processes accompanying air mass movements. Measurements in various parts of the world revealed clear, though not very large, day-to-day variations attributable to changed air trajectories. In the Brookhaven measurements mentioned above (Woodwell et al., 1973), westerly winds carried on the average 322.9 ppm, compared to the easterly mean of 319.1 ppm, a significant difference at the 95% probability level. Similar minor variations were reported by Goldman (1974) from Hawaii and by Spittlehouse and Ripley (1977) in Saskatchewan.

Naturally, these small irregularities as well as the regular diurnal march are eliminated by averaging during preparation of monthly means but the annual amplitude is a notable feature of all long-term CO_2 records, no matter where acquired. Even at the South Pole, minima come very regularly either in March or in April and have been up to 1.65 ppm below maxima which are recorded as early as September but also as late as December.

Long-term monthly averages from Mauna Loa show unfailing minima either in September (about one-third of the time) or in October but the absolute concentration differences for these two months are usually so small that one can talk about a clear bottom of 2 months' duration (Keeling et al., 1982). Even more regular are the maxima, occurring without exception in May (only in 1970 did April and May share the same average). Between 1959 and 1980 the largest difference was in 1966, 6.11 ppm, the lowest in 1970 (4.69 ppm), the mean for the 22 years being 5.51 ppm; no trend is apparent in these fluctuations—the undulation remains regular with regard to both its peaks and troughs as well as its amplitude.

But, of course, the annual mean has risen steadily and this figure, with its inexorable ascent, embodies the justification of all the flourishing CO_2 research. Missing values for four months in 1958 and for three months in 1964 preclude the calculation of annual means for those two years but all other averages and yearly rises are listed in Table 2-5. The average from 1965, the first year of the unbroken record, to 1980 is 1.22 ppm, but variation is considerable and some year-to-year differences are notable, such as a relatively huge jump in 1972–1973 (by far the highest of the 15-year span) followed by an increase of a mere 0.58 ppm, the lowest recorded increment. I shall return to these irregularities shortly.

Comparisons with other GMCC stations are quite interesting: they were published for the first time as provisional monthly means in the program's tenth annual report (Bodhaine and Harris, 1982) and for all stations cover the period from 1976 (the first full year of monitoring on Samoa) to 1981. Table 2-6 lists the 1976 and 1981 annual minima, maxima, and averages for the four stations and Figure 2-2 shows the overlapping and rising curves of monthly records. The most notable features in these comparisons are north-to-south gradients for annual averages (with Barrow's mean up to 4 ppm above the South Pole value), monthly maxima (Barrow's being nearly 8 ppm above those at the South Pole), and the annual differences

TABLE 2-5
ANNUAL AVERAGE CO_2 CONCENTRATIONS AND INCREASES AT MAUNA LOA, HAWAII, 1959–1980[a]

	Average concentration (ppm)	Annual increase (ppm)		Average concentration (ppm)	Annual increase (ppm)
1959	315.66		1970	325.51	
		0.93			0.96
1960	316.59		1971	326.47	
		0.71			1.12
1961	317.30		1972	327.59	
		0.90			2.23
1962	318.20		1973	329.82	
		0.51			0.58
1963	318.71		1974	330.40	
		—			0.60
1964	—		1975	331.00	
		—			1.06
1965	319.99		1976	332.06	
		0.67			1.57
1966	320.66		1977	333.63	
		0.81			1.57
1967	321.47		1978	335.20	
		0.83			1.35
1968	322.30		1979	336.55	
		1.88			1.81
1969	324.18		1980	338.36	
		1.33			

[a]From Keeling et al. (1982).

TABLE 2-6
AVERAGE, MINIMUM, AND MAXIMUM CO_2 CONCENTRATIONS AT THE FOUR GMCC STATIONS IN 1976 AND 1981[a]

	1976 concentrations (ppm)			1981 concentrations (ppm)		
	Average	Minimum	Maximum	Average	Minimum	Maximum
Barrow	334.1	323.6	339.0	340.7	331.1	346.1
Mauna Loa	331.7	328.3	334.5	339.7	336.6	342.6
Samoa	332.7	332.1	333.5	338.9	338.4	339.5
South Pole	330.1	329.5	331.2	337.8	337.2	338.6

[a]From Bodhaine and Harris (1982).

between extreme values. These amplitudes shrink from about 15 ppm at Barrow to around 6 ppm at Mauna Loa and to between 1 and 2 ppm at Samoa and the South Pole, an order of magnitude difference.

These gradients are even better illustrated by data from the GMCC CO_2 flask sampling program which now includes 20 stations dotting the earth: its extreme latitudes are Mould Bay in the Canadian Northwest Territories (76°N, 119°W) and the South Pole, its remote island locations include the Falkland Islands, Ascension, and Terceira (Azores) in the Atlantic, Mahé (Seychelles) and Amsterdam Island in the Indian Ocean, and Tutuila, Hawaii, and Guam in the Pacific (DeLuisi, 1981).

Edited flask data for the years 1979–1981 reveal a stable and characteristic seasonal pattern superimposed on a perceptible secular increase of about 1.4 ppm/year (Bodhaine and Harris, 1982). As might be expected, midlatitudes of the Northern Hemisphere show the highest annual gain (about 1.5 ppm) while their

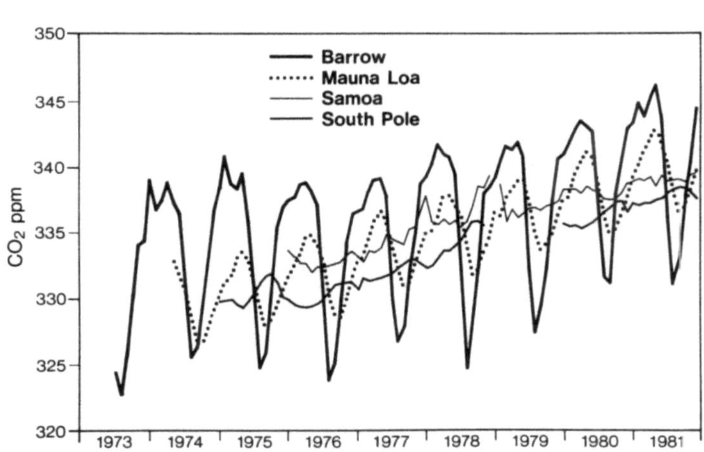

FIGURE 2-2. Monthly means of CO_2 concentrations at the four GMCC stations (Bodhaine and Harris, 1982).

southern counterparts have significantly lower increments (around 1.2 ppm). Latitudinal differences in the seasonal amplitude are best conveyed by a smooth curve fitted to the flask data (Figure 2-3) while the latitudinal dependence of annual CO_2 march is impressively portrayed by a three-dimensional graph of the same data for the year 1981 (Figure 2-4).

When represented by Mauna Loa's record (the best long-term data yet available), atmospheric concentrations of CO_2 have risen by nearly 30 ppm in the 25 years between 1958 and 1983. While in the last two years of the 1950s they stood at around 315 ppm, they now surpass 340 ppm and are increasing by an average of 1.2 ppm/year. Measurements in many other remote locations of both hemispheres suggest that the global mean for current CO_2 concentration increments is 1.4 ppm.

Although no reliable systematic CO_2 measurements are available for the years before 1958 and no reliable predictions can be made about the future levels, the next section will examine the best evidence for the increases since the middle of the 19th century and will look at both theoretical and practical maxima of future CO_2 concentrations.

2.3.2.3. PAST AND FUTURE CONCENTRATIONS Of preindustrial CO_2 concentrations we are not at all certain. The general assumption is that in the middle of the 19th century the levels were not higher than 290 ppm and may have been as low as 270 ppm (Siegenthaler and Oeschger, 1978). Although numerous measurements were made during the last three decades of the 19th century, many of these century-old values are obviously unreliable, as are virtually all measurements made before 1870.

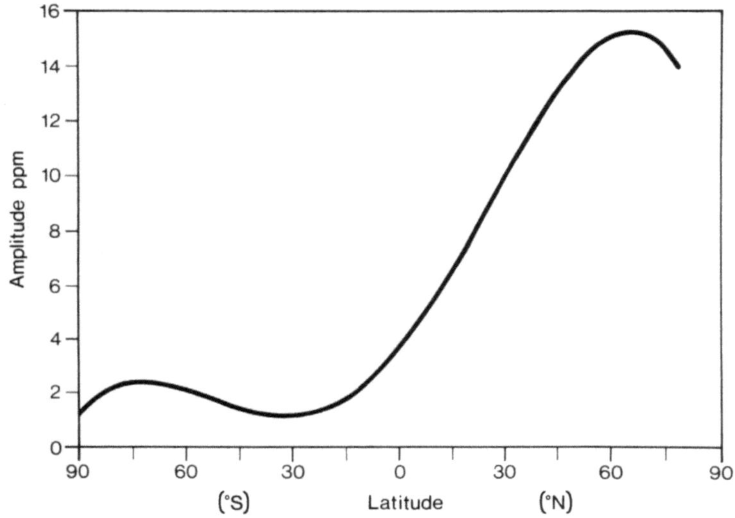

FIGURE 2-3. Meridional distribution of the amplitude of seasonal CO_2 concentration. The line is a fourth-order polynomial fit to CO_2 flask measurements around the world (Bodhaine and Harris, 1982).

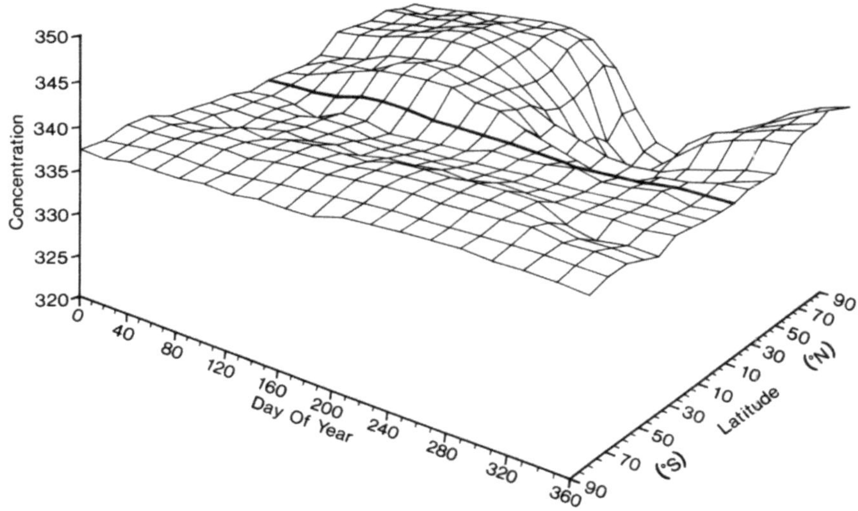

FIGURE 2-4. Three-dimensional graph of 1981 global atmospheric CO_2 distribution from the 20 GMCC flask sampling stations (Bodhaine and Harris, 1982).

The best review of pre-1958 CO_2 measurements, including a variety of 19th century values, was assembled by Callendar (1958) who took care to exclude all suspect values in preparation of what he calls preferred averages. This exclusion criterion did away with all pre-1870 results where, moreover, a relatively crude instrumentation introduced large errors. From five sets of long-term observations encompassing 305 separate determinations made between 1872 and 1901 over periods of 5 to 20 months along the northern coast of France, in a further unspecified countryside location in France, near Belfast, and at Kew (Greater London), Callendar (1958) established a mean of 288.5 ppm for the last three decades of the 19th century.

No upward temporal trend was apparent with the latest values at 286–289 ppm, the earliest at 287–291. Consequently, Callendar (1958) favored 290 ppm as the baseline value for the year 1900. About 60 observations during the years 1882–1883 are available for the South Atlantic between 42° and 50°S; they range between 256 and 273 ppm, about 15–30 ppm less than recorded at the same time in Europe. For the 20th century there are several thousand pre-IGY measurements, above all from New England (1909–1912 and 1930–1935), Scandinavia and its surrounding seas (1917, 1920–1923, 1934–1935, 1955–1956), showing values rising from 293–305 ppm before 1925 to 310–323 ppm in the 1930s and 322–328 ppm in the mid-1950s.

Figure 2-5 charts the best estimates of average global atmospheric CO_2 since 1870 and the diminishing band of uncertainty before 1958; the superimposed curve indicates the increases of atmospheric CO_2 which would have resulted if all the gas

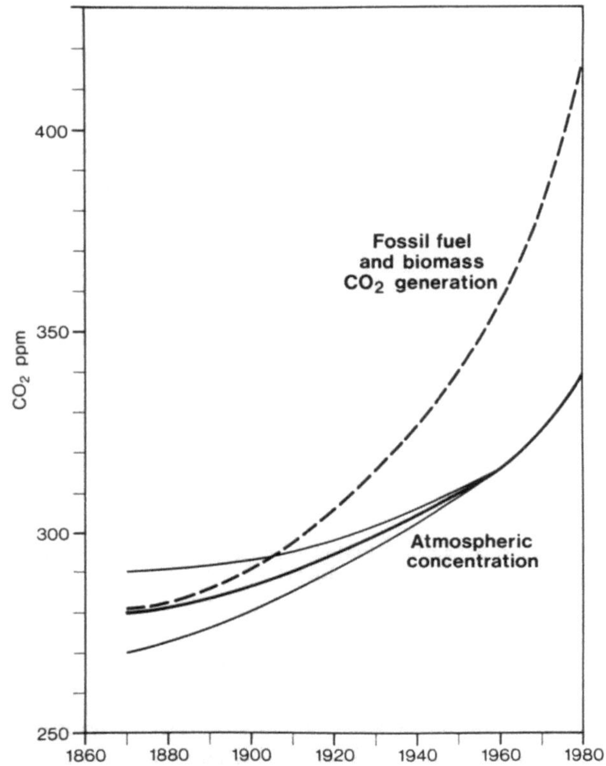

FIGURE 2-5. Increase of average global CO_2 concentrations after 1870 compared with levels resulting from complete atmospheric retention of CO_2 from fossil fuels and biomass.

generated by human activities had remained in the atmosphere after the year 1860. Atmospheric CO_2 levels would have reached about 360 ppm by 1980 and would now be moving up by about 20 ppm/decade.

And if all the CO_2 added to the atmosphere since 1850 by fossil fuel combustion *and* the gas released by ecosystemic change had remained airborne, CO_2 concentrations would have risen by at least 100 ppm and by as much as 180 ppm, with 140 ppm being the most likely increase corresponding to emissions of 1.1 trillion t of the gas. Starting with 270 ppm in 1850, the 1980 level would be 410 (370–450) ppm CO_2 as opposed to the actual level of 340 ppm. This means that about half (extremes being no less than 30% and up to 60%) of all anthropogenic CO_2 was transferred from the atmosphere.

This sequestration can, of course, be followed with greater accuracy after 1960, when both the fuel consumption, and hence CO_2 generation, data and annual increases of CO_2 are known with satisfactory accuracy. During the two decades between 1960 and 1980, CO_2 levels measured at Mauna Loa rose by 21.77 ppm for an average rise of 1.089 ppm/year and during the same period carbon released from fossil fuel combustion, natural gas flaring, and cement production totaled about 80.5 billion t. As an increase of 1 ppm of atmospheric CO_2 requires the presence of an additional 2.13 billion t of carbon in the air, retention of all fossil-fuel-generated carbon would have raised the mean background level by 37.81 ppm, or by an average of 1.89 ppm/year.

As the actual rise was only 1.089 ppm, it is obvious that on the average no more than 57.6% of fossil fuel-derived CO_2 has remained in the atmosphere. However, a closer look shows that this mean is derived from a highly fluctuating sequence rather than from a very regular proportional retention: Figure 2-6 plots the annual increments of CO_2 concentrations at Mauna Loa Observatory together with atmospheric CO_2 increases which would follow total retention of all the gas generated by fossil fuel combustion, gas flaring, and cement production.

A good explanation for such large year-to-year fluctuations is lacking. A perfect knowledge of CO_2 releases from biomass combustion might explain part of some sudden rises or declines but it is unlikely that annual fluctuations in deforestation, grassland and forest fires, and biomass burning for fuel are so large as to account for most of the differences. Undoubtedly, complex climatic, oceanic, and biospheric responses are responsible and no single variable will be found decisive in effecting the observed shifts.

Charting the future atmospheric CO_2 levels in response to various energy consumption scenarios has been one of the favorite modeling activities in the realm

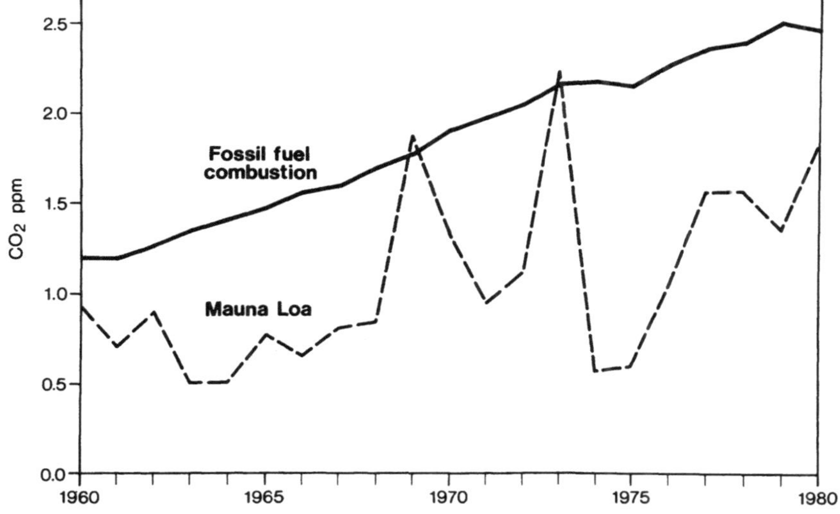

FIGURE 2-6. Annual increases of atmospheric CO_2 concentrations at Mauna Loa and the increase of CO_2 levels which would result from complete retention of all CO_2 from fossil fuel combustion in the atmosphere.

of the carbon cycle. The results range, predictably, from some incredibly high concentrations following scenarios of lavish fossil fuel use to relatively moderate increases brought by soft energy paths eschewing coals and hydrocarbons. The most extreme forecast one can choose is, of course, a theoretical calculation of how much the atmospheric CO_2 concentrations would increase if all of the fossil fuel carbon is burned. This is a patently unrealistic assumption but it fixes the uppermost limit easily when one takes into account the highest available estimates of fuel resources in place (which are, in the case of hydrocarbons, an order of magnitude above the lowest ones!): about 20 trillion t of carbon would then be theoretically recoverable, and if one-half of it stays in the air this would be enough to raise the atmospheric CO_2 content to about 5000 ppm (0.5%) or 15 times the current level.

Restricting the estimate to ultimately recoverable resources is a more sensible approach although it is absolutely certain that far from all of the recoverable mass will be extracted owing to many economic, technological, environmental, and social reasons. A careful estimate of CO_2 releases resulting from eventual recovery of all fossil fuels in the ultimately recoverable category was done by Rotty and Marland (1980). They used the World Energy Conference (1978) expert opinion polls on ultimately recoverable crude oil and natural gas, the Conference's published data on the global coal resources, Culbertson and Pitman's (1973) evaluation of shale oil, and several sources for estimates of tar sands and heavy oil to derive the energy and carbon contents of ultimately recoverable fossil fuels as listed in Table 2-7. For comparison, the last column lists the historic energy consumption and makes clear that the mass of carbon which could be released from burning all conceivable fossil fuel reserves would be almost 30 times larger than the aggregate emissions since the mid-19th century.

Assuming, again, that half of the CO_2 is retained in the atmosphere, this would translate into a rise of nearly 1000 ppm. But I would guess that the eventual

TABLE 2-7

CARBON CONTENT OF ULTIMATELY RECOVERABLE FOSSIL FUELS[a]

Ultimately recoverable resources	Best estimates[b]	Upper limit speculations[b]	Cumulative consumption up to 1980 (rounded to the nearest five)[c]
Coal	3510	6,315	105
Crude oil	230	380	40
Natural gas	143	230	20
Shale oil	173	9,530	—
Tar sands and heavy oil	75	200	—
Total	4131	16,655	165

[a]All values are in billion tonnes of carbon.
[b]Source: Rotty and Marland (1980).
[c]Source: the author's calculations.

cumulative recovery of all fossil fuels will not surpass 3 trillion t of carbon (about 20 times the 1850–1980 total), an equivalent (with 50% retention) of an additional 700 ppm, so that the maximum atmospheric concentration would be roughly three times the current level. In any case, the highest practically conceivable CO_2 increases induced by fossil fuel combustion would not surpass 0.2%, staying well below the Phanerozoic maxima in excess of 0.4% and just about equaling the mean atmospheric CO_2 level of the past 500 million years (Budyko, 1982).

All of the above calculations were done simply to set the proper order of magnitude: releases of such huge quantities of CO_2 would have to be spread over hundreds of years, a time span during which deep ocean absorption would have to be taken into account. Any very long-term modeling of CO_2 levels must thus include a variety of assumptions regarding atmosphere–ocean interactions and a good example of such a work is found in the simulations prepared by Keeling and Bacastow (1977). These authors predict the increases up to the year 3400 and any interested reader's curiosity (akin to a desire of an early medieval monk to know some particulars of the 20th century) will be satisfied (with ultimate benefit much like that monk's) by consulting the original source based on a simplistic six-reservoir model.

But I shall not look so far, just some decades ahead. Here I have to admit that in the early 1970s, when I submitted to the lure of computer modeling and was busily building various models and scenarios, I assembled a moderately complex dynamic simulation of future atmospheric CO_2 levels responding to two extreme and one probable set of assumptions (Smil and Milton, 1974). The most probable of the three sets—with energy growth rates declining to half of their pre-1970 value, nonfossil sources covering about one-third of all needs by the year 2000, about three-fifths by 2020, and no additional carbon storage in biota—yielded a value of 371 ppm by the end of the century, 430 ppm in 2020,

Since then, the literature on future CO_2 levels has grown to scores of entries but until a few years ago most of the forecasts brought much more rapid increases than my most likely model, predicting levels around 600 ppm (i.e., an approximate doubling of preindustrial concentrations) no later than about 2030, only about half a century ahead. This result was inevitable when it was assumed that global energy use would at least continue its post-WW II exponential rise of 4.5%/year and that close to 60% of fossil fuel-generated CO_2 would remain airborne (average, calculated with Mauna Loa values for the years 1958–1975, was about 57%).

But both of these premises had to be altered. The second one fell first as it came to be realized that biotic releases have been far from negligible: if the higher estimates for this flux, on the order of 5 billion t C/year, are included, then the atmospheric retention rate would be around a mere 25% and even when the calculation is done with an annual net flow of 2 billion t of biotic carbon, the airborne fraction for the years 1958–1980 is still much lower than previsouly thought, just short of 40%.

The first premise, long-term continuation of post-WW II exponential growth rates of primary commercial energy, seemed unshaken even after the rapid crude oil price changes of 1973–1974, but after another round of price escalation in 1979–1980 the inevitable started to happen with surprising speed: no exponential growth can continue forever and as the size of an affected system increases the rate must moderate.

The best way to generalize the differences arising from these shifting premises is by plotting doubling times as functions of average annual increases of CO_2 from fossil fuels and average airborne fraction: Figure 2-7 (Clark et al., 1982) shows that an airborne fraction close to 0.6 and energy consumption growth rates in excess of 4% bring a doubling of preindustrial CO_2 by 2030. In contrast, what the authors call typical assumptions of the early 1980s' CO_2–climate models—3% energy growth and 50% retention—postpone the doubling to the fifth decade of the next century while their best estimate, a combination of 2% growth and 40% atmospheric retention, moves that date close to 2080.

I am pleased that a decade ago my most likely scenario implied CO_2 doubling no sooner than after 2070, in perfect agreement with the latest informed consensus, but even slower changes may be set out in plausible combinations: sustained energy consumption growth just in excess of 1% and lower atmospheric retention owing to increases of primary productivity might easily push the doubling dates into the early decades of the 22nd century.

Interestingly, changing assumptions about the rate of net biotic CO_2 generation does not move the doubling date much, as the higher estimates for such fluxes are

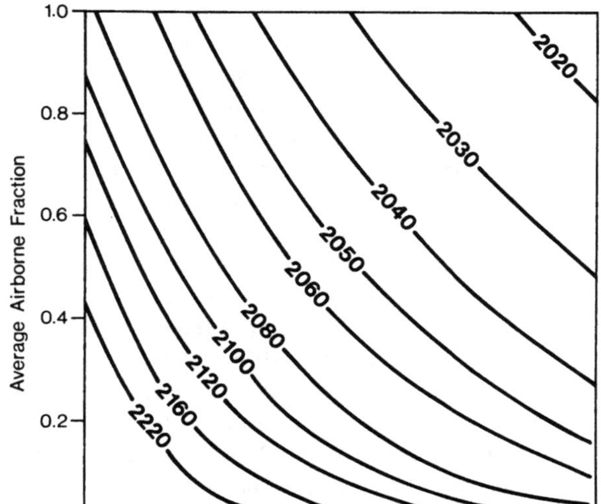

FIGURE 2-7. Time of atmospheric CO_2 doubling shown as a function of average annual increase in fossil fuel consumption and average airborne fraction (Clark et al., 1982).

largely compensated by the lower atmospheric retention implied by such higher values. Clark et al. (1982) are certainly right when they conclude that the contribution of future land use changes is "substantially less important than that of changing growth rates in fossil fuel releases or changing values of the airborne fraction." Only some truly catastrophic vegetation shifts could negate this generalization so that inclusion of future biotic CO_2 releases comparable to the recent rates might bring the doubling date only about 10 years closer. So the best current consensus is that at some time between the years 2060 and 2100, atmospheric CO_2 will reach 600 ppm, twice its preindustrial level, 1.75 times today's value.

Nordhaus and Yohe's (1983) probabilistic scenario supports these conclusions, as it results in the following combination of percentiles and years: 5th and 25th percentile after 2100, 50th percentile in 2065, 75th in 2050, and 95th in 2035. Thus, the study's results give even odds for CO_2 doubling occurring during the years 2050–2100 or outside that period and that doubling before the year 2050 has only one chance in four.

Obvious questions to ask are: how much warming might be expected with CO_2 doubling, and what would be the effects of such global climate change? The remainder of this chapter will be devoted to considering the case—from basic modeling assumptions to climatic variation uncertainties, from ocean behavior to plant response. Some evidence is hard or at least satisfactorily solid but much in the coming sections will deal with assumptions and speculations—at best "tested" by computer models but untestable in nature. Still, I hope to make some clear, though only probabilistic, conclusions.

2.4. CONCERNS AND UNCERTAINTIES

While the presence of undesirable nitrogen or sulfur compounds in the environment takes us in diverse directions and raises many different concerns—some of which, as the subsequent chapters will show, we can face with confident understanding with others remaining controversial, even largely speculative—the CO_2 story is dramatically simple, a perfect feed for catastrophic popular science writings, a message which goes well in two-sentence explanations on television: the threat of a warmer planet.

As usually served, the progression is very linear and often most uncomfortable. Burning of fossil fuels raises atmospheric CO_2 whose greenhouse effect inexorably raises temperatures so that large-scale climatic changes occur and while some places may benefit, temperate latitudes will become drier, crop production will start faltering—but perhaps it will not matter so much in comparison with the ultimate effect, melting of ice caps and worldwide submergence of coastal regions.

These worries cannot be dismissed lightly. Without belaboring what some feel to be the ultimate, and deeply ethical, essence of the problem—do we have the right to permit the CO_2 increase to continue in the face of our awareness of its possibly destabilizing effects?—I believe we must strive for better understanding of the CO_2

puzzle but to do so within the wider contexts of the still so poorly understood climatic change, planetary equilibria, and ecosystem resilience, with deliberate caution and with ready admissions of lasting limitations. The following sections will detail the arguments why it should be so.

2.4.1. Warmer Planet?

First, I will address the key point on which rests the whole carbon dioxide worry: how much warmer would the mean temperatures be? This means engaging in descriptions of long-term climatic models, appraising the relative importance of many constituent variables, and comparing the now abundant predictions. A truly vast literature is now available on CO_2 modeling so I shall be content in capturing succinctly the essence of the basic approaches and outlining the main areas of both consensus and disagreement.

Then I will go on to summarize the variety of possible effects arising from this predicted warming. Again, there is no shortage of envisioned specificities here and I will treat them, just as a matter of convenience, in three categories: first, the effects on global climate, then the postulated changes of terrestrial ecosystems, and last the fate of the oceans on a warmer planet. Only when I am done with these neutral reviews of published models and effects will I focus on the uncertainties which make all of these exercises and conclusions amazingly shaky.

2.4.1.1. Models and Predictions Leaving 19th century estimates aside, efforts to quantify surface temperature response to increases of CO_2 concentrations—usually taking the form of CO_2 doubling exercises—have been going on, with intensified frequency and complexity, since the mid-1950s, encompassing a range from simple calculations to state-of-the art hemispheric computer models. Chronological review of major published models will show best the degrees of disagreement and consensus after a quarter century of effort. Other concise reviews of CO_2 models can be found in Clark et al. (1982) and Schlesinger (1983).

Plass (1956) opened this modern period by using a simple surface energy balance model to calculate a 3.7°C rise of surface temperature upon doubling of atmospheric CO_2. Kaplan (1960) dismissed these calculations for failing to take into account the influence of cloudiness, and argued that with 50% overcast skies and an average cloud distribution at different atmospheric levels the net long-wave radiation of the surface is reduced by 38% and hence the resulting temperature change with CO_2 doubling would be only 2.23°C. Möller's (1963) work was the next step in complexity: he pointed out that both Plass (1956) and Kaplan (1960) neglected the overlap of CO_2 and water vapor absorption between 12 and 18 μm and he argued that the atmosphere tends to conserve relative rather than absolute humidity.

However, as Manabe and Wetherald (1967) pointed out, all of these models, in spite of inclusion of different assumptions concerning radiative transfers, have tried to determine changes in the surface energy budget resulting primarily from higher downward infrared flux brought about by elevated CO_2 concentrations increasing atmospheric opacity—rather than accounting for the whole earth–atmosphere system. Consequently, they included the mixing effects of vertical heat transport in their radiative–convective model; the first of a series of increasingly more realistic simulations.

The principal conclusions of their work were that with fixed relative humidity, cloudiness, and radiative–convective equilibrium, and without horizontal circulation, doubled atmospheric CO_2 would raise the mean surface temperature by 2.35°C. Four years later Manabe (1971) used the same model but with a somewhat different radiation scheme to arrive at a lower mean increment of 1.9°C.

The next step by Manabe and Wetherald (1975) was a fundamental advance: although inevitably simplified—with limited computational domain, idealized terrain, no heat transport by ocean currents, and fixed cloudiness—theirs was the first three-dimensional general circulation model. The principal conclusions of the model were: a tropospheric warming averaging 2.93 K at the surface with CO_2 at 600 ppm, and ranging from 2 K in the equatorial zone to more than 4 K north of 60°N with a maximum of 10 K at 80°N; a rather large stratospheric cooling of 1 K (at 20 km) to 5 K (at 30 km) with little latitudinal difference; an intensification of the water cycle by about 7% with mean zonal rates of precipitation higher everywhere except around 15°N and between 30° and 40°N; predictably, a lower surface albedo everywhere north of 50°N; and a reduction of total poleward heat transport between 30° and 60°N.

A year later Bryson and Dittberner (1976) tried a relatively simple approach in their effort to simulate the past hemispheric response to CO_2, volcanic ejecta, and anthropogenic particles pollution, the three variables assumed to be primarily responsible for variations in atmospheric transmissivity and absorption of cloud-free atmosphere by volcanic emissions and man-made pollution.

Their model yielded a surface temperature sensitivity of 0.08 K for a 1% (3 ppm) change of CO_2 concentration, and hence CO_2 doubling would bring a higher mean temperature increase than indicated by nearly all previous studies. However, Bryson and Dittberner hasten to argue that considering CO_2 variations as the sole parameter of climatic change is unrealistic and that aerosols from fossil fuel combustion and forest and agricultural burning—the processes which are also the principal sources of CO_2—must be made a part of any proper calculation. Accounting for the lowered tropospheric transmissivity then results in a reversal of the CO_2 effect and doubling of CO_2 concentrations would lower the average hemispheric surface temperature by 4.93 K.

Bryson and Dittberner's (1976) work, although not a model of future CO_2 levels but rather a simulation of past climatic events, can be seen as the first of a series of calculations which departed substantially—and all of them downward—from the emerging "3 K warming" consensus. Ohring and Adler (1978) used a simple, zonally averaged numerical model to simulate some features of the annual average climate of the Northern Hemisphere, among them responses to changed cloudiness and solar constant and to doubled CO_2. Without ice-albedo feedback, they obtained a mean hemispheric rise of just 0.52°C; with the feedback of surface temperatur increased, this value went up to 0.8°C, ranging from 0.5°C at the equator to 2.5°C at 75°N.

Augustsson and Ramanathan (1977) concentrated on the sensitivity of the global surface temperature increase in relation to the CO_2 15-μm band absorptance, on the contribution of weak CO_2 bands, and on the effects of some critical assumptions inherent in radiative–convective models. Assuming the constant cloud top altitude, their best estimate of temperature increases for CO_2 bands in the 12- to 18-μm region was 1.9 K for doubled CO_2 concentration, and inclusion of the CO_2 band in the 10- and 7.6-μm region raised the value to about 2 K.

Newell and Dopplick (1979) were the first major "downward" dissenters. Their approach was via a static consideration of radiative fluxes, a much less spectacular method than multilayer dynamic computer modeling. In their short paper they calculated the decrease in infrared flux from the surface to the atmosphere to between 1.0 and 1.6 W/m^2 for CO_2 doubling between 330 and 660 ppm and a net change at the surface, with additional absorption of 0.3 W/m^2 of solar infrared by CO_2, of 0.8–1.5 W/m^2 with cloudy skies and 1.1–2.6 W/m^2 with clear skies, values much lower than Manabe and Wetherald's (1975) 3.5 W/m^2.

They then used 1 W/m^2, the average increase in the energy received at the surface, to estimate how much the tropical sea temperature would change in response to such warming by plugging it into standard sensible and heat loss and radiative transfer equations and assuming a mean wind speed of 3 m/sec. The result, 0.03 K, was also applied to the atmosphere, again much less than previously published values of around 2 K for tropical latitudes. Calibration of this crude model with observations recorded after the Mount Agung eruption in 1963 yielded a sensitivity of 0.1 K for the entire planet; even using the Manabe–Wetherald value of 3.5 W/m^2 with the authors' model yields a warming of no more than 0.24 K for tropical latitudes, still no more than one-eighth of the generally accepted level.

Another dissent, which caused much greater reaction, was also a short paper, a mere four columns in the Reports section of *Science*: Idso's (1980) "natural experiments" established that the surface air temperature response functions to radiative perturbations mimic CO_2 effect. From altitudinal redistribution of atmospheric dust over Arizona he obtained a value of 0.173°C/W per m^2, from an influx of moisture over Phoenix during the rainy summer season he obtained 0.196°C/W per m^2, and from annual variation of solar radiation at 105 United States stations he derived a value of 0.185°C. But in the last experiment the response of 15 West Coast stations heavily influenced by the Pacific was only half of the nationwide total and so Idso derived a weighted upper limit for the whole planet to be 0.113°C/W per m^2.

Good agreement among the responses to three different perturbations then encouraged Idso to look for a proof that the result also applies to time scales on the order of decades to centuries and that it embodies thermal inertia of the oceans. His solution was that an airless planet would have a mean equilibrium temperature of −18.6°C, while the actual mean is 15°C, so that the total greenhouse effect of the entire atmosphere is to elevate surface temperatures by about 33.6°C; as the atmosphere radiates back to the surface 348 W/m^2, the response function is about 0.1°C/W per m^2, a result naturally including all land–ocean–atmosphere feedbacks operating over long periods of time.

This is, of course, identical to Newell and Dopplick's (1979) value and gives a temperature increase of no more than 0.25°C for doubled atmospheric CO_2. Furthermore, Idso argues that when the surface air temperature response function has an average value of 0.1°C/W per m^2 when the atmospheric emissivity goes from 0% (no atmosphere) to the current level of about 90% it would be most unlikely that an additional 10% increase (i.e., all of the surface radiation reflected back to the ground) would cause any drastic change. But even a tenfold increase of CO_2 concentration would close only about 20% of the radiation window, making it impossible for atmospheric emissivity to reach unity and limiting the surface temperature response to only about 0.8°C even with such a huge CO_2 influx.

Idso's (1980) verdict is that "considering the available *evidence*, then, there seems to be no reason to suppose that carbon dioxide, present in the atmosphere only as a 'trace', has any more than a 'trace effect' on Earth's surface air temperature." The reception of Idso's findings by the CO_2 modeling "establishment" was, not surprisingly, negative. Schneider et al. (1980) were "dismayed" by Idso's claim of uncovering a major discrepancy between the prevailing theory and his "natural experiments," faulted him for not considering the positive feedback arising from increased water vapor content, the same omission as found in Newell and Dopplick (1979), and rejected his claim that "serious reconsideration of the whole CO_2–climate problem" is in order.

The latest in the series of Manabe and Wetherald models was also published in 1980 as a modification of their previous effort. The improvements include extension of computational domain to 90°N, proper assignment of −10°C as the boundary of permanent ice or snow cover, and, most notably, a variable treatment of cloudiness rather than the assumption of a fixed state (Manabe and Wetherald, 1980). All other simplifications of the previous model were retained.

Resulting increases of the mean zonal surface and tropospheric temperature (and declines in stratospheric temperatures) are very similar to those of their 1975 model and the rise of the surface air temperature averaged for the whole simulated area is 3.0 K, again virtually identical with the 1975 mean in spite of incorporation of the cloud-radiation feedback. Also identical was the finding of an intensified water cycle with mean precipitation going up by 7%, with latitudes north of 40°N receiving even larger increments. In addition, experiments employing a changed solar constant have shown that doubling of CO_2 and a 2% increase of the constant result in virtually identical changes of zonal-mean surface air temperature, a surprising finding in view of the quite different radiative influence of the two forcing factors.

Similarly, the distribution of cloud cover change with CO_2 doubling is very much like a new pattern arising from a higher solar constant but the shifts, above all the decline of upper and middle trophospheric cloudiness at most latitudes and the increase of lower troposphere clouds north of 50°N, are not very large so that the feedback has little effect on the overall sensitivity of climate.

All of these models were reviewed by two NAS panels set up to formulate state-of-the-art consensus. Their "statistically significant conclusions," identical in both 1979 and 1982 reports (NAS, 1979c, 1982a), follow: First, doubling of current

CO_2 concentrations should increase surface temperature by 3°C with a probable error of 1.5°C. Second, this tropospheric warming will be accompanied by stratospheric cooling with relatively small latitudinal variation. Third, mean global evaporation and precipitation would increase, i.e., the water cycle would intensify.

Fourth, surface temperature changes will vary appreciably, with polar regions warming two to three times as much as the tropics, the Artic much more than the Antarctic, and seasonal variations large in the north (with minimum summer and maximum winter warming) and much smaller below 45°N. Fifth, changes of water cycle would include higher mean annual circumpolar runoffs, later snowfalls and earlier snowmelts, lower soil moisture in midlatitudes of the Northern Hemisphere, and lesser extent and thickness of polar sea ice. The following sections will examine these terrestrial and oceanic effects in some detail.

2.4.1.2. Possible Effects: Land Warming would bring a variety of fairly inconspicuous ecosystemic changes such as gradual zonal shifts of many plant and animal species, accelerated slide toward extinction for many sensitive varieties, possible dominance of weedy growth under new competitive conditions, with pests and diseases spreading more rapidly owing to the loss of former climatic controls. But by far the most significant consequence of global warming on land would be a change of precipitation patterns of obviously nonuniform character: some regions would become wetter, in other areas, including some with already light water supply, precipitation totals would decrease. So the challenge is to foresee which shift will descend on a particular locale. This can be done by assuming a "more of the same" distribution, with coastal rainy regions getting ranier and dry inland areas becoming drier, but the best way to do guesswork on warming effects is to study analogical states.

Past climatic analogues offering suitable comparisons are, in reverse chronology, the well-known medieval warming between about 800 and 1200 A.D., a period to which Greenland owes its name; the Altithermal period of the Holocene (between 4000 and 8000 years ago); the last interglacial (some 120,000 years ago); and the last period when the Arctic Ocean was ice-free, about 25 million years ago (Flohn, 1979). But the applicability of these analogues is made uncertain by different ocean and cryosphere boundary conditions and even more by the fact that our knowledge of temperature and precipitation patterns for those past warm eras had to be gleaned from a variety of indirect information and while it is satisfactory in qualitative terms it is not quantitatively very reliable.

Budyko (1982) goes the farthest in his thermal analogy illustrating possible climatic changes on the territory representing the European part of the USSR and in western Siberia by using the Pliocene model. Assuming that CO_2 doubling will bring to the Northern Hemisphere temperature averages comparable to Pliocene levels, he uses Sinitsyn's paleoclimatic maps to show that surface air temperatures in the Arctic could be up to 20°C higher in January and that the 0°C isotherm for January would shift northward by 10–15° of latitude (Figure 2-8). The thermal regime of the northwestern part of European Russia would then be like central France today and even the north of western Siberia would mellow so much that it

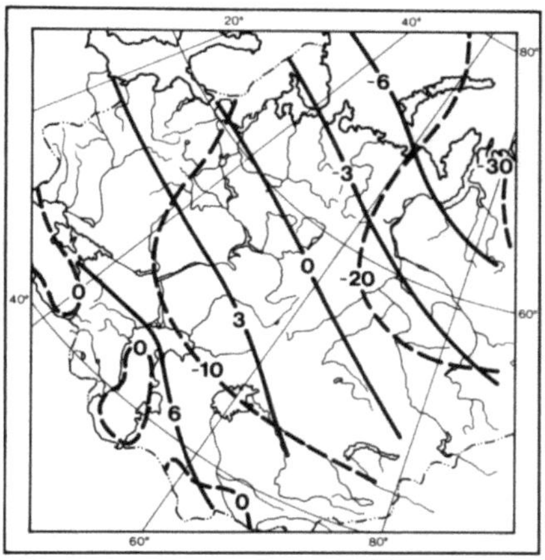

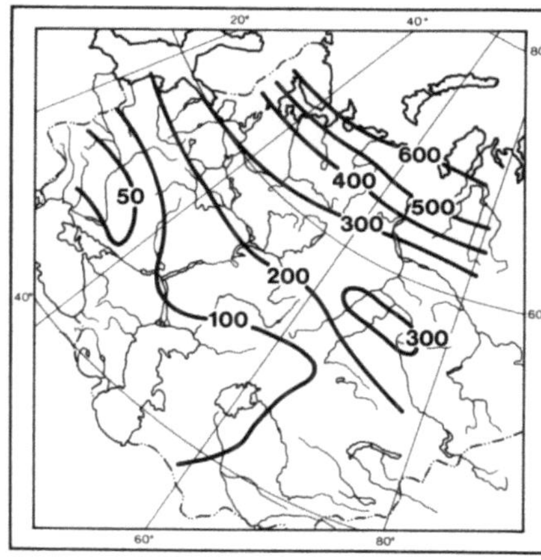

FIGURE 2-8. January isotherms for the USSR (without eastern Siberia). Dashed thin lines show current temperatures, solid thick ones are for possible warming with doubled CO_2 and thermal regime approaching that of the Pliocene (Budyko, 1982).

FIGURE 2-9. Differences (in mm) between current annual precipitation totals and levels resulting from doubled atmospheric CO_2 over the western half of the USSR based on Sinitsyn's paleoclimatic reconstructions (Budyko, 1982).

would resemble current southern Poland. As for the precipitation, Budyko (1982) believes that rainfall over the western USSR will increase, by as much as 600 mm on the Kola peninsula, by at least 200 mm everywhere north of the line Riga–Tselinograd, and by up to 100 mm in central Asia, along the lower course of the Volga, and in central Russia (Figure 2-9).

A wetter Europe and Russia are also predicted by Kellogg (1978) who assembled a variety of paleoclimatological evidence for the Altithermal period which he believes to be an approximate representation of the future climate. Figure 2-10, which indicates the areas of wetter or drier climate than at present, suggests some possible future shifts (the map's blank areas do not necessarily mean that such regions did not deviate in any way from the current precipitation pattern, but that simply our paleoclimatic information is inadequate there).

Perhaps the most notable Altithermal departure was the prevalence of a wetter climate throughout most of Mediterranean Africa, the Middle East, India, and China, an essential precondition for the rise and entrenchment of ancient civilizations in valleys of the Nile, the Tigris and the Euphrates, the Indus and the Ganges, and the Huang He. But Kellogg is also quick to caution that owing to different causes of the Altithermal warming (largely from the variation in the earth's elliptical orbit rather than from changes in radiation balance), one cannot expect the onset of the same circulation patterns with the future CO_2-induced warming.

Although many of the climatic shifts expected from the Altithermal analogue can also be seen in a modern analogy used by Wigley et al. (1980), there are some

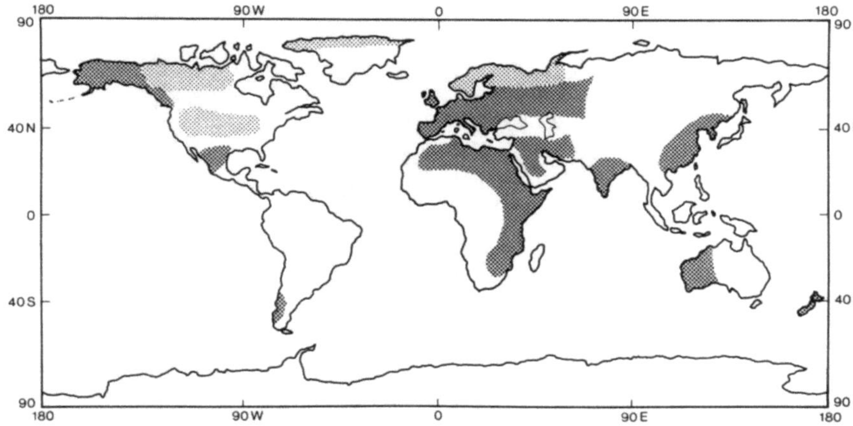

FIGURE 2-10. Kellogg's (1978) reconstruction of the Altithermal (4000–8000 years ago) climate which may be an approximation of future conditions brought by doubled CO_2. Heavier pattern shows areas wetter than today, lighter shading those with drier climate. Blank spaces do not necessarily imply no rainfall change, rather a lack of information.

notable differences. Their approach to presenting a plausible scenario of future temperature and precipitation changes is simple but effective: instead of comparing records for a group of warm years with a long-term mean, they derived a warm-world scenario by contrasting composites of the five warmest and the five coldest years for the period 1925–1974, with the warmest and coldest years defined in terms of temperatures between 65° and 80°N to accommodate the generally accepted opinion that any CO_2-induced warming will be most discernible in such high latitudes.

For these latitudes the average temperature difference between the coldest and the warmest group was 1.6°C, for the Northern Hemisphere as a whole only 0.6°C, considerably less than predicted by consensus models. Consequently, CO_2-induced precipitation shifts may be both more intensive and spatially more extensive than suggested by this scenario (Figure 2-11). The most important shifts are decreases of wetness over much of the United States, most of Europe, the Soviet Union, and Japan; much of Southeast Asia also has a decrease. On the other hand, all but the southernmost tip of India, much of eastern China, and nearly all of the Middle East have higher precipitation.

Revelle and Waggoner (1983) presented calculations based on an empirical relationship among mean annual precipitation, temperature, and runoff and showing that while a slight warming (2°C) and a reduction of precipitation (by 10%) would not have a serious effect east of 100°W, such changes would probably severely reduce both the quantity and the quality of water resources in the western United States. As a result, they favor careful consideration of creating more resilient water supply systems to ease such burdens.

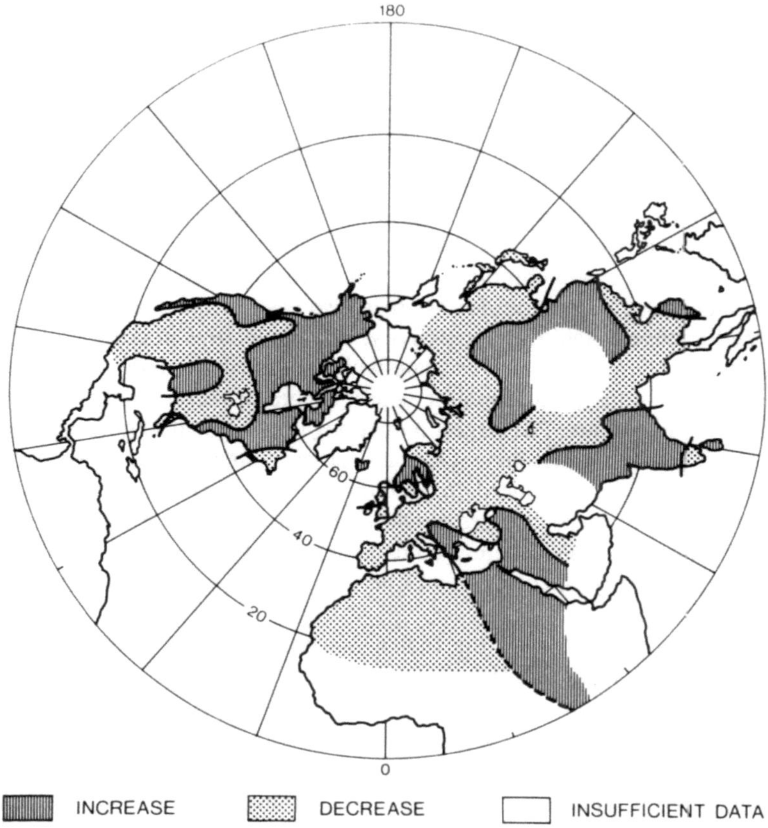

FIGURE 2-11. Mean annual precipitation changes from cold to warm years (Wigley et al., 1980).

Temperature shifts would inevitably influence the duration and extent of circumpolar snow cover. The annual variability of continental snow cover is very large, especially in Eurasia, but during a typical winter about 38 million km^2 in Eurasia and North America may be covered by at least 2.5 cm of snow and appreciable hemispheric warming would obviously eliminate much of this cover, first in those areas of low and middle latitudes where the occurrence of snow rather than rain is already marginal (Barry, 1978). There is clear evidence from Europe that during this century the period was snow on the ground was shortest in the 1920s and 1930s, the decades of greatest mean temperatures.

In contrast, in high latitudes where snowfall is limited by low humidities, warming would most likely mean snowier winters although one cannot expect any simple linear relationship between higher mean temperatures and greater snowfall, and duration of snow cover may be only slightly affected. Similarly, permafrost changes would not be very significant in the high Arctic where the frozen layer is very deep, but general retreat should be expected along the discontinuous southern margins.

Without knowing the intensity, perseverance, and regional distribution of all the discussed climatic changes, any considerations of possible health, social, or economic consequences are speculative and must be confined to general features. In terms of human health risks, milder winters in temperate latitudes would be welcome, but hotter summers everywhere may aggravate the already present dangers of heat stress and strokes which accompany current heat waves and result in much discomfort, debilitation of chronically ill people, increased hospital admissions, and higher incidence of premature deaths. It is also easy to imagine how the already clearly felt intense instability, aggressivity, and intolerance in crowded urban settings during hot summers could escalate with additional warming.

To determine the socioeconomic consequences of these shifts, their seasonal distribution is important. The largest reductions in precipitation generally occur in winter and fall but most affected regions also show summer precipitation decreases, a change especially pronounced in eastern Europe, central and western United States. In any case, these changes would be representactive only of 5- to 10-year mean conditions on which there would be superimposed inevitable annual variability: if these yearly fluctuations will be greater or lesser than today we cannot say.

General intensification of the Indian monsoon and less rain over European Russia, plains of the Kazakhstan—remember, both Kellogg (1978) and Budyko (1982) have it the other way—and the Corn Belt would have perhaps the most undesirable impacts in terms of crop harvests. Although in the Indian case the problems usually arise from late or weak monsoonal flows, a considerable increase of seasonal rains would almost certainly cause detrimental floods; the importance of timely ane recurrent rains over wheat and cornfields of Eurasia and North America is obvious.

Not surprisingly, then, most of the interest in CO_2-induced climatic change and its effects on terrestrial temperatures and precipitation has concentrated on the fate of crop production in major grain-exporting and cereal-consuming nations and on its socioeconomic impacts. Most of the available information on this rapidly growing research field is conveniently packaged in multiauthored volumes by the Climatic Impact Assessment Program (1975) and by the National Defense University (1980). I will summarize the findings of the latter study, the first systematic effort to outline agricultural consequences of different climatic scenarios.

A slight global warming, the most likely climatic change (with a probability of 30%) identified by an international panel of experts, was found to have negligible effect on 15 key food crops. Even a large warming, where mean Northern Hemisphere temperatures would be between 0.6 and 1.8°C warmer than today (considered much less probable in the near term with a probability only 10%), was forecast to have relatively minor effects.

Winter wheat yields were seen to decline in the United States (but only by about 1%), Argentina, Australia, and India (all by about 4%), but to rise in China (over 1%) and the USSR (by 6%); spring wheat yields in Canada and the USSR were forecast to go up by about 7% and remain unchanged in the United States; for rice and soybeans the forecast yield declines were mostly just a fraction of a percent, for American corn 2.5%.

Reductions in corn yield would be in response to both higher summer temperatures and lower rainfall. An increase of just 1°C in average August temperature would alone be sufficient to cause a 2% yield reduction, and 2–3 cm less rainfall in July, during tasseling and silking periods, would bring as much as a 7% yield decline. With wheat, higher yields would be brought above all by slightly higher rainfall in what are now drier farming regions (Great Plains, Kazakhstan). Rice cultivation could also benefit from higher temperatures and higher summer rainfall which would enable its greater northward expansion.

In general, informed consensus does not foresee any worrisome destabilizing shifts in food productivity of major grain-growing nations as a result of (no matter how induced) global warming raising average temperatures by as much as nearly 2°C. And even the postulated minor yield declines may not materialize as fertilized and irrigated crops would be able to respond to higher atmospheric CO_2 by appreciable increases in the rate of carbon fixation (see Section 2.4.2.3 for details). The net effect of global warming may thus well be an overall increase of crop harvests although some regions would undoubtedly experience yield declines.

If an average temperature increase of 1–2°C does not appear to precipitate any unmanageable food production crises, greater concern may then focus on the fate of oceans, especially in relation to the warming of polar regions, and the following section will look in some detail at this threat, probably the most popularized part of the CO_2 story.

2.4.1.3. Possible Effects: Ocean Having described the ocean's critical role in regulating atmospheric CO_2 levels (in Section 2.3.1.2), it should be obvious that any relatively rapid releases of CO_2 from fossil fuels or from terrestrial phytomass will be reflected in important direct and indirect ways as changes of the ocean's titration alkalinity, total inorganic carbon, and pH, while the general warming might eventually result in restriction of phytoplankton productivity, gradual decline of huge polar ice storages, and substantial rise of mean ocean levels.

Basic chemical relationships can be predicted with fair accuracy from our present understanding of the inorganic reactions involved in the process (Baes, 1982). But these "immediate" direct effects, including uptake of excess CO_2 by surface waters and resulting pH decline (a very slow shift of just 0.2 pH unit with doubling of atmospheric concentration), would soon be modified by feedback responses with primary producers assuming a key long-range role.

Oceanic photosynthesis might be adversely affected by decreasing pH and by catastrophic impact of extensive carbonate dissolution—but only very high atmospheric CO_2 increases could push these variables to levels adverse to phytoplankton and hence the changes in water circulation caused by planetary warming would appear to be of much more serious concern.

As the warming would increase with latitude, temperature gradients driving ocean circulation would be reduced and its global patterns may be considerably altered. Lowered circulation would inevitably cut down the availability of nutrients in the euphotic zone, leading to decline of organic carbon in the surface layer and to CO_2 releases to the atmosphere. Biotic feedback may thus be quite appreciable—but as yet it cannot be modeled with any confidence (Baes, 1982).

Possible intensification of CO_2 releases owing to gradual devegetation of the ocean would cause economically far-reaching declines and dislocations in fishing but the most unwelcome CO_2-induced change involving the ocean would obviously be the coastal submergence following polar deglaciation. This is the ultimate threat of popular science "melting of the ice caps" accounts, and it could descend in a milder unipolar form, with just an ice-free Arctic Ocean, or it could involve disintegration of the huge west Antarctic floating ice shelf, a process which would, in turn, trigger a deglaciation of the land ice in the entire adjacent region and elevate the mean ocean level by about 5 m.

Flohn (1982) offered a very thorough account of the first possibility, with arguments based on a combination of paleoclimatic data and physical models. His reasoning is as follows. As there is clear paleoclimatic evidence that the evolution of the two polar ice caps occurred quite independently and at different times and as there was a very long coexistence of an ice-free Arctic Ocean with huge Antarctic continental glaciation, this disparity may arise again when the average temperature increase of 4 or 5°C starts melting the relatively thin Arctic sea ice which is much more sensitive to climatic fluctuations than the thick southern ice shelves or the large continental sheets.

Temperature rise over land or ice will be two or three times higher than the global means in high northern latitudes and hence only a short period of a few decades may be needed to complete the creation of a unipolar glaciated earth. Thus Flohn's suggestion is in sharp contrast to prevailing appraisals that a relatively smooth, long-drawn transition is needed to achieve such formidable changes as complete melting of Arctic ice—and his premises, too, are seen as tenuous by his critics. Perhaps most importantly, they point out that Flohn's Miocene–Pliocene analogue, central to his forecast of future changes, may be faulty as all the evidence for unipolar glaciation is, contrary to his interpretations, equivocal while there are some clear proofs that it in fact never existed.

Moreover, observations show that snow cover which now persists through early summer delays the summer ice melt rather considerably and hence the likely increases in Arctic snowfall may help to preserve the ice. Estimates for Barrow (Alaska) attribute only a 15% thickness reduction upon 5°C warming (Barry, 1978), and sedimentary records indicate the existence of pack ice cover during many glacial–interglacial fluctuations of the past 2 million years.

There is, by now, an almost inevitable pattern to the CO_2 story: that of a future seen through models with clarity and certainty which is truly amazing and which rapidly disintegrates not only owing to general critical distrust of putting fine points on exceedingly uncertain affairs, but also owing to sharp identification of missing or untenable links.

Regardless, melting of the northern ice would not carry the threat of a biblical deluge. According to Flohn's (1982) scenario, an ice-free Arctic would become, once a new thermal equilibrium had been reached, an exporter of surplus heat into adjacent continents, snow-covered in winter and hence much colder. Climatic zones would rapidly shift at least 200 km to the north and the changes could be especially significant around 40°N, where considerable warming and drying might bring longer drought spells.

But unless there is a massive deglaciation of Greenland—and there are no strong indications that this should happen as the melting in lower altitudes may be more than offset by increased precipitation owing to much enhanced cyclonic activity—the sea level would hardly be affected. In any case, any substantial changes in Greenland's ice mass would be slow, measured in millenia rather than hundreds of years or decades, and the concomitant sea level rise may be no more than 2–4 mm/year (compared to the present annual rise of 1.2 mm).

Similarly, the huge land-grounded ice cover of eastern Antarctica could not be subject to rapid deglaciation even with unrealistically high and fast temperature increases (say up to 5°C in a century) and could waste only gradually over several millenia by *in situ* melting. On the other hand, the marine ice sheet of western Antarctica can survive only as long as its grounded part is buttressed by large fringing in shelves, above all the Ross formation facing the Pacific and the Ronne ice shelf opening to the Atlantic, and their melting would lead to relatively large and widespread submergence.

A group of studies published in *Nature* in 1977–1979 looked at the prerequisites, process, and impacts of such a major geophysical change (Clark and Lingle, 1977; Mercer, 1978; Thomas et al., 1979). Warmer temperatures would start weakening the ice shelves by melting at both upper and lower surfaces and by enhancing existing rifts and crevasses so that calving rates would rapidly increase. Calculations show that complete removal of the ice sheet and a 5-m rise in world sea level could theoretically occur in less than a century but more realistic modeling predicts, even with major and sustained warming, a gradual collapse over at least 400 years and most of the associated sea rise only during the final century of the disintegration.

Although we do not know with great accuracy the total volume of the western Antarctic ice sheet that is currently above sea level, the generally assumed share of above 50% would, if totally melted, result in a global sea level rise of 4–5 m. That is a rise not so unusual even in the very recent history of the planet: during a warm interglacial period of some 120,000 years ago, the global mean sea level was at least 6 m above the present one (Hollin, 1972). But this is of little solace in the current situation when the flooding of coastal areas would have profound consequences for the global population distribution and when it would carry enormous economic losses.

But the sea level changes will not be evenly distributed: owing to the earth's immediate elastic response to changed ice and water loading and the change in gravitational attraction exerted by the western Antarctic ice sheet on the surrounding ocean, sea level changes would be nonuniform. As an example, Clark and Lingle's (1977) calculations for uniform removal of 1 m of ice from the entire western Antarctic sheet show an oceanwide average sea level rise of 0.69 cm but in the mid-Pacific the value would be 25% higher, in the central Indian Ocean and central North Atlantic about 20% above the mean, while near the Ross Ice Shelf the sea would actually fall by 40%.

The effect of global warming on ocean fisheries most likely would be rather positive. Recession of ice covers would lead to a diffusion of fish species and to uncovering of new fishing grounds; judging by past experience, catches in such areas would be very good at least during the initial period. Reduction in zonal temperature differences might weaken the trade winds and, as a consequence, El Niño episodes may become less frequent and the upwelling fisheries of the western Pacific may be more stabilized and easier to manage in the long run (Bardach and Santerre, 1981).

As for the effects on coastal settlements, farming, and industries, substantial regional differences in submergence would mean that the actual population and economic displacement could be much more severe, or considerably lighter, than the impact resulting from a uniform rise of, say, 5 m. But even when assuming a uniform rise of 5 m, it would not be possible to evaluate the global impact without fixing the new sea level with the help of topographic maps, a prodigious undertaking on a worldwide scale.

The largest population displacements would be, not surprisingly, in Asia, affecting high-density settlements in the Ganges Delta (already subject to recurrent monsoonal flooding), in the deltas of Irrawaddy and Mekong, and on the low-lying coastal plains of several countries (most notably in China in Jiangsu province and along Bohai Wan). But the highest economic losses would, obviously, occur along European and North American coasts, and a detailed appraisal done by Schneider and Chen (1980) for the United States eastern seaboard is a very good indication of the overall cost.

Consequences of an approximate 5-m rise were evaluated on the basis of topographic sheets for 15- and 25-foot (4.6–7.6 m) contours. Dramatic regional effects of such coastal change are best illustrated by transformation in Florida and Louisiana where even the lower level (4.6 m) would submerge, respectively, about 24 and 28% of the land and displace 43 and 46% of the total state populations. In terms of estimated market value of their immobile wealth, the two states would lose about half of their assets.

Outside Florida and Louisiana, a 25-foot (7.6 m) rise would submerge such places as Savannah (Georgia) and Charleston (South Carolina) and "one could launch a boat from the west steps of the United States Capital . . . and row to the White House South Lawn" (Schneider and Chen, 1980). By contrast, along the West Coast only the Sacramento River floodplain would be heavily affected and for the whole country (excluding Alaska and Hawaii) the lower rise would flood only 1.5% of all land, displace less than 6% of the population, and affect about 6% of immovable assets. For the higher rise the shares would be, in the same order, 2.1, 7.8, and 8.4%.

Rapidity of the change would determine the seriousness of the damage and the responses to it. Should the surge come without much warning in just a year or two—a most unlikely possibility—the sudden economic loss and even more so the relocation of displaced people would obviously be burdensome. But as there would be almost certainly a warning period before the onset of such a change and as the process would be gradual, the overall damage, at least in the North American and European context, could be considerably minimized.

For example, even when Schneider and Chen's (1980) estimate of a \$(1980) 107.5 billion loss associated with a 4.6-m rise is doubled to account for various indirect costs and secondary effects and the sum of \$(1980) 200 billion is spread over an unrealistically short period of just one decade, then the federal compensations would amount to a mere 0.18% of the country's GNP, and about 3% of the annual federal budget, hardly an economically crippling damage. Spreading the payments over a longer, and much more likely, period would bring them to levels of many routine government support programs (such as farm product subsidies) whose existence is not required by natural dislocations—merely by the presence of powerful lobbies: a fine example is the powerful milk lobby which costs United States consumers about \$(1983) 2.5 billion every year (Donahue, 1983).

A rising ocean would also lead to large-scale shifts and transformations of coastal ecosystems: tidal flats and marshes, mangroves and countless beaches would be inundated, wave erosion would establish new shorelines, and eventually new coastal communities would evolve. There would be much change but not necessarily much species extinction or impoverishment. The only marine communities which might cease to exist in the form we know them today would be coral reefs—but only if the ocean rise were relatively rapid.

Coral reefs have a unique ability to precipitate $CaCO_3$ from seawater at rates sufficient to keep them growing as regionally extensive three-dimensional structures with seaward shallow portions producing about 4 kg of the carbonate per m^2 each year (Smith and Kinsey, 1976). Thus, the extreme potential for upward reef growth is about 3–5 mm/year, although for larger formations the typical value may be more likely just 0.6–1 mm. Consequently, the coral reef communities could not survive as three-dimensional structures if the water rises more rapidly than 3–5 mm annually for extended periods.

In sum, the most threatening of all oceanic impacts, coastal submergence, should be seen first on its proper time scale, taking a minimum of hundreds of years even in the extreme case of very unlikely high and sustained warming, and second, with the realization that even its eventual occurrence, unwelcome and costly as it would be, would leave plenty of time for gradual management of the problem whose cost appears to be well within our current capacities to deal with it. And, of course, the uncertainties ruling this realm of theories, speculations, and models must be identified and appreciated so that all of the bad news would be seen in their proper, that is, inherently tentative and tenuous, perspective.

2.4.2. Major Uncertainties

Later in this book it will be seen again and again how uncertain is our understanding and hence how limited are our predictive and normative capacities in studying and managing processes which are subject to relatively easy, controllable, and repetitive experimentation and whose durations are measured in anywhere from seconds to a few decades.

With these temporary, spatially limited, and often easily compartmentalized processes studied in theory, in the laboratory, and in the field by several generations of interdisciplinary effort proving so elusive and turning the relevant policy-making into a contest of sharply divided opinions and vociferous lobbies thriving on the very ambiguity and tenuousness of our understanding—how then can we expect to be guided in a matter of an incomparably more complex process unfolding on a time scale of centuries and embracing the whole biosphere of this planet? The outcome of the CO_2 "problem" rests, in the first place, on virtually unpredictable rates and kinds of future energy consumption and, even should these be known with clairvoyant perfection, the minimum required for any meaningful predictions of future atmospheric concentrations is a confident understanding of the relative importance and mutual interaction of scores of climate-forming factors, ranging from the obvious effects of changed cloudiness and terrestrial albedoes to much less obvious shifts in ocean thermal upwelling and productivity responses of primary producers.

Even if we could correctly identify all of the large and small ingredients of this awesome mix, a baking analogy illuminates quite well our dilemma: depending on the relative importance of initial inputs and their mutual interactions, one can end up either with nearly weightless puffs or with nearly stone-hard lumps. The fact that the "current consensus" ordains certain amounts of particular inputs to behave in given ways and that the CO_2-modeling patisserie stoutly maintains that theirs is the only possible recipe makes for a scientifically poor diet. We must do better than that—but we cannot do it in a hurry. As the following brief reviews will show, the uncertainties are both nearly endless and endlessly deep. The goal—a decent understanding enabling a sensible forecast—is not unattainable but less haste and greater humility are the two essential preconditions for overcoming the very unsatisfactory state of constructing even more complex computerized castles of sand.

2.4.2.1. Atmospheric Response The above chronological review of major models investigating surface temperature changes with doubled atmospheric CO_2 concentrations has revealed a wide variety of approaches—and no smaller variety of results. Below are shown all those results, and a few more, this time ordered in a single table and disaggregated by spatial coverage and, where such calculations were made, also by latitudinal effects (Table 2-8). When presented in this manner, it is immediately clear that, in the first place, the models are not easily comparable: one simply cannot assume that the values given as global means will be identical for the Northern Hemisphere and vice versa.

Indeed, a simulation by Sellers (1974) which attempted to obtain the latitudinally disaggregated surface temperature change for the whole planet shows a

TABLE 2-8
COMPARISON OF CALCULATIONS OF TROPOSPHERIC WARMING INDUCED BY DOUBLING OF AMBIENT CO_2 CONCENTRATIONS

	Surface temperature change (°C)				
	Global	Northern Hemisphere	Tropical (0–20°)	Midlatitude (40–50°)	Polar (70–90°)
Plass (1956)	3.8	—	—	—	—
Kaplan (1960)	2.23	—	—	—	—
Möller (1963)	9.6	—	—	—	—
Manabe and Wetherald (1967)	2.36	—	—	—	—
Manabe (1971)	1.9	—	—	—	—
Rasool and Schneider (1971)	0.8	—	—	—	—
Weare and Snell (1974)	0.7	—	—	—	—
Sellers (1974)	1.32	1.5	1.0	1.5	2.5
Manabe and Wetherald (1975)	—	2.93	2	3	8
Bryson and Dittberner (1976)	—	−4.93	—	—	—
Augustsson and Ramanathan (1977)	1.98	—	—	—	—
Ohring and Adler (1978)	—	0.8	0.5	0.7	2.2
Newell and Dopplick (1979)	0.25	—	—	—	—
Manabe and Wetherald (1980)	—	3.0	2	3.5	7
Idso (1980)	0.26	—	—	—	—
Chou et al. (1982)	—	2.3	1.6	2.5	5.2

considerably smaller effect in the Southern Hemisphere where only the belt between 70° and 80°S would experience a temperature increase equal to or greater than 2°C and that only between June and August during the austral winter; in contrast, all surfaces above 60°N would have such increases for at least 10 months of the year. The absolute simulated values are, of course, irrelevant: the undeniable point is that any temperature change will be distributed with considerable latitudinal variation and while only three-dimensional models can provide such distributions, they remain almost exclusively limited to the Northern Hemisphere and the nearly general omission of the Southern Hemisphere is an obvious fundamental flaw.

Further obstacles to comparability arise from the different model types which range from simple, one-dimensional radiation balances to three-dimensional simulations of general circulation, from very long forecasting runs to matching of the historical record which can be used for predictions. And, of course, even models belonging to the same category in terms of techniques and execution differ significantly as far as the choice of fixed variables and treatment of feedback relationships are concerned.

Still, the models will continue to be compared just in terms of single figures and with the incorrect assumption that the value represents the global average (for example, they are so listed in the otherwise careful review by Clark et al., 1982). Comparisons on that simplistic basis alone show no satisfactory degree of consensus emerging in spite of the rapid increase of modeling efforts since 1975: while I have reviewed only the key published contributions in Section 2.4.1.1, Clark et al. (1982) list 25 published and unpublished estimates, including approximate values derived from works where response to CO_2 doubling was only of peripheral concern or from studies which looked only at quadrupling of CO_2 levels.

Even when eliminating Möller's (1963) solitary (and erroneously derived) peak, the spreads are still considerable. For the most abundant group of one-dimensional models derived in many simple as well intricate ways, the lowest value is Newell and Dopplick's (1979) 0.25°C, the highest is the 3.8°C estimated by Plass (1956); the extreme range is thus about 15-fold, and the average for 19 available studies is exactly 2°C. Results of the six two-dimensional studies range from 0.8 to 2.4°C, a much smaller 3-fold span, and the average is 1.8°C. Three-diemsnional models have the highest average, 2.5°C for eight studies, including two unpublished simulations run by the Goddard Institute for Space Studies, and range from 0.2 to 3.9°C, an almost 20-fold range.

The three groups are too small to offer meaningful statistical evaluation of the sets and, besides, four of the six published three-dimensional models, which should best reflect water cycle intricacies and cloud, snow, and ice feedbacks critical for establishment of new thermal equilibria, originate from a single source, the Geophysical Fluid Dynamics Laboratory (GFDL) of the NOAA. However, it is abundantly clear that matters are not as simple as two National Research Council panels (headed and staffed by researchers from the GFDL and the Goddard Institute, the authors of "three-degree" estimates) make them out to be.

In their conclusions, the panels believe that the best estimate of long-term climatic response to doubled CO_2 is an average surface temperature rise of 3 ± 1.5°C (NAS, 1979c, 1982a). Why 3 as the mean when the average of the six published three-dimensional models is 2.13°C? Even when Gates and Cook's (1980) very low value of 0.2°C is excluded it is still only 2.52°C! And just because they are less complex and less interactive, should some 25 other models be excluded from consensus forming? If all the models are taken into account, the median value is exactly 2°C, the lower quartile 1.4°C, and the upper quartile 2.8°C.

One cannot easily dismiss the feeling that any work yielding lower values is discounted while the higher estimates are welcomed by what might be called the CO_2 modeling establishment at the GFDL and the Goddard Institute. A perfect illustration of this is seen in the contrast between the treatment given to Idso's work and that to the Goddard Institute simulations.

When Idso's (1980) controversial report came out in *Science,* it became doubly contentious as the key reference for his very low response value was listed as "in preparation," a fact which Schneider et al. (1980), writing a defense of "three-degree consensus," found dismaying. But when atypically high estimates of temperature change were proposed by the Goddard Institute group and included in the NAS evaluation, there was, to quote Clark et al. (1982), "little more than disgruntled mumbling" even when the supporting data and calculations were unavailable for independent review. This "implicit double standard" (Clark et al., 1982) is so sadly inappropriate!

Just reviewing the available calculations and estimates illustrates the considerable degree of uncertainty. There is little doubt that virtually all authors believe that the mean surface temperature will rise in response to doubled CO_2 and that the simple mean and median values of all estimates published since the early 1970s are about 2°C and that half of all values fall within ±0.8°C of that average. But even subscribing to such a carefully stated consensus may be wrong. After all, it would not be the first time that predictions based on a very simplified understanding of an extremely complex process permeated with countless nonlinear feedbacks would fall considerably off the mark. And while it may readily be granted that all the earlier calculations and all one- and two-dimensional models are no heuretic match to more recent general circulation computer simulations, the only reason to proclaim this family of models to be superior is an agreement of perhaps a score of researchers that these simulations best account for important feedback processes.

Again, easily granted—but, again, this fails to answer the ultimate questions as to how well these models simulate the real world and what the correspondence has been to date between their predictions and actual atmospheric behavior. There is no evading these questions, "for we live in a *real* world with a *real* climate, and not [in] some computer-generated substitute of unknown or dubious quality" (Idso, 1982). So first we must look closer at how well the models mimic reality, then at the actual behavior of the atmosphere.

The best way to illustrate the numerous simplifications and limitations of even the most advanced models is to look closely at the two simulations by Manabe and Wetherald (1975, 1980) which have certainly been the most quoted predictions of surface warming and which are considered to be the best available estimates by the National Research Council panel which produced the CO_2–climate assessment (NAS, 1982a, 1983b). As complex as they may be numerically, the models are still surprisingly crude substitutes for reality.

To begin with, the computational domain is but a small part of the planet. In the 1975 run it is an area in Mercator projection embracing 120° longitudinally and extending to 81.7°N; cyclic continuity is assumed for the two meridian boundaries while "free-slip insulated walls" are placed at the equator and at 81.7°N. In the 1980 run the domain is a 120° wedge with the same boundary properties. Can a model be considered realistic when it simulates a global phenomenon with complex linkages between atmosphere, land, and ocean—and disregards completely the Southern Hemisphere with its nearly 60% of the planet's seas?

Whatever remains of the global ocean is, in both models, treated in so simple a manner that on this point alone the model abdicates the essential resemblance of reality: the domains are simply divided in half between land and ocean but the models contain no separate ocean computations as their "oceans" are treated merely as areas of wet land able to supply infinite evaporation—"oceans" without any heat capacity and without any heat transport by ocean currents.

Among the basic tenets of the planet's thermal balance are, of course, the facts that the ocean does not cover one-half of the surface, that it is the world's largest heat reservoir, and that it distributes heat through an amazingly complex system of currents whose perturbations may lead to far-reaching climatic changes. For example, the anomalous El Niño of 1982–1983—which first appeared in the western tropical Pacific and then spread eastward attaining an unusually large extent and persisting for 2–3 months longer than the more common events—was associated with catastrophic floods in Ecuador and northern Peru, droughts in neighboring areas, and the virtual disappearance of marine life in the central and eastern equatorial Pacific, and the atmospheric circulation changes it caused were credited with contributing to droughts in northeastern Brazil and to unusual weather over large parts of the United States (Philander, 1983).

Even when one assumes, most unrealistically, that the ocean's heat capacity and currents will have no decisive effect on reaching new global thermal equilibria, there are numerous sufficiently notorious examples to appreciate the enormous regional effects which may substantially modify any minor shifts over areas as large as 10^6 km^2. Other obvious departures from reality in both models are the complete elimination of diurnal and seasonal variations of insolation and, in the 1975 run, the assumption of fixed cloudiness. In the later model, cloud cover is placed wherever condensation is predicted and otherwise clear skies are assumed—a better approach than before but still so obviously simplistic.

This brings me to numerous specific problems with variables and relationships which are included in the GFDL and other three-dimensional models but whose complexity and insufficient understanding make it inevitable that mistakes will result from their simplified representation, mistakes which can shift the final outcomes by uncomfortably large margins. Of the many reviews of these touchy variables and processes, I find the works of Schneider (1975), Dickinson (1982), Idso (1982), Reck and Hummel (1982), and NAS (1982a, 1983b) the best of the lot. So unless specifically referenced, the following paragraphs will largely be brief summaries of the observations and criticisms offered in these publications which, of course, rely on numerous other contributions.

So back to clouds, whose importance in the whole problem is so obvious: at any given time they cover about half of the planet and are thus critical in determining global albedo and governing temperature lapse rate feedbacks. Modelers thus face the extremely difficult task of incorporating these ever-shifting wisps into their equations if their models are to resemble the actual atmosphere—and so far they have not done well at all. Not surprisingly, perturbations in cloud dynamics are ignored altogether, and even the very inclusion of the phenomenon is done clumsily, mostly by assuming fixed presence and fixed cloud top level.

As for the effects that changes in cloudiness will have on global climate owing to changes in cloud cover, optical thickness, and altitude distribution—we simply do not know: "the current capabilities for modelling cloudiness are not yet advanced to the point where it is known how cloudiness might change with climate" (Dickinson, 1982). Consequently, we cannot even say if changes in cloudiness will have positive, negative, or neutral feedback on future surface temperatures (Schneider and Thompson, 1980) or if they will provide a net amplification or an attenuation of the climate's sensitivity (NAS, 1982a).

But we know that changes in cloudiness depend on changes in relative humidity whose levels are controlled by a multitude of variables which never enter today's simplistic models. Evaporation is of critical importance in regulating climate sensitivities. Even small wet surface heating—caused, for example, by an initial reduction in the outgoing infrared radiation—will result in considerably higher evaporation (owing to the nonlinear relationship between evaporation and surface temperature) which will heat the atmosphere by releasing the latent heat.

Kandel (1981) showed how the degree of compensation in these perturbations of latent (and also sensible) heat fluxes and the ways in which the atmospheric humidity evolves determine the surface temperature sensitivity. According to his calculations, only slight changes in humidity lead to large (order of magnitude) differences in surface temperature sensitivity. The other key weakness concerning atmospheric response and the water cycle is the mismatch between relatively complex general circulation atmospheric models and the simplistic treatment of the ocean.

As imperfect as our understanding of atmospheric processes is, it is considerably more advanced than our knowledge of ocean mixing and circulation. Especially uncertain is our appraisal of heat transfer, a critical consideration in any model of a warming atmosphere. Our continuing inability to formulate realistic ocean models and to couple them with atmospheric simulations would, in itself, disqualify all published quantitative predictions. Similarly simplistic has been the treatment of snow cover and sea ice. The NAS (1982a) panel had to conclude that an assessment of the role sea ice will have in the high-latitude response is difficult even in qualitative terms until we understand the controlling climatic factors.

The importance of water cycle variables in this whole effort can also be illustrated by testing the sensitivities of the models. Chou et al. (1982) looked at the effects of various treatments of physical processes critical for feedback mechanisms. The topmost curve in Figure 2-12 represents the latitudinal distribution of temperature changes as determined by the standard model run; the other lines represent cases where one critical value was fixed at the normal CO_2 level. The smallest reduction results from fixing meridian transport of latent heat whereas fixed ice/snow cover or fixed absolute humidity more than halves the temperature change in the high latitudes and the values averaged for the whole hemisphere are, respectively, 32 and 53% lower.

As all these feedbacks are directly related to water, it is obvious that even relatively small errors in delineating these relationships can multiply to differences on the order of several tens of percent. Without the ability to model the water cycle with a much greater sophistication than found in any of today's climatic models, even a rough usefulness of CO_2-forced runs remains elusive.

Using the multilayer energy balance model of Chou et al. (1982) as an example, Figure 2-13 shows the extent to which the downward infrared flux and the

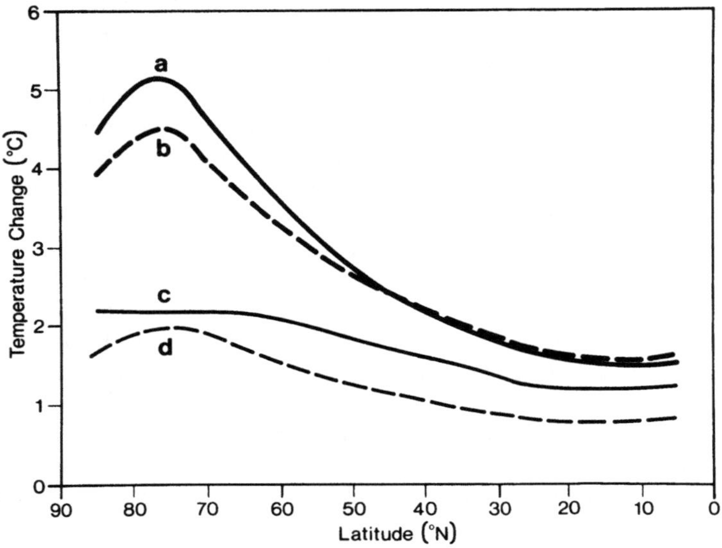

FIGURE 2-12. Sensitivities of the surface temperature to CO_2 doubling for (a) the control case, (b) fixed ice/snow cover, (c) fixed meridional transport of latent heat, and (d) fixed absolute humidity (Chou et al., 1982).

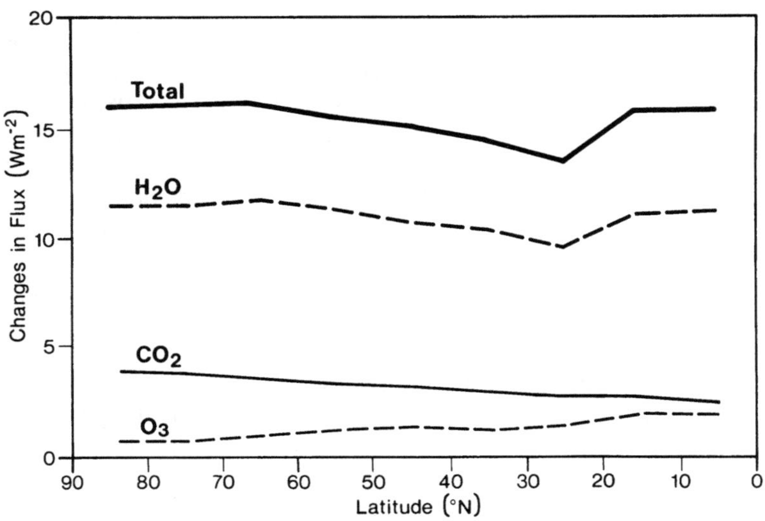

FIGURE 2-13. Changes in the downward infrared fluxes at the earth's surface for the entire spectrum and for H_2O, CO_2, and O_3 bands in response to doubled atmospheric CO_2 content (Chou et al., 1982).

spectral distribution of this change are dominated by water vapor bands. Of the total change in the downward infrared flux averaged over the Northern Hemisphere, 72% was due to water, just over 19% to CO_2, and the rest to ozone.

The only systematic way to estimate the reliability of model results in CO_2 doubling calculations, or in any other instances where the outcome cannot be compared with observation, is by the adjoint method of sensitivity analysis. Hall et al. (1982) demonstrate the application on a radiative–convective climate model where single adjoint calculation, requiring roughly the same computation time as the original model, was done to compute the sensitivities of surface air temperature to all of the model's 312 parameters. Parameter ranking shows that constants in CO_2 transmission functions for terrestrial radiation, CO_2 concentration, surface albedo, solar constant, and relative humidity are high on the list, but by far the largest relative sensitivities are those belonging to constants used to compute the saturation vapor pressure of water.

But this is an endless list of requirements: even with the water cycle well covered, and cloudiness simulated to perfection, fundamental difficulties would remain, none of them perhaps more important than the effect of atmospheric aerosols on long-term climatic trends. Once again, no reliable way exists even to interpret the sense and the magnitude of the effect (Hansen et al., 1981). Even when looking at the tropospheric aerosols alone, we cannot tell if they will shift us toward warming or cooling and if we couple this essential uncertainty with all the previously mentioned lacunae of atmosphere–ocean response, we are in the realm of opinions which will be, of course, appropriately couched in terms of models and inevitable computer simulations.

I cannot resist the ancient analogy of the blind men and the elephant: as the individual CO_2 modelers or groups feel, via jargonistic and highly formalized computerized exercises, various parts of the beast's body they come away with a very definite but very wrong account of what it is *all* about.

The impossibility of producing fine numerical results is obvious and one wonders about all the commotion and the need to defend particular models when, if carefully read, even the authors and appraisers plead ignorance and disclaim practical utility. Manabe and Wetherald (1975), who have devoted years to perfecting general circulation models, warn that owing to various simplifications the readers are not "to take too seriously the quantitative aspect of the results" obtained by their studies, and the National Research Council panel which considered three-dimensional models as the best available evidence admitted that our current capabilities are not realistic enough to offer reliable predictions in the detail necessary to assess the likely impacts (NAS, 1982a).

Perhaps the best perspective is offered by Ackerman (1979). In an effort to test the reliability of the models of Manabe and Wetherald, which used a rather simple treatment of the sensitive CO_2–H_2O overlap in the 15-μm CO_2 band, he compared their results by employing a high-resolution long-wave radiation code—and found that both approaches produced similar changes in the heating rate for doubled CO_2 concentrations. However, he did not interpret this finding as yet another proof of an inevitable 2 or 3°C warming. Instead, he noted that the difference in comparing the estimated accuracy of the calculations with the difference in heating rates is, particularly in the troposphere, "on the order of or less than the absolute accuracy of the models."

Moreover, he pointed out the possibility that the model errors are not systematic and, consequently, the heating rates have a component of random error with significant effect on the values of the heating rate differences. This led him to suggest "that the quantitative results of models used to study the effects of changing CO_2 concentrations should be treated with considerable care . . . while we can have considerable faith in the qualitative results . . . we should remain cautious about the quantitative results. It is indeed unfortunate that so much uncertainty is evident in what is a reasonably straightforward theoretical problem . . . ."

Long as this litany of uncertainties has been, nothing as yet has been said about no less fundamental and no less extensive gaps and inadequacies of our knowledge concerning the general processes of climatic change and, with a brief exception of particulate matter, about specific anthropogenic interferences other than CO_2 generation which can potentiate, neutralize, or negate the effects of higher atmospheric CO_2 levels. The study of climatic change is now such a growth industry that only the outlines can be given here, just enough to further reinforce the already plentiful arguments about the lack of confidence in our best CO_2-doubling models.

2.4.2.2. CLIMATIC VARIATIONS Rapid acceleration of interest in the CO_2 problem should be seen as just a part of a broader and more general explosion of concern about climatic change (for good reviews of this vast topic, see Matthews et al., 1971; NAS, 1975; Gribben, 1978; Miller, 1978; WMO, 1979; Berger, 1981; Budyko, 1982). What used to be a quiet preserve of individual enthusiasts from a variety of specialized fields quite suddenly turned into a highly fashionable quest whose high profile is eagerly kept up by the public media. Anchormen and writers of scientific columns and Sunday supplements started to see—behind every drought, flood, or winter tougher than the last one—unmistakable signs of impending climatic change and, I am saddened to say, not a few scientists offered predictions and warnings where the only honest course was to *plead ignorance!*

The enormous difficulties in understanding global climate—a dynamic system covering the entire planet and joined with countless regional and local feedbacks with the oceans, continents, and plants and influenced by changes ranging from the small but critical luminosity shifts of the sun to many forms of anthropogenic degradation of the environment—must be at least intuitively obvious. These feelings should have been intensified by the discussion of uncertainties permeating our efforts to model atmospheric response in simplistic model systems forced by a single variable so that even a reader who has never pondered these difficulties before should start feeling melancholic about our abilities to forecast the climate's course in future decades and centuries.

Not surprisingly, most of the public interest and much of the research work in studies of climatic change has focused on temperature changes: the variable is obviously critical, it can be easily and precisely measured, and we have more extensive past records and possess indirect analytical methods to derive its ancient trends better than for any other essential climatic parameter. And yet, the observational studies indicate that with our present monitoring network we are unable to determine reliably the average global surface temperature (Williams, 1978b). If we cannot do this, how can we pretend to evaluate its recent changes, much less the future fluctuations?

These fluctuations, we know, are driven by a complex interaction of many factors, natural as well as anthropogenic. Among the natural causes, by far the most influential ones appear to be variations of the earth's surface position relative to the sun and fluctuations in volcanic activity which have been influencing the climate by emissions of CO_2 as well as by releases of large masses of aerosols (Budyko, 1982).

The question of the climatic effect of astronomical factors, most comprehensively developed by Milankovich between 1920 and 1941, has been a controversial subject. As the earth's orbital eccentricity, the inclination of the planet's axis to the elliptic plane, and the time of precession of the equinoxes vary over periods of tens of thousands of years, they change the latitudinal distribution of total radiation in the different seasons (but not the total amount of solar radiation reaching the outer atmosphere as a whole) and may thus bring mid- and high-latitude temperature changes of several degrees.

Uncertainties in identifying which aspects of the radiation budget are critical to climatic change and inaccuracies of geological chronology have made conclusive tests of the orbital hypotheses difficult. Perhaps the most impressive recent proof comes from measurements of three climatically sensitive parameters ($\delta^{18}O$ of planktonic foraminifera, summer sea-surface temperature changes derived from a statistical analysis of radiolarian assemblages, and a relative abundance of *Cycladophora davisiana*) in two deep-sea sediment cores (Hays et al., 1976).

When monitored over the past 450,000 years the climatic variance is concentrated in three discrete spectral peaks at periods of 23,000, 42,000, and 100,000 years and these peaks, corresponding to the dominant periods of the earth's solar orbit, contain, respectively, some 10, 25, and 50% of the climatic variance, with the dominant 100,000 year component being in phase with orbital eccentricity. Extension of these findings to the future indicates that, ignoring other possible effects, the long-term trend over the next 20,000 years is toward extensive Northern Hemisphere glaciation and cooler climate.

During the Phanerozoic era, fluctuations in volcanic activity, which seem to have peaks approximately every 100 million years, and atmospheric CO_2 levels display unmistakable correlations (Figure 2-14). Budyko's (1982) calculations of Phanerozoic variations of atmospheric CO_2 (based on the rough proportionality between atmospheric CO_2 concentrations and the carbon mass in sediments) show that during the past more than 500 million years, the concentrations were around 0.2% with peaks up to 0.4%, and that only in the late Mesozoic (less than 100 million years ago) did they start to decrease unevenly and gradually. Consequently, modern atmospheric CO_2 concentrations reached an unprecedented low level, nearly an order of magnitude below the Phanerozoic mean—and hence the average

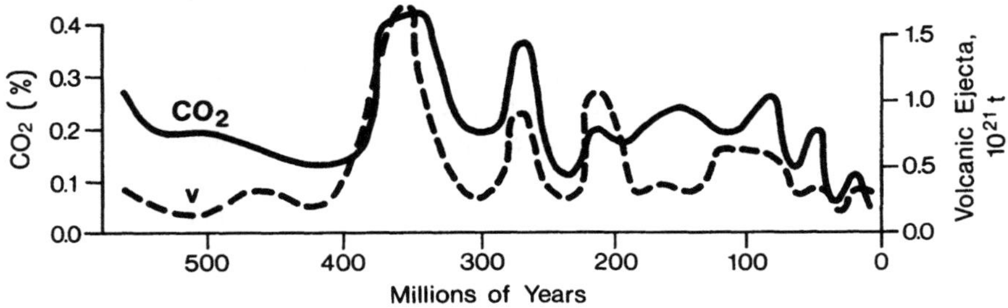

FIGURE 2-14. Correlations of global volcanic activity and atmospheric CO_2 concentrations during the past 500 million years (Budyko, 1982).

surface temperatures during the millenia of emerging and spreading human civilization have been, although obviously warmer than the last ice spell, appreciably cooler than during most of the evolution of life on this planet.

Even when one assumes a lower solar constant in the early Phanerozoic, CO_2 levels six- to tenfold higher than today would still have made up for the lower insolation. Using Sinitsyn's paleoclimatic data, Budyko (1982) calculated the departures from the contemporary mean temperature in the Northern Hemisphere at +4.8°C for the early Pliocene and 6°C for the early and middle Pliocene while in the Cretaceous the difference may have been as much as 11°C.

But for climatic changes of the past millenium it is not the CO_2-induced warming effect of volcanic eruptions but rather the aerosol-precipitated cooling which has received much research attention. The essential role of volcanic dust in controlling short-term declines of surface temperatures has been demonstrated quite convincingly by historical records as well as by actual monitoring for different periods and locations. Lamb (1970) concluded that the middle latitudes would experience a decrease of 0.5–1°C during the year after a large volcanic eruption and there is an abundant climatological literature documenting cooling around the Northern Hemisphere after the eruptions of Tambora (1815), Krakatau (1883), and Katmai (1912). This cooling effect wears off after a few years but long-range effects cannot be discounted.

A detailed historical review of volcanic explosions has shown a dramatic increase of their incidence after the year 1500, but it seems more likely that better reporting rather than a real increase in occurrence is the proper explanation, especially as the number of large, highly explosive eruptions has not grown proportionately (Newhall and Self, 1982).

Perhaps the best way to visualize the CO_2 warming and particulate cooling effect on atmospheric temperature change is to contrast them in one graph with observed temperature change in the Northern Hemisphere (Mitchell, 1975). When normalized to 19th century levels, the expected departures owing to the modeled warming effect (basically a line followed the GFDL model response) and tentative cooling effect [with temperature decline as calculated by Rasool and Schneider

(1971) and considered by Mitchell to be the maximum likely response] were unable to account for an important part of the recorded fluctuations at least until the 1920s (Figure 2-15).

Starting in the 1930s, either of the effects would have sufficed to determine a large part of the recorded temperature change and today these effects could be dominant. However, if the two effects have been additive, the net outcome would have remained negligible until the 1960s and even today the net shift (cooling) would be a small part of the actually observed changes. Obviously, what matters here are not the precise, absolute positions and courses of the warming and cooling lines but the realization that even liberal allowances for temperature responses to CO_2-induced warming and particulate-driven cooling result in net changes which continue to be hidden by the ever-present climatic fluctuations whose existence does not require any specific "forcing" by CO_2 or particulates.

In turn, the spells of intensive volcanic activity may be caused by variations in tidal stresses on the earth caused by the sun and moon so that tide variations may be seen as an important controlling factor in recent climatic change: Roosen et al. (1976) found that at least 13% of the variance in the $\delta^{18}0$ temperature record in the Greenland ice cap might be attributed to long-term (180 years) periodic variations in tidal stresses at high northern latitudes. And there is another 180-year periodicity which has been offered as a major explanation of climatic variation: solar activity corresponding to variations foreseen both by the planetary theory of sunspots and by natural oscillations inside the sun. While mentioning the sun, it is also interesting to

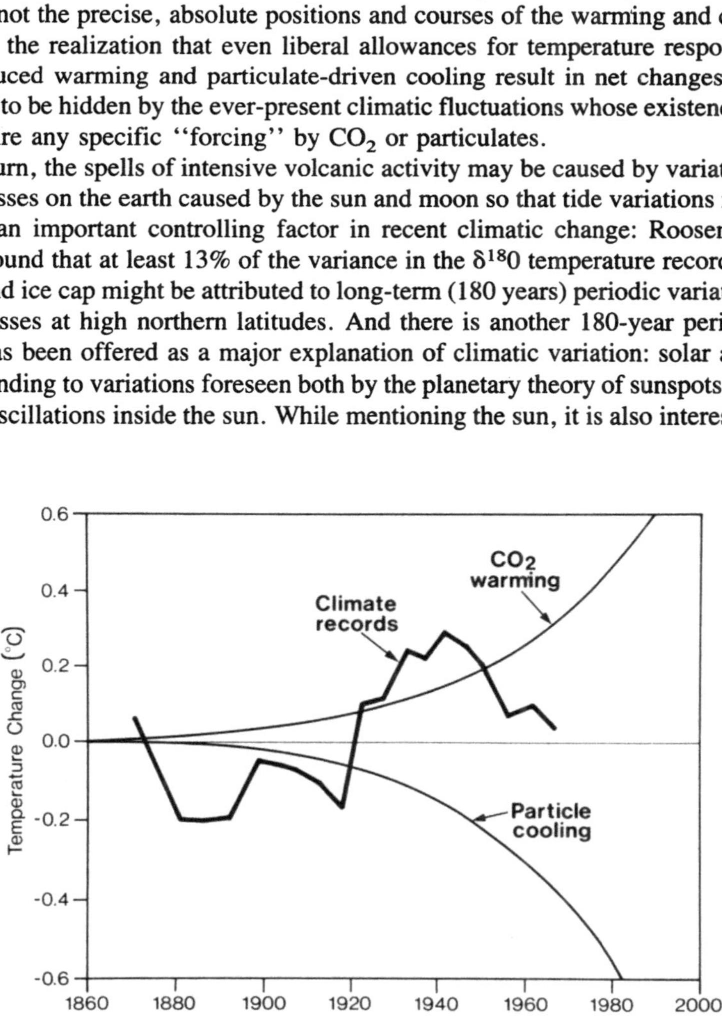

FIGURE 2-15. Fluctuations of observed mean global temperature compared with expected warming effect of CO_2 and probable maximum cooling contribution owing to atmospheric particle loading (Mitchell, 1975).

note that although the radiative forcing is quite different, the response of some atmospheric models to doubled CO_2 was very similar to that following a 1% increase in the solar constant (see, e.g., Chou et al. 1982).

Besides the influences arising from the changes of climate-controlling natural processes, other human interferences must also be considered: treating CO_2 as the only or as the pre-dominant anthropogenic air pollutant with climate-modifying capability is an inadmissible simplification of reality. There are, of course, particulate matter releases not only from fossil fuel combustion where they originate concurrently with CO_2 but also from a great variety of other industrial activities as well as from farming and even aerosols entrained by winds from newly desertified lands may be seen to be of anthropogenic origin. And, no less importantly, there are a variety of air pollutants whose atmospheric levels are very small (mere hundredths to millionths of the current CO_2 level) but whose infrared radiation absorption capacities can combine to bring appreciable temperature changes should their future concentrations increase substantially.

These "greenhouse" gases have received increasing attention since the mid-1970s when Wang et al. (1976) wrote a frequently quoted paper; Chamberlain et al. (1982) and Machta (1983) offer detailed progress reviews. The presence and continuing buildup of these trace gases in the atmosphere enhances a possible temperature rise and confuses expected CO_2-induced changes, especially as there are so many compounds being absorbed in the critical thermal window (7–14 μm) through which most of the long-wave radiation from the surface and lower atmosphere leaves the planet. These trace constituents include gases produced by natural anaerobic decay such as N_2O, NH_3, CH_4, and C_2H_4 (but releases of the first two compounds have been greatly accelerated owing to spreading nitrogenous fertilization; see Chapter 3), SO_2 from fossil fuel combustion and natural oxidation of H_2S, HNO_3 from atmospheric reactions of NO_2 and HO, CH_3Cl from decay of marine organisms, and chlorofluorocarbons (freons) of solely anthropogenic origin, with $CFCl_3$ (CFC-11), CF_2Cl_2 (CFC-12), and CF_2HCl (CFC-22) being the most important compounds. CCl_4 and tropospheric and stratospheric ozone must be also considered.

Concentrations of all these compounds are mostly fractions of 1 ppbv (except for about 6 ppbv for NH_3 and 1.6 ppm for N_2O; on N_2O see much more in Section 3.4.1.3) but they may increase substantially owing to more extensive fertilizing, higher fossil fuel combustion, and continued use of freons for many industrial, commercial, and household purposes. Various calculations of the thermal effects these greenhouse gases would have upon doubling their atmospheric concentrations show that increases may be 0.25–1.1 K for N_2O, 0.2–1.4 K for CH_4, and fractions of 1 K for other trace compounds.

Although it is far from certain toward what extremes of the estimated ranges the actual temperature increases would tend with eventual doubling of concentrations or what the combined effect of these changes would be, the increases could be equivalent to 30–50% of the CO_2-induced rise as expected by the current consensus—and the necessity to take all possible trace gases into careful account becomes obvious, and the task of singling out CO_2 contributions becomes even more difficult.

A notable complication involving CO_2 and one of the greenhouse gases, methane, is the possibility that the tropospheric temperature rise induced by the former will warm up the bottoms of shallow oceans which in turn will release unstable methane hydrate (clathrate) from continental slope sediments. Bell (1982) estimated that over a period of 100 years, 120 million t of CH_4 could be released annually from the Artic Ocean whose waters would be warmed up by about 3°C. Revelle (1983) put the global total at 640 million t/year.

Such huge releases could then be substantial positive feedbacks to further tropospheric warming—but the cited releases are based on estimates of hydrate abundancies whose accuracy is unknown and on chains of assumptions regarding enhanced warming of the Northern Hemisphere atmosphere, heat transfer to the sediments, and liberation of the gas, all major uncertainties not to be resolved soon.

Major uncertainties also prevail with respect to the climatic effects of particulate matter from human industrial, construction, and agricultural activities (Bryson, 1974). Bryson and Dittberner (1976) modeled the mean hemispheric temperature between 1890 and 1960 and achieved an excellent match with observed values by using only three independent variables: atmospheric CO_2, volcanic ejecta, and anthropogenic particulates. They argue that calculation of the thermal effect of CO_2 variation alone is unrealistic as its releases are inevitably accompanied by aerosol production and that a simple observed relationship between CO_2 and anthropogenic aerosols actually brings a slight net temperature decrease with an increase in CO_2 (about −4.93°C for doubled CO_2). Hence, a continued generation of both CO_2 and aerosols may have opposite effects than usually assumed by considering CO_2 alone.

In contrast to this, one can cite Ellsaesser's (1975) review of airborne particulate matter showing estimates for the anthropogenic fraction ranging from just 5 to 45% of all particulates, with the best assessments of individual components putting the most likely share at about 14% and concluding that while an upward anthropogenic trend in airborne particulates was evident in the earlier decades of this century it was halted and most probably even reversed over the past few decades.

Further complications are introduced by special considerations of stratospheric pollution, a topic which rose to great prominence during the 1970s, first with the controversy about supersonic air transportation, a bit later with rather frightening concerns about the effects of chlorofluorocarbons. As with tropospheric pollution, there are large uncertainties—unfortunately not only in magnitude but also in sign—concerning the long-term climatic impacts of various stratospheric pollutants (and even greater uncertainties regarding the consequences of such changes for biota) and it is most unlikely that accurate appraisals of these atmospheric effects will become available in less than one or two decades.

Potential anthropogenic stratospheric pollutants include NO and NO_2 from high-flying aircraft and N_2O from denitrification of nitrogenous fertilizers (for much more on this, see Sections 3.4.1.2 and 3.4.1.3), aircraft condensation trails, radioactive particles from nuclear weapons tests, radon-222 from coal combustion, and, above all, chlorofluorocarbons, whose future production, stratospheric concentrations, and effects on ozone destruction have been one of the most controversial and fashionable research topics of the 1970s.

How then is it possible, amidst all of these natural and anthropogenic influences, perturbations, and uncertainties, to detect the climatic change attributable to higher CO_2 levels? That is, taking the current consensus view and looking for small but detectable temperature increases, how does one go about uncovering and attributing these gentle rises on the background of the well-recognized inherent variability of climate? Obviously, one should know first the typical magnitude of these ever-present perturbations before looking at the feasibility of discerning a CO_2 signal amidst this background noise.

A good example of this detection effort is the work of Madden and Ramanathan (1980). First, they assembled nearly continuous records of mean monthly temperatures for the years 1906–1977 for 12 stations encircling the Arctic roughly along 60°N (actual range was from 50°N to nearly 70°N). This choice was guided by the limited availability of data for still higher latitudes and by the concern to avoid large ocean gaps that would have resulted had more southerly locations been chosen.

After some rather complex statistical manipulation of the monthly temperature means to find the level of noise, i.e., the effect of all variables other than CO_2, comparisons of these perturbations with the most likely CO_2-induced temperature response led to the conclusion that the surface warming predicted by general circulation models should be evident today or that it should become detectable no later than during the 1990s. But how meaningful can this kind of complex statistical manipulation be when our understanding of the causes of climatic change and feedbacks operating in it is so imperfect?

The authors are well aware of the dilemma and the impossibility of offering a satisfactory solution: "While it is reasonable to attempt to account for changes not due to CO_2 by looking at past records as we have done, we cannot be certain that such changes have not masked a CO_2 effect, or conversely that they may indicate in the future that there is an observable effect of CO_2 when there is none" (Madden and Ramanathan, 1980). All the impenetrability of a process complex beyond our understanding is so well summed up here!

Another effort to forecast a clear emergence of CO_2-induced warming is a simulation of the trend of observed global temperature by Hansen et al. (1981). This global model, forced primarily by variations in CO_2 and volcanic aerosols, achieved such an enviable fit that the authors, equipped with this "improved confidence in the ability of models to predict future CO_2 climate effects," offer some rather clear-cut predictions for the next century and foresee the CO_2 warming effect rising out of the noise level of natural climatic variability during this decade when the increment induced by warming will surpass one standard deviation of the observed climatic trend of the past century; 2σ level should be surpassed in the 1990s.

But Idso (1982) points out that by using a purely hypothetical solar luminosity variability as well as questionable volcanic dust values, theirs was basically a guaranteed-to-fit approach; the authors admit that much openly when they write that "other hypothetical solar radiation variations that we examined degrade the fit" (Hansen et al., 1981). Even more surprisingly, their historical global temperature trend was created by merging data for three broad belts of northern (90–23.6°N), low (23.6°N–23.6°S), and southern latitudes (23.6–90°S), but such a procedure hides the fact that between 1935 and 1980 the northern latitudes have experienced a pronounced *cooling* of over 1°C/decade—in the very region where the CO_2-induced warming should have been most discernible.

The incongruity of all of this is almost comic: on the one hand, Hansen et al. (1981) claim that their model shows a global warming of about 0.25°C between 1935 and 1980 and that northern latitudes should be warming two to five times that much, or 0.5–1.25°C—yet their own data show an average drop of 0.5°C for that region. Idso's (1982) conclusion is incontrovertible: "What better refutation of a model's validity could be found than this: dramatic cooling where dramatic warming is predicted?"

The most comprehensive consideration of complexities surrounding detection and monitoring of CO_2-induced climatic changes is a lengthy analysis by Weller et al. (1983). This study reviews roughly a score of studies estimating the degree of present tropospheric warming attributable to CO_2—their results ranging from 0 to 0.6 K since the middle of the 19th century—and by contrasting them with relatively large and far from adequately explained temperature fluctuations of the last century, it concludes that the overall pattern of both global and Northern Hemisphere temperature variations and associated changes of other climatic variables does not yet confirm any occurrence of temperature changes attributable to increases of ambient CO_2 levels.

Although the available evidence does not preclude the possibility that slow CO_2-induced change is already under way, numerous uncertainties make any analyses showing a definite relationship between rising CO_2 and changing temperature merely suggestive. Even when the CO_2 signal does, as generally expected, gradually increase, Weller et al. (1983) rightly point out that relative contributions of the gas and other radiatively active compounds will still have to be inferred from model calculations and precise measurements of radiation fluxes.

Clearly, without extensive monitoring of solar radiance and concentrations and radiative effects caused, above all, by aerosols, chlorofluorocarbons, methane, N_2O, and ozone and careful charting of cryospheric changes (sea ice cover, snow cover, ice cap balances), ocean levels and temperatures and cloud and water vapor parameters, it will be impossible to accurately attribute even an unmistakable temperature increase.

Such an increase will have to be a sustained rise of at least 1°C: after all, a chart of annual surface temperature anomalies for the Northern Hemisphere for the years 1891–1980 shows that the extremes span 1.34°C within what is only a very short period on the time scale of climatic changes (Figure 2-16). Still, theories telling us when we shall see unmistakable signs of global warming are not in short supply.

The tidal–volcanic theory mentioned earlier in this section offers a quite plau-

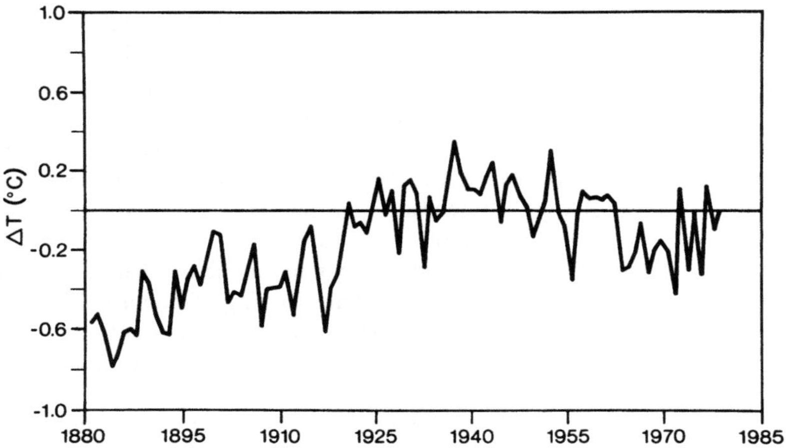

FIGURE 2-16. Annual surface temperature anomalies for the Northern Hemisphere, 1881–1980 (Jones and Wigley, 1980).

sible explanation of why the predicted CO_2-induced warming has not yet been demonstrated although its effect should already have been measurable: the planet is now in a cold spell lasting several decades, a spell which might be even colder save for the CO_2 warming which, however, is not yet strong enough to override the cooling trend. When the tidal stresses ease and the natural trend reverses by 1990–2000, then the CO_2 effect should become abundantly clear.

An almost identical prediction, although one based on different evidence, was offered by Broecker (1975). Using $\delta^{18}O$ records from the Greenland ice core as indicators of long-term climatic changes, Broecker believes that the natural climatic cooling which has since 1940 more than compensated for the CO_2 effect will soon bottom out and that it will swing into a new warming phase by about 1990; global temperature would then rise for at least four decades and this warming would be sufficient to bring, together with the CO_2 effect, a dramatic rise of surface temperatures beyond the range experienced during the past 1000 years.

But Broecker (1975) had to make a critical assumption readily disputed by other researchers—that anthropogenic dust is unimportant for climatic change and can be ignored in construing his prospect—and he had to admit that the exact date of the minimum on the extended natural climate curve as well as the rate of warming beyond the minimum are not known with any quantitative certainty.

Yet another argument is that CO_2-induced warming has been inhibited since 1940 by a downswing in a 76-year solar cycle and that around the year 2010 the onset of an upswing will seriously and rapidly amplify global warming caused by fossil fuel combustion (Gilliland, 1982).

And so we remain in the domain of hypotheses and guesses: where there are no hard facts to offer, the latter ones may be no worse, even better, than forbiddingly computerized theorizing, and the responses gathered from 24 leading climatologists by the National Defense University (1978) in its study of climatic change are quite interesting. The response was sought to five predicted global temperature changes, including large cooling (0.3–1.2°C colder mean in the Northern Hemisphere, mid-point decline of −0.75°C), modererate cooling (−0.175°C), continuation of the last 30-year trend (+0.10°C), moderate warming (+0.425°C), and large warming (+1.2°C).

The outcome was perfectly symmetrical, with 10% each going to the extremes, 25% each to moderate departures, and 30% to continuation of the pattern from the recent past—clearly a consensus of insignificant changes. Another outcome of the survey notable in the CO_2-warming context are the means of annual changes in the total and in the variability of precipitation. Only in the cases of large and moderate cooling did the respondents feel that the annual precipitation would decrease both in the higher midlatitudes and in the subtropics, while remaining the same in the lower midlatitudes. In cases of continued past trend and moderate and large warming, they foresaw either no changes or clearly detectable precipitation gains between 2.3 and 6.8% in all latitudes. And, perhaps even more importantly, the responses indicated an expectation of unchanged precipitation variability with the three nonextreme scenarios and its decline (by about 6%) with large warming in all latitudes.

2.4.2.3. EFFECTS ON PLANTS The only certainty concerning doubled atmospheric CO_2 and plant productivity is that general microbial activity sustaining the photosynthesis will hardly change as CO_2 concentrations in the soil atmosphere are already 10–50 times higher (3000 ppm in aerobic sites, much higher in waterlogged places) than the current ambient levels (Lamborg and Hardy, 1983). Beyond this, the sides are drawn as sharply as in the most controversial disputes concerning climatic effects of higher CO_2 concentrations. The extremes are easily stated: on one side are those who believe that future doubling or tripling of atmospheric CO_2 will greatly increase primary productivity (for perhaps the most vehement assertion, see Idso, 1982; but also Wittwer, 1980), even to such an extent that the role of fossil-fuel-derived atmospheric CO_2 concentrations will be slowed down (Bacastow and Keeling, 1973); on the other side are those who claim that our understanding of CO_2 enrichment is too weak for any sensible speculations on the effect (Marland and Rotty, 1979; Kramer, 1981) and hence that the long-term effects of higher CO_2 levels are as likely to bring either increases or decreases in plant productivity (e.g., Hansen et al., 1981; Cooper, 1982).

At the most general level, one must acknowledge two simple facts: for more than nine-tenths of the time span since the onset of higher life forms about half a billion years ago, the earth's surface temperatures was considerably warmer than during the past five millenia which have seen the rise and then global diffusion of high civilizations and, as already pointed out, CO_2 levels were much higher. Glaciers of the Pleistocene receded just before the oldest civilization had started to evolve its complex socioeconomic structure but we still live in an anomalously cold age. From this perspective—and it is a perspective which cannot be ignored by any natural scientist–there is nothing frightening about 600 ppm of CO_2 in the atmosphere and a rise of 1 or 2°C.

Indeed, to believe that doubled CO_2 and a couple of degrees' warming will imperil life on this planet is ludicrous as that very life evolved, complexified, and diffused while bathed in atmospheres containing up to an order of magnitude more CO_2 and being up to 10°C warmer than today. Unless one wants to discard the evidence of several hundreds of millions of years, then one must see future warming and higher CO_2 levels as a return to long-term normal from ice age aberrations of the recent era.

What is very difficult, and today simply untranslatable into quantitative terms, is to evaluate the extent to which, if at all, crops and forests will respond to the greater availability of CO_2. In spite of decades of sustained progress in plant physiology, uncertainties abound about the effects of higher CO_2 levels on changes, adjustments, and limits of carbon metabolism (Tolbert and Zelitch, 1983). The initial phase of CO_2 fixation is incompletely understood as are the needs for energy and intermediates needed for carbon assimilation at higher CO_2 concentrations. Least known are the influences of higher CO_2 levels on the duration of photosynthesis in the plants' life cycle from germination to senescence, environmental feedbacks that may limit the rise of primary productivity, and effects and allocations of more abundant photosynthate in plant development. Limits on CO_2 fixation also exist owing to chloroplast envelope translocations and there is little information on how C_4 plants, which are less likely to respond to more CO_2 than C_3 species, will react to atmospheric CO_2 enrichment.

Perhaps the most notable physiological benefit would be increased efficiency of water use. Photosynthesis is predicated on a very uneven CO_2–H_2O exchange: gaining a single molecule of the gas means losing, through the same diffusion path, hundreds of molecules of water. As the stomatal conductance would decline with higher CO_2, levels in both C_3 and C_4 plant transpiration would be reduced and water use efficiency, judged by experiments with cotton (C_3 plant) and corn (C_4) in 640 ppm CO_2, would rise by nearly 90% with doubled CO_2 levels (Pearcy and Björkman, 1983).

The beneficial effect of higher CO_2 atmospheric levels on the rates of photosynthesis was first noted by Priestley, an outstanding English scientist of the 18th century, in rather unusual circumstances: his experimental plants grew better in a bell jar in which a mouse had died than in ordinary air. Since then we have had numerous proofs, ranging from simple tests to elaborate, carefully controlled experiments, from de Saussure's classic pioneering trials with peas through Cummings and Jones's (1918) extensive tests of many field crops (peas, beans, potatoes, lettuce, strawberries) to the wide variety of CO_2 yield effect measurements of the past two decades.

Virtually all of the available historical evidence was summarized in an annotated bibliography of 590 papers dealing with the link between plant productivity and CO_2 concentration which was prepared for a Workshop on Anticipated Plant Responses to Global Carbon Dioxide Enrichment in August 1977 (Strain, 1978). And the same topic became the sole focus of the International Conference on Rising Atmospheric Carbon Dioxide and Plant Productivity in May 1982 at Athens, Georgia (Lemon, 1983).

Consequently, there is a large amount of empirical evidence on the effect of higher CO_2 levels on plants whose main conclusions may be summarized as follows: higher CO_2 concentrations generally lead to higher photosynthetic rates, increases in leaf area, dry weight per unit leaf area, branch, fruit, and seed numbers, better germination and earlier and more profuse flowering on the one hand and accelerated maturity on the other, as well as to higher tolerance of some common atmospheric pollutants (most notably SO_2), greater dark fixation of CO_2, and, as already noted, lower losses through water transpiration.

These combined effects appear so overwhelmingly desirable that one should hope for CO_2 increments to come as rapidly as possible to furnish an automatic means of higher food, feed, and wood production in a world that is facing increasingly serious regional shortages of all three biomass commodities. And, indeed, such a world was promised by some participants in the Athens meeting where the doubling of atmospheric CO_2 (i.e., levels close to 700 ppm) was credited with the following possible consequences.

Higher photosynthetic rates may increase yield and dry weight by at least 20% and as much as 45% for C_3 plants; efficiency of water use (i.e., dry phytomass produced per unit of transpired water) may double for both C_3 and C_4 plants; and tuberization of potatoes may rise severalfold. Other positive effects are difficult to quantify but there should be "significant" increases in symbiotic mycorrhizal fixation of nitrogen in legumes, "general" improvement in resistance to drought, and better tolerance of air pollution, soil and water salinity, and high temperatures. The differential response of C_3 and C_4 plants appears especially fortuitous when one notes that 12 of the 15 main food crops have C_3 metabolism and that 14 of the 18 most damaging weeds are C_4 plants (Waggoner, 1983).

Actually, all of these beneficial effects would be translated into higher net productivity so that it is sufficient here to look at the reliability and applicability of the experiments measuring higher yields. The first readily apparent weakness is that most of the available experimental studies were done with high CO_2 exposures lasting for very short periods of time—as little as 1 hr or up to several days—and such measurements cannot be used for reliable extrapolations of long-term productivity response. Moreover, many published reports do not specify the exact levels and variations of experimental CO_2 enrichment.

Kimball (1982) took care of the dual problem of unspecified or variable CO_2 enrichment and short-term exposures by focusing on only 81 of 437 separate experimental observations for 24 different crops and 14 nonagricultural species where carefully monitored CO_2 enrichments of up to 1200 ppm were applied during virtually all the daylight hours of the plants' entire growth period: this choice best approximated the possible ambient exposures with doubled or tripled CO_2 concentrations. The mean result for the wide variety of tested plants remains impressive: a doubling of current ambient CO_2 levels would likely bring a 33% increase in worldwide crop production and tripling would lead to a 67% rise.

Idso (1982) calls these findings "truly amazing, and the phenomenon they portray could well prove to be a godsend in the days and years ahead" because our greatest challenge during the next two decades will be to produce sufficient food.

One may absolve Idso of "days" as a patently rhetoric phrase—but the fact is that even years, even the two decades he specifies, will make very little difference. As shown earlier in this chapter, the time of doubled CO_2 may come only after several generations, too late to aid in increasing the global food production over the next two to four decades—or to enable new plantings on arid lands owing to higher water use efficiency.

In this respect, Kimball and Idso looked at 46 separate observations for 18 plant species where the reduction of transpiration was noted with CO_2 doubling (Idso, 1982). Percentage reduction ranged from just around 10% in wheat and barley to as much as 58–68% in sorghum with the average at 34%. When coupled with the mean 33% rise in plant productivity, the average water use efficiency (yield per unit of transpired water) would go up by about two-thirds and, Idso believes, it "may well double," thus not only helping current farm production but opening large areas of now idle arid lands for cultivation.

He cannot contain his enthusiasm for the combined effect of higher growth rates and lower water use as he foresees "mind-boggling" rewards when the "yields of plants the world over could actually double, with water usage being cut to but a fraction of what it is today" and, with desert blossoming, the higher CO_2 levels would lead us not to a disaster but rather propel "us back to the Eden which we left so long ago."

I noted previously (Section 2.4.2.1) that Idso deserves much credit for rocking the "mainstream" boat of somewhat less than humble climatic modelers who seem to prefer their theories to the exclusion of clearly observable realities and that his arguments, supported as they are by other "revisionists," deserve no less attention than those of his opponents. However, in this case I cannot endorse his enthusiasm for there is much evidence that, no matter what the date of CO_2 doubling, the actual productivity response of crops, forests, and grasslands will not be so great as to reestablish lost Eden. The situation here is very similar to the disparities between the controlled fertilization experiments in greenhouses and on small outside test plots and the actual performance of field crops: our present knowledge of CO_2 enrichment effects on plant growth is based almost solely on laboratory and greenhouse tests where water and mineral nutrition are rarely limiting and where even long-term experiments do not embrace more than one season of annual crop growth.

In the case of nitrogenous fertilizers, we know very well of the large reductions of experimental results—translating into average harvests—and basic ecosystemic considerations demand the same scaling-down of expectations of the CO_2 enrichment effect. Several factors will be critical in moderating the gains. First, there are large differences in CO_2 response among various species, with indeterminate C_3 plants (such as bean) performing much better than determinate and C_4 species (corn and sorghum fit both categories).

This means that simple averaging in terms of global crop productivity is very misleading: soybeans appear to respond better to CO_2 enhancement than any other major field crop—but their annual total global area is only 40% of average plantings of corn, a plant with relatively low productivity increases in response to CO_2

enrichment. Carefully established means for individual species would thus have to be applied to specific crop areas to calculate potential weighted averages of CO_2-induced yield improvement.

On the other hand, Pearcy and Björkman (1983) caution against what they call a common misinterpretation of Liebig's law, namely that suboptimal supply of water, light, and nitrogen will negate nearly all effects of higher CO_2. Undeniably, the stimulation will be lower in absolute terms but its relative effect may actually be greater as, for example, when improved water use efficiency permits crop cultivation, although low in yield, in environments where such growth was previously impossible. Similarly, with lower lighting, increased CO_2 may have a relatively higher stimulative effect than under optimum lighting.

Their conclusions, based on "our present limited information on basic physiological responses to higher CO_2," are that C_3 plant production will increase in most environments; that the relative enhancement will be greater in warm climates and most notable in areas with currently limited water supply; that in nutrient-poor environments, substantial increments will only follow higher N, P, K inputs; and that C_4 plants may respond appreciably only where water supply restricts their current productivity.

As to the level of the primary productivity increases where it matters most—in food crops—Baker and Enoch (1983) concluded that, in the absence of stresses, plant dry weight and yield will go up one-third with CO_2 doubling while the effect on crop yield under stress conditions will be much more variable, ranging from some declines to doubling or even tripling of yields as high CO_2 will generally alleviate water and salinity stresses but aggravate nitrogen stress.

But even if atmospheric CO_2 enrichment has no beneficial effects on farm crops, the repeated gloomy predictions of food production crises arising from temperature and precipitation changes caused by CO_2-induced warming must be discounted as largely sensationalist utterances having no relationship to reality as seen in long-term climatic records.

Surface temperature anomalies for the Northern Hemisphere for the years 1881–1980 show an annual range in excess of 1°C, a summer season range of the same magnitude, and spring and fall ranges (critical times for planting and harvesting) of 1.4–1.5°C (Jones and Wigley, 1980). Thus, during the past century, we have gone, in terms of extreme years, through an equivalent of warming predicted as a lower limit caused by CO_2 doubling (i.e., 3 ± 1.5°C in "consensus" models).

What has been the effect on agricultural productivity? Through the inescapable annual fluctuations (caused by a variety of factors ranging from late frosts to heavy pest infestations to rainy harvesting days), the yields have steadily risen, an upward trend brought by improved seeds, new technologies, better farming practices, a trend traversing periods of fairly steady mean temperatures of relatively rapid warming (1915–1940), or cooling (1940–1970). For all major food and feed crops in the United States, the trend never was downward and the performance not only of European countries and Japan but also of such large and populous nations as China and Mexico shows the same direction, if not the same rates of yield increase.

Consequently, as Wittwer (1980) rightly notes, the past century has provided evidence that United States farming not only can cope but that it can improve during climatic change. His conclusion is that "American agriculture already has demonstrated that it can adapt to a trend of +0.1°C per year, assuming no change in interannual fluctuations."

Many essential food and feed crops, owing to their very wide spatial distributions, would require hardly any adaptation: wheat is grown from the subtropics to 60°N; corn is a major crop in Kenya, Java, and tropical Brazil as well as in South Africa, Argentina, northern China, and Romania, spanning a range from 40°S to 50°N; barley, millet, sorghum, and potatoes reach to even greater extremes; and even rice is now grown at nearly 50°N. As for vegetables and fruits, there is no shortage of varieties able to thrive in climates whose temperatur changes much more than a few degrees Celsius. And a quick perusal of geobotanical source books will show numerous tree species—conifers and deciduous—which thrive over impressively wide latitudinal ranges.

And there is nothing frightening in the often-cited latitudinal shifts of climate: after all, in the absence of climatic change, North American farmers have extended commercial cornfields 800 km northward since the 1930s when the first suitable hybrids started their displacement of traditional corn varieties. Corn is now a major crop as far north as the Red River Valley of southern Manitoba where a farmer can rely on no more than 90 frost-free days in the coldest years.

Similarly, Rosenberg (1982) has shown that the spatial expansion of the hard red winter wheat cultivation between 1920 and 1980, again a result of breeding for hardiness and new farming practices, encompassed temperature and rainfall changes as great or even greater than those predicted by the current "consensus" warming models. Waggoner (1983) has pointed out that increased nitrogen fertilization, following residue recycling, mulching, and reduced tillage methods, doubled water use efficiency on the Great Plains between 1930 and 1980, enabling surprisingly reliable harvests in this moisture-deficient farming region.

Where changes would have to be made, they can be made with rapidity far outpacing any reasonably imaginable warming trend. A few examples will serve to illustrate the rapidity with which a transition can be made (in these cases, owing to economic benefits). In 1970, soybeans were grown on some 17 million ha in the United States, but just 10 years later they were grown on almost 30 million ha in 30 states, or about one-fifth of the country's farmland. The spread of Brazilian soybeans has been even more dramatic, from a mere 1.3 million ha in 1970 to almost 10 million ha in 1980. And because both countries export a substantial part of their soybean crops, animal feeding worldwide has been shifted considerably toward soybeans. Similarly, adoption of new oil-bearing species has been very rapid, with sunflowers sweeping across the plains of the northern United States and parts of the Canadian Prairies in less than a decade; African oil palm cultivation increased so rapidly during the 1970s around the tropics that palm oil production went up nearly threefold in just a decade.

In any case, from the viewpoint of phytomass storages, any conceivable shifts in productivity of agroecosystems will be negligible. Having reviewed the global distribution of standing phytomass (Section 2.1.2.2), it will be easily appreciated that even a doubling of biomass storage in crops would alter the total planetary stores of living matter in barley in a noticeable way.

Similarly negligible would be any effects of CO_2 enrichment on freshwater ecosystems. Botkin (1977) has reviewed the available evidence demonstrating that the fundamental cause of algal blooms is an increase of phosphorus or nitrogen (see also Section 3.4.1.5) and that carbon does not limit biomass storage over the period of a growing season.

But as higher CO_2 could limit the rate at which biomass can increase, it could accelerate growth up to the limits imposed by phosphorus, nitrogen, and self-shading. In any case, freshwater ecosystems represent such insignificant stores of biomass (see Section 2.1.2.2) that even a fabulous enrichment would not turn them into significant carbon sinks. Consequently, it must be concluded that any response of freshwater phytomass will be very small and additional growth of marine phytoplankton appears rather unlikely.

In the case of natural terrestrial plant communities—unlike with many crops grown experimentally under enriched CO_2 conditions—we have virtually no quantitative base for assessing the long-term effect of increased atmospheric CO_2. Consequently, responses of this largest biotic carbon reservoir are just a matter of informed speculation, and the working group on terrestrial plant communities at the 1982 *CO_2 and Plants* meeting offered only very tentative conclusions (Strain and Bazzaz, 1983). Foremost among these is the differential response of various species and subspecies, whose mutual interaction and reactions to other changing environmental variables will cause shifts in the ecological structure of communities; consequently, predicted increases in net primary productivity of individual plants may not automatically translate into net ecosystem productivity gains in lightly managed or unmanaged communities.

Changes in forests would, of course, matter most and Botkin's simulations are perhaps the best indicators of the counterintuitive complexity emerging once some essential dynamic features of ecosystems are taken into account (Botkin et al., 1973; Botkin, 1977). In modeling a typical northern New Hampshire forest following clear-cutting, Botkin distinguished between shade-tolerant or climax community species and shade-intolerant or successional trees: the former ones grow slowly, live long, and use inputs efficiently to dominate steady-state environments wherever resources are limited; the latter species overwhelm newly established or resource-rich formations with their rapid growth for which they pay by inefficient use of inputs and short lifetimes. The computer model of the forest community also defines each tree by its age and physiological characteristics but does not include specific response function to CO_2 increases for each species as these are unknown.

Consequently, Botkin (1977) considered a case of uniform response by all species. If a mixed-aged, mixed-species forest responded uniformly to CO_2 enrichment it might seem obvious that a given percentage increase in the growth of each tree will increase the growth of the entire forest by that percentage. Not so. The total basal area of the CO_2-enriched forest with a 50% increase in actual growth would be larger than in the normal case only during the first two decades with no significant difference between the third and fifth decades. Afterwards the difference would arise again, but after a century of growth the CO_2-enriched forest would grow only 1.4 times faster and this ratio with 95% confidence would be as large as 1.64 but also as small as 1.15!

Two basic explanations are not surprising: stochastic properties of growth found in the forest can obscure even what appears to be a dramatic enrichment: and while shade-tolerant species (red spruce, sugar maple) do very well, shade-intolerant trees such as white birch may actually have a smaller basal area after a few decades as the rapid responders bury them in shade. As one burgeoning group suppresses another, CO_2 enrichment effects on the shade-intolerant group may thus at least be canceled.

Botkin (1977) drew two fundamental interpretations from his simulations. Unless the effect of CO_2 enrichment uniformly raises annual growth by at least about 50% neither productivity nor biomass would change significantly. Two centuries after clear-cutting, both CO_2-enriched and normal stands reach maximum stable storages and there would be no additional sequestration of atmospheric CO_2 at all. Naturally, these simulations are not precise quantitative indicators but there are unmistakable reminders of how misleading it is to extrapolate from healthy, monocultural, protected, and nurtured greenhouse plantings to a mixed-age, multispecies natural community.

For tropical moist forests, ecosystems with by far the largest share of global biomass, it is especially hard to postulate how even major increases of atmospheric CO_2 would do away with limitations imposed by extensive shading and intensive cycling of rare nutrients. And for both forests and fields the inevitable annual weather fluctuations will continue to play their controlling role, with intensity and duration of temperature and moisture stresses determining the variations of year-to-year productivity more than any other single variable.

Moreover, in natural communities, long-term CO_2 environment effects may cause a decreased availability of nitrogen and phosphorus in all relatively infertile sites, a trend which would return more land to an unmanaged state with its usual N and P growth limitations. In unmanaged but heavily grazed grasslands, increases in quantity of forage may well be accompanied by its declining quality as its C/N ratio goes up. A change of general concern would be the more rapid completion of plant life cycles which would influence many relationships between and among autotrophs and heterotrophs.

Lemon's (1983) summary appears to me a very fair evaluation of the best information we have and cautious conclusions we can make and as such it is worthy of an extended quote: ". . . it becomes clear that after the initial metabolic uptake of CO_2 in photosynthesis, there is a hierarchy of increasingly complex processes controlling the production and allocation of end-products . . . . Little is known of what more CO_2 will do to these long-term complex growth processes . . . . If this is true for experimental plants grown under the best of conditions, it is obviously many times more difficult to predict a final outcome in competitive plant communities of either single species or mixed species at the highest level of complexity. As a general rule, however, one can argue that with increasing complexity, the initial advantages of more CO_2 in the photosynthesis process will be increasingly buffered."

Plants, one can conclude with some confidence, will not be a CO_2 sink to such an extent that the additional storage of carbon in the phytomass could be seen as a practical way of controlling future ambient CO_2 levels. The next section will show that the choice of effective control strategies is generally quite limited.

2.5. CONTROLLING CO_2

The challenge is of a very different kind than reduction or elimination of nitrogen or sulfur oxides. The latter two elements are only minor, though troublesome, contaminants of fossil fuels and can be controlled by a variety of precombustion, combustion, and postcombustion treatments, process adjustments, and removal techniques, and combustion modifications can also minimize that part of generated nitrogen oxides which arise from air's natural nitrogen content during fuel burning at high temperatures (see Sections 3.5 and 4.5).

But, of course, fossil fuel combustion is essentially nothing less than rapid oxidation of carbon so the only two ways to prevent CO_2 formation are to stop burning carbon-containing fuels altogether or, a theoretical possibility, to give up the honorable engineering pursuit of maximum oxidation and, instead of converting the fuel into CO_2 and H_2O, to choke the process by roughly halving the conversion efficiency, ending up with alcohols, aldehydes, and acids as final combustion products. Naturally, this would halve our existing fossil fuel reserves but it would enable us to pump the liquid wastes back into hydrocarbon fields or old mines or dump them into the ocean. The thoroughly impractical nature of these measures is, I trust, abundantly clear.

In contrast, gradual reduction of the global reliance on fossil fuels appears not only practical but inevitable as the reserves of coals but especially oils and natural gases will eventually be exhausted anyway and new energy sources will have to take their place. However, even the simplest inquiries into specific courses and costs of any accelerated shift away from fossil fuels designed to lower CO_2 emissions show the enormous difficulties ahead.

The easiest way to partially reduce CO_2 releases would be to juggle the makeup of the current consumption to minimize emission rates. Comparing the typical emission factors (Table 2-4) it is obvious that large-scale displacement of coal by hydrocarbons would lead to an appreciable reduction of CO_2 emissions: if all energy delivered in the early 1980s by coal could be provided by natural gas, global CO_2 releases would be lowered by about 40%. But such substitution, obviously, is just the opposite of the most likely development; while coal may not reach its renewed dominance as soon or to such an extent as predicted by most future energy-use models, the long-term trend is certainly not in favor of hydrocarbons.

Although there would be no major reserve and development constraints, it would be possible to favor those kinds of coal which produce less CO_2. But as higher heat content means higher carbon content, typical lignites, subbituminous and bituminous coals emit just about the same amount of CO_2 per unit of delivered energy in standard combustion. However, there are major differences in CO_2 emissions when different coal conversion methods are compared on the complete fuel cycle basis, i.e., from extraction through conversion to end use.

Several comparisons of this kind are available. Chen et al. (1980) have shown that in delivering energy for space heating, electricity generation using low-sulfur coal, physical coal cleaning, boiler with a limestone scrubber, or fluidized bed combustion produces roughly equivalent (the differences are no more than 10%) amounts of CO_2—some 300 million t CO_2/EJ—while for synfuels the CO_2 emission rises to 360 million t for gasification with combined cycle and to 440 million t for liquefaction with boiler; in contrast, direct furnace combustion of high-energy gas from a coal gasification plant would produce only 220 million t/EJ.

Synfuels, releasing about 50% more CO_2 in complete fuel cycle than standard coal-fired generation, thus appear to be the worst choice, and during the late 1970s when the Carter administration was favoring a massive program to develop large-scale synthetic fuel capability in the United States, this caused additional worries about intensified CO_2 output from such an effort. As those megalomaniacal programs have gone the way of the administration that bore them and as the direct combustion of the synthetic gas would produce less CO_2 than the current use of coal-fired electricity, one must conclude that any modest future synfuel developments, with the gases or liquids used both directly and for electricity generation, will have negligible effect on CO_2 releases.

Curtailments of fossil fuel combustion on a global scale imply universal reduction of growth rates of energy consumption, a change which may be relatively, at least in the beginning, fairly painless for rich industrialized nations with huge conservation potential so well illustrated by the developments in the early 1980s—but which would impose crippling restrictions on developmental options of virtually all Asian, African, and Latin American countries.

The most obvious question to ask in this context is, assuming that fossil fuel alternatives would be available on a commercial scale commensurate with current energy needs, why any other nation except for a dozen of the largest fossil fuel consumers should carry the burden of an inevitably very expensive conversion. Global fossil fuel consumption is extremely skewed with just 10% of the world's nations consuming almost four-fifths of global primary commercial energy.

Looking at coal, the most likely largest contributor to long-term future CO_2 releases, the number of nations concerned shrinks to just three, the United States, the Soviet Union, and China. These countries control some three-quarters of global coal reserves, but that they could "negotiate some kind of three-way agreement, analogous to arms limitation, to limit coal use and apply CO_2 control technology"

(Chen et al., 1980) is a most unrealistic expectation. Analogy with arms control is particularly unfortunate as there has been no such meaningful agreement on any offensive weapons (the very useful Test Ban Treaty of 1963 did not affect the numbers at all) and as China has never even consented to take part in such talks.

In any case, suggestions about accelerated displacement of fossil fuel in order to lower CO_2 generation are purely speculative and shall remain so for decades to come as there are today no conversion technologies ready to turn abundant direct and indirect solar fluxes into commercial fluxes sufficient to cover energy needs on national scales and as development of such technologies must extend over many decades (see Section 1.1).

However, we may try to curb the growth and use of fossil fuel through taxation. Nordhaus (1977) argued that CO_2 emissions should be considered a resource in short supply and as such be allocated among the various sectors of the economy in ways which would maximize GNP under this unusual constraint. In the same way the allocations could be done among major fossil fuel-consuming nations.

Two years later he proposed introduction of stringent taxes on fossil fuel combustion to lower future CO_2 levels (Nordhaus, 1979) and most recently he simulated effects of five different taxations, two temporary, two permanent, and one gradually increasing over many decades (Nordhaus and Yohe, 1983). The most important result of this modeling is that even the most stringent tax considered (a linear increase from zero in the year 2000 to \$6 per t of coal equivalent in 2020, then to \$68 by 2040, and eventually to \$90 by 2060) did not prevent CO_2 doubling before the year 2100 in the most likely simulated case. This heavy tax, approximately equivalent to \$(1983) 10 per barrel of oil, would reduce CO_2 concentrations by the year 2100 by only 15% from the base run—providing, of course, that the taxation would be global!

Although there is no shortage of postcombustion control proposals for the capture of the gas, all those entering the realm of such designs should be warned. This sphere of intellectual constructs, neat theoretical calculations, and proposals ranging from simple chemical separation devices to oceanic "gigamixers" has very few links with real world realities, and the chances for practical applications of these seemingly so simple ideas appear to me exceedingly slim.

Only the large thermal power plants and major industrial boilers with relatively high CO_2 concentrations in flue gases would be sensible choices to concentrate any scrubbing efforts. Numerous scrubbing pathways have been suggested, usually consisting of CO_2 dissolution (in water or other solvent) and subsequent gas release by heating or pressure treatment (Marchetti and Nakicenovic, 1979). Yet such complete cleanup could lower the plant's overall conversion efficiency by 20–25% (excluding energy costs of waste disposal) and easily double the capital cost of a new large facility (Albanese and Steinberg, 1978).

But even greater difficulty comes with the disposal of the captured CO_2. The quantities to be temporarily stored or permanently dumped are enormous. In a 1000-MW coal-fired plant, no less than about 1000 t of CO_2 is generated every hour so that (with a load factor at 60%) the management would be facing a task of disposing of some 5.5 million t of CO_2 annually, a mass about 2.5 times greater than the plant's total coal consumption. And, of course, the complete global control of CO_2 would entail dumping somewhere about 5 billion t of the gas annually.

Exhausted mines and hydrocarbon reservoirs would, obviously, store only a tiny portion of the annual output and so the only way to dispose of such large quantities of generated gas without scrubbing would be by transferring it to the ocean. Mustacchi et al. (1978) proposed three such ways. The first would require cooling and compressing of flue gases to be released at least 240 m below sea level so that some 95% of CO_2 could be absorbed by ocean water; as about 45% of generated electricity would be needed for the gas compression alone, the total cost of such a scheme would, with machinery and pipes, at least double the current investment. The second method involves separation of CO_2 in the gaseous or liquid phase before underwater disposal. Compression energy needs would be much smaller, but a separation plant would not be cheap.

The best choice would be using easily regenerable mono- or diethanolamines so that the absorption step could be carried at atmospheric pressure; process heat requirements are put by the authors at one-fifth of the plant's output and investment per kilowatt is cited as only about half of the first option's total. This third method entails dissolving CO_2 in seawater in a pressurized absorber directly in the power plant and returning the solution to the ocean. Operation costs for this option are estimated to be about the same as for amine absorption, capital investment a bit less than for untreated fire gas bubbling.

The authors have so little critical distance from their evaluations that they not only confidently forecast increases in the price of electricity to be no higher than 12–15% with large generating units but end their paper with a conclusion which makes one wonder why there is any CO_2 problem: "It appears that if and when it is acknowledged that CO_2 cumulation in the troposphere can be dangerous, no major difficulty and only moderate cost will be involved in its removal" (Mustacchi et al., 1978).

These claims are obviously wrong. The investment and operating costs of any such process existing merely as concepts of a few researchers will bear little resemblance to the actual expenditures and as with any similar technology numerous difficulties are bound to arise. Even setting these factors aside, it is a trivial fact conveniently neglected by the enthusiastic proponents that most power stations are not located along seashores, as is not the overwhelming part of other industrial boilers which would have to be controlled as well. And, equally obvious, effects of such concentrated and extensive CO_2 disposal on coastal waters, atmosphere–ocean interactions, and biota are completely unknown.

Yet an even wilder CO_2 ocean disposal scheme was advocated by Marchetti (1977), who during his years at the IIASA delighted in putting forward various altogether unrealistic proposals of how to solve the world's energy problems. He had a giant "energy island" sitting in the Straits of Gibraltar churning out methanol and disposing CO_2 into the ocean to be carried by a "gigamixer" current which dumps warmer but denser (saltier) Mediterranean water into the deeper Atlantic.

According to Marchetti's expectations, CO_2 injected into this giant waterfall would spread at an equilibrium buoyancy level around 1500 m, eventually covering the whole Atlantic and not coming to the surface for hundreds of years.

Other "gigamixer" sites for CO_2 disposal would be the Red Sea near Bab al-Mandab and the Weddell Sea, and Marchetti's "rough" calculations show that 50–90% CO_2 removal would raise the raw fuel cost by merely 10–20%. All of this makes for nice science-fiction reading but I believe that we would sooner see a ban on further use of fossil fuels than Gibraltar and Red Sea "gigamixers" in action!

Considering the costs, technology penetration rates and difficulties accompanying control of sulfur dioxide, the gas generated in amounts two orders of magnitude smaller than CO_2, one simply cannot foresee any rapid and meaningful wet scrubbing removal of CO_2, and the obvious physical impossibilities (à la pipeline carrying the gas from the large Ohio Valley coal-fired stations to the Atlantic shores), high costs, and environmental uncertainties make the appealingly sounding ocean disposal schemes at least equally unlikely.

Even less realistic to practice on a meaningful scale would be use of the gas to enhance primary productivity and thus to extend the temporary storage of carbon in living phytomass. Here the relatively most practical option would be to pump the gas into greenhouses to boost vegetable and fruit production: it is, of course, a perfectly workable option with rewarding yields but the difference between the magnitude of the generated gas and its volumes needed for horticulture in greenhouses located near large boilers is so large that (even when setting aside the far from low expenses for fertilizers, irrigation, additional land, and capital cost of the structures) only a tiny percentage of the gas could be diverted in that way. In any case, this would be just a very short-term rerouting.

As stressed in the preceding section, even fabulous increases in average global crop yields would bind just a tiny amount of generated CO_2 for a very short time and only increased storage in forests could make an appreciable difference, but the only way toward this goal—massive afforestation in tropical regions—has exceedingly slim chances to succeed in the face of nearly universal and worsening tropical deforestation. Total phytomass storage in boreal forests cannot be extended by a significant margin and afforestation in subtropical regions would flounder on serious water shortages. And even where the plantings would be successful, lowered surface albedo brought by new large forest areas would actually enhance the warming effect they were designed to counteract. Clearly, additional carbon storage in phytomass cannot be seen as a meaningful control option.

A thoroughly impractical idea involving higher primary productivity is to dump phosphorus into the ocean to increase the autotrophic production in the presence of higher CO_2 content and thus by boosting the mass of organic litter sinking to the deep ocean, sequester larger amounts of carbon. Even when one assumes the impossible, that is that phosphorus would stay in the euphotic layer long enough to raise the productivity significantly, disposing of the current production of 5 billion t of excess carbon in that way would call, with C/P ratio of marine species at roughly 100, for annual dumping of 50 million t of phosphorus, i.e., about four times as much as is now applied to all crop fields around the world. And as that dumping would have to be spatially uniform, the logistics of such an exercise is obviously overwhelming.

Albedo manipulations have also been proposed as a possible way to control CO_2-induced warming. Perhaps the most *outré* idea to counteract a warming trend is by spreading myriads of small, thin (0.01 mm) highly reflective latex platelets over large areas of the ocean surface and in the stratosphere and thus increasing the planetary albedo (Revelle et al., 1965; NAS, 1977c). Fortunately, the cost of material alone was estimated at tens of billions of dollars annually to cover a mere 10% of the world's surface and thousands of planes flying on this gigantic dusting mission would run up no less forbidding application costs—and so the economic realities, once again, will keep us safe from experiments whose consequences (on ocean fishing, on stratospheric processes) might well cause more environmental damage than the "problem" they were to solve.

I will not describe how albedos could be manipulated by huge land-based reflectors built at a future central solar power plant or, better still, by satellites: there is no end of surreal proposals in this field. To me it seems more sensible to regain the mundane ground by being reminded of a fundamental difficulty with any conceivable control method, a need so well expressed by Zimmermeyer (1978): ". . . it seems unreasonable to take the second step before the first—namely, to carry out research and development work with the aim of reducing CO_2 emission, and to lay down conditions involving high costs for the operation of power plants before the effects of increased CO_2 concentrations have been clearly determined."

Indeed, one could, with some simplification, classify all the CO_2 researchers into two camps according to their advocacy of timing for action. Forecasters of CO_2-induced perils argue that any delays in taking action may have damaging consequences and urge that waiting for just another decade may turn our unprecedented "geophysical experiment"—a characterization first made by Revelle and Suess (1957)—into a callous, irreversible toying with the well-being of future generations. "Humanity may be in great peril" is their cry and their arguments aim to show the possibilities of CO_2-induced climatic risks eventually so enormous and irreversible that we would do best by erring conservatively in planning our future fossil fuel consumption or, as the minimum, start making adjustments needed to cope with such a dramatic coming change. Schneider emerged as perhaps the most often cited advocate of this school of thought (Schneider and Mesirow, 1976; Schneider and Chen, 1980).

He has reasoned that studying the problem for 10 or more years to avoid costly and premature decisions is dangerous on two counts—first, because it may take decades to confirm or to deny the conclusions of the current climatic models and, second, because of the very long lead times required for new energy sources to displace a significant part of the existing infrastructure. This infrastructure inertia means that even the most far-reaching commitment to reduce CO_2 emissions could not, all cost questions aside, lower the emissions dramatically until after several decades, and hence the conclusion (Schneider and Chen, 1980) that "one cannot delay a decision to curtail CO_2 emissions (or other such active policies) by some 10 years while experts study, without some risk!''

I believe that these, in any case unquantifiable, risks are easy to shoulder. Preceding sections have tried to show the enormous extent of our ignorance regarding the causes of climatic change and CO_2 contributions to it. Even a casual reader of this book has been, I trust, persuaded that much more empirical evidence and much deeper understanding is needed before advocating specific action—and hence the ''study more'' attitude is eminently sensible. Rotty and Marland (1980) put it well: ''Our key conclusion is that we believe there is time—disaster is not imminent—to conduct a carefully designed program to understand thoroughly the processes involved and the probable consequences. The need is not for immediate, short-range, and probably ill-conceived crash programs, but rather for long-range, carefully developed programs in the area of alternatives to fossil fuels.''

The latest NAS (1983b) CO_2 volume shares these views. While calling for R & D priorities for long-term energy options not based on fossil fuels, the report's authors do not believe ''that the evidence at hand about CO_2-induced climate change would support steps to change current fuel-use patterns away from fossil fuels. Such steps may be necessary or desirable at some time in the future, and we should certainly think carefully about costs and benefits of such steps; but the very near future would be better spent improving our knowledge than in changing fuel mix or use.''

In fact, with the establishment of the Geophysical Monitoring for Climatic Change and with the Department of Energy's relatively large-scale interdisciplinary CO_2 program, the scientific bureaucracy of the United States government has been engaged in that ''study more'' course—and in this case, unlike in many misplaced outpourings of public expenditures on fashionable but far from fundamental scientific problems with public policy implications, this commitment must be welcome.

One may disagree with the extent of the funding as so many projects do little more than duplicate already available results or add to an intriguing but far from critical aspect of the concern, with often unmistakably prejudiced attitudes, arising from the domination of CO_2 research by a surprisingly small core (about a score) of experts whose presence graces just about every conference and edited volume on the topic. But these are perhaps inevitable accompaniments of all such efforts.

A fundamental consideration which must be kept in mind concerns the very nature of climatic change of which CO_2, however important, is only a part. The shifts, be it rising temperature or rising ocean level, are gradual and their spatial distribution is very diverse. Under such circumstances, adaptation to altered conditions, rather than impractical efforts to remove or mitigate the supposedly causal process, may be by far the best course to follow.

On this score, the adaptive capacities of *Homo sapiens* are nothing short of amazing. Schelling (1983) pays much of the most appropriate attention to the fact that climatic changes experienced voluntarily by hundreds of millions of long-distance immigrants are far greater than any alterations most populations would be subjected to were they to remain in their settlements during the climatic changes of the next century. Perhaps the most extreme example of such voluntary climatic displacements are the migrations of Mennonite farmers from southern Russia to the prairies of central Canada (begun in 1874) and from there to Mexico (after 1920) and to Paraguay (after 1926).

Within less than three generations, thousands of Mennonite families thus chose to locate in some of this planet's most extreme climates still compatible with farming—and establish in these new locations, again and again, prosperous agricultural settlements. Chinese migrations to the cold northeast (Manchuria), Russian settlement of Siberia and central Asia, and peopling of the North American continent (including the recent strong flow from the Snowbelt to the Sunbelt) are better known and larger-scale examples of similar, though not as far-ranging, moves.

Accumulated experience and human ingenuity represent considerable guarantees that, in most instances, farming will adopt and that yield declines, should they occur, will be limited to bearable levels. This, of course, does not imply painless adjustments: a loss of several percent of food production capacities in the United States, Canada, or France would be of no critical long-term consequence for these countries but a similar decline in Egypt, Nigeria, or Bangladesh could have far-reaching consequences.

Similarly, eventual gradual abandonment of some low-lying settlements and industries, protection of other real estate by dykes, and new construction on landfills could be matter-of-fact responses to eventual sea level rise in rich industrialized countries with much experience in such undertakings. The Dutch example comes first to mind, as do considerable construction on landfills in Japan and building of levies and dykes in the United States. In contrast, protecting the delta of the Ganges would be a hopeless enterprise thwarted by the enormous length of required dyking, shortage of suitable local materials, and dangers of internal flooding of protected areas by monsoonal rains.

As with any change—climatic, technological, or political—there would be winners and losers and we know so little to start apportioning places into these categories. Although Baes et al. (1976) concluded that "almost any summation of the possible social cost from excess CO_2 becomes far too great to languish in the realm of academic debate," all we can do now and in the foreseeable future is to offer little more than a few defensible generalizations. Foremost among them is the prediction that for territorially large nations any eventual changes will be easier to absorb, while for some already highly stressed smaller countries suffering brought by less or more precipitation or ocean encroachment may be disproportionately greater.

But an obvious question to ask here is how much lighter would these impacts be if populations of such countries were smaller and better off: where is the real problem? Bangladesh with half of its current population could relinquish some of its land to the ocean without catastrophic consequence; if the country is twice as populous a century from now, any farmland loss would be inconceivable for survival.

Schelling (1983) put this point well: the most perplexing uncertainty is not how much fuel will be burned and how much temperatures and precipitation will change, but what the world will look like. If the planet's population will be richer or poorer, if international relations will be cooperative or confrontational—these will be the key determinants of the response, be it by controls or through adaptation.

I will return to unusual challenges and choices posed by the CO_2 problem—that is, assuming it will be a problem rather than a minor annoyance or even a benefit—in the last chapter of this book, but here I cannot pass up an opportunity to contrast the CO_2 affair with environmental risks and applied and intended controls of anthropogenic nitrogen and sulfur, the topics of the next two chapters. In both of those cases, we have behind us decades of wide-ranging everyday experience as well as rich works of replicable experimentation, and both the nonglobal nature of the sulfur- and nitrogen-induced interferences and the ready availability of effective management and control techniques put the challenges and solutions into a class very different, immeasurably easier, than the CO_2 concerns.

And yet the experience with uncovering the relationships pinpointing the causes and effects and introducing practical control measures has been rich with errors, misunderstandings, uncertainties, procrastination, and emotional policy debates. Why then are many researchers and bureaucrats so anxious to serve those so much more tenuous insights into the CO_2 story as the undoubted slices of reality-to-be, why do they believe that a handful of ''better'' computer models will furnish the answers sufficient to guide our policies? The field, I feel, would greatly benefit from less institutionalized search for consensus and from much less of now tiring repetition of unprovable and unproductive warnings. The research effort needs broadening in terms of more interdisciplinary participation (although it is already one of significant areas of truly interdisciplinary research) and it has to come to terms with the virtually permanent nature of the topic: encouragingly, the latest NAS (1983b) CO_2 report moves the discussion very much in this calmer, sustained direction. The books are not to be closed on this problem for many decades to come—but this chapter closes to make room for in many ways no lesser challenges arising from the releases of anthropogenic nitrogen and sulfur.

3 NITROGEN

The chemical properties of the noxious portion of the atmospheric air being hitherto but little known, we have been satisfied to derive the name of its base from its known quality of killing such animals as are forced to breathe it, giving it the name of azot, from the Greek primative particle α, and ζωή., vita. . . . We cannot deny that this name appears somewhat extraordinary.

—Antoine Laurent Lavoisier
Elements of Chemistry
(Robert Kerr translation, 1799)

And as if all these grand phenomena were destined to be associated with causes which should appear the most remote, we may here observe upon the sources whence the oxide of ammonium and azotic acid, from which vegetables derive a portion of their food, are themselves derived. They are, in fact, produced upon the grand scale by the action of those magnificent electric sparks that dart from the storm-cloud, and, furrowing vast fields of air, engender in their course the nitrate of ammonia. . . .

—Dumas and Boussingault
Leçon sur la statistique chimique des êtres organisés (1841)

When Antoine Laurent Lavoisier was sorting out the composition of air—in those happy days before Marat's guillotine when he was a celebrated scientist self-assuredly posing with his impeccably dressed wife for David—he used what soon appeared to be an astonishing misnomer to christen the gas which makes up nearly four-fifths of the earth's atmosphere: *azot,* "without life." Classical inorganic chemistry provides, especially in contrast to oxygen, easy justification for viewing the gas in such a way: after all, it is not only odorless and colorless but it is also nonreactive under normal biospheric conditions, nonflammable, nonexplosive, and nontoxic, "nonvital air" left behind in Lavoisier's combustion experiments.

Yet it is one of the cornerstones of life, a quality again not caught by its now universal name nitrogen given to it by Chaptal in 1790 to describe the element's presence in *nitre* (i.e., potassium nitrate), with the principal root *ntr* going all the way to salt deposits of ancient Egypt. Yet appreciation of nitrogen's role in life came soon afterwards. When Dumas, whose lines are quoted at the opening of this chapter, lectured on elemental cycling at the Ecole de Médécine in Paris in 1841, nothing was known of nitrogen-fixing organisms and biochemistry was just beginning its explosive acquisition—but he was the first scientist to comprehend the essential relationships of carbon, nitrogen, and solar energy so well that a century later Hutchinson (1944) could remark that "the cyclical character of the migration of the common non-metallic elements had never been demonstrated so lucidly and with such rhetorical skill."

With the subsequent expansion of organic chemistry, nitrogen's importance started to gain by leaps. The element's ability to assume valence states ranging from nitrate (+5, the most oxidized) to ammonia (−3, the most reduced) guarantees its presence in a large variety of organic compounds. Only when two nitrogen atoms are joined in gaseous dinitrogen (N_2) is the element largely filling our air stubbornly unreactive. While all other diatomic molecules—O_2, F_2, NO, CO—are reactive, dinitrogen is uniquely inert and it forms an unusually well-bonded whole: in comparison with O_2, CO, and NO molecules, it has the shortest bond length, the highest ionization potential, the highest stretching frequency, and its dissociation energy is surpassed only by that of CO (0.93 versus 1.06 MJ)—but carbon monoxide is much easier to oxidize as no oxidizing agent is strong enough to react with nitrogen at ambient temperature, not even fluoride (Chatt and Leigh, 1968).

But in other valence states the element readily partakes in life processes, especially in reduced forms in which organic nitrogen exists almost exclusively to serve in several important roles: as a ligand for metal complexation (when in heterocyclic rings), in hydrogen-bonding with other nucleophiles (most notably in DNA and protein helices), in the peptide bond, in numerous chemical reactions.

Considering the element's importance in amino acids, proteins, nucleotides, nucleic acids, B vitamins, enzymes, and chlorophyll, it can be claimed with only a little exaggeration that "every vital phenomenon is due to some change in a nitrogen compound and indeed in the nitrogen atom of that compound" (Needham, 1965). Nitrogen's biospheric role can be perhaps best appreciated by imagining it to be a carbon substitute (it stands just above carbon in the periodic table) whose five valence electrons introduce "an essential distortion into the symmetry of C, providing compounds with additional properties of coordination, basicity, charge, chemical reactivity, oxidation-reduction, and structure" (Clarkson and Hanson, 1980).

Unfortunately, combined nitrogen in forms usable by plants, i.e., as NO_3^- or NH_3, is scarce in surface waters and often no less so in soils and hence the primary productivity in general and the production of food in particular are limited more by the supply of metabolizable nitrogen than by the availability of any other plant nutrient. This scarcity is reinforced by the relative paucity of organisms capable of incorporating ("fixing") the abundant atmospheric nitrogen into compounds utilizable by autotrophs and by relatively rapid losses of useful nitrogen compounds from soils owing to incessant, complex cycling of the nutrient. Yet without this natural recycling, the current rates of primary production would deplete the biosphere's mobile nitrogen in just 4 million years—and ammonia and nitrate carried away by surface water and precipitation would exhaust even the atmosphere's huge nitrogen stores in about 70 million years (Gutschick, 1980).

The most productive traditional farming of Europe and Asia used extensive recycling of organic wastes and crop rotations including legumes to overcome the element's scarcity but, obviously, all such practices run eventually into a variety of limitations ranging from the necessity to forgo a harvest of the preferred grain crop to logistical difficulties of fermenting and applying huge quantities of organic wastes. Synthesis of ammonia in the second decade of this century removed these limits and opened the way, together with other farming innovations, for unprecedented yield increases.

These developments have been of enormous importance for survival and growth of the rapidly expanding human populations of this century. No vertebrates can synthesize all amino acids and vitamins from ammonia and must ingest them preformed as essential dietary proteins. Of 18 common amino acids, 10 cannot be synthesized in our bodies and must be eaten in food proteins. These proteins contain, on the average, 16% of nitrogen and hence the relative nitrogen content of edible plants and animal tissues can easily be calculated by multiplying their protein levels, listed in food and feed nutritional handbooks such as Watt and Merrill (1963) and NAS (1982b), by 0.16 (naturally by 6.25 converting the other way).

Proteins are thus critical limited and limiting resources for human development and their total availability in typical diets is a revealing measure of existential security while the makeup of their intake is a rather reliable index of affluence as animal proteins in meat, fish, eggs, and dairy products largely displace essential amino acids consumed in plants in richer societies. Sustaining this shift which involves considerable consumable protein losses during conversion of plant to animal tissues has required even higher nitrogen applications to croplands and grasslands which must provide the feed.

Yet often surprisingly large shares of these fertilizer applications have found their way into the waters or into the atmosphere where they may cause acute localized as well as chronic and global problems. And although fertilization with synthetic inorganic compounds is by far the largest human interference with biospheric nitrogen balances, it is not the only one. Combustion of fossil fuels is a much smaller source as far as total global flux is concerned but it is spatially more concentrated in industrialized nations of the northern temperate zone and the resulting oxides and nitrates can participate in complex transformations leading to photochemical smog and can be transported by prevailing air flows over considerable distances and contribute to acid deposition. Animal and human wastes are the smallest source in the global aggregate but spatially the most concentrated one.

The scope of nitrogen-associated concerns is thus enormous—from fertilizer denitrification to health effects of peroxyacetyl nitrate, from protein malnutrition to power plant combustion modifications. I shall attempt to bring in all the highlights regarding the fertilization and combustion interferences but I will not deal with urban sewage treatment and disposal (a concentrated but relatively small nitrogen flux compared to fertilization) or with proteins in diets, a concern beyond the already sprawling scope of this book. Suffice it to say that protein malnutrition is still sadly prevalent in many parts of the poor world but that the problem appears to be more that of inadequate energy intake rather than low content of utilizable protein (Sukhatme, 1977). Simply, when people have enough to eat and when their food is reasonably diverse, protein malnutrition disappears.

A systematic evaluation of human interference in the nitrogen cycle is perhaps best started, as in the case of carbon, by an outline of the essentials of biospheric storages and intricacies of principal natural fluxes.

3.1. COMPLEXITIES OF THE NITROGEN CYCLE

Among the always difficult analyses of principal grand biospheric cycles, those attempting to bring clarity and consistency to the study of nitrogen fluxes appear to face the usual frustrations on a larger scale and to a higher degree. Burris (1980) remarked that "the global nitrogen cycle can be drawn in as simple or as complicated a form as you wish." Indeed, in his "simplistic" model he lumps everything into just three big reservoirs and five interchanges (Figure 3-1); in contrast, Delwiche (1977), whose nitrogen cycle work has been widely quoted, splits the cycle into 12 reservoirs and 23 fluxes. (Figure 3-2).

Clark (1981) reviewed the historical development of nitrogen cycle concepts and divided it into four stages: the process diagram stage, the process and compartmental stage, the budgeted flows and compartmental stage, and the simulations stage. The number of diagrams drawn in the first three categories must by now be in the scores, and the multitude of quantitative models is another telling indicator that the nitrogen cycle has probably been researched more than its other two macronutrient companions—but there are still numerous and major uncertainties or outright blank spots. The Amsterdam symposium on nutrient cycling listed among its recommendations for "considerable research required" several key segments of

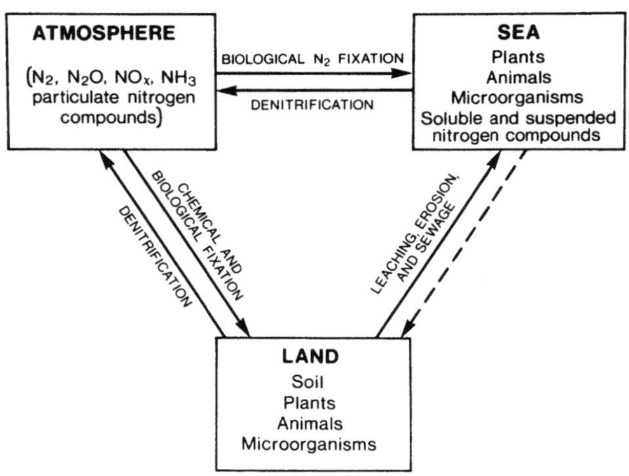

FIGURE 3-1. Simple three-reservoir model of nitrogen cycle according to Burris (1980).

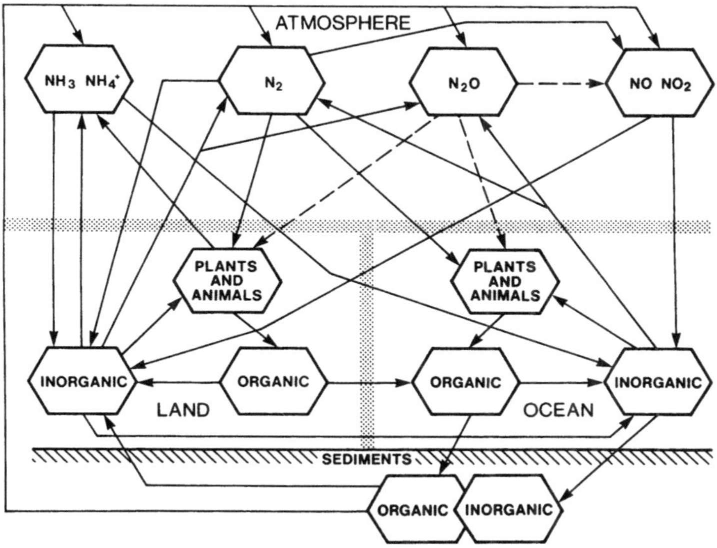

FIGURE 3-2. Nitrogen cycle as outlined by Delwiche (1977). For his estimates of annual fluxes see Tables 3-1 to 3-3.

the cycle, ranging from immobilization transfer by weathering and mineralization to outputs by denitrification and ammonia volatilization (Frissel, 1977a).

The following sections will review in some detail all important reservoirs and fluxes of the element, identifying in each case the extent of uncertainty underlying our quantification attempts and suggesting the most likely values based on the latest available information or consensus.

3.1.1. Major Reservoirs

Virtually all standard discussions of biospheric nitrogen storages in biological texts focus on the element's abundant but inactive presence in the atmosphere and on its large reserves in soil organic matter. But as with carbon, so with nitrogen and, as will be seen, with sulfur: the lithosphere holds the overwhelmingly largest shares of these elements.

Consequently, the following accounts of nitrogen reservoirs will start with quantifications of nitrogen in the lithosphere, including fossil fuels, and only then will more attention be given to atmospheric dinitrogen and nitrogenous compounds residing in the troposphere (above all, NH_3/NH_4^+) and to the nutrient's stores in soils and biota.

3.1.1.1. Rocks and Waters As just noted, the lithosphere holds by far the largest share of nitrogen but unlike with carbon, whose main stores are in sediments, the bulk of the lithospheric nitrogen is contained in primary, igneous rocks, far removed from any effects of weathering and cycling, forever inaccessible for biospheric flows. Earlier inventories based on the information from the 1950s put the lithospheric nitrogen mass as high as 16.36×10^{16} t, with 99% of this total in the mantle rocks, but over 0.5% in the crust and the rest in the fossil sediments (Stevenson, 1972).

These inventories were based on Rayleigh's pre-WW II estimate of nitrogen content in igneous rocks, but by the early 1960s the lithospheric nitrogen estimates were revised substantially downwards. Delwiche (1970) used 1.8×10^{16} t (of this, 1.4×10^{16} t in the crust, the remainder in the sediments) and Rosswall (1981), 5.74×10^{16} t. For analyses of the global biospheric nitrogen cycle, these differences are irrelevant as only a tiny fraction of 1% of whatever the lithospheric nitrogen total may actually be will eventually be transferred into the biosphere via weathering of organic sediments and combustion of fossil fuels.

Weathering contributes less—by at least an order of magnitude—"new" biospheric nitrogen than fixation by symbiotic and free-living organisms (see Sections 3.1.2.1 and 3.2.1) and is important only locally or regionally where there are nitrogen-rich sediments. On the other hand, nitrogen released from the combustion of fossil fuels, largely as nitrogen oxides but also as NH_4^+, represents a major flux, a flow which must not be neglected especially in heavily urbanized and industrialized temperate regions where the consumption of fossil fuels is by far the highest.

Ultimate analysis of coals shows a range of 1.0–1.7% of nitrogen in anthracitic and semianthracitic varieties, 1.0–1.9% in bituminous ones, and 0.7–1.4% in subbituminous coals and lignites (USBM, 1954). Similarly, ultimate analysis of crude oils reveals nitrogen contents of 0.11–1.70% by weight (King et al., 1973), with most frequent values much below 1%. Finally, natural gas has usually less than 5% of dinitrogen but some fields may contain much in excess of 10% and in all such cases nitrogen is removed from the fuel (Mullins, 1977).

Nitrogen present in all fossil fuels is a major source of nitric oxide and nitrogen dioxide generated during combustion: small-scale experiments showed that 50–70% of the total nitrogen oxide flux from coal and residual oil is fuel related, although a larger part of fuel nitrogen is released as dinitrogen (Martin and Bowen, 1979). This conversion appears to be relatively insensitive to temperature, the main factor being the oxygen availability—unlike with the formation of the thermal oxides which is strongly temperature-dependent and which occurs during the combustion of all fuels at peak temperatures present in all diffusion flames.

Because of the presence of these two kinds of oxides in effluent gases (of fuel and thermal origin), it is difficult to estimate reliably the size of reservoir exchange: thermal nitrogen oxides formed by high-temperature oxidation of atmospheric N_2 represent just a transformation within the same reservoir and only fuel-derived nitrogen oxides are the key nitrogen flows between the lithosphere and the atmosphere.

But nitrogen is introduced into the atmosphere during combustion in yet another way, as ammonia from the burning of coal. Usually between 15 and 25% of the total nitrogen in coal is released as ammonia (Söderlund and Svensson, 1976), but in calculating the global output the total must be reduced by ammonia recovered during coking. Of the current worldwide consumption of 3.7 billion t of coal, some 500 million t is used for coking and assuming the average N content of coal to be 1–1.5%, the global NH_3 emissions from coal combustion would be between 5.5 and 13.9 million t N, with the most likely value around 8 million t N.

Compared to the lithosphere, nitrogen in waters represents a tiny reservoir. Besides dinitrogen, ammonium, nitrates, nitrites, and dissolved and particulate organic matter are also present in the ocean but over 95% of the estimated 2.3×10^{13} t of the element is in the inactive molecular form; nitrate holds about 2.5% and organic matter roughly 1.5% of the total (Stevenson, 1972). Although the shares of the listed nitrogen forms may vary locally with depth, season, and biomass production, the total reserve should be considered in a state of quasi-equilibrium with gains and losses balancing each other in the long run.

The two gains are atmospheric deposition (wet and dry in the form of nitrogen oxides, organic nitrogen, and NH_3/NH_4^+) and river runoff (as NO_3^-, NH_4^+, and organic matter); the main loss is through the bottom deposition of organically bound nitrogen: based on the C/N sediment ratio, Holland (1978) puts it at about 10 million t N/year. Atmospheric deposition values have been a matter of substantial disagreements but estimates of river runoff, though far from being very close, show much less disparity with the extremes at 15 and 40 million t/year and with 8 of the 12 available estimates ranging just between 20 and 35 million t of nitrogen annually, an unusually excellent agreement for nitrogen cycle accounts.

Nitrogen residing temporarily in streams and lakes is only a very small fraction of the element's total aquatic stores. Livingstone's (1963) summaries of river water composition list a mere 1 ppm of NO_3^- as the global average, with extremes between 3.7 ppm for European rivers and 0.05 ppm for Australian streams, all very low quantities compared to more than 10 ppm for sulfates and nearly 60 ppm for bicarbonate; in fact, all strongly ionized components are more abundant in river waters than NO_3^-. However, where the low NO_3^- water concentrations rise through human interference, nitrate in water may become a serious concern owing to its potentially dangerous health effects. Both the hydrosphere and the lithosphere are thus relatively inert reservoirs of nitrogen: the busiest flows of the element's cycle involve soils and living organisms with the atmosphere being the key supplier of nitrogen for life on the planet.

3.1.1.2. NITROGEN IN THE ATMOSPHERE The atmosphere contains considerably more nitrogen than does the hydrosphere—3.8×10^{15} t versus 2.3×10^{13} t, a 165-fold difference—and virtually all of this huge mass, comprising 78% of the volume and 75% of the weight of all atmospheric gases, is in a stable triple-bonded dinitrogen form. Quantities of other nitrogenous compounds in the air appear negligible in comparison but these gases and aerosols are the most active links in the atmosphere–biosphere cycling of the element and I shall discuss them in turn.

Present in the smallest amount are nitric oxide and nitrogen dioxide whose interconvertibility in photochemical smog brought their combined designation as

NO_x. Their total atmospheric mass does not add up to more than about 6 million t of nitrogen and their background concentration is measured in parts per billion. However, their local and regional levels are much higher as the two oxides are ubiquitous urban and industrial pollutants released by combustion of fossil fuels in both stationary and mobile sources. A large portion of NO_x may eventually get converted to nitrate and the aerosols and gases are returned to the pools of terrestrial and oceanic inorganic nitrogen by wet and dry deposition. A detailed look at these processes will be given in Section 3.3, a review of NO_x emission controls in Section 3.5.1.

Here will be given just a brief mention of three natural processes converting dinitrogen to NO_x. The first conversion is caused by a huge mass of meteoroid material—on the order of millions of tonnes—entering the earth's atmosphere annually with velocities between 11 and 72 km/sec. These are mostly micrometeoroids but a significant number of larger bodies penetrate the upper atmosphere where their high entry velocities cause high temperatures (around 10^5 K) which dissociate and ionize the atmospheric gases. Menees and Park (1976) were the first to offer an estimate of annual global production rates of nitric oxide owing to this phenomenon: about 40,000 t/year.

The other two processes are much more important: global contributions of ozonization in the upper atmosphere and direct ionization of air by lightning. Ozonization was estimated by Burns and Hardy (1975) to contribute at least 10 million t N annually and lightning activity is almost certainly an even more generous source. Several theoretical calculations and laboratory simulations have attempted to establish the magnitude of atmospheric nitrogen fixation by lightning: Noxon (1976) estimated the NO_x production rate at 2×10^{26} molecules/flash, Griffing (1977) at 4×10^{26} molecules.

Using the generally accepted global total of 100 cloud-to-ground flashes/sec (Dawson, 1980), this average production rate translates to 30 million t N/year. NO_x balance considerations put the upper limit at 75 million t while some theoretical calculations give totals of only a few million tonnes. As the lightning frequency shows a marked increase in lower latitudes, nitrogen fixed in recurrent tropical thunderstorms and deposited in rainfall may bring surprisingly high inputs to some equatorial forests, perhaps up to several kilograms of nitrogen per hectare per year (Turman and Edgar, 1982).

Nitrous oxide is a much more abundant compound than NO and NO_2 combined, with typical background levels around 300 ppb and a total atmospheric burden of about 1.8 billion 5. N_2O is released to the air as the intermediate product of denitrification and this flux has risen with higher applications of nitrogen fertilizers, leading to concerns about potentially worrisome declines of stratospheric ozone levels should the fertilization growth rates keep increasing in the coming decades.

In addition to the three oxides, the only other nitrogenous compound encountered regularly in the atmosphere is ammonia, in both gaseous form and as suspended matter. Harrison and McCartney's (1979) review of NH_3 measurements shows values of 0.9–5.7 $\mu g/m^3$ for rural and suburban sites in England and the Unted States and 1.4–33.7 $\mu g/m^3$ for urban locations. During airborne measurements over a rural area in West Germany, Lenhard and Gravenhorst (1980) measured ammonia fluxes totaling 118 ± 49 $\mu g/m^2$ per hr for NH_3-N and 93 ± 47 $\mu g/m^2$ per hr for NH_4^+-N in summer and only 7 ± 14 $\mu g/m^2$ and 23 ± 34 $\mu g/m^2$ for the same two compounds in winter and explained these largely as a result of volatilization from domestic animal excrements; volatilization of fertilizers and coal combustion are other important sources.

As NH_3 concentrations are higher over land than over the ocean and as they increase with higher soil temperature and over soils with higher pH, there can be little doubt that the soils are a major, if not the principal, source. But substantial amounts of NH_3-N can be also absorbed by soils, water surfaces, and even plant leaves from localized industrial and farm sources. Available figures show annual rates of 21–83 kg NH_3-N/ha in New Jersey, up to 13 kg in the Rothamsted (England) area, and up to 150 kg near Midwestern feedlots (Legg and Meisinger, 1982).

The other sinks are wet and dry depositions, photolytic destruction in the stratosphere, and reactions with OH radicals converting NH_3 to N_2 and NO. The latter two processes are considered to be large sources of nitrogen oxides but the possibility that atmospheric destructions of NH_3 result in NO_x loss cannot be excluded either. In the first case, McConnell's (1973) assumptions would imply tropospheric generation of NO on the order of 100 million t annually; in the other instance, a comparable loss would have to be balanced by a natural source of the same magnitude, most likely by releases from the soil, and in either case natural NO sources would be four or more times the anthropogenic flux. Relatively smaller but still considerable uncertainties also characterize our understanding of nitrogen storages in soils and living organisms.

3.1.1.3. SOILS AND BIOTA Only organic matter can store relatively large amounts (typically some 95% of the total) of soil nitrogen without rapid losses: nitrogen oxides and ammonia can be readily volatilized, nitrates can be leached, but nitrogen in plant and animal residues can be retained and accumulated in soils in considerable quantities. However, in some soils, especially in deeper subsoil horizons from where volatilization is impossible, relatively large shares (20–40%) of total nitrogen may be held as NH_4^+ within the lattices of clay minerals, but in the topmost layers the share of ammoniacal nitrogen is only 4–9% of the total (Stevenson, 1972).

Relatively higher values of NH_4^+-N in deeper soil and subsoil strata are not caused by its higher concentrations but result from sharply declining amounts of total organic matter, and hence of organic nitrogen, with progressive depth: the drop is greatest between the two topmost horizons which means that the large soil organic nitrogen reservoir is quite shallow, just a few tenths of a meter; it includes everything from fresh plant litter and rapidly multiplying and dying microorganisms to well-rotted soil humus, the final product of decomposition. While in mineral soils organic matter may be just a fraction of a percent of their weight, in humus soils 10 or even 15% may be organic (in peats up to 30%), a difference of two orders of magnitude.

And the differences in nitrogen content of organic materials are also large, although within the same order of magnitude: humus has a C/N ratio of about 10 so that in the richest organic soils nitrogen may constitute more than 0.5% of the total mass. But in fresh and decaying plant residues, C/N ratios range from 20 to almost 100 so that typical nitrogen content of most of the soils is usually only between 0.05 and 0.2%.

Published estimates of soil organic matter nitrogen are very disparate: for example, Delwhiche (1970) put the reservoir's total at 760 billion t, and later changed the value to 175 billion t (Delwiche, 1977); Rosswall (1981) gave the total as 307 billion t, virtually identical with 300 billion t N in Söderlund and Sevensson (1976) or in NAS (1978a). Stevenson (1982b) estimated total nitrogen in the world's soils on the basis of Bohn's (1976) tabulation of soil organic carbon by assuming an average C/N ratio of 10 for mineral soils, 30 for organic soils (histosols), and an average of 10% of all nitrogen in mineral soils to be clay-fixed NH_4^+. This reasoning leads to 220 billion t N in soil organic matter and 20 billion t N in clay-fixed NH_4^+. Histosols, accounting for less than 4% of the world's soils, are the richest repository of nitrogen, containing at least 12% of the element's soil shares to a depth of 1 m; as they are often much deeper, their share of the world's soil nitrogen content is actually even more important.

Yet on the basis of actually determined soil nitrogen contents, one can arrive at plausible global totals which are much lower than all these estimates. For example, old central European farming soils whose fertility has been continuously replenished not only by synthetic fertilizers but also by residue recycling and manure application hold between 1500 and 5000 kg N/ha in the top 20 cm (Vömel, 1965). Throughout the United States, there is a large variation, with Mollisols (prairie, chernozem, and chestnut soils) containing the highest average amounts of nitrogen, 3300–5000 kg/ha in the top 15 cm and 12,000–17,900 kg/ha to a depth of 1 m (Stevenson, 1982b) but with the typical values being much lower: most of the Great Plains, Great Lakes, and Northeastern states have between 4000 and 8000 kg N/ha in the surface soils and virtually the entire Southeast has soils with less than 4 t N/ha (Aldrich, 1980).

The nitrogen content in soil organic matter of grasslands is between 640 kg N/ha in semiarid formations of India and 14,000 kg N/ha in mesohalophytic meadows of the USSR, with values between 3000 and 6000 kg N/ha being most typical (Coupland, 1979). In the top 60 cm of forest soils in the United States, the nitrogen content ranges from 1100 kg/ha in a poor jack pine formation in Florida to about 17,000 kg/ha in a rich temperate coniferous forest in Washington, with most values between 2500 and 6000 kg N/ha (Keeney, 1980a).

Satisfactory global order of magnitude estimates can thus be made by multiplying the total dry land area of 13 billion ha (this excludes extreme hot or ice deserts) by a conservative average of about 3000 and a liberal mean of 8000 kg N/ha to get no less than 40 and as much as about 100 billion t N. Taking the average of 0.1% of nitrogen in the top 50 cm of all soils would result in 6500 kg N/ha (assuming a mean soil specific gravity of 1.3) and a global soil total of 85 billion t N.

How much the undecomposed and decomposing litter holds is more difficult to estimate as its mass ranges from nearly 50 t/ha in boreal forests to a mere couple of tons per hectare in tropical rain forests. Even when one generously assumes 10 t of litter per ha of forests and woodlands (about 6 billion ha) and 2 t/ha for the rest of however sparsely vegetated dry land (roughly 7 billion ha), the total nitrogen mass, assuming 2% of nitrogen in the litter, would be a mere 1.5 billion t—a number too small to matter compared to the soil nitrogen total.

Making allowances for relatively small areas of organic soils extremely rich in nitrogen would also not boost the calculated global total in any significant way. Clearly, the global nitrogen content of the soils may not be much higher, if at all, than 100 billion t. But all of the above estimates face the unsurmountable obstacle of using averages where none easily apply as variability of soil nitrogen content even within a single field can be quite considerable, with coefficients of variation ranging commonly up to 30–50% (Biggar, 1978).

In comparison with nitrogen reserves in and on the soils, the aggregate mass of nitrogen bound in the living biomass must obviously be much smaller. In herbaceous plants, seeds may have as much as 7% of nitrogen but leaves, stalks, and stems nearly always have just between 0.5% (e.g., rye straw) and 1% (e.g., corn stover) of nitrogen; only in legumes do leaves have 2.5–5.0% and whole plants average 2–3% of nitrogen. In trees, which hold about nine-tenths of all global phytomass, the nitrogen content is much lower in wood (a mere 0.03–0.2% in temperate species, typically 0.5–0.7% in tropical stem wood) than in leaves (usually 0.9–3.0%) and the content changes with age: spring leaves may have nearly twice as much nitrogen as those in the fall (Heinsdijk, 1975).

The easiest estimate of total nitrogen in phytomass can be derived from the global aggregates: taking Olson's (1982) conservative value of 560 billion t C (i.e., 1.25 trillion t of dry biomass) sequestered in all terrestrial vegetation (see Section 2.1.2.2) and an average of 0.75% of nitrogen in the phytomass results in 9.5 billion t. Using the earlier, and higher, total terrestrial biomass estimates by Whittaker and Likens (1975) would bring the nitrogen total in plants to nearly 14 billion t. Delwiche's (1970) first estimate was 12 billion t, later scaled down to 8 billion t, while Rosswall (1981) put the total at 12.2 billion t. The differences between these estimates are, in comparison with usual nitrogen cycle uncertainties, negligible and one can confidently say that 10 billion t of terrestrial biomass nitrogen is a very good consensus total.

Needless to say, standing marine biomass, being considerably smaller than the terrestrial one, contains much less nitrogen. Delwiche's (1970, 1977) estimates decreased from 970 to 340 million t, both an order of magnitude below his terrestrial totals. Using Whittaker and Likens's (1975) marine biomass estimate of roughly 4 billion t and 10 as the C/N ratio of ocean phytomass implies no more than 180 million t of nitrogen in phytoplankton and algae. And nitrogen in zoomass appears to be approximately of the same magnitude: Delwiche's (1970) total is 200 million t, Rosswall's (1981) 280 million t. The animal nitrogen pool is rather insignificant in comparison with other living storages but, as Burris (1980) acidly remarked, "studies of animal metabolism receive most of the research largesse in this country and elsewhere because government agencies are run by mammalian chauvinists."

Plants and animals thus contain only about 1/400,000th of the atmospheric nitrogen so that the complete cycling of atmospheric nitrogen through the biota is an affair measured in 10^5 years. And, of course, to cycle the atmospheric nitrogen through anthropomass would be a much lengthier affair: the nitrogen content of mankind is—with nearly 4.5 billion people, mean weight of 50 kg, and average protein content of 15% of live weight—a mere 5.5 million t. Yet it is for the maintenance and enlargement of this minuscule storage that we manipulate and influence much greater pools and flows of the element. When we incorporate nitrogen from amino acids and other organic compounds synthesized by plants and animals, our tissues become tiny temporary reservoirs of the nutrient after a long journey from the air through the microorganisms, fertilizer factories, soils, and crops (or animals) and at these flows, determining the availability of nitrogen assimilable by humans, I shall look next.

3.1.2. Natural and Anthropogenic Fluxes

By far the most important transfers involving human interference in nitrogen cycles are the complex biotic processes going on in the soils. These microorganism-driven processes have all been subjects of intensive research and experimentation and the monumental monograph published by the American Society of Agronomy, *Nitrogen in Agricultural Soils* (Stevenson, 1982a), is *the* source.

Here only the principal processes will be briefly introduced in the coming section and the key supply and loss fluxes, fixation and denitrification, will then be dealt with in much greater detail during the discussion of soil fertilization (Section 3.2.1) and nitrogen losses in agroecosystems and N_2O effects in the atmosphere (Sections 3.4.1.2 and 3.4.1.3).

Three other important fluxes through which considerable amounts of nitrogen are lost from soils—NH_3 volatilization and NO_3^- leaching and erosion (all abiotic)—will be treated in considerable detail later (Sections 3.4.1.1 and 3.4.1.4) and will not be mentioned among the soil nitrogen flows in the following pages—except in the summary of all major nitrogen transfers.

3.1.2.1. Nitrogen in Soils Although soils contain only a minuscule fraction of the lithospheric nitrogen, it is this small reservoir which is essential in sustaining life through its supply of the element to plants as NH_4^+ and NO_3^- ions, and through its sheltering of myriads of microorganisms participating in intricate processes which transform and cycle nitrogen compounds. These processes can be divided functionally into three categories: conversion of inactive inorganic atmospheric dinitrogen into organic forms, the transfer generally known as fixation; provision of plants with assimilable nitrogen from the element's soil reserves through mineralization (aminization, ammonification, and nitrification); and return of the soil nitrogen to the atmosphere through denitrification.

Jansson and Persson (1982) proposed a useful functional division of these processes into three interdependent partial subcycles—elemental, autotrophic, and heterotrophic. The first subcycle forms the outermost "envelope" returning atmospheric dinitrogen via fixation, mineralization, nitrification, and denitrification, the second one involves primary producers building up and using the pool of soil organic nitrogen, and the last one is governed by the activities of heterotrophic microorganisms immobilizing mineralized NH_4^+ (Figure 3-3). Functioning of all three subcycles depends on this mineralized nitrogen with nitrification, assimilation, and immobilization "driving" the flows.

Beginning with mineralization the three subcycles are in competition which is intensified at the NO_3^- pool where denitrification, plant uptake, and immobilization further split the available nitrogen flux. Production of all unfertilized plants is thus determined by the net outcome of these successive splits and the fate of cycling nitrogen can be traced quite conveniently within these three interdependent cycles just by throwing the "switches" at NH_4^+ and NO_3^- sites.

Following this model I shall start my brief descriptions with mineralization processes. As the soil nitrogen is primarily stored in organic compounds, above all in proteins, it must first be released by proteolysis (aminization) and then reduced to

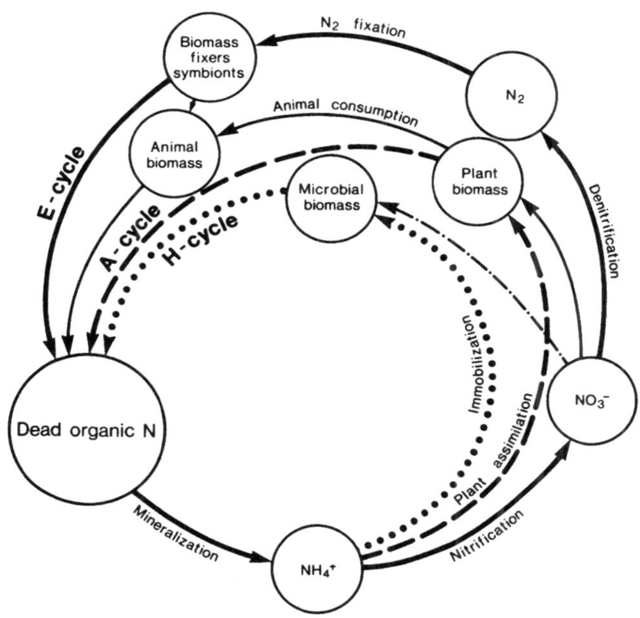

FIGURE 3-3. Three interdependent subcycles of nitrogen flows in soils (Jansson and Persson, 1982).

NH_3 (ammonification) to become available to plants. On a global basis, ammonification is the principal source of assimilable nitrogen owing to numerous bacteria, yeasts, and other microbes which utilize organic carbon as their energy source and carry on under aerobic as well as anaerobic conditions in virtually all environments. However, higher temperature, hydration, and soil aeration are the principal factors promoting faster decay rates and aerobic destruction results in more assimilable compounds than the anaerobic process.

Ammonification comprises not only enzymatically catalyzed reactions during which proteins, nucleic acids, and other soil nitrogenous substances are transformed to yield NH_3 (which then equilibrates to NH_4^+) but also decomposition of urea, added in the animal wastes or in fertilizers, by ureases. Ladd and Jackson (1982) offer a detail review of the biochemistry involved in these processes.

Ammonia ions resulting from ammonification can be easily oxidized, first to NO_2^-, then to NO_3^-:

$$2NH_4 + 3O_2 \rightarrow 2HNO_2 + 2H_2O + 2H^+$$

$$2HNO_2 + O_2 \rightarrow 2H^+ + 2NO_3^-$$

The purely biotic nature of nitrification had been established by Winogradsky's isolation of nitrifying bacteria in the late 1890s and many excellent reviews of the process are available: those by Focht and Verstraete (1977) and Schmidt (1982) are perhaps the most useful. The first step is mediated mostly by *Nitrosomonas* bacteria, the second one by *Nitrobacter,* both obligate aerobes with preference for neutral to slightly acid pH. This nitrification process can proceed quite rapidly under optimum temperature and humidity and just 2 weeks may suffice to convert virtually all NH_4^+ into NO_3^-.

Soil organic matter, especially in calcareous sites, has the capacity to react with NH_4^+ and to bind it in a form resistant both to chemical hydrolysis and to microbial attack; however, during normal NH_3 applications where the fertilizer is distributed only through a fraction of the top soil volume, the extent of this immobilization will not exceed 5% of nitrogen added (Nommik and Vahtras, 1982).

Under field conditions, especially in waterlogged, nearly anaerobic soils, in arid environments, or in very acid soils, nitrification proceeds much slower or even stops completely. This has important implications for the efficient utilization of synthetic fertilizers: nitrifiers oxidize NH_4^+ applied in the fertilizer in the same way as they convert NH_4^+ released by ammonification and if the transformation proceeds at rates higher than the uptake rate of the plant, considerable amounts of NO_3^- may be lost through denitrification or leaching. In contrast, NH_4^+ ions adhere so strongly to clay minerals in the soil, especially to illites, vermiculites, and montmorillonites, that drainage waters usually contain about 100 times less NH_4^+ than NO_3^-. Consequently, moderating nitrification rates would result in a more efficient use of synthetic nitrogen—as well as in less environmental pollution.

The net equation for nitrification shows that the process is a large source of hydrogen ions and hence a major contributor to soil acidification:

$$2NH_4^+ + 4O_2 \rightarrow 2NO_3^- + 4H^+ + 2H_2O$$

The rapidity and extent of these processes should be kept in mind for the discussion of soil acidification caused by atmospheric deposition of sulfurous compounds in the next chapter.

Efficient utilization of the applied nutrients is also initially dependent on the assimilation and incorporation of inorganic nitrogen by soil microorganisms. Soils with high C/N organic matter ratios will have lower rates of mineralization than soils with nitrogen-rich organic matter: in the former case, part of the added nitrogen will be taken up by the microorganisms thriving on the available carbon. This process of nitrogen immobilization (or biotic nitrogen fixation) is especially notable after incorporation of a large mass of crop residues, above all straws which have very high C/N ratios. Naturally, the immobilized nitrogen is not lost as it will be released gradually through subsequent proteolysis and mineralization, but the rapid initial decline after residue incorporation may result in yield depression and so additional synthetic nitrogen application may be needed to cover the temporarily higher microorganismic demand.

With so many variables affecting the outcome, net mineralization rates are not a subject of meaningful generalizations: even in the same field during one warm and fairly rainy growing season, twice the amount of nitrogen may be made available to plants than during a following colder and drier summer. The reported values range from just a few kilograms to a bit over 100 kg N/ha per year and in old farm fields of central Europe the rates are from about 10 on sandy soils and loessial loams to around 40 kg N/ha per year on alluvial loams and clays (Vömel, 1965).

These values imply mineralization of about 0.5–1.0% of total nitrogen contained in the upper 20 cm of the soil (ranging from 1600 to 5000 kg/ha). For comparison, Delwiche (1977) uses a rate of 1% in his global model but as he started with a larger organic nitrogen pool in the soil this value would prorate to nearly 150 kg N for every hectare of the planet's dry land.

Whatever the actual rate of mineralization, the process could not continue without continuous replenishment of organic nitrogen in the soil. Only a small part, most likely less than one-tenth, of this supply comes from weathering of organic sediments and the rest moves into soils through nitrogen fixation. Fixation is a collective term for natural and anthropogenic processes through which the inert atmospheric dinitrogen is converted to nitrogen compounds (NH_3, amino acids, nitrogen oxides) which may be, directly or after ammonification and nitrification, used by plants.

By far the most important among the fixation processes is the conversion done by marine and above all by terrestrial microorganisms living freely or in symbioses with higher plants. This fixation (Section 3.2.1) is critical for agricultural production and on a global scale may supply at least 100 million and as much as 175 million t of nitrogen annually. The second largest contributor to the soil nitrogen pool is now undoubtedly the chemical synthesis of nitrogenous fertilizer, the leading form of human interference in the nitrogen cycle to whose genesis, expansion, pathways, consequences, and controls will be devoted Sections 3.2.3, 3.4.1, 3.4.2, and 3.5.2. The annual global rate of this process now surpasses 60 million t of nitrogen.

Atmospheric chemical reactions are the third largest imput of fixed nitrogen to soils, now probably led by anthropogenic combustion of fossil fuels and biomass, followed by natural fires and direct ionization of air by lightning, falling meteorites, and cosmic radiation which can deliver the large energy inputs required for the fixation. Anthropogenic combustion, the second largest human contribution, fixes about 20 million t of nitrogen a year and will be discussed in detail in Section 3.3 and its controls in Section 3.5.1; as noted, lightning contributes most likely about 30 million t annually, falling meteorites no more than 100,000 t, and ozonization could produce up to 15 million t of nitrogen in oxidized compounds.

Finally, the process opposite to fixation is denitrification, the principal closing arm of the nitrogen cycle as nitrates are reduced through bacterial action to nitrogen oxides and ultimately to dinitrogen in the following steps:

$$NO_3^- \rightarrow NO_2^- \rightarrow NO \rightarrow N_2O \rightarrow N_2$$

The only organisms able to carry on the denitrification are common heterotrophic bacteria (except for *Thiobacillus*) belonging to nearly 20 different genera with *Pseudomonas* and *Alcaligenes* being the most numerous ones; other common denitrifiers are among the genera *Achromobacter, Bacillus, Micrococcus, Neisseria, Rhizobium,* and *Azospirillum.*

These biochemically and taxonomically very diverse bacteria have denitrifying capabilities owing to their possession of requisite reductases (though some microorganisms may not have all of them) which drive the dissimilatory reduction from NO_3^- to NO_2^- to NO to N_2O and finally to N_2. In some cases the outlined reduction path is not followed as several species start with nitrite rather than nitrate and others stop at N_2O rather than at N_2. Principal controlling environmental factors are the levels of oxygen (all NO reductases are repressed by the gas) and water (for a particular O_2 level, denitrification increases with higher soil water content), availability of organic carbon to feed the heterotrophic bacteria, NO_3^- concentrations (high levels inhibit the process), pH (in acid soils the reduction drops; optima occur at pH 7.0–8.0), and temperature (denitrification being higher at higher temperatures between 10 and 35°C) (Knowles, 1981).

Yet another form of gaseous nitrogen loss from soils is through nitrite reactions in a process termed chemodenitrification. Nelson (1982) summarizes the available evidence in four points: first, substantial gaseous losses have been noted under conditions not conducive to microbial denitrification or NH_3 volatilization; second, aerobic incubation experiments after addition of NO_3^- or NH_4^+ show much higher recovery of the former than the latter; third, high gaseous losses are found in circumstances leading to NO_2^- accumulation, i.e., after copious applications of NH_4^+ or NH_4^+-forming fertilizers; and, finally, nitrite added to sterilized or unsterilized acidic soils decomposes rapidly to dinitrogen.

Quantifying denitrification in the field is not easy owing to measurement problems: trapping the evolved N_2 in the atmosphere which itself contains 78% of dinitrogen makes the possibility of assay errors resulting from contamination exceedingly high. N_2O would be a better indicator of the rate of denitrification as its atmospheric concentrations are very low and a way to accomplish this is to block the reduction of N_2O by addition of acetylene gas (Balderston et al., 1976).

Information on the total denitrification flux for a variety of environments is so insufficient that the leading students of the process would not even want to guess at global rates but they put the nitrogen losses owing to the process between 15 and more than 50%, depending on the conditions of the soil, and they believe them to be only slightly lower than the rates of nitrogen fixation (Payne et al., 1980; Payne, 1981; see also Section 3.4.1.2). Obviously, any way to slow down denitrification would be immensely beneficial in raising the global nitrogen availability to crops. A soil additive which would selectively inhibit *de novo* synthesis of *c*-type cytochromes in bacteria (the key links in the reduction) could slow down denitrification rates.

However, such a solution may remain impractical for a very long time. Not being able to stop these losses of fixed nitrogen, the only way open to us in managing agroecosystems for higher yields is to keep supplying them with increasing quantities of artificially fixed nitrogen in fertilizers, a flow supplemented by formerly decisive but now nearly everywhere uncomfortably declining inputs of nitrogen through natural fixation of legume symbionts and free-living bacteria and through recycling of organic wastes.

Before summarizing all of these nitrogen fluxes on a global scale, a few remarks on nitrogen in flooded soils—which are so important for feeding about two-fifths of the world's population—are necessary. Waterlogging causes fundamental changes in soil properties which greatly influence nitrogen's behavior: above all, the root zone is changed from an aerobic to a virtually anaerobic layer and this switch retards decomposition of organic matter and promotes denitrification as facultative anaerobes substitute nitrate for oxygen; nitrogen can also be lost by NH_3 volatilization but leaching is not usually severe (Patrick and Mahapatra, 1968). For rice cultivation, the better stability of ammonium ions in waterlogged soils means that ammonium nitrogen is a preferred choice, especially when applied a few centimeters deep into the soil before flooding to avoid oxidation to nitrate.

In flooded soils, organic nitrogen tends to accumulate considerably more than in well-drained soils, partially as a result of higher nonsymbiotic fixation in wet fields but perhaps mostly owing to the slower decomposition of organic compounds under near-anaerobic or anaerobic conditions: slower mineralization thus leaves a larger part of the fixed nitrogen to remain in organic form (Buresh et al., 1980).

3.1.2.2. SUMMARIZING THE TRANSFERS Describing the calculations and assumptions presented for most nitrogen fluxes in all of the currently available global mass balance models would be both too tedious and too space-consuming. Instead I shall present the information in tabular form, proceeding by logical subcycles and offering separate tables for fixation and denitrification, ammonia fluxes, and releases of nitrogen oxides. Detailed discussions of data bases, estimating procedures, and uncertainties for most of the listed individual fluxes will be found in appropriate sections throughout the remainder of this chapter.

But one generalization applies readily: data on nitrogen fluxes are hard to gather not only because the exchanges and storages operate over vast areas but also owing to the heterogeneity of even seemingly uniform areas (demonstrable in soil conditions within a single field), and large seasonal differences in primary productivity of extratropical regions (i.e., most of the world's cultivated land) make most large-scale annual averages mere indicators rather than true measures of the ever-changing flows.

First, fixation–denitrification transfers (Table 3-1) will be reviewed. Here the only value known with high accuracy is nitrogen fixation by the Haber–Bosch process for fertilizers and industrial uses (Section 3.2.3): there is no reason why some of the late 1970s' totals should be so low. Estimates on fixation by fossil fuel combustion are in excellent agreement owing largely to application of identical or very similar emission factors to consumed fuel totals but these multipliers are not universally applicable with a very high confidence.

A similar situation prevails with terrestrial biotic fixation estimates: although most of these range between 140 and 200 million t N annually, the actual value may be appreciably lower (Section 3.2.1); marine fixation values are obviously very uncertain, essentially mere guesstimates. Also far from reliable are all figures for fixation by lightning, fires, and combustion of biomass, although in the last case we are getting onto a bit firmer quantitative ground (Section 3.3.1.3). And the return part of the cycle is not quantified with satisfactory certainty either: terrestrial denitrification estimates range over roughly an order of magnitude (Section 3.4.1.2), and an even greater range can be seen for oceanic denitrification values.

Estimates of ammonia/ammonium fluxes are scarcer than values for the fixation–denitrification cycle (Table 3-2). Only NH_3 volatilization is given in all of the reviewed works while values for dry and wet NH_3/NH_4^+ deposition are mostly lumped together or joined in a grand total with all other nitrogenous compounds (NO_x, organic nitrogen). Robinson and Robbins's (1970) estimates are a gross overestimate: when uniformly spread over the planet's dry land, this flux would average almost 90 kg NH_3/ha, i.e., some 75 kg N/ha per year and it would not only easily surpass total annual soil nitrogen additions through mineralization and fixation in all but the richest organic soils or legume-dominated ecosystems but it would also be much higher than the global mean of nitrogen fertilizer applications to cropland. However inadequate our knowledge of the nitrogen cycle, fluxes of such magnitude would not elude us.

According to Dawson's simple cycling model, emissions from uncultivated soils amount to only 47 million t NH_3/year with 32.5 million t (70%) originating in the Northern Hemisphere. But how large a source of ammonia are cultivated,

TABLE 3-1
ESTIMATES OF GLOBAL NITROGEN FIXATION AND DENITRIFICATION FLUXES[a]

	Industrial fixation	Fossil fuel combustion	Biotic fixation		Lightning	Biomass combustion	Denitrification	
			Terrestrial	Marine			Terrestrial	Marine
Eriksson (1959)	—	15	104	—	—	—	65	87
Delwiche (1970)	30	—	44	10	8	—	43	40
Robinson and Robbins (1970)	20	19	118	12	—	—	—	—
Burns and Hardy (1975)	30	20	139	36	10	—	140	70
CAST (1976)	57	20	149	1	10	—	70–100	70–100
McElroy et al. (1976)	40	40	170	10	10	—	243	106
Söderlund and Svensson (1976)	36	19	139	30–130	—	—	107–161	25–179
Delwiche (1977)	40	18	99	30	7	50	120	40
Hahn and Junge (1977)	40	—	180	85	—	—	150	165
Liu et al. (1977)	40	20	200	40	10	—	140	130
NAS (1978a)	70	21	139	100	30	10–200	197–390	0–120
Sweeney et al. (1978)	35	15	100	15–90	0.5–3	—	90	50–125
Bolin (1979)	40	20	140	20–120	—	—	63–245	35–330

[a]All values are in million tonnes of nitrogren per year.

TABLE 3-2
ESTIMATES OF GLOBAL NH_3 AND NH_4^+ FLUXES[a]

	Ammonia volatilization	NH_3/NH_4^+ deposition	
		Wet	Dry
Eriksson (1959)	99	99	—
Robinson and Robbins (1970)	957	796	175
Burns and Hardy (1975)	165	140	—
McElroy et al. (1976)	150	—	—
Söderlund and Svensson (1976)	113–244	38–85	72–151
Delwiche (1977)	75	79	
Hahn and Junge (1977)	170	—	—
Liu et al. (1977)	190	—	—
NAS (1978a)	36–90	—	—
Bolin (1979)	110–250	110–240	

[a]All values are in million tonnes of nitrogen per year.

fertilized soils and volatilization from animal wastes and plant remains in comparison with these uncultivated soil emissions? Delwiche (1977) put the global NH_4^+ soil volatilization at 22 million t N/year and that from plants and animal wastes at 53 million t N/year. The first value would prorate to about 1.7 kg N/ha of all dry land, in contrast to Dawson's (1977) average of 3.4 kg N/ha of nonfarming soils only, and the second one to 4 kg N/ha annually.

Söderlund (1977) prepared a detailed ammonia mass balance for the atmosphere over northwestern Europe (accounting for 0.44% of all dry land) showing a total loss of 1.7 million t N/year from domestic animals and 700,000 t N/year from synthetic fertilizers. These totals average to roughly 30 and 12 kg N/ha annually, and combustion of nearly 1 billion t of coal in the area adds about 1 million t of NH_3-N each year, or another 17 kg N/ha per year.

My range estimates will be based strictly on the most likely percentage losses from the approximate total outputs or reservoirs of ammonia-generating substances. The minimum NH_3 volatilization losses from synthetic fertilizers are 10%, average maxima are unlikely to surpass 30% (see Section 3.4.1.1) for the range of 6–18 million t N/year, or between 3.4 and 12.9 kg N/ha per year with a conservative average of 5 kg N/ha (i.e., 7 million t N). If the total output of nitrogen in animal wastes is about 75 million t (Section 3.2.2.2), and volatilization losses are at least 25 and as much as 50%, the animal total would then be between 19 and 38 million t NH_3-N (25 million t will be used as the single most representative value).

Owing to a large range of decomposition rates in various communities, and to no smaller differences in soil retention of the generated NH_3, volatilization from natural ecosystems may range from just small fractions of 1 kg N/ha per year in temperate grasslands (Woodmansee, 1978) to many kilograms in some forest soils.

Minimum rates could be calculated assuming an average nitrogen content of 5000 kg N/ha, a mineralization rate of 0.5–1.0% (i.e., 25–50 kg N/ha per year), and NH_3 volatilization losses not surpassing 10% of this amount, or 2.5–5.0 kg N/ha per year, globally 29–58 million t N/year (roughly 45 million t as an average).

Adding up the contributions from fertilizers, animal wastes, and nonfarming soils results in an aggregate range of 54–114 million t of nitrogen volatilized as NH_3 each year, with about 75 million t as the "best" single value which, although derived in quite different ways, coincides accidentally with Delwiche's (1977) total.

The third logical group embraces fluxes of nitrogen oxides, but NO_x combustion fixation estimates have been listed in Table 3-1 so that in Table 3-3 only the figures for biogenic NO_x production and NO_x deposition are summarized. Both sets have, once again, a considerable range of uncertainty. Finally, estimates of nitrogenous compounds in river runoff are rather abundant and in fair agreement, the lowest value being 15 and the highest 40 million t N/year, while an insufficient number of estimates for various exchanges of organic nitrogen precludes a tabular comparison; Söderlund and Svensson (1976) discuss the available fragmentary evidence.

These authors also list approximate residence times for nitrogenous compounds. With the single exception of N_2O, whose turnover time appears to be at least 100 years, all common nitrogenous substances cycle very rapidly: tropospheric turnover for NH_3 is a mere 1–4 days, for NO_x 1–20 days, for NH_4^+ 1–3 weeks, for organic nitrogen about 10 days. In contrast, dinitrogen's huge atmospheric reservoir makes the elemental turnover very slow: assuming that some 300 million t of N_2 is fixed by biotic and anthropogenic actions, it would take some 13 million years to process each molecule of atmospheric dinitrogen.

To close these summaries I offer a graphic review of all important fluxes and reservoirs of biospheric nitrogen in a comprehensive cycle model (Figure 3-4) which is also handy for comparisons with its carbon and sulfur counterparts (Sections 2.3 and 4.1, respectively). The following sections will highlight, in an inter-

TABLE 3-3
ESTIMATES OF GLOBAL NO_x FLUXES[a]

	Biogenic NO_x production	NO_x deposition	
		Dry	Wet
Eriksson (1959)	—	—	48
Robinson and Robins (1970)	234	22	83
Burns and Hardy (1975)	—	—	60
Söderlund and Svensson (1976)	21–89	25–70	18–46
Delwiche (1977)	—	34	
NAS (1978a)	22–66	—	
Bolin (1979)	20–90	40–110	

[a]All values are in million tonnes of nitrogen per year.

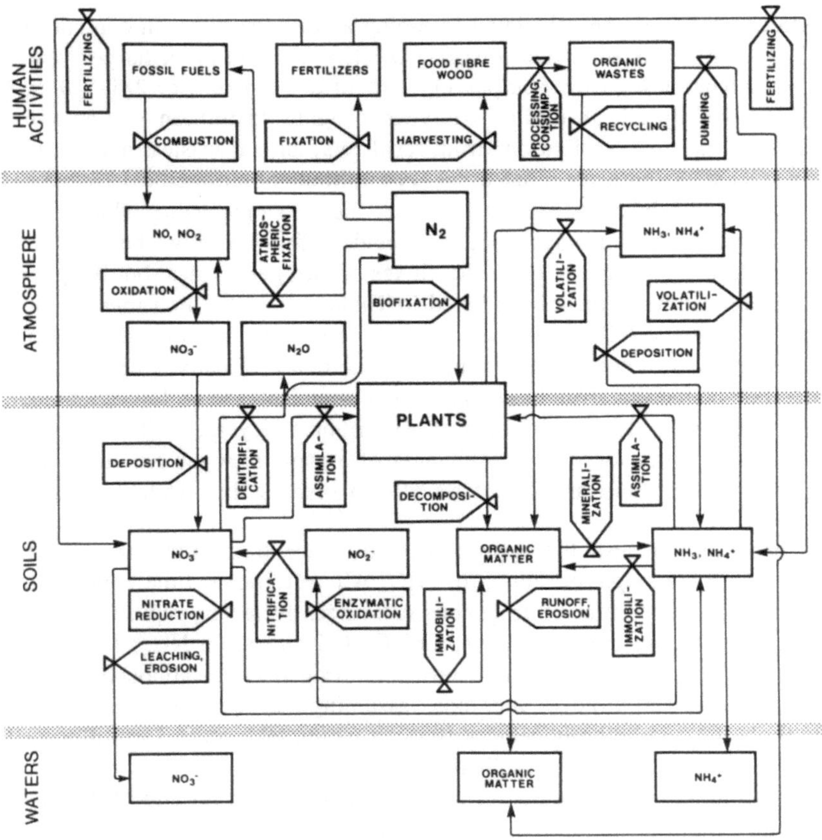

FIGURE 3-4. All important flows of the biogeochemical nitrogen cycle are traced in this graph which focuses on complex fluxes of the element in soils.

disciplinary fashion, many fundamentals as well as some intriguing specifics of all major nitrogen fluxes involving human interference. Both by their extent and magnitude of utilization, fertilizers must come first.

3.2. FERTILIZATION

As noted at the start of this chapter, nitrogen's role in plant life is essential in many ways, from the synthesis of enzymes controlling the metabolic processes to the formation of the chlorophyll molecule. The other two macronutrients and a host of micronutrients can, and not infrequently do, seriously affect plant growth, but nitrogen requirements are by far the highest and hence its deficiencies are most often encountered in natural communities lacking sufficient numbers of leguminous species or in unfertilized crops of traditional subsistence farming.

Nitrogen—the element making plants as luxuriously green and thriving as in Rousseau's jungle paintings—is the foundation of vigorous vegetative growth, taller stems, larger leaves. And in the staple grain crops it is the element largely determining the grain size and protein content, and thus the nutritional quality of the harvest. Shortages of nitrogen are manifested by both stunted growth and yellowing of leaves (chlorosis) proceeding from the lower layers upward, in many species from the leaf tip along the midrib.

Many highly productive natural ecosystems have developed efficient ways of recovering and recycling the limited amounts of nitrogen available to them from precipitation, soil reserves, organic decay, and symbiotic or free-living nitrogen-fixing organisms. Until the beginning of this century, our efforts in sustaining the fertility of agroecosystems on which we depend for our survival were limited to emulating nature through cultivation of leguminous species and recycling of organic wastes.

These two options, involving relatively mild interference with the existing biospheric mass and energy flows and staying fully within solar (renewable) energetics, will be discussed first in two separate sections. Invention of ammonia synthesis introduced an "extrasolar" intervention whose scale has now approached the magnitude of the principal natural fixation processes. This necessitates not only a bit more detailed look at nitrogenous fertilizers and at crop responses to their applications but later (Sections 3.4.1–3.4.3) also a discussion of their effects on ecosystems and human health.

3.2.1. Nitrogen Fixation by Living Organisms

Biotic reduction of atmospheric nitrogen to ammonia is done largely by prokaryotic bacteria and blue-green algae (collectively termed diazotrophs) living alone or, more often, in symbioses with some fungi, liverworts, ferns, gymnosperms and with thousands of species of angiosperms, above all among Leguminosae. Although it will be seen (Section 3.2.1.2) that our knowledge of actual field fixation rates is far from adequate, there is no doubt that on the global scale nitrogen-reducing organisms are by far the most important suppliers of the element to the soils, surpassing the synthetic inputs at least twofold.

Clearly, the primary production of natural ecosystems is critically dependent on biotic fixation and, in spite of the increasing mass of synthetic nitrogen, productivity of human agroecosystems would drop much below the current levels without bacterial and algal contributions. Not surprisingly, the recognition of the essentiality of biotic nitrogen fixation for the functioning of agroecosystems—the process started with Hellriegel and Wilforth's identification of rhizobia–legume symbiosis in 1888—brought considerable scientific attention which has been greatly aided since the 1940s by many fundamental advances in analytical methods, biochemistry, and bacteriology.

Consequently, today's information on biotic fixation, especially on *Rhizobium*–legume symbiosis, is very rich, encompassing numerous scientific fields from agronomy and plant breeding to molecular chemistry and genetic engineering. For the almost endless details now available, the following summary and review volumes will serve best: Quispel (1974), Burns and Hardy (1975), Evans (1975), Stewart (1975), Newton and Nyman (1976), Nutman (1976), Hoveland (1976), Silver and Hardy (1976), Ayanaba and Dart (1977), Hardy and Gibson (1977), Newton and Orme-Johnson (1980), Havelka et al. (1982), Vincent (1982).

3.2.1.1. NITROGEN-FIXING SYSTEMS The two most basic divisions of the variety of nitrogen-fixing arrangements are (1) heterotrophic and autotrophic setups and (2) processes carried out by symbiotic and free-living organisms (Havelka et al., 1982). Certainly the best known and the most important mode of natural nitrogen fixation, involving not only many common farm crops (soybeans, alfalfa) but also numerous tropical and subtropical trees, is a nodule-forming symbiosis of heterotrophic *Rhizobium* bacteria with leguminous species. Other nodule-forming macrosymbiont–diazotroph arrangements are between nonleguminous angiosperms (*Alnus, Myrica*) and *Frankia* spp.; a singular *Rhizobium*–nonlegume link with *Trema cannabina;* and a symbiosis between a nonlegume (*Gunnera*) and a blue-green alga (*Nostoc*).

Nonnodulated heterotrophic symbioses involve associations of tropical grasses and bacteria (*Paspalum* and *Azotobacter* is the best known example) and various angiosperms with *Azotobacter, Beijerinckia,* and other bacteria. Autotrophic symbioses are very common in blue-green algae (above all *Nostoc*) and lichens, liverworts, mosses, ferns as well as gymnosperms (*Cycas*) and angiosperms.

Many of the just described diazotrophs are also free-living. Among heterotrophs the leading species are aerobic (*Azotobacter, Beijerinckia, Rhizobium*), facultative anaerobic (*Bacillus, Klebsiella*), and anaerobic (*Clostridium, Desulfovibrio*) bacteria; among blue-green algae above all heterocystous species (*Anabaena, Calothrix, Nostoc*), less frequently filamentous or unicellular ones (*Trichodesmium, Gloeocapsa*). Some species of photosynthetic bacteria—*Rhodospirillum, Rhodopseudomonas, Chromatium*—are also diazotrophic.

Rhizobium–legume symbiosis has been, and undoubtedly will remain, by far the most important natural nitrogen-fixing arrangement diffused and encouraged by men in search of higher food production, improved soil fertility, or better supply of some raw materials. Consequently, the next two sections will be devoted to that symbiosis and to leguminous plants while in this section I will introduce in some detail all other important nitrogen-fixing systems.

I shall start with nitrogen fixation by symbiotic actinomycetes (*Frankia* spp.) attached to roots of many dicotyledonous plants, an arrangement which used to be considered much less important than *Rhizobium* symbiosis. Only through more extensive root sampling did we come to realize the considerable extent of this form of nitrogen fixation: about 160 species in 15 genera of seven families form root nodules following actinomycete infection and the total amount of nitrogen fixed in this way is at least comparable to that in legume symbiosis (Bond, 1976; Torrey,

1978). Belated appreciation of the actinomycetes fixation stems mostly from the great diversity of plants supporting the nodules and even more from the fact that none of the species involved has been an important farm crop.

The largest number of actinomycete-nodulated species are among Eleagnaceae(*Eleagnus, Shepherdia,* and *Hippophaë* species), Rhamnaceae (above all *Colletia*), Betulaceae—among them, 33 of 35 species of *Alnus* growing on poor soils and on wet sites throughout Eurasia, North America, and the Andes have actinomycetes symbiosis—and Casuarinaceae. *Casuarina equisetifolia* is planted for erosion control, soil improvement, timber, pulp, and fuelwood in India and Southeat Asia; *Alnus, Myrica,* and *Eleagnus* have been cultivated for the same uses in Europe and Japan. Alders could be the first trees cultivated on a larger scale as new timber or fuelwood crops (Smil, 1983).

As there are no reliable estimates of the total amount of nitrogen fixed annually by either rhizobia or actinomycetes, quantitative comparisons of aggregate contributions are impossible but the available fixation rates are sufficient to conclude that the nitrogen reduction proceeds with the same order of magnitude in both symbioses: annual values for *Alnus rubra* are reported to be 140–300 kg N/ha, for *Casuarina equisetifolia* 60–200 kg N/ha. The obvious disadvantage of actinomycete symbiosis is its restriction to woody dicotyledons which makes annually only a part of the fixed nitrogen available to other plants through leaf litter but the long-term nitrogen contributions to their ecosystems are at least equal to those from leguminous symbiosis on a global basis.

Associative nonnodulating symbioses (or biocoenoses) with higher plants appear most frequently in tropical C_4 grasses, i.e., more photosynthetically efficient, nonphotorespirative species with Hatch–Slack metabolism. Neyra and Döbereiner (1977) offer a thorough review of these arrangements, including a detailed bacteriological appraisal.

The first tropical associative symbiosis between a C_4 grass and bacteria investigated in detail was the *Paspalum notatum–Azotobacter paspali* link in Brazil (Döbereiner, 1977). Sugarcane (*Saccharum officinarum*), grown in many areas in monoculture for over a century without any external nitrogen additions, has nitrogen-fixing *Beijerinckia indica* in the rhizosphere or in the surrounding soil. Other tropical C_4 grasses (*Panicum, Digitaria, Andropogon*) have associations with *Azospirillum* and the same bacterium is also found in relative abundance in corn and sorghum roots (both are C_4 crops). Rhizosphere nitrogen fixation is also important in two C_3 grasses supplying the bulk of human grain—in rice and wheat. *Beijerinckia* and *Enterobacter* are the most common genera found in the rice rhizosphere, while *Enterobacter, Bacillus,* and *Azospirillum* can be isolated from roots of wheat.

Among the at least 25 genera of free-living bacterial fixers, heterotrophic aerobes are most abundant: as they obtain energy through the breakdown of plant residues, they are highly active only in fertile soils. *Azotobacter, Azotomonas, Azospirillum, Azotococcus, Beijerinckia, Pseudomonas,* and *Rhizobium* are all in this category. Facultative anaerobes (*Bacillus, Enterobacter*), anaerobes (such as

Desulfovibrio), and photosynthetic bacteria (*Rhodopseudomonas, Rhodospirillum*) are much less important. The overall contribution of free-living bacterial fixers to soil fertility remains uncertain: very likely they do not make enough of a difference to raise crop yield above a statistically significant level (NAS, 1979a).

No doubts can be had about the distribution of blue-green algae which are the most universal of all nitrogen-fixers. These photosynthesizing autotrophs can survive temperatures in the range 0–60°C and thus can be found in almost all kinds of terrestrial, aquatic, and marine environments, including Antarctic waters and hot springs and only low acidity checks them (Peters, 1978; NAS, 1979a). Moreover, blue-green algae are among the oldest photosynthesizing life forms preserved in ancient rock formations, indicating that nitrogenase has been around long enough to become more widely available to primary producers. That it has not is one of the most intriguing anomalies of the biosphere, and one for which we have no good explanation.

Of the roughly 60 genera known to fix nitrogen, heterocystous filamentous forms, led by *Anabaena*, are most abundant; nonheterocystous filamentous forms are represented by less than a dozen of genera, unicellular forms just by *Gloeocapsa* [for detailed classification and characteristics of nitrogen-fixing blue-green algae, see FAO (1981a)]. Blue-green algae have been critical suppliers of nitrogen in many stressed environments but nowhere has their contribution been so crucial as in maintaining the fertility of Southeast Asian paddy fields before the introduction of synthetic fertilizers. International Rice Research Institute (IRRI) experiments in the Philippines showed that the algae were able to replace all nitrogen removed from the wet fields by as many as 24 consecutive crops grown on otherwise unfertilized soils—and do so without an apparent decline of the total soil nitrogen content (Buresh et al., 1980). The FAO (1981a) now recommends broadcasting dry algal matter (*Nostoc, Anabaena, Ailosira* spp.) over the standing water in paddy fields at a rate of at least 10–15 kg/ha 1 week after transplanting and seeding to gain benefits on the order of 20–30 kg N/ha.

The actual rates of nitrogen fixation by blue-green algae in rice fields have been extensively investigated at the IRRI in the Philippines and also in Thailand, Ivory Coast, and Louisiana. Buresh et al. (1980) survey these studies which show a wide range of values determined mainly by the C_2H_2 method: from as little as 0.5 kg to as much as 75 kg N fixed/ha per year. Similarly, in the photic zone of various coastal habitats the rates range from only a few kilograms in Massachusetts creek banks and bottoms to 462 kg N/ha per year in stagnant pools of water in England.

In tropical paddy fields the highest fixation rates can be achieved when the blue-green alga *Anabaena azolla* is present in symbiosis with the small floating fern *Azolla pinnata*. The alga lives in a large cavity of the upper lobes of *Azolla* leaflets, forming a floating nitrogen-fixing unit which appears to grow in unison and which can provide, as long as the fern does not cover more than half of the water surface to avoid excessive shading, high rates of nitrogen additions. In Indonesia, south China, Vietnam, and Thailand, *Azolla* is now being used fairly extensively as a green manure (during the hot season its stocks are kept for distribution and multiplication in cooler weather), as well as an excellent duck and pig feed (up to 24% crude protein in dry mass).

The benefits of *Azolla–Anabaena* symbiosis are restricted to flooded farm soils and are best seen in rice cultures; in dryland crops supplying most of the world's food and feed, the only practical natural way to increase significantly the soil fertility is by *Rhizobium*–legume symbiosis.

3.2.1.2. *Rhizobium* NITROGEN FIXATION The association between legumes and rhizobia has found very widespread practical applications which may be greatly improved by inoculation: many soils do not harbor the proper nodulating bacteria or else contain ineffective strains, so that inoculation of seeds before planting with proper rhizobia results in excellent response with less than 1 kg of high-quality inoculant replacing more than 100 kg of synthetic nitrogen per ha (NAS, 1979a).

Actual field fixation rates are surprisingly uncertain. One might have expected to find hundreds of published values for a few dozen of the principal leguminous plants of the world, but a recent review by La Rue and Patterson (1981) shows a surprising paucity of such information, with most figures repeatedly cited in fixation reviews and biological texts derived largely from two several decades-old sources using inappropriate experimental methods and overestimating the actual field fixation rates! Another popular figure, an annual average of 140 kg N/ha calculated by Burns and Hardy (1975), was shortly after its publication appraised as perhaps twice too high by a group consensus of workers engaged in fixation research. And answering the question about actual field rates of nitrogen fixation is difficult not only because of the scarcity of available assessments but, no less importantly, also because of difficulties with the assessment methods themselves.

Estimates of nitrogen fixation by leguminous crops have been done by a variety of methods, none of them quite satisfactory. Estimates based on total nitrogen accumulation overestimate the crop performance as all of the plant's nitrogen is not derived from symbiotic fixation. Lysimetric studies may be biased owing to nitrogen volatilization. Each of the three commonly used difference methods (comparisons of a legume with a nonlegume, a legume with a nonnodulating legume, and an inoculated legume with an uninoculated one) rests on questionable assumptions: the first method implys that there are no significant differences in soil nitrogen absorption caused by growth patterns and root morphology, the second approach ignores the fact that nodulated plants may absorb more nitrogen than nonnodulated ones, and the third technique assumes that root structure and function are not changed by the nodules.

Direct isotopic methods have problems of high cost and measurement errors (dilution method based on ^{15}N added as a small amount of nitrate or ammonium) or are seen as giving only qualitative, at best semiquantitative, results (the technique based on natural isotopic abundance, or very slight increase of ^{15}N produced by symbiotic fixation). The very widely used acetylene reduction (where nitrogenase reduces acetylene to ethylene) is sensitive, fast, and cheap but there is no adequate way to calibrate ethylene formation with nitrogen fixation, the results are much affected by diurnal variation in fixation (as immediate incubation is essential) and by plant-to-plant variability so that an estimate for the whole growing season requires numerous replications.

To these numerous methodological and measurement biases, difficulties, and errors must be added the no less numerous differences and uncertainties owing to climate, management practices, cultivars, stand densities and compositions, and growing seasons. The results: very wide ranges of nitrogen fixation values have been reported in the literature for all important leguminous crops (Table 3-4). Values for soybeans, obtained from every important soybean-growing region of North America, appear to be the most reliable but they, too, are quite wide-ranging and are representative only of experimental plots in specific locations so their extension to actual field fixation is speculative.

This, of course, is a problem common to all extrapolations of experimental figures to field conditions: test plots almost always yield much more than the field

TABLE 3-4
RANGES AND MEANS OF PUBLISHED *Rhizobium* FIXATION VALUES FOR THE MOST COMMON LEGUMINOUS FOOD, FEED, AND FUELWOOD CROPS[a]

		Nitrogen fixation (kg N/ha per year)	
Latin name	Common name	Published ranges	Calculated averages[b]
Arachis hypogaea	Groundnut	72–240	140
Cajanus cajan	Pigeon pea	96–280	150
Cicer arietinum	Chick-pea	65–140	100
Cyanopsis tetragonolobus	Guar	41–220	130
Glycine max	Soybean	15–331	120
Lens culinaris	Lentil	62–114	90
Leucaena leucocephala	Leucaena	110–500	300
Medicago pratense	Red clover	154–300	230
Medicago sativa	Alfalfa	128–600	280
Melilotus alba	Sweet clover	125–140	130
Phaseolus vulgaris	Bean	10–64	30
Pisum sativum	Pea	17–77	55
Trifolium alexandrinum	Egyptian clover	62–235	150
Trifolium repens	White clover	117–189	150
Trifolium subterraneum	Subterranean clover	58–207	150
Vicia faba	Broad bean	45–648	260
Vicia villosa	Vetch	111–184	150

[a]Sources: Ayanaba (1980), Burns and Hardy (1975), Evans and Barber (1977), Högberg and Kvarnström (1982), La Rue and Patterson (1981), NAS (1977b), Neyra and Döbereiner (1977), Nutman (1976), Peters (1978), Phillips (1980), and Silver and Hardy (1976).

[b]Averages are simple means of the listed extremes; modal fixation values may be quite different.

average, as soils in experiments are often quite unlike field soil and there are no valid ways to extrapolate from a few small plots to over several orders of magnitude and to account for differences between a carefully managed test plot or a greenhouse experiment and water, pest, pesticide, or fertilizer stresses in the field. Consequently, even extensive experimental data can be used only in estimating upper limits of actual field fixation.

In view of the all too clear importance of leguminous crops for global food supply and maintenance of soil fertility, the conclusion reached by La Rue and Patterson (1981) is almost incredible: "There is not a single legume crop for which we have valid estimates of the nitrogen fixed in agriculture." With this harsh caveat in mind, all the figures listed in Table 3-4 must then be seen as mere magnitude indicators, most likely biased toward maximum possibilities.

Even if the general rates were assessed accurately, the specific field fixation values would have to vary rather substantially as there are numerous environmental factors affecting the host plant, the *Rhizobium*, and the development and functioning of their symbiosis. The ecology of *Rhizobium* is relatively poorly known compared to the enormous literature on the biochemistry of nodules, inoculation problems, and agronomic aspects of symbiosis: in fact, Alexander (1977) went so far as to compare our knowledge of the bacteria's ecology to that of the art of ecology "four or even seven decades ago." What we appreciate fairly well are the numerous nutritional deficiencies that can exert a limiting influence on the fixation (Edwards, 1977).

Higher soil concentrations of inorganic nitrogen clearly decrease nodulation and reduce fixation in most instances. Even small nitrate concentrations inhibit the initial rapid phase of nodulation in some species while nitrates applied during fixation periods can cause fast decline of activity and nodule breakdown. However, some strains can function well even in the presence of relatively high soil nitrogen levels, especially at later stages of development, and sometimes small amounts of nitrogen can actually stimulate early seedling growth and nodulation. Deficiencies of several other nutrients essential for hig her plant growth—potassium, calcium, magnesium, phosphorus, sulfur, iron, boron, zinc, copper—are reported to reduce fixation. Cobalt and molybdenum requirements of N_2 fixation are higher than those for the host plant growth so that deficiencies of these two micronutrients are especially restrictive. Molybdenum is especially important as it is a constituent of at least five different enzymes catalyzing reactions of nitrogenase and nitrate reductase; consequently, the element is an essential micronutrient both for nitrogen-fixing bacteria as well as for higher plants in nitrate reduction.

Among the abiotic environmental variables, both shortages and excesses of moisture are crippling, as is the soil acidity. Although certain strains may survive longer dry spells, reversible and irreversible damage afflicts most nodules during drought, especially in lighter, sandy soils. Yet periods of waterlogging also lead to declines in both nodule tissue production and nitrogenase activity as the supply of oxygen is reduced. *Rhizobium* fixation is, of course, an aerobic process with maximum rates at 0.2–0.3 atm O_2 (Bergersen, 1977), and as the oxygen level in soils is usually well below that range, both nodule formation and fixation are often suboptimal even in soils of normal moisture.

Soil acidity has direct restrictive effects as well as secondary influence through Ca and Mo deficiencies and Mn and Al surpluses (for much more on acidity in soils see Section 4.4.3.2). Low pH (generally below pH 5.0) restricts infection and nodulation and increases the need for scarcer calcium; high Al levels appear to reduce the potential number of nodule sites and limit both nodulation and fixation more than the host-plant growth (Edwards, 1977).

The efficiency with which *Rhizobium* cells reduce nitrogen is difficult to calculate in the absence of information about carbon flux from plant cells to bacteroids but the energy cost to the whole plant (including maintenance of the nodules, reduction of atmospheric nitrogen, and assimilation of NH_4^+) has been determined by extensive research, mainly by measuring CO_2 evolution by nodules relative to nitrogen accumulation by the whole plant. The review by Phillips (1980) shows extremes of 0.3 and 20 g C/g N, the values being, respectively, for *Trifolium subterraneum* and *Glycine max*. Typical values for common pulses (*Pisum sativum, G. max*) and forage legumes are 1.5–8 g C/g N and if a single approximate value is needed, 6.5 g C/g N (mode).

In spite of many studies of natural *Rhizobium* occurrence, few generalizations are possible (Alexander, 1977). Densities are nearly always low in acid soils but otherwise populations vary widely depending on the previous soil treatment, season, and planting of *Rhizobium* symbionts (although the bacteria do not persist in soil in the absence of the host for long). Initiation of the symbiosis is also poorly understood and the basis for the specificity of the plant–bacterium interaction remains largely unexplained.

Many of these information gaps and uncertainties will undoubtedly be removed by the continuing intensive research in *Rhizobium* fixation, an effort without which the greater contribution of legumes to global food, feed, and raw material production would be seriously slowed down.

3.2.1.3. LEGUMES Extensive as it has been, human intervention in the nitrogen cycle through cultivation of leguminous crops represents only a fraction of the natural nitrogen fixation and, more importantly, only a small part of the potential manipulation. Leguminosae, split into three large subfamilies (Papilionoideae, Mimosoideae, and Caesalpinioideae), is the second largest family of plants (with Gramineae first) with 650 genera and 13,000 described species—yet only about 100 are grown commercially, and a mere dozen (soybeans, groundnuts, lentils, three *Phaseolus* beans, *Vicia faba* beans, peas, pigeon peas, cowpeas, red clover, alfalfa) are cultivated extensively for food or feed (Delwiche, 1978; NAS, 1979a).

Legumes grown for direct consumption of their seeds are commonly called pulses (most notably, FAO statistics use this term for annual areas sown, yields, and production) and are, in the aggregate, the world's second most important source of vegetable protein. And although the yields of common cereal and root staples are usually much higher than the harvests of all important leguminous crops, the latter require much less management, often much less water, and owing to high nitrogen content their protein production per hectare easily equals and mostly surpasses the protein harvests in grain and fibers.

Yet, incredible as it may seem in view of the widespread food shortages and nutritional deficiencies, legume cultivation worldwide has grown very sluggishly during the past generation: FAO statistics put their current global production at 42 million t, the same level as two decades ago. In contrast, during the same period, the global output of cereals increased roughly 2.5% a year and that of sugar by over 3%.

However, not all edible legume production fell or stagnated during the past two decades: worldwide cultivation of soybeans, above all for livestock feed, expanded rapidly, from just about 30 million t annually in the early 1960s to nearly 90 million t two decades later (annual growth rate in excess of 6%). Of this total, the United States produces nearly two thirds, and Brazil next with about 15 million t, a result of an extremely rapid expansion started only in the early 1970s. Yet Brazil, now the world's second largest exporter of soybeans, has become an importer of one of its traditional staples—black beans, as irrational a legume trade-off as imaginable.

Overall, this is a clearly discouraging development, even more so when one notes stagnation or only marginal expansion of many leguminous food crops grown by subsistence farmers of the three poor continents. A wide variety of these cultivars include not only such nearly universal species as *Phaseolus vulgaris*, *Vicia faba*, or *Pisum sativum* but less known plants whose cultivation is mostly restricted to a part of one or two continents.

Table 3-5 lists a dozen of the most important food legumes, and the detailed information on their botany, ecology, agronomy uses, and food values can be found in FAO (1959), Aykroyd and Doughty (1964), Rachie and Roberts (1974), Rachie

TABLE 3-5
IMPORTANT FOOD LEGUMES[a]

Latin name	Common name	Main areas of current activation	Perennial or annual	Maturing period (months)	Yield range (t/ha)
Arachis hypogaea	Groundnut, peanut	Tropics and subtropics	A	3.5–5	0.4–1.5
Cajanus cajan	Pigeon pea	Tropics	P	5–6	0.5–2.5
Cicer arietinum	Chick-pea	Near East, India	A	3–7	0.4–1.6
Glycine max	Soybean	East Asia, USA, Brazil	A	3.5–7	0.9–2.0
Lens culinaris	Lentil	Near East, north Africa	A	2.5–4	0.5–1.5
Phaseolus lunatus	Lima bean	Tropical America	P	3–9	1.0–3.0
Phaseolus vulgaris	Bean, kidney bean	Global	A	2–5	0.3–2.5
Pisum sativum	Pea	Temperate climates	A	2–4	0.4–4.0
Vicia faba	Broad bean	Global	A	3–7	3.0–6.0
Vigna mungo	Black gram	Tropical Asia	A	2.5–5	0.5–1.5
Vigna radiata	Green gram	Tropical Asia	A	2–5	0.3–2.0
Vigna unguiculata	Cowpea	Tropics	A	3–8	0.4–3.0

[a]Source: Aykroyd et al. (1982).

(1977), NAS (1979b), and Aykroyd et al. (1982). Seeds are obviously by far the most important food in edible legumes—eaten cooked (most beans, lentils, peas), fresh when immature (soybeans, peas, Bambara groundnuts), fermented or otherwise processed (soybeans turned into Indonesian *tempeh,* East Asian *doufu* and soy sauces or into oil)—and they are the justification of further substantial extension of such crops as an essential part of strategy of meeting future global nutritional needs. However, herbaceous legumes also have edible leaves (cowpeas) and roots (yam beans, winged beans), the latter having potential for surprisingly high yields even with unimproved cultivars (NAS, 1979b).

The other large group of cultivated legumes are forage crops with alfalfa (lucerne, *Medicago sativa*) and clovers (*Trifolium pratense, T. subterraneum, T. repens*) being the dominant species. Grown in pure or mixed stands, these cultivars greatly enhance the productivity of range land for long periods of time, and a single planting can, in milder climates, establish many years of continuously photosynthesizing cover allowing repeated harvests. Pure or mixed stands of alfalfa now cover about 11 million ha in the United States, Australia has over 100 million ha in legume pastures, and *T. repens* is the foundation of New Zealand's pastoral economy. In Egypt, rotations with berseem (*T. alexandrinum*) clover have traditionally provided essential forage for maintenance of indispensable draft animals as well as fertilization for subsequent food crops.

At one time, legumes were extensively planted just to be tilled back into the soil as green manures. However, this practice, essential for maintenance of soil nitrogen content and tilth in both Asian and European (by extension also North American) traditional farming, has declined considerably owing to the easy availability of low-cost nitrogen fertilizers. Such crop rotations are now less common not only in North America and Europe but also in China where winter planting of Chinese vetch (*Astragalus sinensis*) has been widely relied on to fertilize subsequent grain crops and to improve soil structure.

In addition to food, feed, and fertilizer, some leguminous crops in traditional farming societies have other, locally important, uses: their stalks can be used as stakes (e.g., pigeon pea, yam bean), as a most welcome fuel (nearly all common species) in energy-sparse rural areas [see Smil (1983) for extensive discussion of rural energy shortages in the poor world], or as an excellent mulch (unlike cereal straw mulch, the leguminous ones cause much less nitrogen deficiency during decomposition). Many legumes, edible or not, serve as outstanding ground covers to control erosion or weeds: *Pueraria phaseoloides* (tropical kudzu), *Coronilla varia* (crown vetch), *Lespedeza bicolor, Indigofera spicata, Arachis prostrata,* and *Lablab niger* are so far the most extensively tested examples.

Although much less cultivated than pulse crops and forage grasses, many leguminous trees and shrubs have outstanding nitrogen fixation capacities and can be used in ecosystemically appealing multipurpose ways as excellent sources of fast-growing fuelwood, luxury timber, natural fertilizer (either through intercropping or by incorporating nitrogen-rich foliage into soils, organic chemicals, animal feed, and highly nutritious food). Moreover, many of these trees can be grown in poor soils and in arid climates where they can be used effectively in checking environmental degradation. Of many hundreds of suitable species, only a score or so have achieved more extensive diffusion and forestry or agroforestry attention (Smil, 1983).

Clearly there is no shortage of both traditional and innovative approaches to further extend the use of legumes either as annual crops or in tree plantations. However, a closer look will uncover more than a few important limitations which will tend to restrict considerably the practical applications.

3.2.1.4. Limitations and Potentials In view of the relatively large number of biotic modes of nitrogen fixation and their already high contribution to the fertility of ecosystems on the one hand, and the rising need to supply adequate food for growing populations while conserving finite resources of fossil fuels or just lowering the cost of supplementary fertilization on the other, it is not surprising that during the mid-1970s, spurred by the rapid rise of fossil energy prices, a wave of enthusiasm arose, a multifaceted effort aimed at further increasing our dependence on biotic nitrogen fixation as a double solution to the "crisis" problems of food and energy.

A decade later, the energy situation in the rich industrialized nations is considerably more relaxed and as our policies are guided almost solely by short-term perceptions, the feelings of the mid-1970s—when hopes were stimulated that "substantial grant funds may become available" and when "both experienced and inexperienced research groups" were gearing up with new approaches (Evans and Barber, 1977)—have swiftly changed: in the mid-1980s, basic research in nitrogen fixation occupies much lower ranks on the list of generally perceived research and development priorities than a decade ago.

These calmer attitudes must be welcome as long as one remembers that the poor nations (i.e., three-quarters of mankind) *do* continue to suffer through chronic shortages of both energy *and* food and that synthetic nitrogen accounts for the largest share of their farming energy subsidies (for more see Section 3.2.3.3). Consequently, any improvements and extensions of biotic nitrogen fixation should prove especially beneficial in their case. Yet the scope and pace of such desirable changes will have to be modest.

Spectacular solutions have repeatedly been outlined in the literature of fixation research: transfers of nitrogen-fixing genes directly to higher plant cells or into harmless bacteria invading these cells so that crops, especially cereals and forage grasses, and trees would become self-sufficient fixers; protoplast fusions to create new bacteria–angiosperm symbioses; development of new associative nitrogen-fixing bacteria for nonnodulating crops; genetically engineered rhizobia insensitive to synthetic nitrogen in soil which inhibits nodulation; and so on (see, among many, Child, 1976; Newton and Nyman, 1976; Gutschick, 1977; Evans and Barber, 1977; Postgate, 1977; MacNeil et al., 1978). As long-term research possibilities, all of these options are undoubtedly interesting to follow, but linking these conceptual proposals with soon-to-come relief for farm fertilization is pure wishful thinking.

The search for new root symbionts would be a long-term task for intensive research and many decades may elapse before successful symbionts are found for principal food crops. Yet such matches will inevitably decrease yields by about 6%, making these arrangements far from easily superior to synthetic fertilizers as long as costs of ammonia synthesis rise in crude proportion to net food costs. Prospects for leaf symbionts are even less promising as only crops in rainy tropics could be considered for such arrangements.

Transfer of nitrogen fixing genes to all major food and feed crops would be the best solution but the amount of basic and applied research to be done in this area is enormous. Gutschick (1977) believes that "100 years, if ever" is the likely prospect. Complexities in the path of genetic engineering applied to nitrogen fixation easily justify pessimistic prospects. For example, successful transfers of nitrogen-fixing genes or whole nitrogen-fixing organisms to a higher plant would be only the beginning of a viable process: without protection of nitrogenase from oxygen inactivation, without complex provision of host cell metabolites to the fixers, without efficient means of conveying the nitrogen away from the fixation site, and without continuous maintenance of fuctioning nitrogenase, there will be no reliable, sustainable fixation (Stewart, 1977).

Hence, it is not possible to disagree with Evans and Barber (1977) who argue for improvements of nitrogen-fixing capabilities of well-known nodulated legumes and nonlegumes and of algal systems as the most sensible way of bringing tangible benefits in a relatively short span of time. Inevitably, even with these "simpler" approaches, difficulties abound and the past experience has sometimes been very discouraging.

The best example is the use of *Rhizobium* inoculation: there is no greater limitation to efficient fixation in leguminous crops than the lack of viable, effective inocula for many legumes (NAS, 1979a). Even when the legume root system is well nodulated by rhizobia present in soils previously planted to common leguminous crops such as soybeans or alfalfa, the fixed nitrogen may be insufficient to provide the total nutrient yields, and introduction of superior *Rhizobium* strains appears to be the best solution. Although many companies now produce various types of inoculants, their field performance has been quite unimpressive. For example, Dunigan et al. (1980) concluded, after testing 11 different commerical soybean inoculants during a 4-year period in five different locations in Louisiana, that they "generally did not result in statistically significant increases in nodule number, nodule dry weight, N_2-fixing activity . . . or in seed yield." Simply, the commercial strains of *Rhizobium japonicum* advertised as high performers are not any better than the native strains already residing in the soils.

Improving and diffusing other biotic nitrogen-fixing systems appears to be even more difficult. As already pointed out, nodulating nonleguminous species are almost exclusively perennial woody plants and hence cannot be integrated into annual farming rotations. This limits substantially their adoption. But the problems with diffusion of more efficient biotic fixation go much beyond the microbiological, biochemical, or agronomic innovations and improvements. For the foreseeable future, only arrangements involving leguminous crops could make any perceptible difference in the effort to supplant a part of synthetic fertilizers by rhizobia-fixed nitrogen but a wider integration of leguminous species into farming of both rich and poor nations will be resisted owing to several reasons.

First, in many densely populated poor nations, a more frequent cereal–legume rotation, even should the legumes be associated with rhizobia of much improved nitrogen-fixing capacity, will be resisted as the leguminous crop could displace one of the now common two or three staple grain plantings grown with synthetic fertilizer in response to shortages of arable land and increasing population density. Even where these pressures are less evident, it is hard to find enthusiasm for food legumes supplying a larger share of everyday diets.

Unfortunately, legumes contain a wider variety of toxic substances than any other plant family: in their leaves, pods, and seeds are found, for example, antitryptic and estrogenic factors, hemagglutinins, goiterogens, toxic amino acids, vitamin antagonists (Rachie, 1977; Aykroyd et al., 1982). Although many of these substances can be neutralized by seed soaking, proper cooking, fermentation, or other processing, the presence of toxins, leading to digestive disorders and to occasionally severe reactions in susceptible individuals, is clearly a limiting factor to wider acceptance of legumes. But still more important limitations to a wider use of legumes are their far from easy digestibility, and less than outstanding palatability in comparison with other high-protein foods (meats, fish, or eggs): given a choice, most people have readily forsaken legumes for animal proteins to complement cereal staples.

Food preferences and avoidances are guided by complex, often hardly rational, traditions and fixed perceptions, and changing the legumes' image is not going to be easy even should an effort be made; so far, the decline in food legumes appears to be largely left to continue. And a similar trend has been noticeable in recycling organic wastes, an important way of closing the nitrogen cycle in a most beneficial fashion.

3.2.2. Organic Fertilizers

Until a few decades ago, most of the world's farmers had to rely on photosynthesis to replenish the fertility of their croplands or pastures: they could grow legumes in intercropping, rotation, or as a green manure; or they could return to the soil portions of the harvested phytomass, either directly by plowing-in the often abundant crop residues, or indirectly as animal and human wastes, most often after suitable fermentation or composting—or, naturally, they could combine legume cultivation with organic waste recycling.

This traditional organic farming, perfected over centuries especially in eastern Asia and Europe, is capable of sustaining impressively high yields but only with much tedious and heavy labor as the organic fertilizers often have diffuse origins (necessitating extensive everyday collection), very low nutritional content (necessitating large volume applications), and must be properly treated before field spreading. Yet, the following pages will show that, in the aggregate, organic fertilizers contain a very large amount of nutrients and their maximum practicable recycling would easily rival current applications of synthetic nitrogen and bring other agroecosystemic benefits as well. But the following sections will also show how the actual recycling potential is rather limited and that the crop residues and organic waste can no longer suffice as the cornerstones of farming—although they certainly deserve continued promotion rather than neglect.

3.2.2.1. Crop Residues Besides the plant parts harvested for food, feed, and fiber, all field crops leave some kind of residues and the yield of these straws, stalks, stems, leaves, and vines is often considerably greater than the production of grain, tubers, or pulses. Using approximate residue rates (expressing dry matter residue yield in relation to field height of harvested parts) to multiply the outputs of the world's major cereal, tuber, leguminous, sugar, and fiber crops results in an annual global output of some 2.3 billion t/year (Smil, 1983). Cereal straws and stalks account for 80% of this total and wheat and rice straws alone provide some two-fifths of it; sugarcane tops and leaves are a distant second.

The total crop residue mass is split almost exactly between the poor and rich countries but as the poor world contains three-quarters of the global population, per capita availability in rich nations is three times larger than in poor ones, a fundamental inequity significantly limiting contributions and uses of crop residues throughout Asia, Africa, and Latin America where they are needed to fill many needs as feed and bedding, fuel and raw material for construction and manufacture (thatching, household utensils). Consequently, in all poor populous countries, only a small fraction of residues is returned directly to the soil.

Although reliable data on crop residue use are hard to come by, the available estimates show that in China about half of all residual crop phytomass is fed to draft animals and pigs and only then recycled in manures and the rest is burned as fuel (Smil, 1984b). In India, Bangladesh, and Pakistan, most straw is used as fodder but unlike in China the dung is not composted but burned as household fuel instead. In Egypt, where berseem clover is available for feeding, about three-quarters of all crop residues, including all rice straw, corn stover, and cotton stalks, are used for fuel (el-Din et al., 1980).

Throughout most of the densely settled areas of the poor world, usually no more than one-fifth of all produced residues is thus returned directly to the soil; moreover, in some countries this recycling takes the unfortunate form of postharvest burning during which the nitrogen is lost (at 500°C the loss is virtually complete; only a lower temperature smoldering at around 300°C can preserve about half of the nutrient). This practice is common in rice fields of Southeast Asia and it is still surprisingly frequent even in some rich nations: for example, in the early 1970s, nearly 40% of all English cereal straws and 15–20% of French straw were burned in the field (Staniforth, 1979).

In the United States the situation is quite different: about 70% of all residues are returned to the soil by far the most important category, feeding being a distant second (USDA, 1978). Although the water conservation and antierosion protection are, together with maintenance of soil organic matter, undoubtedly the greatest benefits derived from crop residue recycling in North America's fields, recycling of nutrients is a far from negligible service.

On a national basis, America's crop residues contain nearly 5 million t of nitrogen (assuming an average N content of 0.7% in dry matter), and with the current rates of recycling, some 3.5 million t of this essential nutrient is returned to the fields each year, an equivalent of one-third of all nitrogen applied annually in commercial fertilizers. Supplanting this input by chemical fertilizers would thus call for a substantial expansion of current fertilizer production and would cost United States farmers roughly an additional $(1983) 1.5 billion each year. In energy terms—assuming an average cost of 60 MJ/kg synthetic N (see Section 3.2.3.3 for details)—the currently recycled residues are worth roughly 5 million t of crude oil equivalent when only nitrogen is counted (6 million t when the energy costs of inorganic P and K are also added). Clearly, even complete recycling of crop residues would fall far short from providing the nitrogen needed to sustain the current crop production levels in the United States and other rich nations—but without it, resource, financial, and environmental costs associated with higher applications of synthetic fertilizers would be substantial.

To make an analogical estimate in global terms is much more difficult as there is only fragmentary information on the rates of recycling. My best estimate is about 17 million t of nitrogen, equally split beteeen rich and poor countries; if all of these residues were recycled, they would supply nitrogen equivalent to nearly 30% of current fertilizer applications in rich nations and to over 50% of inorganic fertilizer use in poor countries. Competing uses described earlier in this section cut the direct recycling shares very significantly so that the actual average contributions in rich and poor countries are most likely equivalent to no more than 10% of synthetic nitrogen applied in both the rich and the poor world.

Not only are substantially larger national shares hidden in these global estimates, but also much higher contributions made by residue recycling to some crops: whereas eastern Asian rice, western European wheat, and American corn are all heavily fertilized, subsistence rice, millet, corn, and tubers in many Asian, African, and Latin American countries must rely on recycled phytomass as the only major source of managed fertilization. And as in many of these cases the residues may contain about as much (or even more) nitrogen as the harvested grains or tubers—for example, data on nutrients removed by leading food crops in Sri Lanka's dry zone show 38 kg N/ha in rice grain and 37 kg N/ha in rice straw, 31 kg N/ha in sweet potatoes, and 58 kg N/ha in their vines (Amarasiri, 1978)—their maximum possible recycling is most desirable. Moreover, in the humid tropics the slow release of nitrogen from crop residues is preferable to fast-acting soluble inorganic fertilizers better suited to moderately rainy temperate environments.

Among the problems associated with crop residue recycling, nitrogen immobilization can be quite severe: straws, stalks, and leaves have very high C/N ratios (about 100) while the decomposer microorganisms have C/N ratios around 5 and their rapid need for nitrogen exhausts most or all of the stores present or mineralized in the soil until the substrate is exhausted and until the nutrient is released in eventual proteolysis (Lynch, 1979).

Eventually, the mineral N content will surpass the preapplication soil N concentration but to avoid the initial massive immobilization, nitrogen fertilizer must be added in quantities sufficient to support the microbial growth. For example, incorporating 1 t of straw per ha would require about 10 kg of inorganic N, a modest amount in comparison with current average nitrogen applications (see Section 3.2.3.1). Sufficient fertilization will thus ease or eliminate the competition between the crop and the microorganisms which the decomposers would otherwise win (Allison, 1973). Yet in fallow the immobilization of nitrogen is actually advantageous as it saves the nutrient from possible erosion or leaching losses.

Neither nitrogen immobilization nor some difficulties in tillage, effects on soil temperature, or occasional temporary phytotoxicity of crop residues are detrimental enough to outweigh the advantages of moisture retention, erosion prevention, and nutrient recycling. Incorporation of straw also greatly increases the activity of heterotrophic nitrogen-fixers; some Asian studies showed counts of nitrogen-fixing *Clostridium* more than four times higher than in control plos with no straw addition (Lee, 1978). And the combination of residue recycling and compost applications can sustain record yields in organic farming practices for long periods in densely settled lowlands of Asia and Europe, and is now looked upon with a renewed interest in the United States (USDA, 1980). Most of the composts used in organic farming contain, of course, a high percentage of animal wastes which are much more concentrated sources of nitrogen than crop residues.

3.2.2.2. Animal Wastes Quantification of animal waste output is even less reliable than that of crop residues as the production rates are affected by breed, amount and quality of feed, water consumption, sex, age, and health of the animals, and in females by pregnancy and lactation. Average figures representing nationwide, regional, or global output are thus good only as order of magnitude estimates (Smil, 1983). The global output is now around 2 billion dry t annually with cattle and water buffaloes producing some 70% of the total, pigs less than a tenth, and sheep, goats, and poultry the rest. Poor countries account for about three-fifths of the sum and China, with close to 20o million dry t annually, is the world's largest producer, the United States second with between 160 and 170 million dry t (USDA, 1978; Smil, 1983).

What share of the voided manures is recycled to croplands depends on how many animals are reared in confined conditions where the wastes are economically recoverable. The USDA's (1978) national estimate for all livestock and poultry is that only 39% of animal wastes originate while in confinement and that 73% of this is actually applied to croplands. Thus, less than one-third of all American animal wastes are managed as fertilizer: of course, all the wastes excreted on pastures and rangelands—about 40–50 million dry t/year—are automatically recycled. In China, a nation with an ancient tradition of careful waste recycling, almost three-quarters of all output—virtually all manures outside of those voided on the extensive grasslands of the Northwest and Tibet—are fermented for fertilizer. In India, where dried dung is used heavily as fuel, about half of the manure produced by draft oxen and water buffaloes is used as field fertilizer.

As to the nutritional content of manure, variation in these estimates is relatively no smaller than the total production. In fact, Lauer (1975) notes that even the variation itself is difficult to estimate. Consequently, the nutrient value of a particular manure can be accurately known only by analyzing the waste at the time of field spreading (Hobson and Robertson, 1977), an exceedingly rarely taken measurement. In general, nitrogen is the most abundant macronutrient in animal wastes, its content (in percent of fresh weight) being 0.6–4.9% for beef cattle, 1.5–3.9% for dairy cows, 2.0–7.5% for swine, and 1.1–11.0% for poultry (Gilbertson et al., 1978).

Not surprisingly then, estimates of the total quantities of nutrient voided by the world's livestock can be done only as order of magnitude exercises. My heavily rounded estimate (Smil, 1983) is 75 million t N, and if this total were recyclable to fields and pastures without losses, it would supply more nitrogen than all chemical fertilizers now applied globally. In reality, no more than two-fifths, or 30 million t, of this nitrogen is returned to cropland or managed pastures. For the United States, these kinds of estimates are on a firmer quantitative ground. The USDA (1978) put manurial nitrogen applied to fields at about 2.1 million t, an equivalent of about 20% of the nutrient supplied in synthetic fertilizers, while White-Stevens (1977) put it at 1.7 million t.

In any case, these theoretical calculations of manure's nutrient values are far removed from actual nutrient contributions owing to very large volatilization and leaching losses between voiding and application to soil and eventual mineralization. For example, the USDA (1978) estimates an average loss of 63%, shrinking the initial N content in applied manure from 2.1 to 0.8 million t. This loss of about 1.3 million t N prorates to about 6 kg N per capita per year, i.e., almost exactly as much nitrogen as is actually consumed each year in plant and animal protein by the average American (see Section 3.4.2.2).

The losses may be even larger as indicated by Lauer's (1975) sequence for dairy manure. Freshly voided, the manure has a nitrogen content of 5.7%, three-fifths of which is present in ammonia; 24 hr later (at daily cleaning) the two shares drop, respectively, to 3 and 36%; after a few weeks in unprotected piles, the values decline to 1.8 and 25%; and several days after field spreading a mere 1.5 and 4.5% are left to be mineralized and so available for uptake by crops. Usually between one-half and four-fifths of the remaining nitrogen is eventually mineralized at annual rates not surpassing 3–5% in temperate regions (USDA, 1978).

Consequently, the principal benefit of manuring may not be its low and slow nutrient contribution but rather its humus-renewing, aeration-imposing, and water-holding capabilities (Salter and Schollenberger, 1938), benefits similar to those arising from the recycling of crop residues. In the United States, even the maximum possible field spreading of confined manure production would be almost everywhere grossly inadequate to support current yields. Van Dyne and Gilbertson's

(1978) calculations show that on the average the country could expand the manuring rate fivefold and still remain within the minimum rate.

On the other hand, there are clear drawbacks from excessive manure application. While applications of up to 40 t/ha can be made annually in most cases with benefit, and while in the Netherlands pig manure is spread at rates of 50 t/ha for sugar beets and 70 t for potatoes (Voorburg, 1974), rates exceeding 125–150 t/ha cause salt problems, poor seed germination, and reduced yields (McCalla, 1974). Even otherwise acceptable applications may bring soil and water contamination by copper, boron, zinc, arsenic, and manganese concentrated in the manure, and improper spreading leading to serious leaching losses can contaminate waters and affect aquatic biota by increased oxygen demand and progressive eutrophication and render waters unfit for drinking, swimming, or shellfish production owing to high concentration of nitrates and pathogens.

In practical terms, there are also the disadvantages of large mass. For example, wheat requiring 60 kg N/ha will need 73 kg NH_3 or 283 kg ammonium sulfate to supply the requisite nitrogen but if applied as a 10% slurry manure with 6% N in solids, close to 30 t will have to be handled and distributed when volatilization losses (about two-thirds) are taken into account—a mass difference of two orders of magnitude! The cost of this distribution will obviously limit the radii of manure distribution.

There are other side effects of manuring which appear to be more troublesome, perhaps none more so than the buildup of inorganic salts of sodium, potassium, calcium, and magnesium in soils in drier climates. This becomes an even greater problem when manures are applied to irrigated fields because irrigation water also contains soluble salts. Thus, application of an adequate amount of nitrogen should be carefully weighed against detrimental salt buildup and formulas are now available for calculating the benign limits for organic waste fertilizing (Powers et al., 1975).

Yet in spite of some inevitable and some potential disadvantages and in spite of the fact that manures are now insufficient to cover nutritional needs even in poor countries with lower crop yields, proper applications of well-fermented animal wastes should be used as much as practicable as an integral part of sustainable farming, with closing of the nitrogen cycle being only a part of wider ecosystemic considerations. In comparison with these contributions, the recycling potential of human wastes is much more limited.

3.2.2.3. Human Wastes and Other Organics Most nitrogen in food is excreted in urine, which has an average nitrogen content of about 1%; feces have about 1.5% (Jaiswal, 1971). Assuming typical daily per capita outputs of 140 g of feces and 1.3 liters of urine results in an annual global output of 2.5 billion t of fresh waste for 4.4 billion people; solids comprise nearly 150 million t and nitrogen totals 25 million t, one-third of the domestic animals' total output and one-half of the current synthetic fertilizer applications. One-quarter of all human wastes is produced in rich countries where it is for the most part mixed with very large volumes of water and released to streams, lakes, or the ocean, usually after some, though often inadequate, treatment.

In the United States, nitrogen from human wastes, calculated at a per capita average of 5.4 kg annually (NAS, 1972), totals only 1.2 million t for the whole population and about three-quarters of this total moves through sewage plants where secondary treatment removes almost one-half of the nitrogen. Consequently, nitrogen directly released to waters in untreated sewage or after only primary treatment, amounts most likely to no more than 800,000 t/year, a negligible sum in comparison with other anthropogenic nitrogen fluxes. However, both these uncontrolled discharges as well as the field applications of digested sewage sludge may be troublesome point sources of ammonia and nitrate pollution.

Nor is the nitrogen content of solid wastes a significant total. Assuming a daily average of 2.5 kg of garbage (N content about 0.5%) per person, the annual total buried throughout the country's disposal sites would be only 1 million t N. But, where the landfills are in contact with groundwater, considerable local nitrogen leaching is possible.

Even complete recycling of easily disposable wastes would make a negligible contribution to cropland fertility; for example, total reuse of sewage sludge in the United States would replace less than 1% of the current fertilizer nitrogen (USDA, 1978). Complete recycling is impossible owing to the restrictions on feasible sludge transport distances around cities and the often poor quality of sludges; moreover, urban sewage sludges often contain very high concentrations of heavy metals and toxic substances which would greatly restrict the volumes of sewage applicable to farmland in the long run.

Throughout the poor world, with the notable exception of the Chinese cultural realm, recycling of human wastes is either absent or much restricted owing to socioreligious principles. In eastern Asia the practice has supported some outstanding yields and today remains quite extensive. Traditional Japanese farming depended almost solely on careful reuse of night soil; between 1908 and 1917, this produced annual rice yields of 3.3 t/ha (Takahashi, 1978), an average which even today is surpassed in Asia only by China, Taiwan, and the two Koreas.

After World War II, this labor-intensive practice rapidly declined in Japan but it continues in China, although there have been reports—from the richer parts of the country where the more convenient synthetic fertilizers are on the rise—of a decreased interest in removing urban wastes to suburban fields (Smil, 1984a). Still, I have estimated (Smil, 1981) that in the late 1970s, recycled human wastes annually contributed 800,000 t N or about 15% of the nitrogen recycled from all organic sources. Of course, any extensive use of night soil as fertilizer requires proper collection, storage, and handling both for hygienic reasons (to eliminate most pathogenic bacteria, protozoan parasites, roundworm and hookworm eggs) and to minimize nitrogen losses. Anaerobic fermentation in biogas digesters is an especially effective method of night soil treatment but possibilities for its significant diffusion outside eastern Asia are rather limited.

Thus, the global outlook for efficient recycling of human wastes is quite limited, and most of these wastes, especially throughout the poor world, will continue to be abandoned, improperly treated, or untreated, often a public health hazard and serious source of pathogens in polluted waters. Nor are the prospects good for recycling of a wide variety of other organic wastes which should be composted before field application: composting on farms in the rich world, quite common two generations ago, declined precipitously with increasing specialization and with the spread of large meat and dairy complexes. It is more common in Europe than in North America but remains, in spite of recent declines, most extensive in China where it has been a part of farming practice for millenia.

A wide range of composting practices—aerobic or anaerobic, at field sites or in the farmyards, at different temperatures, and with a wide variety of organics—have evolved to suit different conditions and needs (FAO and UNDP, 1977). The biochemistry of composting is well understood (Poincelot, 1975) and the resulting product—depending on the ingredients a slightly acid to fairly basic (pH 6–9), moist (40–90% liquid) matter with a nitrogen content of about 0.5% (typical range 0.1–1.0%)—is not only a valuable fertilizer but even more an excellent soil conditioner activating microbial processes and improving filth. But this beneficial practice is highly labor-intensive in all of its stages, and thus even the Chinese are abandoning it as synthetic nitrogen is becoming increasingly available.

The worldwide decline of composting is just a part of a broader shift away from organic fertilizers to synthetic nitrogen. Organics have several undisputed benefits which should secure their place in any sustainable agroecosystem but in comparision with chemical fertilizers they are not a more efficient, cheaper, and flexible source of nitrogen. Thus, regrettably but not surprisingly, they are being supplanted by synthetic nitrogen in one of the most fundamental technological transformations of this century.

3.2.3. Synthetic Fertilizers

Nearing the end of a century which has seen the emergence of so many technological wonders which have transformed the ways of our civilization, it is not surprising that even well-informed people would rarely place the synthesis of nitrogenous fertilizers among the great post-1900 technological innovations.

For most of those living in great cities of the Western world and removed from the land for several generations, the glamor or curse of high technologies overwhelms something so mundane as fertilizing fields—yet that act remains one of the few unalterable bases of our survival and it has been made much easier with the synthesis of nitrogenous fertilizer (i.e., anthropogenic fixation) at the beginning of this century. Standard comprehensive texts such as Tisdale and Nelson (1975) or Mengel and Kirkby (1978) should be consulted for a wide variety of chemical and agronomic facts concerning all types of fertilizers. In the following sections I will focus only on the basic techniques and requirements—in both financial and energy terms—of nitrogen fertilizer production and its continuing improvements.

3.2.3.1. NITROGENOUS FERTILIZERS: EMERGENCE AND DIFFUSION Until the second decade of this century, the only significant sources of nitrogen in farming were the natural fixation processes, including the cultivation of legumes, recycling of organic residues and wastes—and the mining of rich Chilean nitrates. Their large deposits (predominantly $NaNO_3$), discovered in 1809, became the most important source of inorganic nitrogen during the 19th century, augmented by the recovery of ammonium sulfate from coking ovens. Cyanamide process introduced in Germany in 1898 was the first commercially successful synthesis of inorganic nitrogen fertilizer compound but its high energy requirements and the need for large volumes of coke to react with CaO precluded its widespread use. Nor was the arc process, where oxidation of nitrogen in an electrical arc to NO is followed by further oxidation to NO_2 and production of HNO_3 by its dissolution in water, a success.

Total consumption of fertilizer nitrogen remained negligible until the third decade of this century as shown by available historic statistics for the United States, where just 34,000 t of inorganic N was used in 1890 and 56,000 t a decade later, although the two other macronutrients, phosphorus and potassium, were already being applied in larger quantities. In 1913 when the world's first synthetic ammonia plant became operational in Germany, American farmers spread only 150,000 t of nitrate nitrogen together with 225,000 t of phosphorus (in phosphates) and 180,000 t of potassium (N : P : K ratio 1 : 1.5 : 1.2) (USDA, 1971).

Parenthetically, whereas I have given the ratio of these macronutrients in elemental form, the most common practice in expressing fertilizer compositions and ratios is still to quote only nitrogen in elemental form with phosphorus and potassium given, respectively, as P_2O_5 and K_2O; multiplications by 0.4369 and 0.8302 are needed to reduce these to their elemental shares.

Although nitrogen applications did surpass those of phosphorus and potassium by the late 1920s, the use of phosphates grew as fast and that of potassium even faster during the 1940s so that as recently as 1949, the national N : P : K ratio was 1 : 0.92 : 0.97 (i.e., nearly identical), with about 800,000 t of each nutrient spread on America's farmlands. However, rapid changes followed: by 1960 (almost 2.5 million t N), the ratio was 1 : 0.41 : 0.65; a decade later (6 million t N), 1 : 0.29 : 0.46; and currently (over 10 million t N), 1 : 0.21 : 0.45. This exponential rise best illustrates nitrogen's irreplaceability in sustaining the country's record harvests (Figure 3-5).

The global trends have been similar. By 1950, 3.8 million t of synthetic nitrogen was applied worldwide, 6.6 million t 5 years later, and nearly 11 million t at the beginning of the 1960s. Although the three regions of traditionally high use—western Europe, the United States, and Japan (collectively, 66% of the 1961 global total)—maintained their primacy, synthetic nitrogen application in Asia (excluding Japan), Africa, and Latin America rose rapidly (2% in the late 1940s to 17% in 1961). By 1970, global nitrogen consumption surpassed 30 million t, and reached 60 million t in the 1980–1981 statistical year. Consumption by the poor nations now represented 40% of the global total.

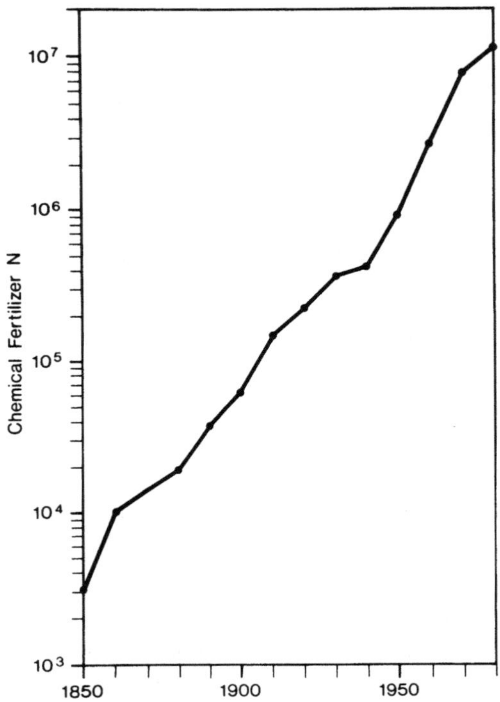

FIGURE 3-5. Exponential rise of agricultural nitrogen applications in the United States between 1850 and 1980 [data from USDA (1971) and Hargett and Berry (1981)].

Although the global difference between rich and poor nations has narrowed considerably in terms of the absolute consumption of synthetic nitrogen, the gap in relative applications remains considerable. Global nitrogen consumption at the beginning of this decade prorated to just over 40 kg N per ha of arable land (including permanent crops), but for poor nations this average was just over 30 kg while for rich nations it exceeded 50 kg N/ha. However, both of these averages come from considerable regional extremes.

Among the rich nations of the northern temperate zone, western Europe averages nearly 110 kg N/ha and Japan 130 kg but North America only about 50 kg and the Soviet Union just 35 kg. Among the poor nations, China's application surpasses 100 kg N/ha and although the rest of the Far East averages only about 25 kg, even this is significantly above that of the Near East (20 kg) and Latin American (17 kg); the African average is a pitiful 4 kg (excluding the Republic of South Africa).

In terms of national levels of synthetic nitrogen use, the extremes range from zero or small fractions of 1 kg N/ha in many countries of Sahelian and black Africa to an unparalleled 560 kg in the Netherlands. Ranking behind the latter as high users are Ireland and North Korea (over 250 kg each), Belguim (220 kg), West Germany and South Korea (200 kg), Egypt (190 kg), and the United Kingdom (180 kg). European countries (except Finland, Sweden, Poland, Rumania, Yugoslavia, Italy,

Spain, and Portugal), Israel, and a few small Caribbean island nations (Martinique, Saint Lucia, Saint Vincent) have levels over 100 kg.

Approaching the level of 100 kg N/ha are three Central American nations (Costa Rica, Cuba, El Salvador) and, with the exception of Portugal, Spain, and Yugoslavia, all European nations listed in the preceding paragraph. American and Soviet levels (55 and 33 kg) are still those of extensive agroecosystems relying on large areas rather than on high yields (however, American corn is highly fertilized). And even lower levels of fertilization prevail in three of the world's high-ranking food and feed grain exporters: Canada (20 kg), Australia (5 kg) and Argentina (below 2 kg). Most African national totals are just around 1 or 2 kg.

Application levels are thus extremely uneven: of the approximately 150 nations and territories for which statistics are available, 50% use less than 20 kg N/ha per year, 25% use less than 60 kg, and only about 15% use in excess of 100 kg. About the high correlation between the levels of nitrogenous fertilization and agricultural production there can be no doubt (with the notable exception of the land-rich countries mentioned above). Nor should there be any doubt that future applications will rise to at least match the population growth; while saturation levels for some crops in some countries may have been approached or even reached, most Asian, Latin American, and African crops yield much below their potential and can absorb multiples of the current nitrogen applications.

And even complete fertilization of all common farm crops would not exhaust the potential demand for nitrogenous fertilizers. After decades of extensive harvesting practices, forestry in both temperate and tropical countries is moving toward more intensive management methods, ranging from aerial fertilization of natural stands to short-rotation silviculture very much akin to field cropping. As a result, use of fertilizers in forestry has been spreading.

Fertilizing of natural stands has been done at relatively low application rates: because of their economic importance, coniferous species in the Pacific Northwest, North America's leading producing region, were the first ones to receive aerial urea fertilization. These applications started in the early 1960s and after temporary uncertainty and slowdown in 1973–1974 the practice resumed with a rapid expansion; by the late 1970s, more than 500,000 ha of young to middle-aged stands (mostly Douglas fir and naked fir hemlock) in Washington and Oregon was fertilized with urea (Bengtson, 1979). During the 1980s, about 200,000 ha has been fertilized annually, the application rates being between 150 and 225 kg N/ha 5 to 15 years before harvesting.

The spreading practice of short-rotation, intensive tree plantation, so far composed largely of fast-growing pines for pulping, requires much more intensive fertilizing. Shoulders and Wittver (1979) have shown that harvesting of two crops in a decade would require applications of up to 700 kg N/ha. Clearly, this represents a new market of huge potential for nitrogenous fertilizers. And increasingly, these fertilizers have come to be supplied as just a few major compounds. But future applications will require a wide variety of nitrogenous compounds.

3.2.3.2. Major Fertilizer Compounds As data reliably disaggregated on a global scale are lacking, I shall use as a quantitative example United States trends accurately documented by USDA statistics.

In the early 1950s, ammonium nitrate represented the largest single source of all nitrogen used (about 20%); anhydrous ammonia supplied only about 10%, the same share as ammonium sulfate and sodium nitrate; urea provided a mere 1% and nitrogen solutions even less. Altogether, only half of all nitrogen came from nitrogenous fertilizers, the other half being incorporated in phosphatic compounds. In contrast, anhydrous ammonia today delivers roughly 50% of all nitrogen used in American crops, and urea and nitrogen solutions each supply about 20%; ammonium sulfate has retained its approximately 10% share.

Globally, the rapid expansion in fertilizer use since the early 1950s ensured substantial increases in the applications of virtually all of the compounds then on the market, but the relative contributions have undergone great shifts and more changes will follow with attrition of old fertilizer plants and opening of new facilities.

The rapid and massive American shift toward anhydrous ammonia has been repeated only in Canada but the trend worldwide toward ammoniacal fertilizers is unmistakable, especially among new producers where urea is the compound of choice. By 1980, about 25% of Chinese nitrogen output was in the form of urea and the shares were much higher elsewhere in Asia: 65% in South Korea, about 75% in both India and Pakistan, 97% in Indonesia. Even Japan, with its traditionally highly diversified production, now puts out about 45% of its nitrogen as urea.

In Europe, Italy and the Netherlands lead the shift with, respectively, 47 and 37% of their nitrogen produced as urea. East and West Germany, France, and Belgium, traditionally large fertilizer producers, continue to turn out a wide assortment of nitrogenous compounds but ammoniacal fertilizers (nitrate and sulfate) are again dominant, as they are in countries such as Mexico and Egypt. Clearly, a more detailed look at today's leading global nitrogen fertilizers should concentrate just on ammoniacal compounds, preferred above all for their high nitrogen content and low cost.

Ammonia (NH_3), a colorless pungent gas containing 82.25% N and 17.75% H, has been derived for over 100 years as a by-product (ammonia liquor) from coking of metallurgical coke with subsequent conversion to ammonium sulfate, more recently also to phosphate. But the direct synthesis from atmospheric nitrogen and from hydrogen derived from hydrocarbons was discovered only at the beginning of this century by Fritz Haber and Karl Bosch, originally to supply Germany with nitrates for explosives to overcome any possible blockade of Chilean nitrate deliveries in the event of war.

The original Haber–Bosch synthesis has undergone many modifications and efficiency improvements during the past seven decades but, in principle, the process follows the same main routes today as originally. For detailed technical descriptions, interested readers should consult, among many others, Markham (1958), Slack (1966), Quartulli (1974), Turner (1974), Greenberg et al. (1977); above all, Slack and James (1973–1979). The latest patented technical innovations and energy-saving processes are reviewed by Brykowski (1982). Here I will present just the essential facts.

The first step in the synthesis is gas preparation, most commonly accomplished by steam reforming of methane (i.e., basically desulfurized natural gas): the resulting mixture of H, CO, CO_2, unconverted CH_4, and excess steam then undergoes secondary reforming by addition of air to reduce the methane and to obtain the 3 : 1 H : N ratio necessary for the following synthesis. Before this synthesis, however, the gas must be purified by catalytic methanation to reduce CO concentrations to below 10 ppm. Afterwards the gas is compressed and catalytically reacted to form ammonia.

Most of the ammonia produced worldwide is used as feedstock for other ammoniacal fertilizers either in integrated plants or after pipeline transport. In the United States, between 75 and 85% of the ammonia synthesized annually is used for fertilizing, the rest going to various chemical industries. Elsewhere, the industrial shares are similar or somewhat lower. As the nitrogen in NH_3 used for subsequent industrial chemical syntheses either enters long-lived compounds which do not disintegrate readily (a wide variety of plastics, above all) or goes into explosives from whose detonation it is rapidly returned to the air, I will ignore this part of anthropogenic nitrogen fixation; it does not burden soils, waters, biota, or atmosphere with any notable nitrogen compound fluxes.

Where anhydrous ammonia is used directly as field fertilizer, special equipment are required for storage, handling, and application of the compound, a complication limiting its use throughout the poor world as well as on small farms in rich nations. Bulk storage at atmospheric pressure requires refrigeration to liquefy the volatilized gas; storage as a liquid must be in high-pressure tanks, which also must be used for field application during which the ammonia is injected into the soil through a series of tubes. But where the infrastructures are in place, distribution and application of NH_3 are no problem: owing to their economies of scale, the largest ammonia plants in the United States now ship their products over an average distance of about 1000 km, and shipping distances of 500–1000 km are common for urea and nitrogen solutions. For smaller farms, custom application obviates the need to purchase the special field and storage equipment.

Ammonia dissolves readily in water forming (exothermically) ammonium hydroxide (NH_4OH), commonly called aqua ammonia. The solution contains 22.2–24.7% nitrogen, is strongly alkaline, and is the base for producing ammonium phosphates, sulfate, and nitrate by reaction with strong acids (H_3PO_4, H_2SO_4, and HNO_3).

Many of the Kellogg or Topsøe ammonia plants recently built around the world feed their product directly to urea synthesis. As noted, urea is gaining an ever larger share of the global nitrogenous fertilizer market, owing to the fact that it has the highest nitrogen content (45%) of any solid fertilizer now used.

In current synthesis, ammonia and CO_2 are reacted to produce ammonium carbamate, which is then dehydrated to yield urea and water:

$$2NH_3 + CO_2 \leftrightarrows NH_2COONH_4$$

$$NH_2COONH_4 \leftrightarrows NH_2CONH_2 + H_2O$$

Depending on the process used, 1 t of urea requires feedstocks of 570–580 kg of NH_3 and 745–760 kg of CO_2, 0.9–2.4 t of steam, 60–160 t of cooling water, and 93–179 kWh of electricity (Payne and Canner, 1969). A 1975 survey of ten United States urea plants showed that in actual operation, the pre-1963 plants producing less than 150 t/day consumed 590 kg of NH_3, 780 kg of CO_2, and 147 kWh per t of product, while the newer plants with capacities of up to 400 t/day required 600 kg of NH_3, 744 kg of CO_2, and 202 kWh per to of urea (Paul and Kilmer, 1977).

Ammonium nitrate (NH_4NO_3) has the second highest nitrogen content (35%) among solid fertilizers and, in spite of its hygroscopicity and resulting susceptibility to caking, it remains a widely used compound worldwide. Manufacturing process is based on reaction of anhydrous ammonia with nitric acid:

$$NH_3 + HNO_3 \rightarrow NH_4NO_3$$

with final concentration to above 99% achievable in falling film evaporators.

Nitrogen solutions comprise a large group of liquids first introduced in the 1930s and greatly expanded along with the growing NH_3 synthesis after 1950. They are usually produced in ammonia, ammonium nitrate, or urea plants by adjusting the concentration of the process liquors. Less frequently, their production is undertaken as an independent process. The nitrogen content of the solutions is usually between 28 and 32%, i.e., lower than ammonium nitrate or urea, but, unlike with solid fertilizers, solutions are easier to incorporate in irrigation water and present no coking or dusting problems. Solutions containing ammonia are pressurized to eliminate volatilization, whereas other liquids are stored and distributed at normal pressure.

The latter solutions are, naturally, the easiest to use and high application rates can be achieved with tractor-drawn booms (more than 40 ha/day) or with planes (40 ha/hr). Repeated testing has found no difference between gases, solids, and solutions as nitrogen sources for farm crops, and with urea and anhydrous ammonia prices being cheaper per unit of nitrogen than other fertilizers, the two compounds will undoubtedly keep gaining in importance. Cost has been, naturally, the key factor in the rapid adoption of all ammoniacal fertilizers and their relative cheapness has had, in turn, much to do with the impressive efficiency improvements introduced since the 1950s, innovations which have considerably reduced the energy needed to produce these compounds. Both costs, financial and energy, will be discussed in the next two sections.

3.2.3.3. ENERGY ANALYSIS During the 1970s, energy analyses of individual crops as well as national agricultural systems became an important part of the overall evaluation of farming's efficiency, a revealing comparison with, and a complement to, traditional economic appraisals. Although farming is not a highly energy-intensive activity as far as its total claim on fossil fuels and electricity is concerned, it uses several high-energy inputs of which none is more important than nitrogenous fertilizers. Pesticide synthesis needs more energy per kilogram of active ingredients but, unlike fertilizers, these products are used in very low densities (fractions of a kilogram to a few kilograms per hectare); and phosphates and potash, the largest sources of P and K, have energy requirements up to an order of magnitude lower than synthetic nitrogen.

Of the total global energy consumption in agriculture (in the United States, nearly 4% of all energy uses), fertilizers account for some 45%, of which nitrogen compounds command about 90%. The high energy requirements of nitrogenous fertilizer production stem not only from the high temperatures and high pressure of the production processes, but also from the fact that natural gas (or, less frequently, crude oil or coal) is also the feedstock to the synthesis. As the bulk of all nitrogen fertilizers are ammonia compounds, a careful appraisal of ammonia's energy cost must be the foundation of all subsequent analyses.

Not surprisingly, the largest amount of detailed information on ammonia's energy cost is available for the United States where natural gas is by far the leading feedstock as well as the fuel. Inputs of natural gas as feedstock are only slightly influenced by the plants' size or age: based on a 1975 ammonia survey of 47 plants, the average consumption per tonne of NH_3 is a mere 1% higher for the smaller plants put into operation in the 1970s while the small (less than 200 t of daily capacity) units built in the 1960s may consume up to 15% more than the large (over 1000 t a day) facilities; the 1975 average for all years and all plants was 27.6 MJ/kg NH_3, the range being 26.4–32.8 MJ/kg (Paul and Kilmer, 1977).

Similarly, gas consumption for heat and power is also fairly uniform, with the older and smaller units actually consuming less than the newest and largest ones: the weighted average is 17.5 MJ/kg NH_3 (range 12.1–21.1 MJ/kg). Combined consumption of natural gas thus ranges from roughly 39 to 49 MJ/kg (plants with the highest feedstock use are not concurrently the highest users of gas for heat and power, hence the top value is only 49 rather than 54 MJ/kg as it would be when adding the two maxima), with feedstock accounting for about two-thirds of the total.

The great consumption differences arise only with regard to electricity use. All pre-1963 plants, as well as newer units producing less than 550 t NH_3/day, have piston compressors operated by electric motors, whereas all larger post-1963 plants use centrifugal compression powered by steam turbines. Consequently, the oldest small plants use up to 700 kWh/t NH_3, the newest a mere 40 kWh, a staggering saving of roughly 95%! Converted to energy units (assuming 33% generating efficiency for thermal electricity), this translates to as much as 7.5 MJ or as little as 0.43 MJ/kg NH_3.

Total production costs are thus about 45 MJ/kg NH_3 for the largest new plants and up to 53 MJ for small pre-1969 units, the weighted average for all plants being 47.4 MJ/kg NH_3; in terms of nitrogen this converts to 55–65 (average 58) MJ/kg.

There is no shortage of other values in the literature: Leach and Slesser (1973), 48.5 MJ/kg N (British experience); Blouin (1974), average for a modern plant with 600–1000 t NH_3/day (based on the TVA performance), 51 MJ (the same value found by Sherff, 1975); Hayes (1976), 47.3 MJ as the United States average; and Mudahar and Hignett (1981), 57.2 MJ as the North American mean for the late 1970s.

Process description for Haldor Topsøe plants published in *Hydrocarbon Processing* in November 1981 (p. 129) states that typical requirements for their new large facilities are merely 29.3 MJ of feed and fuel and a minuscule 0.095 MJ of electricity (just 9 kWh) for each kilogram of NH_3, or a total of only 35.6 MJ/kg N. The range from the best to the worst operating plants is thus about twofold (35–65 MJ/kg N) and a value of 60 MJ/kg N would perhaps be still a slightly optimistic average on a global scale.

Naturally, fertilizers based on NH_3 must cost more. For urea, assuming inputs of 60 g NH_3/kg $(NH_2)_2CO$, an average NH_3 cost of 45 MJ/kg, and 5 MJ of gas and electricity for synthesis and prilling (Paul and Kilmer, 1977), the total is around 32 MJ or roughly 70 MJ/kg N. Similarly, ammonium nitrate will cost at least 72 and as much as 90 MJ/kg N, while the cost of nitrogen solutions will be just a shade above the energy cost of the compounds used for their production.

Improvements in ammonia synthesis have brought down the energy cost exponentially between the 1920s and 1970s, from roughly 400 MJ/kg to just over 30 MJ/kg for the best technology, but this very success will make the future gains much more difficult as the thermodynamic limit for NH_3 synthesis, 17.5 MJ/kg, is not far away. Although the best ammonia plants may not get much better, the national averages should continue to improve. As in many such instances, United States operations have until recently been considerably less efficient than those in Europe and Japan where the feedstock and electricity prices have historically been much higher (up to three times during the 1960s) and where the energy conservation measures have been evaluated on the basis of a 4–5 year payback, rather than 3 years or less as in the United States.

Whatever the actual rate and extent of new-term improvements, two conclusions are very safe to make. First, the Haber–Bosch process will remain the foundation of the nitrogen fertilizer industry for decades to come as none of the long-term alternatives (discussed in Section 3.2.3.5) can be commercialized on a significant scale within one or two decades. Second, the costs of NH_3 synthesis, so closely tied to prices of fossil fuels, will not return to their low pre-1973 levels, a fact of major consequences for farming in both rich and poor nations. But the technological improvements will help in keeping the money cost of nitrogenous fertilizers at acceptable levels.

3.2.3.4. Financial Costs The continuous decline of fertilizer prices during the two decades preceding the early 1970s had been a key factor in the rapid diffusion of nitrogen applications. With prices of feedstock and process energy declining or, at worst, steady in real terms, the productions costs were pointed downward even without major engineering changes but their decrease was potentiated by technological innovations. The most important of these changes, the switch to centrifugal compressors, forced both the large size, as the turbine-driven compressors are not economically viable in plants producing less than 600 t NH_3/day, and higher load factors, as the efficiency of these new large plants drops rapidly with production rates below 70% of the installed capacity (Paul et al., 1977).

Significant economies of scale resulted from these developments. Detailed comparisons of 400, 600, and 1000 t NH_3/day plants show that the total investment and capital costs drop nearly 12% per t of installed capacity with the first step-up and then decrease by 15% with the second (Paul et al., 1977). Similarly, for larger urea plants, total investment and capital costs go down almost 19% with doubling of the facility size from 300 to 600 t NH_3/day and then another 8% by 1000 t/day.

Variable and fixed operation costs in ammonia plants decline 18% upon shifting plant output from 400 t to 600 t NH_3/day, and a further 9% by 1000 t/day (Paul et al., 1977). Savings on electricity by eliminating piston compressors are impressive (95%), as are the lower labor costs: each plant requires just five operators per shift so that the biggest facility saves 60% on salaries per tonne of NH_3.

As a result, anhydrous ammonia prices in the United States began an impressively rapid decline after 1962; when they reached their bottom in 1970, they were, even in terms of current prices, nearly 50% lower than at the beginning of the 1960s. After 1970, prices adjusted for inflation remained virtually identical for the next 3 years, a stability followed by an instantaneous rise in 1974 caused by OPEC's crude oil price increases mirrored by higher costs of natural gas. Even when adjusted for inflation, the 1974 price was 1.8 times above the 1973 level, and the 1975 price in constant dollars rose again by about 30% over 1974. Since then, United States ammonia prices continued this upward trend with the highest rise coinciding with the second period of rapid hydrocarbon price increases in 1979–1980 and with much more moderate increases (even temporary declines) before and after—but in constant monies, ammonia in the early 1980s is still cheaper than in the early 1960s!

While there appears to be no parallel to ammonia's rapid price decline in the United States in the 1960s, other fertilizers—in the United States, western Europe, Japan, and also many poor nations—were selling for less in the early 1970s than in the early 1950s. For example, USDA's annual surveys of prices paid by American farmers show that ammonium sulfate cost about 20% less in current, and 40% less in constant dollars in 1970 than in 1950—while prices of farm machinery, the other rapidly rising input during that period, grew appreciably even in real terms. And FAO's annual *Fertilizer Reviews* show either slow but continuous price declines or steady prices for a variety of nitrogenous compounds in countries as diverse as Japan and Egypt.

And, again, the reversal after 1973 was universal although the increases in prices actually paid by farmers differed substantially: between 1972–1973 and 1977–1978 when the prices generally, and temporarily, settled to new highs, Philippine farmers had to pay 3.5 times more for their nitrogen whereas Indian farmers spent only 55% more (FAO 1981b). These differences reflect, above all, various subsidies given by the governments of at least half (and more likely almost two-thirds) of all Asian, African, and Latin American countries to raise the demand for fertilizer.

But even generous subsidies cannot eliminate many inherent deficiencies or disadvantages hampering a more extensive use of fertilizers in most poor nations. Most importantly, infrastructural deficiencies are widespread and affect the entire delivery and application chain. To begin with, fertilizers are too often packaged shoddily with many bags tearing and spilling during transportation and storage (losses of 10–15% are common), and rebagged fertilizer may be adulterated or underweight. Shipping the fertilizers into farming areas remote from the plants so that numerous transshipments are necessary increases the likelihood of losses and/or delays so that the chemicals may arrive after planting time.

As for application, only the larger land-owners and state farms have the proper spreading equipment and thus nearly all of the fertilizer must be broadcast by hand, usually with no subsequent incorporation and hence higher potential for losses. Extension services are either nonexistent or inadequate so that the applications are too often untimely and improper. Still, the main problem throughout the poor world is the farmer's poverty and his resulting much less than optimum use of fertilizers ensures a limited income, a vicious circle so commonly encountered in developing economies.

In contrast, in the world's most productive farming region—the United States Corn Belt—restrictions on the use of chemical fertilizers were seriously contemplated during the late 1970s in the wake of concerns about leaching of nitrates into the region's water (see Sections 3.4.1.4, 3.4.1.5, and 3.4.3.2). Obviously, such restrictions would have had serious economic consequences.

Clearly, no matter if the problem is inadequate use of synthetic nitrogen or its possibly harmful overapplication, both demonstrate the critical dependence of global food production on continuous use of man-made fertilizers, a dependence whose future options will be examined in the next section.

3.2.3.5. LONG-TERM OPTIONS As noted previously (Section 3.2.3.3), no substantial changes in the currently dominant Haber–Bosch synthesis based overwhelmingly on natural gas as the feedstock (with naphtha supplying the small remainder) can be expected at least for the remainder of this century. Even should hydrocarbon prices resume the upward spiral of the 1970s, the inertia of the existing technology, the very high cost of plant conversion, and the unavailability of cheaper-priced coal in many locations would militate against any rapid change. And as hydrocarbons used as feedstock and fuel in nitrogen fertilizer plants are only such a small fraction of the total gaseous and liquid energy use and as their product is so essential for the survival of modern societies, there is little doubt that priority allocation, rationing, and government subsidies would help the industry to cope even with temporary supply shortages or rapid price fluctuations of essential feedstocks.

Coming up with alternatives is thus a very long-term exercise. When we do have to face the necessity of change, no sooner than sometime in the first two decades of the next century, the conditions, perceptions, and abilities will be much unlike today: it helps to remember that this is equivalent to having to forecast the performance and needs of today's nitrogen fertilizer industry in the 1940s or early 1950s. Hence, the only sensible thing which can be done now is to outline the options for long-term strategies of nitrogen supply as we currently see them and to offer critical comments on their advantages and drawbacks. No quantitative or temporal forecasts will be given here, just probabilistic qualitative appraisals.

There appear to be four technological alternatives, all being methods which need no fundamental scientific breakthroughs but whose successful diffusion on a national and global scale is currently precluded above all by their prohibitive costs. The least promising of the four is thermochemical dissociation of water to produce feedstock hydrogen. At present, there are no commercial processes for the thermal conversion of water. Direct single-step splitting requires a temperature in excess of 2500°C for good dissociation yields: costs aside, we have neither a reliable source of such heat nor adequate process equipment.

Consequently, thermal decomposition must rely on sequential chemical reactions, and many of these multistep methods (initiated with reactions involving compounds such as Fe_3O_4, CdO, $CrCl_2$, HBr, HCl, CuCl) have been tested in laboratory. However, unless they proceed at temperatures in excess of 600°C, their efficiency is no higher than that of water electrolysis (including the cost of thermal electricity generation), or about 25–30% (Gregory and Pangborn, 1976). Some cycles working with temperatures over 800°C can reach efficiencies of 45–60% but no realistic cost estimates are possible in extrapolating from laboratory tests to reliably operating full-size plants.

As the use of nuclear process heat would be required to run these decompositions, the fate of thermochemical hydrogen generation from water is also closely tied to the now rather dim fortunes of nuclear generation. If only for this reason, Gutschick (1977) believes that this approach, even if successfully commercialized, will be limited to a fraction of its potential. Also, some of the cycles use volatile iodine or mercury and require hundreds of kilograms of toxic reactants for each kilogram of water decomposed per cycle, a potential source of both air and water pollution.

Obtaining hydrogen from electrolysis is a technologically much easier proposition. Thousands of electrolytic hydrogen plants, some very large ones consuming over 100 MWe, have been operating successfully for a variety of special industrial applications—but rarely to produce feedstock for nitrogen plants owing to the high capital and operation costs. Only where there are no other feedstock sources and where electricity is very cheap does electrolysis not look to be impractically expensive. Nichols et al. (1980) present data (based on Norsk Hydro experience) showing that the total capital investment in a 900 t NH_3/day plant is 1.5 times higher than in a natural gas reforming facility, but the production cost, at \$(1978) 0.027/kWh, is 2.8 times higher than for ammonia based on natural gas costing \$(1978) 70/1000 m^3.

Attempts to revive the electric arc method on a small scale also have a very long way to go. They were initiated at the Kettering Laboratory in the 1970s with experiments using a 3-kW unit which could produce 1 t of nitrogen per 55,000 kWh (Franda, 1980). But the costs [on the order of $(1980) 5000], relative inefficiency of the operation, and far from negligible maintenance and repair requirement make it most unlikely that such units will be available when they would be most welcome and affordable to those who would most benefit—remote areas endowed with hydropower and settled by poor subsistence farmers (Nepal and Bolivia are two outstanding examples).

Thus, the best chances for any large-scale replacement of hydrocarbon-based synthesis are in switching to coal. The three countries that currently dominate nitrogen fertilizer production with about half of the global output—the United States, USSR, and China—also have the world's largest coal deposits and lead in annual extraction. The greatest advantage is clear: no alterations of Haber–Bosch plants would be needed as the gasification of coal would deliver "pipeline" gas, a synthetic equivalent of the natural one. Future major discoveries of natural gas and the increase of international trade in liquefied natural gas may, together with numerous environmental concerns surrounding massive coal conversion, postpone any such significant developments for decades, but the process is both appealing and much less expensive than thermochemical or electrochemical water dissociation.

Consequently, the choices shrink to the well-known but so far generally shunned process of partial coal oxidation, basically also coal gasification but in small multiple units specifically matched to the ammonia plant's needs. Principal reaction produces CO_2 and H_2 as does the reforming of methane but the balance of the two gases is different so that a better way is to follow the initial gasification by a partial shift conversion with removal of the generated CO_2 to achieve the same gas ratio as in reforming.

Disadvantages of this process compared to natural gas or naphtha reforming are obvious. Limited capacity of individual gasifiers, high steam and heat requirements, necessity to cope with often great variability in coal quality and sulfur content, and higher capital cost for the gasifiers, gas purification units, and liquid air plant make coal-based NH_3 facilities significantly more expensive than their hydrocarbon-based counterparts. In their conceptual designs and cost estimates standardized for a 900 t/day ammonia plant newly built in 1978, Nichols et al. (1980) found the same total investment difference (1.9 times more expensive) between a partial-oxidation coal-based plant and natural gas reforming but the shift in feedstock and fuel prices to $(1978) 70 per 1000 m^3 of gas and $29.70 per t of coal translated into a production cost just 1.23 times higher. Moreover, at that level, ammonia from coal was also marginally cheaper than ammonia from naphtha or from partial oxidation of fuel oil, although in 1978 the international crude oil price was still only $(1978) 13/barrel.

Another 5 years later, the basic capital cost relationship remains unchanged, but continuing fossil fuel price changes have pushed the total production cost even further in the direction clearly discernible by 1978. In 1983, average cost estimates had to be computed with up to $(1983) 128 per 1000 m^3 of gas (United States ceiling price for new discoveries), around $40 per t of coal, and, of course, $28 per barrel of crude oil. With these costs, crude oil-based processes are most expensive and the production cost gap between natural gas reforming and partial coal oxidation almost disappears. This still does not signal any rush to coal-based plants but it makes for a better long-term outlook for ammonia synthesis. And this outlook will be shaped not only by the availability and prices of feedstock, technological innovation, and demand for nitrogen fertilizers, but also by a multitude of choices, changes, and adjustments made in an effort to increase the efficiency of fertilizer use. The best possible knowledge of crop response is an obvious prerequisite for such better management.

3.2.4. Crop Response

Raising the yields is the obvious goal of fertilization and hence it is not surprising that a truly enormous number of studies have investigated the response of many crops to applications of various levels of macronutrients in the form of dozens of different direct and mixed fertilizers. Still, there are a great many things we do not know very well. Gutschick (1981) lists 11 puzzling plant responses to nitrogen, including nearly general counteraction of fertilization on maturation; frequently documented repression of fixation by fertilization of grain legumes; the plants' preference for nitrate over ammonia although the latter is an almost obligatory decay intermediate and its oxidation presents a considerable expense of energy; and reduction or complete exclusion of legumes and other fixers from temperate climax ecosystems. He also offers plausible explanations for most of the puzzles but in many cases satisfactory explanations remain elusive.

Extensive reviews and summaries of what we have learned are available in FAO publications (FAO, 1958, 1966a, 1981b) and only a brief survey of major findings from these and other sources can be presented here by discussing the removal of nitrogen by different crops, the recommended fertilization levels, and the yield response functions.

3.2.4.1. Nitrogen Requirements of Crops Although plants may take up nitrogen as urea and nitrite, these two flows are dwarfed by the uptake of ammonium and nitrate ions, with the latter dominating even in legumes which fix symbiotically only about 25–30% of their nitrogen needs directly as amino acids not affected by nitrification. The study of ammonium and nitrate nutrition of plants has involved many branches of agronomy, plant and soil science, and biochemistry and extensive reveiws can be found, among so many others, in Kirkby (1970), Beevers (1976), Nielsen and MacDonald (1978), Haynes and Goh (1978), Hageman (1980), and Stevenson (1982).

The quantities of nitrogen removed by harvests of annual and permanent crops are, obviously, not a matter of easy generalizations: their variations will be determined by yields which, in addition to the effects of fertilization, are a complex function of rainfall (its total, distribution, intensity), energy characteristics of the growing season (its length, available heat, soil temperatures, last and first frost dates), soils (depth, nature of root-restricting layer, texture, structure, mineral and organic content, drainage), previous production history (years under cultivation, preceding crops), production practices (planting densities, weed and pest control), and the cultivar's potential. Agronomists are still trying to optimize the path through this thicket of largely uncontrollable and unpredictable factors but scientific assurance will always be hard to come by.

Greenwood (1976), in his extensive discussion of nitrogen stress in plants, shows that it is not feasible to measure the nutrient's deficiency in a precise way similar to water stress and that we have to rely on qualitative evaluation, a situation he finds to be "an extraordinary weakness within the discipline of plant nutrition." He proposes five plant parameters—leaf nitrogen, dry weight, leaf elongation, leaf area, and carbon dioxide exchange rate—to be used singly or in any combination, depending on the objective and expertise available, to estimate the nitrogen stress.

A farmer who would take such elaborate steps with each crop would be rare indeed. The best he can do is to set realistic yield targets on the basis of the planted variety's known potential performance in given soil under standard management practices (i.e., assuming that all controllable factors will be nearly optimal) and to fertilize accordingly. For example, on a large Illinois corn farm, this would mean applying 150–170 kg N/ha; however, in a dry year the optimum would be below 120 kg, in a wet year the crop would profit from nearly 200 kg (Figure 3-6): overfertilization or underfertilization in expectation of average weather translates into inevitable losses.

However, the two kinds of losses differ. While there is no way to recover the yield lost from underfertilization in a favorable year, overfertilization in a dry year will result in partial carryover of unused nitrogen for the following crop. Consequently, in the Corn Belt, or in other rich agricultural regions where the expenses for nitrogen are but a part of the total farming cost (though they may easily be as much as half of all variable expenditures), the farmers are not going to jeopardize their return on other production inputs by intentional underfertilizing. On the other hand, for a small farmer in a poor country, fertilizer may be by far the single largest expenditure and he will have to weigh the returns differently. Aldrich (1980) put it well: "In summary, there appears to be no feasible substitute for the judgment of individual farmers based upon the best information they can get from agricultural scientists and their fertilizer suppliers."

Among the principal crops widely cultivated for food, feed, or fiber, high-yielding corn hybrids in the United States, short-stalked European wheats, and eastern Asian *japonica* rices remove the highest quantities of nitrogen—in excess of 100 kg/ha—and hence must be supplied with the largest amount of the nutrient. For example, nitrogen guidelines for the Corn Belt based on the optima from experimental plots call for 170–225 kg N/ha and actual applications in central Corn Belt states now average nearly 170 kg N/ha (Aldrich, 1980). Möhr and Dickinson (1979) recommend 135 kg N in light soils, 149 kg in medium, and 170 kg N in heavy soils for a yield target of 8000 kg/ha.

Table 3-6 lists nitrogen removal values for more than a dozen of the world's major crops: as these values reported in various agricultural publications come

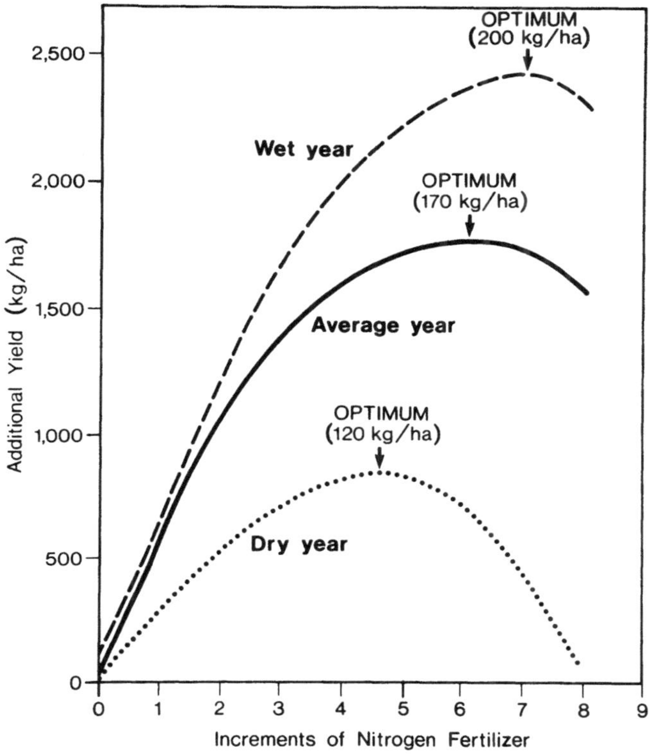

FIGURE 3-6. Nitrogen fertilization optima of central Corn Belt grain corn as a function of mean annual rainfall (after Aldrich, 1980).

mostly from experimental plots and hence pertain to yields which are significantly higher than most national averages and, of course, the global mean, they must be prorated to lower harvests to derive the annual worldwide total of nitrogen removed from the fields. Needless to say, such an estimate can only be a rough approximation but a very useful one for comparison with the total nitrogen mass applied in synthetic fertilizers.

All the relevant figures are listed in Table 3-7 for 14 principal grain, tuber, sugar, and fiber crops grown annually on some 850 million ha; annual nitrogen removal for these crops totals roughly 40 million t but this sum must be augmented by the removals relating to legumes, vegetables, and permanent crops. In the case of legumes, it is not difficult to make an approximate calculation of nitrogen incorporated in the harvest but it is much harder to decide what portion of this withdrawal is not covered by *Rhizobium* fixation. Assuming that 10% of the nitrogen in soybeans, peanuts, and pulses comes from the soil, the total would increase by at least 1 million t, and the removals by leguminous forage would add another 1

TABLE 3-6
NITROGEN REMOVED BY MAJOR CROPS AND THEIR RESIDUES[a]

Crop	Yield (t/ha)	Nitrogen removal: Harvested phytomass	Residues (kg N/ha)	Total
Rice	2.5	55	15	70
	4.0	90	25	115
Wheat	2.5	45	20	65
	5.0	80	40	120
Corn	4.0	60	35	95
	7.0	130	70	200
Barley	2.5	40	15	55
Oats	1.8	35	10	45
Sorghum	1.0	20	5	25
Potato	20.0	120	10	130
Sweet potato	20.0	80	—	—
Cassava	9.0	35	30	65
Soybeans	1.7	140	—	—
Peanuts	1.5	100	—	—
Sugarcane	55.0	80	—	80
Sugar beets	25.0	145	—	—
Cabbage	30.0	100	—	100
Onions	20.0	40	—	40
Apples	20.0	30	—	30
Oranges	40.0	100	—	100
Bananas	10.0	20	20	40

[a]Sources: FAO (1958, 1978, 1981b), Sanchez (1976), and Smil (1983).

million t. Vegetables, fruits, nuts, coffee, and tea, most likely do not remove in excess of 3 million t.

Total removal of nitrogen in global agriculture thus appears to be roughly 45 million t annually or, considering the uncertainties inherent in estimates of this kind ($\pm$ 10% as the minimum), most likely somewhere between 40 and 50 million t, or about 15–30% less than the total applied annually in nitrogenous fertilizers. However, this comparison is no indicator of more than sufficient replacement of the removed macronutrient as there are considerable losses of the applied nitrogen through volatilization, leaching, and runoff which not infrequently surpass half of the total (for details see Section 3.4.1); moreover, as discussed earlier (Section 3.2.3.1), nitrogen applications in most of the world's nations are extremely low. So in spite of the seemingly positive global application/removal ratio, there is little doubt that for most of the world's crops, fertilizer use is too low and that this inadequacy is especially worrisome in view of the long-term maintenance of soil fertility needed to support future yields for larger populations.

TABLE 3-7
AGGREGATE NITROGEN REMOVAL BY THE WORLD'S MAJOR FIELD CROPS AROUND 1980[a]

Crop	Harvested area (10^6 ha)	Average yield (t/ha)	Nitrogen removal (kg/ha)	Total nitrogen removal (10^6 t)
Wheat	230	1.9	45	10.4
Rice	145	2.6	60	8.7
Corn	120	3.0	60	7.2
Barley	95	2.0	45	4.3
Millet	55	0.7	25	1.4
Sorghum	50	1.3	40	2.0
Cotton	35	1.3	35	1.2
Oats	30	1.9	25	0.8
Potatoes	20	15.0	65	1.3
Rye	15	2.0	60	0.9
Sugarcane	14	55.0	75	1.0
Cassava	13	9.0	35	0.5
Sweet potatoes	12	9.0	35	0.4
Sugar beets	9	30.0	60	0.5
Total	843	—	—	40.6

[a]Sources: harvested area and average yield according to the FAO's *Production Yearbooks*; nitrogen removals adjusted from data in Table 3-6.

Indeed, without the greatly increased applications of nitrogenous fertilizers since the 1950s, it would have been impossible to bring in the higher harvests which still have not eliminated widespread malnutrition but without which the frequency and extent of regional or national famines would have been much higher. Crop response to higher nitrogen applications thus remains a critical consideration.

3.2.4.2. RESPONSE TO FERTILIZATION While most of the varieties of commonly cultivated staple food crops will respond to increased nitrogen applications, the rates and extents of these responses vary appreciably. By far the greatest differences can be seen between the responses of the traditional long-stemmed varieties of wheat and rice and the new short-strawed cultivars introduced during the past three decades largely through the efforts of the Centro International de Mejoramiento de Maiz y Trigo (CIMMYT; International Maize and Wheat Improvement Center) in Mexico and the International Rice Research Institute (IRRI) in the Philippines. Introduction of these high-yielding varieties highly responsive to nitrogen extended intensive staple cereal cropping, previously the domain of temperate latitude farming, to subtropical and tropical countries in one of the most significant and most rapid transformations of cropping practices in history.

Dramatic illustration of the high and sustained response of new varieties is presented in Figure 3-7 comparing an IRRI-bred high-yielding cultivar with a local

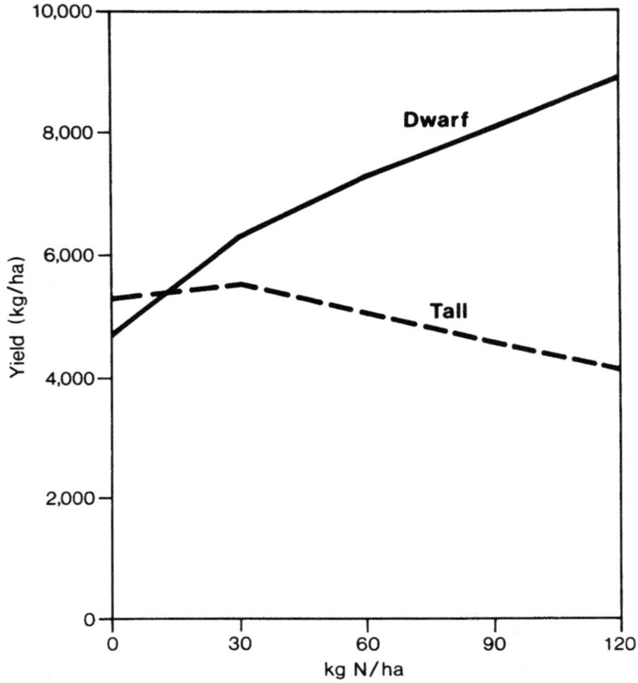

FIGURE 3-7. Differences between response of a traditional Philippine rice variety and a modern short-stalked cultivar to nitrogen fertilization (FAO, 1981b).

Philippine variety. Not all responses are as impressive, but of nitrogen's key role in rice production there can be no doubt. Higher phosphorus applications are needed with increased nitrogen fertilization of high-yielding varieties but the yield response to phosphorus is both more erratic and less universal than to nitrogen, and the response to potassium is even less significant. Doyle's (1966) survey of data on 385 experiments in 20 major rice-growing countries showed that the crop, especially its short-stemmed *japonica* varieties, responds to nitrogen on all soils except those previously supplied with the element by legumes and that on a worldwide basis, a linear relationship between fertilization and yields applies to the whole range of application rates from 10 to over 200 kg N/ha.

However, most of the data in this set are for applications between 20 and 80 kg N/ha and using just these it is evident that some leveling-off sets in at around 40 or 50 kg N/ha. Another examination of more than 400 response figures (FAO, 1966b) showed the same average gain as Doyle's (1966) study—12–13 kg of unmilled rice per kg of N—but put the onset of leveling-off at between 30 and 40 kg N/ha. Returns in most trials of the past two decades clustered around 12 kg of grain per kg of N but considerable departures from this typical value have been common, with many lower values clearly indicating not only performance plateaus but also significant yield declines with higher nitrogen use.

For example, the FAO tests in Bangladesh in 1976 found returns of 18.7 with 75 kg N applied to summer (*aus*) crop while in 1346 Indian experiments between 1971 and 1974, returns with 120 kg N applied to *kharif* crop were only 6.4, and for *rabi* crop (1974 trials) even lower at 5.5 (FAO, 1981b). Similarly, recent information from China shows an unmistakable decrease of yield gains with increasing application of chemical fertilizers in many intensively farmed and relatively heavily fertilized regions. Yet when other environmental factors influencing the yield are well controlled, response has continued in some Japanese and American experiments all the way up to applications of 200 kg N/ha.

Another interesting, indeed a more revealing, way to look at the response of rice to nitrogen is to compare yield increases achieved in different locations or countries after application of the same amount of fertilizer, preferably just the amounts recommended as a general practice. Perhaps the most extensive set of this kind consists of 11,202 experiments from 12 Indian states where applications at the recommended rate of 22.4 kg N/ha resulted in yield increments averaging 10.84 kg of grain per kg of N, the range being 5.5–13.0 kg (FAO, 1966a).

Wheat varieties respond to nitrogen in the same general way as do the rice crops: Figure 3-8 shows the classic response curves for an unimproved tall Indian cultivar and for Sonora 64, a semidwarf wheat developed by the CIMMYT. With the traditional crop, the response plateau is reached at around 100 kg N/ha, while the high-yielding strain continues to respond to an additional 70 kg N. With adequate moisture, semidwarfs will commonly produce 15–20 kg, even up to 30 kg, of additional grain for each kilogram of nitrogen up to the first 50–70 kg N, while the

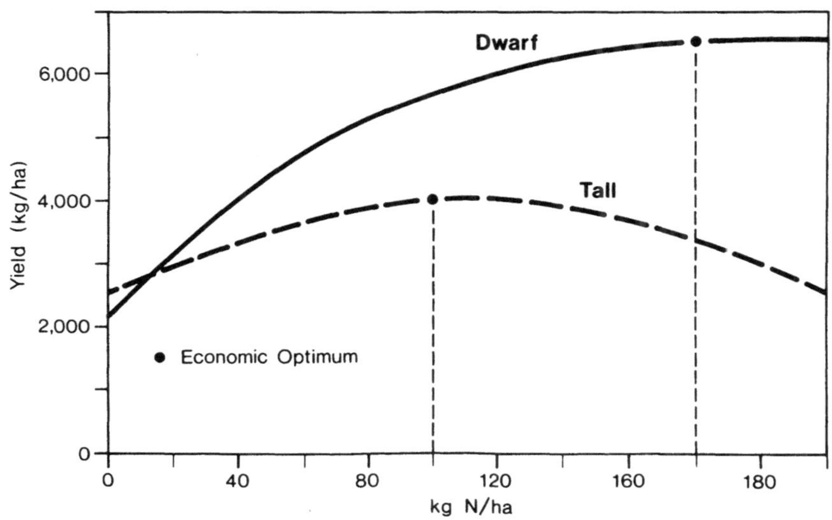

FIGURE 3-8. Responses of traditional Indian tall and modern Mexican dwarf wheat varieties to nitrogen application (FAO, 1981b).

tall varieties will respond to the same fertilization with only 8–12 kg of additional grain (Hanson et al., 1982). Above 50–70 kg of applied nitrogen, the response of semidwarfs slows down considerably (to as little as 5 kg of additional grain above 100 kg N) but it continues, as with some rices, up to 200 kg.

Hybrid corn varieties respond to nitrogen fertilizer more vigorously than any other major staple grain crop as is best illustrated by the success of American grain corn farming. In 1950, each hectare of harvested corn received just 8 kg N and yielded 2400 kg of grain; three decades later, the average hectare was fertilized with 140 kg N and yielded 6200 kg, an increase resulting in a return of nearly 30 kg of grain for each additional kilogram of N (Smil et al., 1982). In contrast, the response of sugar crops is a more complicated affair. Sugarcane almost always produces more phytomass with fertilization but sugar percentage drops with increased applications so that a trade-off must be sought especially as excess nitrogen also lowers drought resistance (Williams, 1975). Another complicating consideration is the tremendous varietal difference in response to nitrogen which makes leaf nitrogen tests useless as guides for fertilization unless they are designed for a specific cultivar (Nickell, 1977).

In sugar beets, nitrogen has by far the greatest influence on root quality and sucrose production although the plants must be deficient in nitrogen about 4 weeks before harvest to retard the utilization of sucrose for growth and hence increase its accumulation in roots. Moreover, the plant appears to be unusually variable in its response to nitrogen. In long-term field trials conducted in the 1970s in California, the amounts of nitrogen required to produce maximum sugar yield varied from none to 270 kg/ha, with the most common optima between 70 and 200 kg/ha (Hills et al., 1978).

Similarly, cassava, the staple food for hundreds of millions of poor people in Latin America, Africa, and Asia, and a crop with high nutrient requirements, has an extremely variable response (Williams, 1975; Onwueme, 1978). On poor sandy soils, nitrogen's effects are small; in the presence of adequate potassium, cassava responds well to moderate applications of nitrogen; and excess nitrogen brings luxuriant shoot growth at the expense of roots and tubers.

Before closing this section, brief discussion must be made on the general contribution of nitrogen fertilizers to rising crop yields. The agronomical literature has repeatedly stressed the essentiality of the complete modernizing package: new high-yielding crop varieties supplied with sufficient water and adequate pest control and cultivated in improved ways are as essential as higher fertilizer use and to isolate nutrient contribution as the causative factor of better harvests is thus quite difficult, if not impossible. Still, such attempts have been made at least for all macronutrients combined (N + P + K) and between 25 and 40% of the long-term crop harvest increase has been ascribed to their rising use (FAO, 1981b).

In a more dramatic perspective, this contribution might be assessed as White-Stevens (1977) did: "To deny American farmers the use of artificial fertilizers . . . would require the removal of 100 million Americans, some of whom, in a democracy it would seem, are likely to vote against the idea. It would also become necessary to open up some 250 million acres of new lands, pressing wildlife into further extinction; to return some 20 million nonrural people back to the toil of the soil and stoop labor; to reduce the present standard living by at least 50% and to retreat to the way of life a century ago."

One may quibble about the figures but not about the essence: fertilization must be extended globally. Even the Chinese, the world's most prolific recyclers of organic wastes, are now relatively as dependent on synthetic fertilizer (i.e., mostly on nitrogen) as the Americans are. Even the most extensive recycling of all organic wastes conceivable spells massive hunger with current populations. Nitrogen fertilization appears to be the least replaceable of major anthropogenic interferences with a key biospheric cycle. Its effects on ecosystems and people thus deserve careful attention. But before I turn to these problems, I must introduce the other great flux of anthropogenic nitrogen—oxides from the combustion of fuels.

3.3. NITROGEN FROM COMBUSTION

Although fertilization is undoubtedly the most widespread and potentially the most worrisome human interference with biospheric flows of nitrogen, the emissions from combustion of fossil fuels—while much smaller in global terms than the applications, in fact smaller than the losses, of nitrogenous fertilizers—are dominant in many industrialized and urbanized regions and their influence may extend, largely through nitrates formed during complex atmospheric reactions, for hundreds of kilometers downwind and may be of considerable importance in "acid rain" phenomena (analyzed in detail in the next chapter on sulfur flows).

Estimates for NO_x emissions from both stationary and mobile fossil fuel-fired sources will be presented first, followed by an appraisal of releases from biomass combustion and from other minor sources not accounted for in any previously discussed categories. Next, a review of emission trends and background and urban concentrations will precede a brief outline of major atmospheric reactions and removal processes involving the oxides of nitrogen.

3.3.1. Sources

As noted in the discussion of nitrogen reservoirs (Section 3.1.1.1), all fossil fuels contain minor amounts of nitrogen; with the exception of natural gas, this nitrogen is not removed before the combustion and can be oxidized and enter the atmosphere as NO and NO_2. And with higher combustion temperatures required to sustain better conversion efficiencies, increasing volumes of the atmospheric dinitrogen are also oxidized increasing the total NO_x emission rates.

One of the most notable final fuel use shifts of the past four decades has been the steady growth in the share of fossil fuels (above all, coal) burned at high temperatures to generate steam in large power stations. Together with a great variety of smaller stationary sources (mainly industrial boilers), this combustion is the leading, and increasing, source of NO_x emissions.

Contributions from mobile sources have also expanded rapidly since 1950 and although the ownership of passenger cars will not increase substantially throughout the OECD nations, the demand for cars is still far from saturated in the richer Communist countries and among the elite of the poor nations where truck and bus numbers will also grow considerably. NO_x emissions from all of these sources will be estimated in the following three sections together with the until recently neglected but surprisingly large flux from combustion of biomass.

3.3.1.1. STATIONARY FOSSIL FUEL COMBUSTION Emissions of nitrogen oxides from stationary combustion sources have a twofold origin—in oxidation of nitrogen present in the fuel (see Section 3.1.1.1) and in oxidation of atmospheric dinitrogen. No single generalization can be made regarding the typical shares of these two processes. In power plant boilers burning heavy oil or coal, oxidation of fuel nitrogen can be a significant contributor, but in all high-temperature combustion processes the combination of atmospheric molecular nitrogen and oxygen is the dominant source of nitric oxide owing to two reversible reactions:

$$O + N_2 \rightleftarrows NO + N$$

$$O_2 + N \rightleftarrows NO + O$$

As nitric oxide formation is much slower than the combustion process, most of the NO is actually formed after completion of combustion (NAS, 1977a). NO is thermodynamically unstable at lower temperatures but it decomposes so slowly that its concentrations remain almost constant long after the effluent gases have cooled. Critical variables increasing NO formation rates are excess air present in the furnace combustion zone and in the postcombustion gases, flame temperature, and residence time of the fuel and air at the combustion temperature. During the cooling of combustion gases, a small amount of NO is oxidized to NO_2 (for most combustion equipment the rate of emitted NO/NO_2 is 95 : 5) and, as noted earlier, these emissions are commonly referred to in the air pollution literature as NO_x.

The most practical way to estimate the releases of NO_x or other gases or particulates on large—global or national—scales is to use standard air pollutant emission factors. Naturally, these factors are average estimates of the rates at which individual pollutants are released per unit weight of a fuel (or a product) and as such there are no precise indicators of emissions for a single source even in those cases when their means have been established on the basis of numerous measurements and are considered highly representative. Often, only minor operation differences can result in relatively large differences in NO_x emissions even from identical sources. Still, when properly applied (i.e., in large-scale inventories), the emission factors allow rapid appraisals of the total pollutant burdens and where they are known with excellent or above-average degree of accuracy—as is the case with fossil fuel combustion in stationary sources—the final value can be seen as highly reliable.

At the beginning of the 1980s, some 3.7 billion t of coal was consumed worldwide with 500 million t used for coking, 100 million t in cement production, and the rest burned in power plant, industrial, and commercial boilers and household stoves (see Section 2.2.1). No reliable global data are available to break down coal combustion into these categories required for applying the appropriate emission factors. However, good approximations can be made on the basis of available OECD, Soviet, and Chinese materials.

OECD countries now burn about two-thirds of their coal consumption in power plants, just over 10% in industrial and commercial boilers, and a mere 5% for household heating (OECD, 1982). For the USSR, these shares are, in the same order, roughly 37, 20, and 20%, and for China, 25, 35, and 20% (Wu, 1980; Tsentralnoye Statisticheskoye Upravleniye, 1982). In all cases the remainder is used mostly in coking.

Applying these shares to coal consumption of the early 1980s means that in the above countries approximately 50% of all coal went for electricity generation, 20% for industrial and commercial boilers, and 12% for household stoves. As these nations account for over 70% of the total global coal use (and for some 85% of bituminous coal combustion), these shares can be safely extended to represent worldwide usage and, with average emission factors in the three combustor categories taken as 9, 7, and 2 kg NO_x per t of coal, the annual global NO_x generation from coal burning would be nearly 23 million t.

For liquid fuel consumption, breakdowns for the three categories are roughly (all in percent) 12–22–18 for OECD (transportation consumes most of the rest), 50–22–5 for the USSR, and 25–30–2 for China. Again, as these countries consume nearly 80% of the world's liquid fuel, I shall take their combined consumption shares (roughly 20, 23, and 14% for the three categories) to be representative of the global use and apply average emission factors of 12.0, 7.0, and 1.5 kg NO_x per t of liquid fuel. Worldwide NO_x emissions from oil-fired stationary sources in the early 1980s would then be (with total crude oil consumption of 2.7 billion t) about 11.5 million t.

Finally, applying the estimating procedure used for coal and oil to natural gas—OECD, Soviet, and Chinese consumption splits in the three combustion categories being, respectively (all in percent), 18–31–27, 25–30–25, and 20–60–1, the combined weighted percentages of the total natural gas use at roughly 20, 30, and 25% applied to global consumption of about 1.4 trillion m^3 and NO_x emission factors of 9.6, 2.8, and 1.6 kg per 1000 m^3 of gas—yields annual worldwide emissions of almost 4.5 million t. All stationary fossil-fueled combustion sources would then be emitting about 40 million t NO_x/year, with almost 60% coming from coal, 30% from oil, and the rest from natural gas combustion.

If all of these emissions were in the form of nitric oxide, the total nitrogen flux from stationary fossil-fueled combustion would be roughly 19 million t (with some 95% of this total released in the Northern Hemisphere), a contribution much larger than NO_x emissions from mobile sources.

3.3.1.2. Mobile Sources In countries with large numbers of motor vehicles, transportation has become a source of NO_x about as large as stationary combustion after gaining steadily in importance since the end of World War II. While correct stoichiometry eliminates CO emissions and nearly minimizes total hydrocarbon flux, the more efficient combustion with higher flame temperatures results in a near-maximum production of NO_x. Higher combustion efficiencies thus translate into higher NO emissions and emission factors must thus be related to the typical performance of specific car populations rather than to the absolute quantities of fuel burned.

In the United States, nitrogen oxide emissions from new motor vehicles have been regulated together with other automobile pollutants since 1968. The USEPA's (1978) review of the exhaust emission rates for light-duty vehicles and trucks and heavy-duty gasoline and diesel trucks shows impressive declines in a relatively short period of time for new light-duty vehicles (i.e., passenger cars). While the pre-1968 emission factor of 2.24 g NO_x/km actually rose to 2.77 g between 1968 and 1972, it declined to 1.86 g in 1973–1974, 1.51 g in 1975–1976, and to just 0.94 g by the end of the decade.

Shull (1979) reviews in detail the measures, control equipment costs, and fuel economy penalties or gains involved in these emission reductions which led to nearly total controls by 1980 when 92% of all passenger cars in America were manufactured to comply with the Clean Air Act Amendments of 1970. Similar decreases occurred in light-truck emissions but, in contrast, heavy-duty gasoline and diesel trucks had been emitting during the 1970s on the average at least as much and often more NO_x per kilometer than in the 1960s.

Historical review of United States transportation NO_x emissions shows continuous increases from 3.2 million t in 1940 to 11.7 million t in 1970 followed by a decline to 10.7 million t by the mid-1970s and a further drop below 10 million t by the end of the last decade (Cavender et al., 1973; Hunt, 1976; USEPA, 1980a). The most notable changes have been the decline of passenger car contributions (from 90% in 1940 to just over 60% by the late 1970s) and rapid rise of emissions from diesel vehicles (from a negligible share to some 15%) and from off-highway machines (from 6 to some 20%).

But gasoline-fueled passenger cars will remain the largest mobile NO_x source even with stricter controls in the future (although naturally a source of highly varying local importance), with diesel-fueled trucks and buses in a distant second place. Estimates of global NO_x emissions from all of these highway vehicles cannot be made with high accuracy owing to the absence of even rough nationwide average emission factors for most countries. And as the actual emissions can range from just below 1 g NO_x/km for a new passenger car to over 20 g for an old diesel truck, it is not possible to pick a single representative mean.

Perhaps the best approach is by extending the United States values at the beginning of the 1980s. Passenger cars, trucks, and buses registered in the United States accounted for just below 40% of the respective global totals (MVMA, 1980). Straight extrapolation from the United States NO_x emissions of about 8 million t/year would then result in a worldwide flux of 20 million t NO_x annually. Statistics available for OECD nations show fewer vehicle-kilometers outside North America (and these totals are even lower in non-OECD countries) but this may be largely negated, or even outweighed, by the lack of NO_x controls and higher NO_x emissions from diesel vehicles. Consequently, I shall stay with the approximate value of 20 million t NO_x annually which, assuming the total to be expressed as NO_2 (NAS, 1977a), translates into some 6 million t of nitrogen. However, totals differing by up to ±25% cannot be excluded.

Compared to highway vehicles, aircraft are a negligible source of NO_x. For example, in the United States they were estimated to contribute less than 400,000 t NO_x (two-thirds from commercial, one-third from military operations) a year during the 1970s. Only in places with very large airports can this contribution be on the order of a few percent: for example, in the Atlanta area, with what is now the busiest airport in North America, just over 3% of all NO_x comes from the planes (Naugle anf Fox, 1981) and at Chicago's huge O'Hare International Airport emission densities are approximately the same as elsewhere in the urban area (Jordan and Broderick, 1979).

In the absence of the detailed global aircraft data required to use specific emission factors, the best approximation of worldwide NO_x emissions can be derived simply by extrapolating from the United States flux by assuming, on the basis of performance statistics, that the country's aviation accounts roughly for one-third of all flights in the world: the global aircraft NO_x emissions would then be between 1.0 and 1.5 million t annually. Emissions from trains and river, lake, and ocean vessels would, again by extrapolation from the United States estimates, add at least another 1 million t.

The last mobile sources category to be considered is the broad class of off-highway vehicles ranging from snowmobiles and dune buggies to farm and forestry machinery and huge dump trucks used in surface mining. The USEPA's estimates put these nonhighway NO_x emissions at about 2 million t during the 1970s and the global total might then be about 5–6 million t.

3.3.1.3. OTHER NO_x SOURCES Compared to stationary combustion of fossil fuels and transportation, the other sources of NO_x emissions are rather negligible in most industrialized countries: for example, the best available estimates for the current United States contribution are no more than about 5% of the total (see Section 3.3.2.1). Considering the uncertainties involved in estimating any air pollution emissions on a nationwide level, the inevitable error in quantifying the NO_x pollution from the two leading sources in the United States is almost certainly greater than the estimated total for emissions from assorted industrial processes, solid waste disposal, or biomass burning.

Petroleum refining and chemical industries are the two leading sources in this minor category. Applying an average value of 0.2 kg NO_2/1000 liters of fresh feed to the global refinery input of roughly 2.9×10^{12} liters results in worldwide emissions of about 600,000 t; in the United States the total is now 200,000 t NO_2/year. Among common chemical processes, it is the production of nitric acid, now done overwhelmingly by high-pressure catalytic oxidation of ammonia, which is the major source of NO_2. However, the global output of 30 million t HNO_3 would add up to a mere 30,000 t NO_2. From the portland cement kilns, about 1.3 kg NO_x is released per tonne of the product so that the global total is now very close to 1 million t. Incineration of wastes generates about the same amount of nitrogen oxides per tonne of burned material (1.5 kg) but a global emission total is hard to establish. Even liberal allowances could not top 10^5 t NO_x/year. Coking releases NO_x (as well as NH_3) in such small quantities (a mere 0.005–0.015 kg/t) that it may be completely neglected.

So it is the burning of biomass which creates by far the largest NO_x flux among these miscellaneous sources. As was seen in the discussions of biomass and CO_2 generation in the preceding chapter (Section 2.1.2.3), any quantifications associated with burning of crop residues and woody phytomass are highly uncertain but a proper order of magnitude will be offered here with confidence. For establishing an approximate global total, I shall assume that some 500 million t of crop residues (this total includes dung) and 1.9 billion t of wood are burned worldwide (for derivations and discussions of these values see Section 2.1.2.3), that, using a conservative estimate, phytomass has on the average a nitrogen content of only 0.5% and that half of this will be lost in combustion: this chain of assumption leads to an annual release of 6 million t N. Applying the USEPA's average emission factor of 5 kg NO_2 per t of wood and bark to all the burned phytomass would give about 3.6 million t N/year, a close agreement in calculations of this kind.

If one assumes that there are only 20 million families of shifting farmers each burning just 1 ha of forest per year (Myers, 1980), only 20 million ha would be affected annually. As much of the burned forest is a poor secondary growth, one should not put the average standing biomass at more than 20 t/ha; if 75% of it gets burned releasing virtually all bound nitrogen, the annual total would be (with 0.5% N in the phytomass) some 1.5 million t N. Furthermore, assuming that wildfires annually destroy 0.5% of the world's forest area—the share is derived from the United States annual wildfire rate—some 20 million ha would be affected; should each hectare contain on the average 30 t of standing phytomass of which three-quarters would be burned releasing all bound nitrogen as NO_x, the annual global flux of No_x-N from forest fires would be 2.25 million t.

Thus, using far from conservative assumptions results in roughly 6 million t of NO_x-N from household and industrial fuelwood and crop residue combustion, 1.5 million t from shifting agriculture, and over 2 million t from forest fires, roughly 10 million t of NO_x-N each year from all burning of phytomass.

The order of magnitude for total NO_x releases from all sources of phytomass combustion is thus undoubtedly 10^7 t, the actual value being more likely just a couple of tens of millions rather than five tens or more. The rate of 20 million t of NO_x-N would put annual emissions from phytomass burning at roughly the same level as those arising from combustion of fossil fuels and it would make the process a major contributor to locally and regionally elevated ambient NO_x concentrations.

3.3.2. Emissions, Concentrations, Reactions

Discussion of nitrogen oxide sources made clear the impossibility of preparing accurate emission inventories but the available information is sufficient for presentation of a global summary of the right order of magnitude to be compared with NO_x biogenic flux. More accurate totals are, of course, available for the United States and they will be presented. Concentrations arising from these emissions have been measured systematically for several decades in many urban areas and a brief review of commonly encountered levels will follow a few paragraphs on background concentrations.

Reactions involving nitrogen oxides are the centerpiece of the involved photochemistry of polluted atmospheres whose unraveling took about two decades after the initiation of the first Los Angeles studies. Here only the bare essentials will be mentioned before noting our surprisingly inadequate understanding of atmospheric removals of nitrogen oxides, especially their dry deposition.

3.3.2.1. Emissions and Concentrations Global emissions of nitrogen oxides from stationary combustion of fossil fuels (about 20 million t N), transportation (8 million t N), and chemical and other industries (at least 2 million t N) total about 30 million t of nitrogen annually. To this must be added at least two other major anthropogenic sources: domestic and industrial combustion of fuelwood and crop residues (at least 4 million t N) and recurrent burning of vegetation in shifting farming (no less than 1.5 million t N) for an annual man-made flux of around 35 million t NO_x−N. This total should be contrasted with annual biogenic releases of nitrogen oxides put most often at between 20 and 90 million t N/year (see Table 3-3).

Thus, anthropogenic NO_x emissions appear to be rather large in comparison with biogenic production but as the latter values are very uncertain (derived basically just to balance the atmospheric deposition of nitrogen compounds), no proper quantitative comparisons are possible. That the human contributions have been rising is obvious, and the global consumption of fossil fuels and the expanding number of cars worldwide are two simple but appropriate measures of the general trend.

For the United States there are detailed disaggregated estimates of air pollutant emission trends since 1940 and their review is quite interesting (Cavender et al., 1973; Hunt, 1976; USEPA, 1980a). NO_x emissions totaled just over 7 million t in 1940 and rose by 34% during the 1940s and by another 34% during the 1950s; then came the rapid jump of the 1960s—a 61% rise—followed by a slight decline in the 1970s as stationary combustion contributions kept increasing but automotive pollution control spread rapidly. Stationary fuel combustion was a larger contributor until the early 1950s and it regained its primacy during the 1970s owing to accelerated expansion of coal-fired power-generating capacities.

The USEPA's (1980a) projections assign only about 8% of all NO_x emitted by the year 2000 to cars, trucks, and planes; utilities may emit, owing to projected higher uses of coal, over one-third of all emissions; and contributions by industrial combustion could nearly double—accounting for almost 20% of the projected 23.5–26.5 million t NO_x by the year 2000, totals representing an increase of roughly 20–40% compared to the early 1980s.

As for the regional changes, emissions in the Southeast, South Central, Mountain, and Northwest regions are projected to go up faster than the national average, while those in the Northeast should grow at the slowest pace. The current distribution of all major NO_x sources is available as part of the United States National Emissions Data System of the Environmental Protection Agency. This inventory of point sources of air pollution (every stationary source emitting at least 90.7 t/day of any of five primary criteria pollutants) holds roughly 200,000 records which contain information on location and strength of the sources. Because cars and trucks are by far the largest source of NO_x area emissions, their distribution basically reflects population density and the largest cities will dominate the pattern; the distribution of many stationary NO_x sources will be essentially that of large power plants.

Mapping of both stationary and area sources within 80-km grid squares was done for the United States and southern Canada east of the Rocky Mountains by Clark (1980). The highest values within single squares are in Indiana, Ohio, and Pennsylvania, representing the large thermal power plants; the distribution of large point sources of NO_x is thus virtually identical with that of large SO_2 sources shown in Figure 4-3 in the next chapter. However, unlike SO_2 emissions, a very large part of the NO_x flux comes from transportation, basically an area source, whose distribution coincides with the most densely populated parts of the country (Figure 3-9).

The best way to appraise the magnitude of atmospheric NO_x contamination arising from these anthropogenic contributions is, as with any pollutant, to compare the background concentrations with those measured in urban and industrial areas.

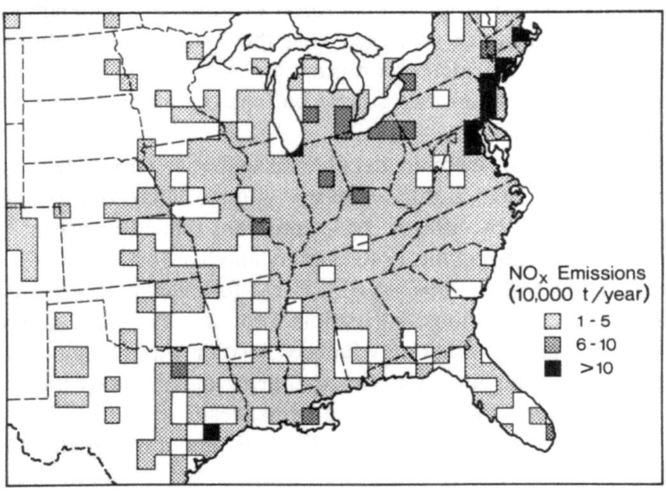

FIGURE 3-9. Areal NO_xL emissions in the eastern United States and Canada (Clark, 1980).

The earliest published monitoring of ambient NO_x levels in "clean" locations was done in Hawaii (Junge, 1956), Panama (Lodge and Pate, 1966) and the mountains of southern Appalachia (Ripperton et al., 1970). Hawaiian observations averaged 2.4 $\mu g\ NO_2/m^3$, Panamanian and later also Brazilian measurements showed virtually identical volumetric results for NO and NO_2, with a range of 0.25–1.3 $\mu g/m^3$ and respective means at 0.5 and 0.75 $\mu g/m^3$ (Lodge et al., 1974). Appalachian monitoring at the top of Green Knob, 1573 m above sea level, in North Carolina found summer values of 6.4 $\mu g/m^3$ for NO_2 and 2.75 $\mu g/m^3$ for NO, with NO_x values higher at times when O_3 levels were lowest.

In contrast to background values of fractions of a microgram to a few micrograms, average NO_x concentrations in urban and industrial areas are usually two orders of magnitude higher. A very large volume of NO_x monitoring data, many going back to the late 1950s, are available for about 100 sites throughout the United States, with California having by far the densest coverage. Unfortunately, an earlier reference method for NO_2 analysis (modified Jacobs–Hochheiser technique for collecting 40-min samples in a fritted glass collector using 0.1 N NaOH solution with added butanol) has been shown to be inaccurate (with errors up to 20%), but even when all values acquired by it are discarded there are still abundant data sets obtained by continuous colorimetric Saltzman method or its modifications (its main drawbacks are inaccuracy at low concentrations and interference from high O_3 levels) and, more recently, by measurement of chemiluminescence, currently the best automated procedure. All these methods are described in detail in WHO (1976) and in Katz (1980).

Long-term analyses of the United States measurements (USEPA, 1971a; NAS, 1977a) show some clear secular, seasonal, and diurnal trends. Review of SO_2 concentrations in the next chapter (Section 4.2.1.2) will show that during the past three decades, they have been declining in most urban and industrial areas of the rich nations, as have the concentrations of particulate matter. Not so with nitrogen oxides, whose urban levels have been, at best, relatively stable since the early 1960s or, more frequently, rising, in some cases rather spectacularly.

These increases were most notable during the 1960s when the six Continuous Air Monitoring Project stations (Chicago, Cincinnati, Denver, Los Angeles, Philadelphia, and St. Louis) had an average increase of 9% in mean NO_x levels. Most of these urban NO_x increases were caused by rapidly growing car use—and Los Angeles County was affected more than any other place in the country: its maximum summer NO_x concentrations doubled between 1960 and 1970 to over 550 $\mu g/m^3$, owing not only to the surge in car use but also to the rapid installation of new power generation capacity and to the partial controls of hydrocarbon and CO exhaust emissions (Chass et al., 1972).

The increases moderated considerably in the 1970s but violations of the national annual NO_2 standard in the greater Los Angeles area continued to be so common that the compliance projected by an air quality management plan for 1982 had to be postponed to 1987 (Horie and Mirabella, 1982). In Los Angeles County, all five measurement sites have been above the standard (the downtown area by as much as 25%) while the two Orange County sites already meet it. There are also many violations of California's 1-hr standard of 470 $\mu g/m^3$ but in relative terms their frequency is a mere 0.3% of the total station hours. Chances for complying with both national and state standards at all sites by 1987 appear to be excellent. United States primary standard prescribes an annual average below 100 μg NO_x/m^3, and most medium- and smaller-sized cities have found no problems in complying, while even outside California several large cities with heavy commuter traffic (most notably Washington, D.C. and Philadelphia) still come close to the mean.

Hourly maxima recorded in southern California used to be as high as nearly 2.5 mg NO/m^3 with values above 1 mg/m^3 for both NO and NO_2 fairly common; in the East such levels have been reached only during high air pollution episodes. In all large cities, hourly NO_x maxima are produced by morning and evening rush hours but the distinct diurnal variations of the two gases are also dependent on the amount of atmospheric mixing and on the intensity of ultraviolet radiation. Nitric oxide concentrations also show pronounced seasonal variation with higher values during late fall and winter (increased heating emissions) and summer minima (when NO is most rapidly converted to NO_2). However, NO_2 concentrations do not show such distinct seasonal patterns.

The two locales with continuously high NO_x concentrations are areas downwind from heavily traveled multilane highways and large airports. Measurements at large airports (such as O'Hare or Los Angeles International) show annual average NO_x concentrations exceeding 300 $\mu g/m^3$ along principal runways and around the terminals and 100 $\mu g/m^3$ isopleths extending for up to 5 km (Jordan and Broderick, 1979).

Measurements available from Canada and some parts of Europe do not indicate secular increases of nitrogen oxide concentrations comparable to the United States trends. For example, monitorings in the center of Rotterdam show monthly NO_2 concentrations fluctuating between 30 and 60 $\mu g/m^3$ with no consistent seasonal variations and with a very steady mean of just over 40 $\mu g/m^3$ between the late 1960s and the mid-1970s (Evendijk and Post van der Burg, 1977). In Canada, the National Air Pollution Surveillance Network has reported on NO_2 only since 1977 when the average for all stations was 60 $\mu g/m^3$; in 1978 it decreased to 54 and in 1979 to 48 $\mu g/m^3$ so that no sites exceeded the maximum acceptable annual air quality standards (Environment Canada, 1981). In contrast, measurements recently available from China show annual NO_x concentrations averaging 80–170 $\mu g/m^3$ in the northern cities and 20–130 $\mu g/m^3$ in the South with the daily maxima surpassing 700 $\mu g/m^3$ (Siddiqi and Zhang, 1982).

This review of NO_x levels would not be complete without a brief discussion of indoor concentrations. While indoor SO_2 levels are usually dramatically lower than the outdoor values—in the cities of modest to high SO_2 concentrations, the reduction is between 50 and 70%, in airconditioned homes 90%—NO_2 exposures may actually be higher indoors than outdoors. Spengler et al. (1979) monitored both levels for 1 year in six American cities and found NO_2 concentrations in homes with electric cooking to be only about 6% lower than the ambient air values, but in homes with gas cooking the indoor NO_2 levels were on the average 6% higher than outside and in houses with no kitchen venting, nearly 10% higher.

Yocom (1982), in his comprehensive review of indoor–outdoor air quality studies done during the 1970s, offers more examples of this relationship with the following elaborations: NO_2 concentrations respond rapidly to gas stove use; there is a significant downward gradient in NO_2 concentrations between the kitchen and the rest of the house; NO_2 indoor/outdoor ratios may be as high as 5.6 in the kitchen and 2.2 in the bedroom; on many days NO_2 kitchen levels can exceed the national ambient air quality standard of 100 $\mu g/m^3$; half-lives of NO and NO_2 are, respectively, about 110 and 40 min, with most NO_2 decayed through reaction and deposition.

3.3.2.2. REACTIONS AND REMOVALS Nitrogen oxides—in the outside air and inside—have fairly short residence times. For example, Chang et al. (1979), using NO_x/CO ratios from the Los Angeles Basin, determined the annual average removal rate of NO_x during daylight hours to be 4%/hr. Other observations from Los Angeles and St. Louis areas confirm this rate, and thus the gases usually reside in the air less than 2 days.

Scavenging processes which limit their buildup embrace a great variety of oxidative and, more importantly, photochemical reactions which have been the subject of lively research interest called to life and supported by that quintessential air pollution classic—Los Angeles smog. Consequently, the field of chemical interactions of nitrogen oxides in the atmosphere has been plowed so thoroughly and repeatedly that an interested reader may ignore literally thousands of papers and conference talks and rely on numerous reviews summarizing the state of our knowledge of these complex phenomena; chronologically, one must recommend Leighton's (1961) and Haagen-Smit's (1962) pioneering works, then volumes by Calvert and Pitts (1966), Tuesday (1971), Spicer (1974), and Innes (1979), to stay within half a dozen choices; NAS (1978a) offers a fine concise summary.

Here I will present just a very brief summary of essential reactions involving nitrogen oxides as established in smog chambers and by field monitoring since the early 1950s. Rapid conversion of nitric oxide to nitrogen dioxide, so clearly recorded during typical photochemical smog buildup in Los Angeles, initiates the changes. What follows is a complex process that cannot be explained either by simple thermal oxidation ($2NO + O_2 \rightarrow 2NO_2$), which proceeds rapidly only at very high NO levels, or by much more intricate chains of photochemical reactions involving various nitrogen oxides, ozone, unstable excited molecules, and free radicals—although these reactions all occur in polluted atmospheres.

But the key path for accelerated transformation of NO to NO_2 leads through reactions between the transient species, most notably the hydroxyl radical formed by collision of singlet-*D* oxygen with water or by photodecomposition of nitrous acid. In polluted sunny skies, these reactions can remove all NO_x within 16 hr. Other reactive molecules, above all carbon monoxide, various hydrocarbons, and aldehydes, are also invariably involved. The reaction chains here become circles, that is at least as far as the hydroxyl radical is concerned. For example, for propylene the events are:

$$C_3H_6 + HO \rightarrow C_3H_7O$$

$$C_3H_7O + O_2 \rightarrow C_3H_7O_3$$

$$C_3H_7O_3 + NO \rightarrow C_3H_7O_2 + NO_2$$

$$C_3H_7O_2 + O_2 \rightarrow CH_3CHO + HCHO + HO_2$$

$$HO_2 + NO \rightarrow HO + NO_2$$

Similar HO-regenerative chains can be written for CO or aldehydes so that many molecules of NO may be oxidized to NO_2 for each HO radical formed.

Photooxidation of NO + NO_2 determines ozone buildup which reaches its peak after considerable depletion of NO and also after the NO_2 levels have started to decline. Ozone can arise from molecular dissociation of NO_2 and subsequent rapid combination of atomic and molecular oxygen with photostationary state established at normal concentrations owing to rapid oxidation of NO by O_3 but this cycle alone is not sufficient to explain large O_3 increases. The second process of O_3 formation involves hydrocarbons oxidized by free oxygen released from NO_2 photodecomposition: as they compete with O_3 for nitric oxide, with NO_2 as an end product, both ozone and nitrogen dioxide concentrations rise at the expense of NO, as the three pollutants peak in sequence.

This strongly oxidizing atmosphere then starts rapid conversion of hydrocarbons whose concentrations plummet and as the smog "matures" and the intensity of solar radiation wanes, both hydrocarbons and NO_2 are removed largely as aerosols (the major cause of reduced visibility) and ozone levels also return to nocturnal minima—before more NO, CO, and hydrocarbons are infused into the air and irradiated during the next morning.

Certainly the most interesting twist to this complex story is the fact that owing to ozone's rapid reaction with nitric oxide, controls of the gas are counterproductive in reducing ozone levels. Innes (1979, 1981) offers the clearest, best supported arguments for this important conclusion appreciated by air pollution control officers in Los Angeles whose standard joke years ago was, as J. P. Lodge, Jr. tells us in his review of Innes's book, that "in the event of a serious ozone episode, the best cure was to get as many motorists on the road as possible, on the theory that the emitted nitric oxide would remove far more ozone that the hydrocarbons would form" (Lodge, 1980).

Clearly, nitrogen compounds are of essential importance in starting the complex photochemical reactions, in sustaining them, and in removing their products. The main photochemical atmospheric pollutants derived from nitrogen oxides are nitric acid, peroxyacetyl nitrate (PAN), and peroxy-propionyl nitrate. PAN and

HNO_3 are the major initial products of atmospheric NO_x reactions accounting, respectively, for almost 70 and nearly 20% of reacted NO_x (Spicer, 1977).

Removal of reacted nitrogen is mostly in the form of nitrites, nitrates, and NH_4^+: these three compounds on the average account for 70–85% of aerosol nitrogen. Particulate nitrites, however, are observed only at very low concentrations during urban monitoring and so it is via NH_4^+ and NO_3^- ions that virtually all deposition occurs. NH_4^+ usually dominates both in aerosols and in precipitation, especially in small sizes (below 1 mm): in large sizes (over 7 mm) the two ions are often about equal (Marsh, 1978). As ammonia fluxes have been discussed in Section 3.1.2.2, here just a few paragraphs on nitrates will suffice.

In comparison to extensive studies of SO_2 oxidation processes and sulfate formation (see Section 4.2.2), much less attention has been given to the atmospheric generation of nitrates. The principal homogeneous reaction involving nitric acid is $NO_2 + HO \rightarrow HNO_3$. With hydroxyl radical concentration at 10^{-7} ppm, the rate of formation of HNO_3 is 8.5×10^{-5} ppm/min with NO_2 at 0.05 ppm, 8.5×10^{-4} ppm/min with NO_2 ten times higher (Orel and Seinfeld, 1977) so that in the absence of any removal, HNO_3 concentrations would rise to 646 $\mu g/m^3$ in 5 hr. Other gas phase reactions, involving NO_2, NO_3, N_2O_5, and NH_3, are much slower, some by several orders of magnitude.

Heterogeneous formation of nitrates is also important, proceeding with complex SO_2–NH_3–CO_2–NO–NO_2–NaCl–H_2O system where nitrate exists in solution as a result of reaction:

$$2NO_2 + H_2O \rightleftarrows NO_3^- + H^+ + HNO_2$$

At coastal sites the formed aerosol may be predominantly $NaNO_3$ or, as it is nearly always inland (and especially in areas of high NH_3 emissions), NH_4NO_3.

The National Air Sampling Network began measurements on suspended particulate nitrates in 1958 so that there is now over a quarter of a century to review. The general distribution of NO_3^- levels shows two expected differences: between large urban areas and small, remote places and, even more dramatically, between urban areas in sunny climates and cities under cloudy Northeastern skies. In the late 1950s, southern California had average 24-hr concentrations of 5–10 $\mu g/m^3$ while large Northeastern cities had only 1.5–2.5 $\mu g/m^3$ (NAS, 1977a).

Since then, there have been general increases in the Northeast and Midwest owing above all to the large expansion of coal-fired power generation located in the two regions. Consequently, most of the Midwest, Appalachia, and Northeast now have levels of 2–5 $\mu g/m^3$ with 24-hr maxima of up to 20 $\mu g/m^3$; concentrations in California also rose with increased traffic and power generation to as much as 10–15 $\mu g/m^3$ with maxima of up to 25 $\mu g/m^3$.

Removal of nitrogen compounds proceeds, as with all similar atmospheric pollutants, by both dry and wet deposition. Dry deposition—gravitational settling, inertial impaction, molecular diffusion, and turbulent eddy transport—of nitrogen compounds is a largely unexplored area. For particles, typical deposition velocities can be calculated once the sizes are known: ammonium ions usually have small diameters (below 2 μm) while nitrates are about equally split between small and large, the latter having, naturally, shorter lifetimes. Particle sizes and deposition velocities will be discussed in greater detail in Section 4.2.2.

Dry deposition of gases has been studied extensively with sulfur dioxide (see Section 4.2.1.3) and ozone, but very little is known about its rates for various nitrogen species. Söderlund (1981) has summarized the few available studies: deposition velocities (all values in m/sec) are less than 0.1 for NO onto soil, 1–3 for NO_2 onto soil and up to 8 onto grass, and about 10 for HNO_3 and NH_3 onto forested or grassed surfaces. As most of the nitrogen compounds present in the atmosphere are highly soluble in water, deposition onto wet surfaces, with the exception of NO, will not be limited by surface resistance. The pH of the solution is another important controlling variable, as is the stomatal behavior of the plants.

In contrast, wet deposition of nitrogen compounds has been measured, episodically as well systematically, in numerous locations. Nitrate concentration in precipitation and nitrate deposition are highest in industrialized regions of the Northern Hemisphere. As for the deposition of ammonium nitrogen, global mapping shows the highest rates to be in Europe (virtually the whole continent in excess of 3 and eastern Europe in excess of 5 kg NH_4^+-N/ha per year) and in Southeast Asia (more than 2 kg). High farming activity and the large number of livestock in these areas are the most likely explanations.

Disaggregated monitoring shows that of the total atmospheric nitrogen deposition ranging from 5 to 25 kg N/ha annually over most croplands and forests of the northern temperate zone, only 5–20% is in nitrates; the bulk is made up largely by dry deposition of NH_4^+ with wet ammonia additions being of similar magnitude as NO_3^- flux (Rosswall, 1977).

Naturally, there are many locations within the Northern Hemisphere's temperate zone where the annual nitrogen deposition will be outside the typical 5–25 kg/ha mentioned above: for example, deserts in the southwestern United States have very low deposition rates, while precipitation alone can bring down in excess of 30 kg N/ha near Midwestern feedlots. And as the emissions of nitrogen oxides have been steadily increasing, so have the average NO_3^- concentrations in rain and annual deposition means.

The longest available data span from the Rothamsted agricultural experiment station in England shows gradual increase from 4.4 kg of nitrogen precipitation per ha during 1889–1903 to 8.6 kg N/ha in 1969–1970 (Legg and Meisinger, 1982), while the weighted annual average of NO_3^- in bulk precipitation at Ithaca (New York) rose from levels of 0.25–1.0 mg/liter between 1915 and 1945 to nearly 3 mg/liter by the late 1970s (Galloway and Cowling, 1978).

In the United States, only the Northwest Pacific Coast and the Southwest now have nitrate precipitation levels below 1 mg/liter; cleaner parts of the Great Plains and Great Lakes area have between 1 and 2 mg, while more polluted Midwestern, Appalachian, and Northeastern locations have in excess of 2 mg NO_3^-/liter (Bodhaine and Harris, 1982).

This increased nitrate presence leading to much higher relative contributions of HNO_3 to the acidity of precipitation—in parts of the eastern United States, the importance of H_2SO_4 decreased by about 30% relative to HNO_3 and that of HNO_3 increased by about 50% relative to H_2SO_4 between 1964 and 1979 (Galloway and Likens, 1981)—will be examined in some detail in the next chapter during the discussion of the roles played by nitrates and sulfates in acid rain while the following sections on the environmental effects of nitrogen compounds will concentrate above all on the consequences of fertilization.

3.4. ANTHROPOGENIC NITROGEN IN THE ENVIRONMENT

With widespread fertilizer applications and combustion emissions, there are many routes and opportunities for unwanted entry and undesirable, objectionable, or outright harmful residence of anthropogenic nitrogenous compounds in the environment. The effects range from clearly acute life-threatening situations (infant methemoglobinemia caused by excessive nitrates in water) to highly uncertain, future alterations (N_2O-mediated destruction of stratospheric ozone), from strong localized presence (ammonia volatilization in huge animal feedlots) to spreading areal impacts (possible health consequences of nitrogen oxides in cities).

First, I shall look at the principal ways in which nitrogen escapes into the environment during farming, then at the element's overall balances in agroecosystems, and finally at the health effects of various nitrogenous compounds, including those of nitrates and nitrites which are deliberately introduced into some food products during their processing. But before all of these, at least brief mention must be made of the effects of nitrogen oxides on animals and plants, a topic marked by a considerable paucity of information and hence not worthy of a separate section.

Perhaps the best way to appreciate these information gaps is to look at the NAS (1977a) nitrogen oxide volume which abounds with statements stressing our very limited knowledge of NO_x influence on plants and animals. Although specific effects of NO or NO_2 in low concentrations are very difficult to dissociate from the effects of other air pollutants, the few available studies did not find that "the current levels of nitrogen oxides have much influence on functioning animal communities" or that there is any ecosystem, or even species, damage. Similarly, "nothing is known about the influence of NO_x on microbial activities in soils, waters, or other ecosystems in which microorganisms multiply."

NO_x effects on crops and trees have been better studied, and the evaluation is that, unlike with SO_2, any direct effects are virtually nonexistent at commonly encountered concentrations. By far the greatest impact of nitrogen oxides on vegetation comes indirectly through ozone in whose production the oxides act as key precursors (see Section 3.3.2.2). Nationwide economic loss from ozone-induced yield reductions is most likely 2–4 billion dollars. Continuing experiments with standardized open-top chambers (3-m diameter, 2.5 m high, equipped with a fan to provide about three air changes per minute through a manifold near the ground) at several locations around the country will eventually result in better understanding of dose–response relationships for major crops and hence in more reliable estimates of potential harvest losses.

In addition to its role in ozone generation, NO_2 can cause direct damage to mesophyll cells of middle-aged leaves. This damage appears as irregular white or brown collapsed lesions on intercostal tissues and near leaf margins, injuries closely resembling those ascribed to SO_2; for acute symptoms, concentrations of nearly 5 mg/cm^3 for at least 4 hr appear to be needed. However, continuing exposure to only one-quarter to one-fifth of these levels may inhibit plant growth or cause marked chlorosis. Dose–response relationships for NO_2 concentrations and reductions of major crop yields are also being studied by the National Crop Loss Assessment Network's open-top chamber experiments. But it will take a very long time before our knowledge of NO_x effects on plants will start rivaling our current appreciation of nitrogen losses in farming.

3.4.1. Nitrogen Losses in Agroecosystems

In spite of impressive advances in analytical techniques (particularly in gas chromatography, ^{15}N-labeling and field measurement of gaseous nitrogen compounds), introduction of computer modeling, and extensive experimental studies of nitrogen flows in various ecosystems, we still have no simple, rapid ways to determine a variety of soil nitrogen losses and our attempts to construct reliable balance sheets can still be done with no more than only mediocre success.

To appreciate the ranges of uncertainties before discussing individual pathways, it is instructive to quote ranges suggested by The Fertilizer Institute (1976) for estimates of potential losses in calculating nitrogen budgets in soils. The values (all in percent) are as follows: crop uptake 40–70, denitrification 5–35, erosion 0–20, immobilization 10–40, leaching 0–20.

Nitrogen losses in many managed grasslands and forest plantations can often be kept very low, but in arable farming the nutrient escapes the fields in several hard-to-curtail ways which can broadly be divided into "up" and "down" categories: return to the air in volatilized ammonia and in nitrous oxide and dinitrogen from denitrification—and removal into waters with soil erosion and leaching. All of these loss routes will be looked at in turn in the following five sections which will also include discussions of nitrous oxide's possible effects on concentrations of stratospheric ozone and of some unwelcome consequences arising from nitrogen's excessive presence in waters.

3.4.1.1. Volatilization Volatilization loss of nitrogen as ammonia from surface-applied fertilizers, organic and synthetic, is a ubiquitous phenomenon easily noticeable in a sty or stable or near a pile of fermenting manure. While not neglected, volatilization has not received a large amount of research attention, especially in terms of field measurements. The best recent surveys of ammonia volatilization are those by Terman (1979) and Nelson (1982). Only field experience will be reviewed here for a few major fertilizers as well as for organic wastes.

Reported values on NH_3 losses from anhydrous and aqua ammonia differ appreciably. For anhydrous ammonia, some studies found only negligible loss even when injection was not very deep (5–7 cm), whereas others reported volatilization of up to 60%; for aqua ammonia the values range similarly, from more than 50% for surface application to alkaline soils to no appreciable loss with even shallow injection of low-pressure solutions. Nelson's (1982) detailed summary of field measurements of NH_3 losses indicates that anywhere between 3 and 50% of added nitrogen may evolve as NH_3 after application of NH_4NO_3, $(NH_4)_2SO_4$, or urea. Typical losses appear to be between 5 and 20%, but are smaller when fertilizers are incorporated into soils and fairly low even when ammoniacal fertilizers are surface broadcast at acidic or neutral sites.

In contrast, large amounts of NH_3 are readily released after surface application of animal manures or sewage sludges, especially to alkaline soils. Naturally, pH levels are a key controlling factor in NH_3 volatilization: while at pH 5.0 just 1–2% of the NH_3-N may be lost in 1–2 weeks, at pH 8.0 50% of the NH_3-N may be lost in 1 week and 80% may be gone in 2 weeks (Van Veen et al., 1981).

The most controversial problem concerns the losses from urea. Some experiments in the 1950s and 1960s indicated that NH_3 volatilization from surface-applied urea was considerably higher than for other solid ammonia fertilizers, and hence the yields were slightly to appreciably lower, while other tests showed equal effectiveness. The problem is complicated by the fact that urea requires contact with urease for hydrolysis so that the losses after application may be delayed if little soil urease is present or until the fertilizer is dissolved and moved deeper into the soil.

Inhibitors of urease activity and additives, including concurrent application of urea and ammonium nitrates, decrease the rate of volatilization: while the losses with surface-applied urea were in some tests as much as 77% of the original nitrogen, a 50/50 mixture with ammonium nitrate cut these losses in half. Yet in other tests urea volatilization losses were fairly low at 12–16% and the addition of ammonium nitrate actually led to their increase to up to 25% of all nitrogen. Clearly, no generalizations at this time are possible.

Volatilization losses from animal manures are usually much greater and much more rapid than from synthetic fertilizers. Nearly half a century ago, Salter and Schollenberger (1938) reported losses of up to 50% and these levels have been confirmed by more recent measurements, above all by Lauer's (1975) extensive field experiments with ammonia volatilization from dairy manure spread on the soil surface. In the nonwinter experiments, mean half-lives were just 1.86 and 3.36 days for, respectively 34 t and 200 t/ha application with more rapid drying of the thinner layer responsible for faster loss in the lower application. Although the rate of volatilization slowed down after the initial fast decline, the mean overall losses were 85% of the NH_3-N contained in the manure at spreading.

In flooded soils, denitrification losses as N_2 and N_2O, described in the next section, may be substantial but NH_3 volatilization is not infrequently even greater and it occurs mostly during the first 10 days after nitrogen application. Reported values range from 3 to 19% and are strongly dependent on the alkalinity of soils and waters. An important factor here is the algal productivity in the flooded fields which causes diurnal rise of water pH: up to 3.5 pH units may be gained in midday when photosynthesis withdraws CO_2 and when waters may reach highly basic pH 10.

Ammoniacal fertilizers broadcast onto such high pH waters, a common practice in rice fields of Southeast Asia, are highly susceptible to direct volatilization losses (Engelstad and Russel, 1975). Although there is great variation owing to water pH, nitrogen source, rates, times, and mass of application, up to 20% of fertilizer nitrogen can be lost within just 4 days after application (Mikkelsen et al., 1978). Typical cumulative losses during the first week are 5–15% for ammonium sulfate and urea broadcast onto water but incorporation of the surface-broadcast nitrogen with a mechanical stirrer can reduce the loss to below 1%, as does placement in the soil at depths of 10–12 cm. Where the fertilizer is merely broadcast, waters poor in aquatic phytomass would be desirable (to preclude diurnal pH rise) but the algae may be *Anabaena* hosts and thus important contributors to the field's nitrogen balance!

Although the reported extreme values for NH_3 volatilization range over nearly two orders of magnitude, the process is, in all cases where the soils have insufficient sorption capacity to hold NH_4-N released from fertilizers, a fairly large avenue for loss of valuable nitrogen. But often this loss must be negated, or at least partially counterbalanced, by soil absorption of NH_3 from the atmosphere: the rates of this transfer have been reported at 10–80 kg NH_3-N/ha annually, with higher values in acid, fine-textured soils (Terman, 1979; see also Section 3.1.1.2).

As these soils will also retain more of the evolved NH_4-N from applied fertilizers, they may display a net gain of ammonia nitrogen, while more basic, coarse-textured soils may have a substantial net loss. However, no quantitative generalizations are possible to assess the magnitude of the net transfer—after all, we are not often even sure of its direction nor can we say how much ammonia is absorbed by vegetation and water surfaces which must be important sinks for atmospheric NH_3.

Finally, gaseous losses may also include volatilization of ammonia, other amines, dinitrogen and nitrogen oxides from the plant tops (Wetselaar and Farquhar, 1980). All of hundreds of tested plants contain ammonia in their tissues and volatilize it at rates of up to about 1 kg N/ha per day. Measurements of the total evolution of nitrogenous compounds from leaves (by pyrochemiluminescence) typically yield about 0.5 kg N/ha per day and increase with temperature to about 1.5 kg N/ha per day. During the late summer weeks just before harvest, these fluxes alone can account for a major part of nitrogen losses from plant tops.

3.4.1.2. Denitrification Denitrification is a ubiquitous soil process which makes nitrogen available at all locations by returning it to the air, and its total annual global flux appears close to the rates of fixation (Payne et al., 1980). On local scales, denitrification can be seen as a "commission" to soil microflora faced with excess N/C ratios at the end of the litter cycle when some nitrogen must be dumped both for microbial viability and for prevention of NO_2^-–NO_3^- toxicity to macrophyta (Gutschick, 1980). Most of the time, the microorganism-mediated process reduces nitrate to dinitrogen with nitrous oxide as the last obligatory intermediate (see Section 3.1.1.3). However, the reduction is not often carried all the way as N_2O escapes from soils and waters into the atmosphere. Published values for N_2O show fractions from fertilizer denitrification ranging from 0 to 1 (i.e., no N_2 production), with high soil NO_3^- levels favoring higher N_2O releases while lower soil NO_3^- and highly anoxic environment (rice paddies, marsh soils) lead to very low N_2O production. CAST (1976) assumed an average field fraction of 0.06, while Rolston's (1981) experiments provided a range from 0.03 to 0.25.

Logical expectation is then that the increased use of nitrogenous fertilizers, as well as wider use of legumes, should increase N_2O releases to the atmosphere. This higher flux would obviously be too dilute to have any direct ecosystemic or health effects but it might possibly contribute to reduction of stratospheric ozone, a potentially dangerous consequence to be discussed in some detail in the next section. However, while denitrification losses during fertilization have come to dominate most practical discussions of this bacterial process, it should at least be noted that there is an important beneficial facet as well, namely use of denitrification in rapid decomposition of organic wastes from feedlots as well as in general wastewater management (Schroeder, 1981; Payne, 1983).

Quantifying denitrification is no easy matter. Maximum rates in experimental plots are not difficult to establish but long-term averages applicable to large areas are nearly impossible to fix. The maximum denitrification rate measured by Rolston (1981) was 70 kg N/ha per day for a plot nearly saturated with water and treated with 34 t of manure incorporated into the upper 10 cm of soil 3 weeks before nitrate was applied to the wet soil surface; in a cropped field plot (perennial ryegrass) the highest denitrification rate was about 10 kg N/ha per day, and for uncropped plots the rate was just 2.5 kg N. In all of these cases no denitrification could be detected after 20 days. A summary of 40 different agroecosystems for which denitrification estimates are available shows losses ranging from just 1 kg N/ha per year to nearly 200 kg (Frissel and Kolenbrander, 1977).

Averages of N_2O releases from unfertilized uncultivated soils range from just a few to a few hundred micrograms (N/m^2 per hr), with most values in the range of 5–20 μg (CAST, 1976; McKenny et al., 1978; Freney et al., 1979; Conrad and Seiler, 1980). Rolston (1981) reviewed a variety of N_2O loss studies where the maximum N_2O fluxes were 9 kg N/ha per day in the wettest, manure-treated soils and minima just 0.0003 kg in an irrigated cornfield in Colorado, and concluded that the maximum N_2O fluxes from cropped farmland without large manure additions, maintainable generally for only short periods during irrigation or rainfall, may range from 0 to 1 kg N/ha per day.

Choice of representative averages is thus open to errors of a couple orders of magnitude. Taking 10 μg N/m^2 per hr (i.e., roughly 0.9 kg N/ha per year) as an average just to establish one of the possible totals and multiplying by 1.1×10^{10} ha (the world's land area minus arable land and extreme cold and hot deserts) results in an annual flux of roughly 10 million t N.

Longer-term measurements of the fertilizer-derived N_2O losses are rather scarce. Rolston et al. (1978) found values of 0.01–0.63% on uncropped soil and 0.05–1.5% on cropped land for sodium nitrate applied to irrigated loam in California at rates of 300 kg N/ha with losses dependent on the season (higher in summer). Ryden et al. (1979) reported higher values, 1.8–2.7% of the applied nitrogen. Denmead et al. (1979) measured the flux just for 18 days in a flooded rice field which initially contained 4 g NO_3-N/m^2 in the top 8 cm of the soil and found a diurnal cycle in phase with water temperature and N_2O nitrogen accounting for a mere 0.8–1.4% of total NO_3 lost.

Conrad and Seiler (1980) measured the N_2O emissions from uncultivated well-aerated sandy soil near Mainz (West Germany) following applications with ammonium nitrate, ammonium chloride, and sodium nitrate, all at a rate of 100 kg N/ha. They established preapplication emissions at 2–6 μg N/m^2 per hr in fall months and 5–13 μg in summer, and the N_2O flux reached peak values 4–12 days after fertilization and then declined, rapidly with ammonium nitrate (in just 7 days the levels were identical to those of unfertilized plots), much slower with NH_4Cl (70 days elapsed before the return to preapplication levels).

Depending on the soil, season, and nitrogenous compounds used, N_2O losses after fertilizing appear to range from a mere 0.01% to nearly 3% of the applied nitrogen, again too broad a range to choose a representative mean with any confidence. Using, perhaps too liberally, an average of 1%—i.e., annual losses of 0.4–2.0 kg N/ha—would result in annual global soil-to-atmosphere flux of about 600,000 t of nitrogen as N_2O. This is much less than the flux assumed in several past models which postulated losses of up to 7%. And this total appears negligible compared to the fluxes from drained organic soils which prior to their drainage represented huge geological sinks of carbon and nitrogen.

Intense microbial oxidation in such soils mineralizes large quantities of nitrogen most of which is subsequently lost through denitrification. Globally, such soils are not extensive but their N_2O production may account for a large share of the annual N_2O soil–atmosphere flux. These conclusions are based on measurements made by Terry et al. (1981) in southern Florida's Everglades where nitrous oxide fluxes range from 4 g during dry periods to 4.5 kg N/ha per day following rains. Annual emissions from fallow land were up to 165 kg N_2O-N/ha while sugarcane fields released the oxide at about 50 kg N/ha per year.

These rates are 100-fold greater than the usually encountered averages and assuming that each of the 300,000 ha of the Everglades agricultural area emits 100 kg N/ha each year, the annual flux would be 30,000 t of nitrogen as N_2O. The total denitrification loss from Everglades soils is estimated at 1100–1300 kg N/ha per year, implying that N_2O-N is, at 50–150 kg, a mere 3.8–13.6% of the total flux. This agrees well with other recent appraisals of total denitrification in farming soils.

Rolston et al. (1978) measured the total denitrification flux from both cropped (ryegrass) and uncropped plots treated with 300 kg N/ha as KNO_3 and maintained constantly wet as well as from manured test sites, all on loam soil. At water contents close to saturation and at 23°C, about 70% of fertilizer nitrogen was denitrified for the manure-treated sites, 14% for cropped and 3% for uncropped plots; at 8°C these rates were, in the same order, 11, 6, and nearly 0%,

Ryden et al. (1979) found a total denitrification loss of 51.2 kg/ha during 123 days following application of 325 kg N/ha irrigated farmland (i.e., a loss of 15.3%) with N_2/N_2O ratios ranging from less than 1 during the first day of the denitrification cycle to 10–20 toward its end, with an average value between 5.6 and 7.4. In colder, drier regions such as the Canadian Prairies, denitrification diminishes considerably the effect of fertilizers applied in the fall: experiments with ^{15}N-labeling resulted in losses of up to 41% for potassium nitrate and 30% for urea.

Although I have cited (Section 3.1.2.1) Payne and colleagues' (1980) conclusion that one cannot even guess at global denitrification rates, I shall at least try to fix the most likely extremes. To begin with, denitrification can be simply estimated just as a fixed share of total fixation: with the fixation aggregate (biotic, atmospheric, anthropogenic) at 150–250 million t N/year and with at least 15 and as much as 50% of it lost to denitrification, the annual total would be somewhere between 20 and 125 million t. The lower bound is clearly too low as that amount of nitrogen may be released as N_2O alone. Thus, one can proceed from the available denitrification measurements, starting with total N_2O-N losses of about 12 million t (10 million t from uncultivated soils, 1 million t each from cultivated mineral and organic soils), and assuming that nitrous oxide is about 10% of the overall denitrification flux, the total would be some 120 million t annually. Combining the two approaches, I would choose a level of around 100 million t N/year.

For comparison, Delwiche (1970) put the denitrification total at 40 million t (assuming total fixation of only 80 million t in the late 1960s) and subsequently (Delwiche, 1977) raised this value substantially to 120 million t N/year (with total fixation estimated at nearly 200 million t). Other available estimates are listed in Table 3-1. In any case, it is the release of N_2O, rather than the aggregate denitrification flux composed largely of N_2, that is of concern. Released dinitrogen merely rejoins its huge atmospheric pool, whereas N_2O can cause undesirable stratospheric changes. Indeed, Bolin and Arrhenius (1977), summarizing the reports of the 30th Nobel Symposium, listed foremost the decrease of atmospheric ozone by increasing denitrification and wrote that "this is a global problem, and total emissions, integrated over the whole earth, must be evaluated and possibly controlled."

3.4.1.3. Nitrous Oxide and Stratospheric Ozone Stratospheric ozone is an irreplaceable screen shielding the biosphere against higher influx of ultraviolet radiation whose increase would have considerable effects on human health and biota evolved under relatively low UV radiation. Increases in UV radiation would appear to have a wide range of effects on important crops, plant pathogens, insect pests, nitrogen fixation, and rates of photosynthesis (NAS, 1982c; Forziati, 1982). Although reliable quantification of impacts is not available, it is clear that increased UV level would cause reductions in leaf area, plant weight and height, delays in germination, and higher respiratory losses in many sensitive species (mainly C_3

plants). This would translate into reduced crop yields and higher pest and disease susceptibility in some crops.

Higher human exposures to UV radiation would cause increased incidence of skin cancer among which the two most common types, basal cell and squamous cell carcinomas, are in general easily detected and successfully treated whereas melanoma is a very serious disease with a high incidence of mortality. Less-pigmented populations of the northern temperate latitudes are especially at higher risk. As a 1% decrease in ozone concentration leads to a 2% increase in UV radiation and as the best available epidemiological models suggest a 2% rise in skin cancer for each 1% of additional UV flux, reduction of the stratospheric ozone level by just 5% may cause as much as a 20% increase in skin cancers in a susceptible population, a worrisome increase. However, considerable uncertainties prevail with regard to the molecular and cellular pathways of this damage, identification and quantification of ecosystemic response, and appraisal of increased incidence of melanoma and other skin cancers (NAS, 1982c).

These uncertainties make it impossible to quantify any damage caused by higher levels of UV radiation (not even for melanoma can we make a satisfactory prediction of increased incidence following higher exposures based on current epidemiological data!) and numerous other gaps in understanding the complex stratospheric reactions do not allow any confident forecasts of ozone reduction as a result of anthropogenic influences which arise from several kinds of activities. Foremost are the releases of chlorinated carbon compounds led by chlorofluorocarbons and methyl chloroform; next, N_2O emissions from farm soils and waters and N_2O flux from combustion of fossil fuels and biomass; and last, higher CO_2 concentration owing to fossil fuel combustion as well as to spreading deforestation.

A closer look at chlorofluorocarbon effects is beyond the scope of this book—the most recent of the NAS reports on the controversy (NAS, 1982c) and a Resources for the Future appraisal (Cumberland et al., 1982) contain all of the facts and hypotheses—but an appraisal of N_2O emission effects is obviously in order, especially the releases of the gas from denitrification resulting from increased application of chemical fertilizers.

First, however, the distribution, concentrations, and sources of the gas must be examined. Hundreds of available short-term atmospheric measurements narrow the tropospheric N_2O concentrations to 250–350 ppbv with relatively uniform distribution throughout the world (Hahn and Junge, 1977). Three long-term series from West Germany and Massachusetts acquired during the 1960s showed some considerable but unexplained irregular fluctuations but more recent reliable continuous N_2O measurements—1-year monitoring at a ground-level site in eastern Washington—found no significant seasonal or diurnal variations with virtually all 3-hr averages being 324–332 ppbv (Pierotti and Rasmussen, 1978).

Regular N_2O measurements have been available since 1977 as part of the Geophysical Monitoring for Climatic Change at five of the program's baseline stations: the last available data, means for 1981, confirm the negligible difference between the hemispheres—302.5 ppbv at Barrow on the northern coast of Alaska and 301.1 ppbv at the South Pole—as well as between continental and oceanic sites—303.2 ppbv in Colorado and 302.0 ppbv on Mauna Loa (Hawaii) and 304.1 ppbv at American Samoa (Bodhaine and Harris, 1982).

The first extensive measurements of N_2O in seawater made during three Atlantic cruises by Hahn in 1969–1971 resulted in surface water concentrations of 0.4–0.5 μg N_2O/liter indicating supersaturation of up to more than 100% in the subtropics and tropics and making the ocean a large net source of atmospheric N_2O (Hahn, 1974). As the highest N_2O levels were measured in layers with the relatively lowest oxygen concentrations, Hahn (1974) concluded that denitrification was the source of excess N_2O, while later N_2O measurements in the Atlantic found the rate of N_2O production to be proportional to the rate of O_2 utilization, leading to the opposite conclusion, namely N_2O being formed by nitrification (Yoshinari, 1976).

Many new details on nitrous oxide distribution, formation, and destruction in seawater reviewed by Hahn (1981) still do not add up to a clear understanding of the ocean's role in the atmospheric N_2O cycle. The total global marine net production of N_2O remains uncertain: Hahn (1974) initially put it at 135 million t/year and when more information became available, he reduced the figure to 45 million t, the range of uncertainty being 12–120 million t N_2O (Hahn, 1981).

This means that oceans are about as large a source of N_2O as natural releases of the gas from soils, and both of these fluxes appear to be substantially higher than the atmospheric inputs derived from nitrogen fertilizers (see Section 3.4.1.2). But releases of N_2O from denitrification, although considered the sole important source of the gas as late as in 1976 (see CAST, 1976), are not the only notable anthropogenic contribution. To begin with, nitrification of fertilizers can also be a source of the gas and Bremner and Blackmer (1981) give the best review of this evidence. The most likely path is via production of an enzyme (nitrite reductase) by *Nitrosomonas* that can convert nitrite to N_2O under both anaerobic and aerobic conditions with concomitant oxidation of ammonium to nitrite although some N_2O may also be formed by nonbiological decomposition of hydroxylamine or nitrite produced during microbial oxidation of NH_3 to NO_3.

Highly questionable assumptions and extrapolations do not enable any reliable assessment of nitrification-derived nitrous oxide, but Bremner and Blackmer (1981) believe that the process may be a more important source of the gas than denitrification in ammonium-treated soils and that in general "a significant portion" of the nitrous oxide that is released from soils is present owing to the nitrifying rather than the denitrifying microorganisms.

Combustion of fossil fuels and biomass have also been identified as two major contributors. Weiss and Craig (1976) measured N_2O emissions from coal and an oil combustion in a large power plants and found nearly identical values (25 ppm) for both fuels. Using their average N_2O/CO_2 ratio in combustion gases of coal- and oil-fired power plants (above 2.1×10^{-4}) and 1980 worldwide CO_2 generation from coals and liquid fuels at about 16 billion t results in annual global N_2O emissions of 3.4 million t (i.e., 2.2 million t N).

Catalytic convertors may be yet another relatively important source of combustion N_2O. Weiss and Craig (1976) estimate that complete conversion of gasoline engines to the platinum catalyst would have produced—with an air/fuel ratio of 15, combustion gases containing 1.88 g/m^3 of NO, and a removal rate of 90%—about 2.1 million t N_2O-N in 1976. Obviously, the actual flux was just a fraction of this value as convertors have been far from universally used, but a potential, together with emissions from catalysis of industrial effluents, for 2–3 million t N_2O from this source in the near future is undeniable.

Crutzen et al. (1979) assumed that N_2O emissions are about 8% of the total volume of nitrogen oxides liberated during combustion of biomass (including forest fires) to derive an average global flux of about 13 million t N_2O/year. As at least three-quarters of this total comes from combustion fuelwood and crop residues and from fires set by shifting cultivators and colonizers of forests, the anthropogenic total is some 10 million t N_2O/year, comparable to (1) the total N_2O flux from soil denitrification, (2) lower estimates of N_2O flux from soil denitrification, and (3) lower estimates of N_2O removal rates in the stratosphere (10–30 million t/year), the only well-documented sink of the gas. Tropospheric photolysis and reactions may remove another 1–5 million t N_2O annually but the main sinks of biospheric N_2O remain unknown.

Comparisons of the three direct anthropogenic sources—fertilizers and burning of biomass and fossil fuels—show quite persuasively that combustion may be as large, or even much larger, a source of N_2O as nitrogenous fertilizers (Table 3-8). In view of the evidence on the share of N_2O released by denitrification of nitro-

TABLE 3-8
UNCERTAINTIES OF N_2O SOURCE AND SINK ESTIMATES[a,b]

	Most likely flux	Range of uncertainty
Sources		
Oceans	45	12–120
Soils	25	10–100
Fertilizers	20	1–30
Biomass burning	10	5–15
Fresh waters	5	0–40
Fossil fuel burning	4	2–6
Lightning	—	0–90
Total (rounded)	110	30–400
Sinks		
Stratospheric reactions	20	14–28
Tropospheric reactions	3	1–5
Unknown	—	15–367
Total (rounded)	23	30–400

[a]Source: according to Hahn (1981), except for biomass burning and fertilizer denitrification.
[b]All values are in million tonnes of N_2O per year.

genous fertilizers presented in Section 3.4.1.2, I find Hahn's (1981) "most likely" total of 20 million t N_2O released from this source annually much too high—but even so the combined contribution of biomass and fossil fuel burning would be almost of the same level.

As for the sinks, there may be a need to find additional placement for about 11 million t N_2O and if higher totals are needed, then the values in the lower half of the given range appear much more likely. Water-logged soils and swamps seem to be the best candidates for absorbing large amounts of N_2O but earlier estimates of this sink's magnitude have been scaled down and soils are now generally considered a considerable source of the gas (Matthias et al., 1979): puzzles remain unsolved.

Owing to its low solubility, N_2O accumulates in the troposphere, and eventually also in the stratosphere where it has a controlling effect on ozone through the transformation to nitric oxide which is one of the principal participants in the ozone-destroying complex of stratospheric reactions. Although there is certainly no imminent danger of substantial ozone reductions resulting from higher fertilization (increased natural fixation should *not* be seen as a significant contributor in view of the barely increasing legume cultivation), the rate of post-1950 application growth (see Section 3.2.3.1) led to worries about long-term effects best exemplified by the N_2O and O_3 report by CAST (1976), by models presented by McElroy et al. (1977), and by several papers published after 1970 by Crutzen (1970, 1974, 1976a). Crutzen's latest review will be used here to show the tentative nature of estimates involved in assessing the effect (Crutzen, 1981).

Although some release of N_2O will follow rapidly the application of nitrogen fertilizers and their subsequent denitrification (or nitrification), this will most likely not happen with that portion of fixed nitrogen which will reach natural reservoirs whose size and transformation rates could be affected by anthropogenic inputs only over very long periods (no less than several hundreds of years).

Consequently, to assess the short-term effects of N_2O releases from fertilizer, the fraction of fixed nitrogen input that does not reach the natural ecosystems must first be estimated. Crutzen (1981) assumes that 80% of NH_3 volatilized from fields and feedlots is dumped on forests, lakes, and oceans where it may increase primary productivity and organic matter buildup. With 20% of fixed nitrogen relatively rapidly denitrified and with generous estimate of N_2O/N_2 shares, the maximum additional source of N_2O from the current applications of synthetic nitrogen would be about 6 million t/year, which means that atmospheric N_2O content would rise by no more than 0.4%/year. Should the anthropogenic fixation reach 200 million t N/year by the first quarter of the next century, the N_2O content of the atmosphere would not double before its end.

The consequences of N_2O doubling are uncertain as the rate coefficients of many stratospheric reactions are not known with sufficient reliability, but the total decrease in ozone content would be as little as 2% or as much as 4% with doubling of atmospheric N_2O. Crutzen (1981) also calculated the consequences of N_2O doubling in an atmosphere continuously perturbed by chlorofluorocarbon releases at about the present rate for another 50 years and found that N_2O (and NO_x) additions might be "beneficial" by causing greater ozone formation if stratospheric concentrations of chlorine are assumed to be large; the total ozone column would be still depleted by about 12% owing to the combined effect of N_2O and chlorofluorocarbon pollution.

But in his earlier work, Crutzen (1976a) assumed that denitrification does not lag significantly after fixation and put ozone reduction owing to increased fertilization at 1.5–7.0% by the year 2025. Similarly, McElroy (1975) assumed that an increase to 200 million t of annual nitrogen fixation in fertilizer would be followed by an essentially simultaneous increase to 200 million t in the denitrification rate; his model postulated a 20-year lifetime for atmospheric N_2O, a stratospheric sink of 10 million t N/year, oceans as a net sink of 25 million t N_2O-N, and the continents as a source of 35 million t N_2O-N, and these resulted in a 20% reduction of O_2 level by the year 2025. A year later he advanced the date of 20% ozone reduction to the year 2013 (McElroy et al., 1976).

The latest NAS review of ozone depletion models (NAS, 1982c) concluded that calculations prevalent at the end of the 1970s indicated that if the production of two major chlorofluorocarbons (CF_2Cl_2 and $CFCl_3$) were to continue at the 1977 rate, the ultimate steady-state ozone reduction would be 15–18% while the latest models foresee only a 5–9% decline under the same conditions and that the best consensus regarding N_2O inputs is that their doubling, in the absence of other perturbations, would reduce the total ozone level by 10–16%.

Although sharing the basic assumption of nitrogen fertilizer applications rising to 200 million t/year within the next four decades, results of the cited ozone reduction models range from 20% even before the end of the first quarter of the next century to a mere 3% 50–70 years later, a very wide span covering fundamentally dissimilar outcomes from a rapid and deep decline almost warranting some emergency measures to a barely noticeable disturbance maturing a century from now when it might become altogether hidden within natural fluctuations or other anthropogenic perturbations.

For several reasons, our assessments will not become clearer any time soon. To begin with, we have no way of reliably forecasting the levels of anthropogenic fixation—but in the mid-1980s it appears almost certain that the mid-1970s' predictions of 200 million t of fertilizer nitrogen by the year 2000 are too high and hence the assumptions of doubled atmospheric N_2O concentrations before the end of the next century appear unrealistic. Even should the fixation be doubled so soon, there is little reason to expect that the return of denitrified N_2O to the atmosphere will attain the same rate, or close to it, with just a short delay. This point, first raised in detail by Liu et al. (1977), is worth stressing. Attainment of steady state will take a long time and short-term (several decades) effects may not even be noted. However, once developed, the consequences of higher anthropogenic fixation may persist for many decades regardless of any corrections then undertaken.

Continued research attention on chlorofluorocarbons and nitrogen oxides seems then to be most desirable as there are undeniable possibilities of unwelcome long-term stratospheric disturbances caused by their accumulation. But the fundamental uncertainties characterizing our understanding of both the complex stratospheric reactions and future rates of N_2O, $CFCl_3$, and CF_2Cl_2 and the rates of natural and anthropogenic NO_x emissions make all current quantitative conclusions merely suggestive warnings rather than reliable foundations for introducing any substantial modifications to our way of farming.

The first 5 years of systematic N_2O measurements at five GMCC stations had shown a gentle rise of what must be considered "background" N_2O levels (Bodhaine and Harris, 1982). Regression analyses for the years 1977–1981 resulted in average growth rates of 0.5 ppbv/year at Barrow (Alaska), 0.7–0.8 ppbv at Mauna Loa and at the South Pole, 1.8 ppbv at Nivot Ridge (Colorado), and an inexplicably high 3.7 ppbv at American Samoa. Yet Samoan data, as well as those from Colorado, show a small but statistically significant decline in annual growth rates.

Taking 1 ppbv as the mean growth rate, the average annual increase for the first 5 years of the monitoring would be 0.33%, or a doubling time of 212 years. Considering the relatively small contribution of fertilizer-derived N_2O to the total generation of the gas, fundamental gaps in our knowledge of its sinks, and the gentle rate of its monitored increase, one must conclude that no worrisome rapid changes lie ahead.

With better understanding of stratospheric chemistry, we shall eventually get more confident appraisals of possible ozone-destroying contributions by N_2O but even now we may conclude that the threat of nitrous oxide from denitrification of synthetic fertilizers was considerably overplayed during the 1970s. And what effect this increment would have on stratospheric ozone cannot be given without considering other atmospheric pollutants with potential stratospheric impacts, a task full of multiple, cumulative uncertainties. Consequently, in practical terms we have all the reasons to worry more about nitrogen losses via leaching and erosion rather than about highly uncertain, remote possibilities of N_2O stratospheric interference.

3.4.1.4. LEACHING AND EROSION Whereas gaseous losses of nitrogen involve three principal substances—dinitrogen, nitrous oxide, and ammonia—leaching losses are almost exclusively a matter of soluble nitrate moving downward through the soil. The obvious variables determining the outcome are the kind, rate, and placement of applied fertilizer, water-holding capacity of the soil, amount of precipitation, and plant uptake. Many field tests have been conducted to elucidate the complexities of leaching processes, especially in the Corn Belt and California's San Joaquin Valley, two of North America's most intensively fertilized farming regions where leaching (in the first instance owing to fairly high natural rainfall, in the second caused by irrigation) became a controversial environmental problem in the 1960s. This brief review is based largely on excellent summaries by Allison (1966) and Aldrich (1980).

Ammonium fertilizers, now the dominant form of nitrogen applied to crops worldwide, are relatively leach-proof as the NH_4^+ attaches readily to clay particles and humus and will not leach through or wash off the surface of the soil and is lost heavily only when the soil is eroded during periods of heavy precipitation. In contrast, nitrate ions in solution will move readily through direct channels in the soil, although the actual transfer is usually composed of a multitude of complex pathways so that months, even years and decades, pass before the fertilizer nitrates appear in the streams and groundwater.

This appears counterintuitive as it might logically be assumed that the completely water-soluble compounds would be washed away rapidly by rains following their application. But this is almost never the case: fertilizer is spread on fields only when they are dry enough to permit the tractors on them and during the first few minutes of rain they are dissolved and are carried into the soil so that if the rain continues to the point of surface runoff the dissolved nitrogen has been overwhelmingly sequestered in the soil. Consequently, virtually all nitrogen reaching surface waters does so through the soil instead of across the surface. The major exception is the movement of soluble nitrogen compounds from manure spread on frozen soil.

Winter spreading of manure on frozen soil has long been considered one of the most hazardous nutrient-leaching practices as a rapid thawing of the accumulated snow and early spring precipitation unable to soak into the ground might carry away most of the nutrients—yet it is impractical to hold all of the winter manure for spring spreading. Fortunately, the hazard may not be as high as formerly believed. In 3 years of experiments on sloping land receiving manure and chemical fertilizer and planted to corn, Young and Holt (1977) found higher total nitrogen loss in runoff from winter surface-manured plots (3.2 versus 2.4 kg N/ha for unmanured land) but nutrient losses owing to erosion were much higher than those in runoff, and as manure applications are very effective in reducing soil erosion, the total nitrogen loss from winter-manured plots was consistently much lower than from unmanured land (3-year average of 27.9 versus 78 kg N/ha). Thus, although the runoff losses from winter-manured plots were 33% higher than those from the unmanured ones, the latter had total nitrogen losses nearly three times as large. This is an excellent example of the necessity to look beyond runoff at erosion-prevention measures (see Section 3.5.2.1).

As the leaching process is so critically dependent on the variable environmental factors, offering representative averages for typical losses is no easier than for denitrification. For example, a listing of 17 leaching loss studies in Thomas and Gilliam (1977) has its lowest average entry at 4 kg N/ha per year for an Ontario mixed crop farm and the highest at 46 kg for North Carolina corn. A survey of estimates for 40 agroecosystems from different parts of the world indicates that leaching losses for farm inputs below 150 kg N/ha per year are scattered around 10% those for inputs above 150 kg cluster around 20% with the highest value (83 kg) reported for irrigated California cotton (Frissel and Kolenbrander, 1977). Nitrogen leaching losses measured in seven different farming areas in Denmark, Sweden, and England ranged from just 3 to 23 kg N/ha per year with typical values between 15 and 20 kg (Edens and Soldberg, 1977).

Consequently, only qualitative generalizations can be unexceptional. The most universal of such generalizations concerns the precipitation–evapotranspiration balance. Throughout the temperate latitudes, the latter surpasses the former during the main part of the growing season between May and October and, moreover, vegetation readily absorbs dissolved nitrogen in the root zone so that the net movement of water and nutrients is upward and there is little possibility of nitrate loss-except when the annual rains exceed 1250 mm, in the regions of heavy summer downpours which exceed the fields' moisture capacities or on highly sandy soils.

In contrast, during the fall, when vegetation stops growing and rainfall starts surpassing evapotranspiration, nitrate can start its downward movement—a typical traveling bulge responding to rainfall accumulation. Naturally, the likelihood that the convexity of the bulge will be greater goes up with the quantity of applied fertilizer. The obvious precaution is to minimize the accumulation of soil nitrates in the late summer and fall months by avoiding excessive spring fertilizer applications which cannot be absorbed by the harvested crop and by abstaining from any fall applications. Different soil structures and textures will also be critical in determining the loss rates although it appears that there is little difference in the amount of rain needed to remove nitrate from surface layers of both light and heavy soils.

Placement of nitrogen is another decisive factor in runoff losses of the nutrient although there are disagreements as to the best courses to follow. For example, in experiments by Whitaker et al. (1978), broadcasting 30% of the annual nitrogen application to corn rather than banding the same amount 10–12.5 cm into the soil resulted in doubled losses. Yet when the moisture content exceeds field capacity, nitrates present in bands can move downward very rapidly (''dropping-out'').

Continuous planting of row crops on sloping land—in the United States, above all corn cultivation in the rolling parts of the Corn Belt—had led to erosion rates found to be incompatible with permanent agriculture (Brink et al., 1977), and to nutrient losses which greatly surpass those of simple leaching. In spite of the high awareness of the problem and many corrective measures taken to slow down the losses, soil erosion remains an acute environmental worry throughout the world, a process whose considerable contribution to nutrient removal is often underappreciated. Naturally, all factors governing the soil loss also determine the nitrogen losses (i.e., rainfall amount, distribution, intensity, soil erodibility, slope length, management practices), but Morisot (1981) criticizes the use of Wischmeier's universal soil loss equation in accounting for nitrogen losses mainly because the nitrogen contents of sediment and soil are different and because nitrogen can also be in soluble form.

In nationwide terms, nitrogen losses through erosion can add up to huge totals. Larson et al. (1983) calculated the total nitrogen loss in eroded sediments from United States cropland at 9.49 million t—nearly the same amount as annual applications of the nutrient in synthetic fertilizers!. Estimates for China show that the eroded soil currently carried by the country's rivers contains more nitrogen than the nation's total fertilizer output (Smil, 1984a).

The advantages of vegetation-covered surfaces are thus obvious. In the world's most erosion-prone region, the Loess Plateau in the middle Huang He basin of China, soil losses from cultivated fields (up to 3570 kg/ha per year) can be cut to only 93 kg by seeding a pasture (Tong and Bao, 1978). And vegetation cover will also lower dramatically the leaching losses. Dutch lysimeter studies (Kolenbrander, 1977) show nearly zero losses under permanent grassland and the highest leaching after a leguminous crop.

When grasslands are converted to fields, subsequent nitrogen releases from soil organic matter may determine the leaching for a long time to come. If a soil has an organic nitrogen content of 0.2% and 1–3% of this is mineralized annually, between 22 and 667 kg N would be added each year per hectare. As this release is stimulated by cultivation and as the movement of percolate waters is very slow in many areas, high NO_3^- levels in some aquifers may merely reflect the onset of farming some decades, or even a century, ago (Keeney and Gardner, 1975).

Finally, leaching losses from plant tops must also be mentioned as they can remove surprisingly large amounts of nitrogen which is transferred to soils. Evidence of these losses was only recently compiled and discussed for the first time in a systematic way by Wetselaar and Farquhar (1980). Significant proofs of this phenomenon are contained in dozens of experiments and trials done on all continents and involving various grain crops (wheat, rice, sorghum, barley, corn), annual and perennial pasture species (e.g., ryegrass, clovers, timothy, bluegrass), as well as cotton, hemp, and tobacco.

Wetselaar and Farquhar (1980) thus had to conclude that decline in nitrogen content of tops is not a rare occurrence and that the phenomenon is a highly significant one owing to its ubiquitous distribution and to the magnitude of reported losses whose maxima range from 16 to 70 kg N/ha representing commonly 20–40% of the peak nitrogen levels.

In comparison with a large number of nitrogen leaching studies on farms, fewer accounts are available for fertilized forests and the observations are usually over shorter periods. Still it appears to be possible to generalize that applications of urea, by far the leading fertilizer used in forestry, have little likelihood of significant nitrogen transport into groundwater. The main reason is, of course, that ammonium ions, the hydrolysis product of urea, are rapidly immobilized in forest soils.

Of more immediate concern are the losses of nitrate from forests after clear-cutting and complete tree harvesting. Although some forests disturbed in this manner respond with only a very small additional NO_3^- releases (just a fraction of 1 kg/ha), other NO_3^- loss studies found a massive and rapid loss with annual losses reaching as much as 97 kg N/ha compared to 2 kg NO_3-N/ha in control plots (Likens et al., 1970). Although the destructive action has the same effect everywhere—increased temperature and moisture availability accelerate mineralization while nitrogen uptake by plants is severely reduced or completely eliminated—nitrogen losses could be prevented or delayed by three kinds of processes (Vitousek et al., 1979).

First, accumulation of NH_3 in soils can be prevented or delayed by nitrogen immobilization, NH_3 fixation in clays, NH_3 volatilization or intake by re-growing plants. Second, accumulation of nitrate may also be prevented or delayed by a lag in nitrification, rapid denitrification, or reduction to NH_3. Finally, nitrate can accumulate in the soil but its leaching into waters can be prevented or moderated by absorption on iron and aluminum oxides, denitrification, or lack of sufficient percolating water (i.e., enough moisture for nitrogen mineralization and nitrification but not enough to transport NO_3^- vertically into streams or groundwater).

No matter how nitrogen enters the waters, it has the potential to cause serious alterations in ecosystems which, before the element's addition, were nearly always marginally deficient in the nutrient, and it can accumulate to the point of being a health hazard in drinking water. Eutrophication of aquatic ecosystems and nitrates in water have thus become serious environmental concerns.

3.4.1.5. NITROGEN IN WATERS Nitrates leached from soils are not the only form of nitrogen in waters: their typical lake and river concentrations ranging from fractions of 1 mg to 4 mg/liter are rivaled by those of ammonia (usually 0–5 mg/liter), and dissolved organic nitrogen (amines, amino acids, peptides, purines) can also reach a level of a few micrograms per liter (Forsberg, 1977). And there are also small quantities of dinitrogen in solution, nitrite nitrogen (no more than 0.01 mg/liter in surface waters), and particulate nitrogen (as a constituent of aquatic biomass). As nitrogen compounds in water are readily interconvertible, the presence of all forms must be considered in evaluating the potentially harmful loading.

Naturally, research on aquatic nitrogen has always concentrated on the cycling of nitrate and ammonia and on the role these compounds play in limiting the primary productivity in different tropic states. Comprehensive reviews by Keeney (1972), Martin and Goff (1973), and Harremoës (1975) were published at a time of growing concern about the effects of anthropogenic nitrogen compounds in aquatic environments, the concern brought to North American public attention most prominently through Commoner's (1968, 1971, 1975, 1977) writings.

Commoner started from a nationwide nitrogen balance, arguing that while the annual nutritional turnover of nitrogen in the United States is 7–8 million t, anthropogenic additions are some 10 million t, and that a large proportion of the unused nitrogen, converted to nitrate, finds its way to streams and rivers, resulting in increased eutrophication and potential hazard from nitrite poisoning (methemoglobinemia).

Commoner's principal examples of locally and regionally serious interference in the nitrogen cycle were taken from the Corn Belt, the country's most intensively fertilized farmland: his leading example was the historical trend of nitrate concentrations in "most of the rivers of Illinois" showing "appreciable rise in the average nitrate concentrations" which were eventually approaching the limit set for drinking water (10 mg NO_3^-/liter). He also showed that the total nitrate-nitrogen entering some Illinois rivers from urban sewage was an insignificant fraction of the total NO_3^- burden, that these concentrations were above those leading to heavy algal growth (0.3 mg NO_3^-/liter), and that the eutrophic pollution of surface waters in Illinois "is now critical—largely due to fertilizer."

These messages of dire environmental threats were rapidly picked up not only by the popular press but also by scientific journals, publishing accounts embarrassing in their eagerness to sound as gloomy as possible. The tenor of the time—the peaking of the environmental movement in the early 1970s—was certainly responsible for much of the ready acceptance but a critical evaluation of the information on which the outcry was based shows how grossly exaggerated the whole case was.

For this critical evaluation we must thank Samuel R. Aldrich, a former Assistant Director of the Agricultural Experiment Station at the University of Illinois, whose book (Aldrich, 1980) is undoubtedly the best available treatment of environmental implications of nitrogen in American agriculture. The following paragraphs will present the highlights of his impeccable analysis.

To begin with, Aldrich pointed out that Commoner's claim of the highest temporal NO_3^- level rise (about threefold) in an Illinois river (the Kaskaskia) is an artifact resulting from sampling at two different locations; comparisons at the same site show no significant change at all! However, in many Midwestern rivers, nitrate content did rise but the rates were nowhere as high as erroneously reported for the Kaskaskia by Commoner. Aldrich (1980) comments that there "is no effective means for correcting errors that become the basis for the conclusions reported in *Fortune, Time, The New Yorker* and *The Wall Street Journal.*"

Reviews of all the available measurements of NO_3^- levels for 11 Midwestern rivers offered by Aldrich (1980) show no strong trends between the mid-1950s and the mid-1970s, the time of the most rapid expansion and intensification of fertilizing. Most of the smaller Illinois rivers with watersheds draining some of the world's most productive soils high in organic matter and still heavily fertilized had slight NO_3^- increases between the mid-1950s and the late 1960s but during the 1970s the NO_3^- concentrations either leveled off or slightly decreased.

Aldrich (1980) offers three possible explanations for the unexpected break in the upward NO_3^- trend after the late 1960s: first, a large increase in the amount of nitrogen removed from the land by harvested crops (total for the 1970s nearly doubled in comparison with the 1950s); second, a leveling off of the row crop areas; third, a temporary leveling off of nitrogen applications between 1966 and 1974. Even at the times of the highest recorded concentrations in the late 1960s when nitrates in Midwestern rivers exceeded the health standard, it was so only at 5 of the 12 sampling sites and usually only slightly above the level of 10 mg/liter and hence well within the built-in safety margin.

United States Geological Survey monitoring of nitrate concentrations in rivers elsewhere in the country shows that with the exception of the Santa Ana River in California (which had up to 7 mg NO_3^-/liter), only three among many hundreds of readings at 88 stations over a period of 43 years exceeded 3 mg/liter. Only the San Antonio River (Texas), the San Joaquin and Santa Ana in California, and the Schuylkill River in Philadelphia show clear upward trends; in other cases there is either little change or gentle decline.

In 1975, data for large rivers put only the Mississippi in Missouri at over 1 mg NO_3^-/liter (about 1.5), with most other streams having less than 0.5 mg/liter. Clearly, only some of the small to medium Midwestern streams and a few rivers in intensively farmed parts of California have in the recent past experienced either almost constantly or seasonally elevated NO_3^- surface water concentrations which were significant fractions of the health standard—or may have even topped it for short periods. This is a far cry from Commoner's (1975) conclusion about "the nitrogen cycle seriously out of balance."

Moreover, Aldrich's (1980) detailed review of studies relating agricultural practices to nitrates in water presents numerous uncertainties—such as the impossibility of relying on $^{14}N/^{15}N$ tracer technique to quantify the portion of NO_3^- reaching the waters from fertilizers or the lack of any reliable estimates of nitrogen lost by denitrification—and makes it clear that the earlier reports categorically ascribing precise percentages of nitrates in surface waters to leaching fertilizers should be properly seen as only qualitative, at best roughly quantitative, appraisals.

Nitrate content in many wells in farming regions used to be considerably above the current health standard and it led to many cases of infant methemoglobinemia (including some deaths) in the 1940s (see Section 3.4.3.2). Studies in Missouri, Illinois, and New York were able to trace most of the contamination to animal wastes (in barns, feedlots) or to domestic waste-disposal systems in whose vicinity the wells were dug—and often improperly lined and maintained. Properly placed, well-cased, and deeper wells are the obvious solution.

Rajagopal and Talcott (1983) provide an unusual series of average groundwater NO_3^- concentrations for Iowa for five decades between the 1930s and the 1970s. The means are based on data from over 6000 wells and show that the concentrations in deep public wells have remained virtually unchanged at around 2 mg NO_3^-/liter, those in nonpublic deep wells declined from over 5 mg to about 3 mg/liter as did the levels in public shallow wells, from as high as 18 mg in the 1940s to about 12 mg/liter in the 1970s.

The authors' survey also confirms that private shallow wells continue to be the most likely sources of high NO_3^- water. Of the 44,000 private wells listed, 18% had nitrates above the maximum recommended level of 45 mg NO_3^-/liter but among the shallow wells this share was nearly 30%. Besides the average high values, shallow private wells also have large seasonal fluctuations with peak NO_3^- levels in May exceeding 50 mg/liter as a result of spring runoff from surface sources.

This study made no attempt to specifically identify and quantify the sources of high NO_3^- levels and hence it is impossible to state what parts of the problem in Iowa farms in the early 1980s are attributable to fertilizer leaching, poor management of feedlot and barnyard wastes, or natural mobilization of nitrates. This last category of nitrate origins should not be neglected as some unexpected rises of NO_3^- concentrations can be explained by mobilization of organic stores: when the organic-matter-rich land at the bottom of Hula Lake and the adjacent swamps were drained and reclaimed in the late 1950s, rapid decomposition of organic matter and subsequent nitrification led to considerable NO_3^- leaching into the Sea of Galilee in Israel (Aldrich, 1980). And, on the other hand, some of the world's most intensive fertilization appears to have made virtually no change in NO_3^- levels in drinking water.

This conclusion comes from Kolenbrander's (1972) comparisons of NO_3^- contents in Dutch drinking water over a 40-year period, perhaps the most important evidence regarding the long-term effects of very intensive fertilization on water supplies; this study established that only one-third of the waterworks examined had marginally higher NO_3^- content (up by just 0.57 mg/liter), and the rest remained unchanged although the annual nitrogen applications over the studied period rose by about 150 kg N!

If the accumulated evidence does not support any alarmist attitudes concerning the trends and levels of nitrates in surface waters and their effect on human health (see Section 3.4.3.2), there is still the concern with the effect of these compounds on aquatic ecosystems. Paralleling the warnings about nitrate health risks during the late 1960s and early 1970s, a good deal of attention was focused on the contribution of fertilizers to eutrophication of lakes. This degradative process is frequently initiated by unusually high productivity of surface algae responding to influx of nutrients; soon the anaerobic decomposers at the lake bottom have a surfeit of substrate from the dead algal matter and produce a large amount of toxic substances which may overwhelm the filtering capacity of submerged photosynthetic bacteria, reach the surface algae, and reduce their photosynthetic oxygen output. The lake gradually turns anaerobic, endangering fish and other higher aerobes.

The key factor in initiating the eutrophication process is a boost in the supply of the limiting nutrient, and for years aquatic biologists argued about the relative roles of phosphorus and nitrogen. There is now a general agreement that phosphorus is the most frequent growth-limiting macronutrient but nitrogen assumes the role in shallower eutrophic lakes at annual loading rates in excess of about 2–3 g N/cm^2; phosphorus governs the primary productivity in oligotrophic and most mesotrophic lakes (Vollenweider, 1968).

For example, the often-mentioned goals of 0.3 mg NO_3^-/liter for lakes and ponds and 1 mg/liter for flowing waters appear to be inherently unrealistic in the overwhelming majority of circumstances where *no* nitrate inputs from fertilizers are involved. As mentioned in Section 3.3.2.2, average nitrate concentrations in rainwater are most of the time above 0.5 mg/liter (and up to 3 mg/liter), and the studies of Illinois streams before the general use of nitrogen fertilizers found that waters flowing through areas of soils naturally rich in organic matter had NO_3^- levels of 1.5–2.0 mg/liter even in areas of low population and livestock density (Aldrich, 1980). Similarly, the Mississippi in Illinois averaged 1.5 mg/liter before the start of universal fertilizing and before the introduction of large feedlots.

Certainly, eutrophication can be an unsightly, bothersome, and in some cases eventually outright harmful process, but any generalization blaming higher and more widespread nitrogen fertilizer applications as *the* cause of the phenomenon is clearly misplaced. Genesis and process of eutrophication is a multifactorial affair with phytoplankton and algae responding to levels of several macro- and micronutrients and a host of other environmental variables; for example, algal growth in the Midwestern rivers with relatively high nutrient loadings is greatly restricted by the severely limited light penetration owing to large sediment loads.

While nitrogen leached from fertilizer must be considered as an important potential contributor to eutrophication, and while this degradative process certainly warrants continuous attention and control measures in some heavily afflicted areas, the surface waters of rich Northern Hemisphere nations practicing intensive fertilization are not being turned into swamps at any perilous rate and most of the problems encountered in the past three decades can be traced to phosphorus or to urban waste releases rather than to leaching of fertilizer nitrogen.

Nitrogen in waters is thus certainly unwelcome and nitrate inputs from any source should remain an object of caution, attention, further research, and effective control wherever practically possible—but in the mid-1980s we must view the problem with considerably less apprehension than was shown in the 1970s. Before looking at the health hazards, I shall review interesting efforts to quantify nitrogen cycling in agroecosystems.

3.4.2. Mass Balances in Agroecosystems

Quantification of nitrogen flows on meso- and microscales started to advance rapidly at about the same time as the global nitrogen budgeteers began publishing their estimates reviewed in detail earlier in this chapter. Consequently, virtually all published national, regional, and local agroecosystemic nitrogen balances originate in the period since the mid-1970s and are thus readily comparable in many ways. Moreover, these comparisons were facilitated by the Amsterdam symposium on nutrient cycling (Frissel, 1977a) which assembled an unprecedented data base on macronutrient flows in a great variety of arable as well as livestock production units.

I will begin this review with some approximate calculations of overall nitrogen balance and utilization efficiency for the United States and with suggestions based on such accounts. Then a separate look at the tropics is necessary as most of our knowledge on nutrient balances is derived from managing intensive temperate latitude fields and cannot be extrapolated to the rainy and hot tropics—a region that will have to supply much more food in the future, a task impossible to sustain without greater understanding of nitrogen's behavior.

3.4.2.1. Nitrogen Budgets The most appropriate departure for an inquiry into nitrogen budgets in agroecosystems is to assess first the nutrient's fate in natural ecosystems which were displaced by fields and grazing land and then to look at reservoir and flux changes brought by cultivation. There has been no shortage of studies of nitrogen budgets and the published diagrams vary quite considerably, partly owing to the necessity of incorporating specific reservoirs and flows as well as owing to preferences and biases of individual researchers. Clark and Rosswall's (1981) collection contains individual papers on nitrogen cycling in all important ecosystems.

Here I will only point out in a simplified manner several generalized relationships between nutrient cycling and successional state as they were formulated over the years. Odum's (1969) influential model assumes continual reduction of nutrient losses with the successional progression and hence a continuous accumulation of nutrients. Vitousek and Reiners (1975) argued that ecosystems store nutrients during the middle successional stages but with maturity the losses increase to equal inputs. Woodmansee (1978) proposed a concept of "abiotically controlled pulse stability" when large, even catastrophic, nitrogen losses occur as a result of water or wind erosion, leaching, volatilization, and denitrification and are subsequently sharply reduced by new plant growth: Figure 3-10 compares the three concepts.

Although Woodmansee's (1978) concept was derived from study of semiarid grasslands, he believes it is applicable also to forests where fires can act as strong abiotic nutrient pulse generators. Retarded losses of nitrogen are then basically seen as a function of primary production, and associated processes of microorganismic, soil colloid, and dead matter storage *regardless* of successional stage—not a function of the ecosystem's maturation.

The usual net effects of cultivation on soil nitrogen content and cycling have been the reduction of sink size, including both the labile and the more resistant organic nitrogen stores, and acceleration of transfer rates (Power, 1981). The presence of nitrogen in the upper soil layers is a good indicator of fertility and its changes mirror very well the decline and recovery of agroecosystems.

Jenny (1941) generalized the typical trend for temperate regions as follows: first, a rapid decline of soil nitrogen content during the first 20 years of cultivation when one-quarter of the initial mass may be depleted through denitrification, volatilization, and leaching; loss of another 10% of soil nitrogen in the subsequent 20 years with an additional 7% departing between the 40th and 60th year of cultivation.

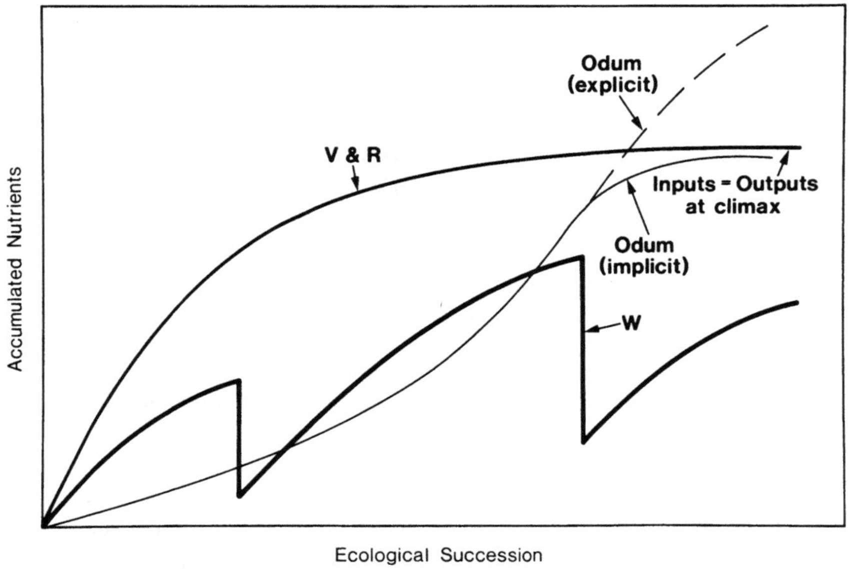

FIGURE 3-10. Three theoretical outlines of nutrient accumulation during ecological succession (Woodmansee, 1978). W, Woodmansee; V & R, Vitousek and Reiners.

Soil nitrogen levels in the Morrow plots at the University of Illinois illustrate how the rates of decline varied with different rotations and fertilizer treatments (Figure 3-11).

Comparisons of the Morrow plots clearly show that proper management can arrest nitrogen losses at levels surprisingly close to the original equilibrium or to establish a new equilibrium, albeit at a fraction of the original level. North American field farming appears to be now, after a century of rapid soil nitrogen losses, generally near steady state or is asymptotically approaching zero nitrogen loss; small declines are still occurring in some agroecosystems, especially in wheatlands, while in a very few cases the nutrient may already be accumulating in soil (Thomas and Gilliam, 1977). Consequently, a new equilibrium at about 50% of the original level is almost in place.

But proper field management may not only set a new equilibrium—it may also lead to steady soil nitrogen increases. In classical long-term cropping and fertilization trials at the Rothamsted Experimental Station in England, soil nitrogen levels have nearly doubled since 1852 with annual manuring at 35 t/ha per year (Stevenson, 1982b). On a national level, a historical appraisal of long-farmed German soils from about 1800 to 1978 indicates an approximately threefold increase in plant-available nitrogen, with soil reserves (humus, mineral weathering) rising from 40 to 85 kg N/ha, organic manure applications going from 8 to 40 kg N/ha and chemical fertilizer from 0 to 92 kg N/ha (FAO, 1981b).

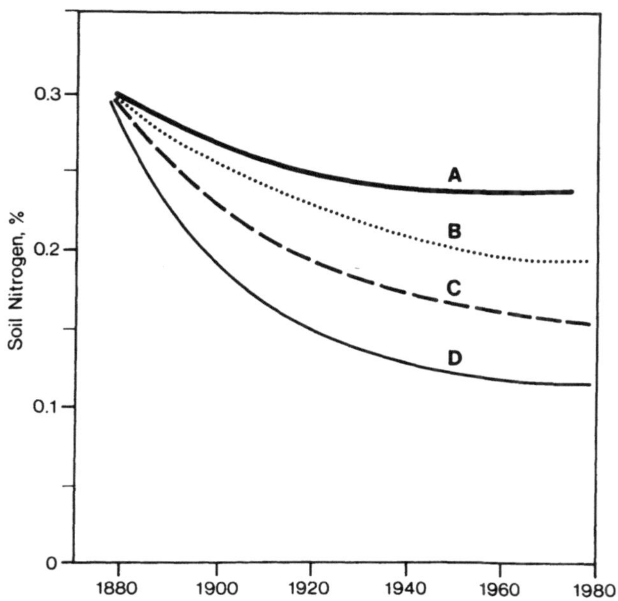

FIGURE 3-11. Soil nitrogen during 100 years of the University of Illinois (Urbana-Champaign) Morrow plot (rotation experiments: A = corn, oats, corn with manure, lime, and phosphate; B = corn, oats, clover; C = corn with manure, lime, and phosphate; D = corn only) (Stevenson, 1982b).

Similarly, traditional Dutch arable farming around the year 1800 had total inputs of some 80 kg N/ha per year, most of these coming from roughly equal contributions by manuring and natural fixation; current inputs are about five times as large (around 380 kg N/ha per year) with chemical fertilizers providing some four-fifths of the total (Frissel, 1977b). Outputs have shifted appropriately with plants now taking up around 200 kg compared to 50 kg N/ha per year in 1800, leaching losses going up about twofold to nearly 60 kg, and denitrification rising from some 20 to 70 kg N/ha per year. Fluxes are thus much more intensive but the approximately zero-loss state during the early decades of the 19th century had been changed into steady nitrogen gains of several tens of kilograms per year (55 kg as calculated from the balance).

No generalizations on a global scale are possible but extensive data presented at the Amsterdam symposium on nutrient cycling (Frissel, 1977a) allow some tentative conclusions. Analysis of 38 agroecosystems for which data on the changes in the soil organic pool are available revealed no change in 13 cases, gains (averaging 35 kg N/ha per year) in 22 cases, and losses (mean at just 9 kg N/ha per year) in only 3 cases (United States corn and wheat farming). Because such relatively low losses are easily replaceable by legume cultivation, Floate (1977) concluded that concerns about exhaustion of organic nitrogen reserves in farming soils are not warranted by the best data currently available for a variety of agroecosystems.

But these organic pool gains are apparently associated with available nutrient pool losses: of the 18 systems without fertilizer 7 were steady, 2 gaining, but 9 losing an average of 11 kg N/ha per year; of 19 fertilized pools, 7 had no change, 6 were gaining, and 6 losing (mean of 38 kg). How long such losses could be sustained is, however, impossible to say without information on the sizes of individual pools.

Information gathered for the Amsterdam symposium also made possible approximate calculations of nitrogen utilization efficiency. Comparisons of consumable nitrogen output and farm inputs (in fertilizers, manure, irrigation, rain and by fixation but disregarding mineralization) showed that arable systems with inputs up to 150 kg N/ha per year have typical efficiencies close to 66% while for higher inputs the performance appears to be somewhat lower at around 50% (Frissel and Kolenbrander, 1977).

Yet Frissel and Kolenbrander's (1977) approach to efficiency of nutrient utilization may lead to dubious results: considering only the nutrients present in consumable farm output measures neither the plant uptake nor the recycling of nutrients which become available for subsequent crops (Karlovsky, 1981). Such a measure is not dependent on soil properties or the plants' absorption capabilities and it leads to incongruities of treating large storages of applied nutrient in soil as net losses—while a major part of nutrients so sequestered will eventually become available to succeeding crops.

Karlovsky's (1981) approach defines utilization efficiency as the sum of the total amount of nutrient taken up by the land and gains (or losses) in available nutrient in the soil divided by total nutrient additions. When recalculated in this fashion, the Amsterdam symposium data on various agroecosystems show that more than three-quarters of these have efficiencies between 50 and 100%, in great contrast to Frissel and Kolenbrander's (1977) results that two-thirds are less than 50% efficient in their use of nitrogen.

Settling the question of nitrogen utilization efficiency in the fields is obviously critical for any attempts at approximating the overall efficiency with which we interfere in the element's cycle in our efforts to grow more food. Detailed nitrogen budgets on a regional or national level are another difficult precondition for such calculations: discussions of uncertainties permeating the quantification of most nitrogen fluxes should have made it clear that with the exception of chemical fertilizer inputs, not much reliance can be placed on any flow estimates even when operating on a meso or microscale rather than on a global level. Still, I shall try to present a fairly comprehensive account of nitrogen flows in the United States agroecosystem and then to discuss its implications.

3.4.2.2. NITROGEN IN UNITED STATES AGROECOSYSTEM Perhaps the best point to start the accounting is with the final products of the whole effort—with foodstuffs consumed annually by the American population (assumed here to total approximately 230 million people). Calculations of nitrogen flows in the human food chain can be done with great accuracy for plant foods but major uncertainties enter with accounting of animal protein. Table 3-9 presents United States per capita consumption of all major foodstuffs and lists their protein contents: Americans consume annually about 11 kg of plant and 23 kg of animal protein, i.e., a total of some 5.5 kg of nitrogen in their diets, or 1.2 million t N for the whole country.

But, obviously, the total amount of nitrogen required to produce animal protein is far greater than the 800,000 t of nitrogen contained in meat, dairy products, and eggs as most of the protein fed to animals is spent in their growth, maintenance, and reproduction and only a fraction is passed on for human consumption. Plant-to-animal protein conversion factors should take care of these inevitable inefficiencies but their published values differ quite substantially.

The difference is greatest for beef which is by far the leading contributor of animal protein in American diets: Sims and Johnson (1972) use a conversion factor of 6.7, NAS (1972) 9.0, and Cook (1979) 13.7. For pork the three sources use 5.0, 9.0, and 8.3, for poultry 4.0, 6.0, and 5.9, for milk 3.3, 4.0, and 4.5, all being much better agreements. Depending on which set of the multipliers one uses, it takes 4.4 or 6.0 or 7.5 million t of plant nitrogen to produce the animal protein eaten in 1 year by the United States population—but the totals should be enlarged by at least 20% to account for growth and maintenance of sire and dam animals and for disease losses to yield values between 5.3 and 9 million t N.

There are three alternatives to check these uncertain totals. One can, naturally, gather the available figures on feed consumption published annually in the USDA's *Agricultural Statistics* and convert them to nitrogen equivalents. In recent years, United States domestic animals were fed annually about 16 million t of high-protein oilseed meals and cakes (largely from soybeans and containing about 1.1 million t N), 2.5 million t of animal protein (almost 200,000 t N), 10 million t of mill

TABLE 3-9
UNITED STATES PER CAPITA CONSUMPTION OF DIETARY PROTEIN AND NITROGEN FROM VEGETAL AND ANIMAL SOURCES AROUND 1980[a]

Foodstuffs	Annual per capita consumption (kg)	Average protein content (%)	Annual protein consumption (kg/person)	Plant-to-animal protein conversion factor	Total plant protein (kg/person)	Total nitrogen flow (kg/person)
Plants			11.3	—	11.3	1.8
Wheat flour	50.0	12.0	6.0	—	6.0	
Corn	7.0	9.5	0.7	—	0.7	
Rice	3.0	6.7	0.2	—	0.2	
Legumes	8.0	22.0	1.8	—	1.8	
Potatoes	40.0	2.1	0.8	—	0.8	
Vegetables	95.0	1.4	1.3	—	1.3	
Fruits	65.0	0.7	0.5	—	0.5	
Animal foods			25.4		119.4	19.1
Beef	55.0	16.0	8.8	6.7	59.0	
Pork	25.0	10.0	2.5	5.0	12.5	
Poultry	25.0	18.0	4.5	4.0	18.0	
Milk	135.0	3.5	4.7	3.3	15.5	
Cheese	7.0	25.0	1.8	3.3	5.9	
Eggs	15.0	11.0	1.7	5.0	8.5	
Fish	8.0	18.0	1.4	—	—	
Total	538.0	—	36.7	—	130.7	20.9

[a]Sources: consumption levels from the USDA's *Agricultural Yearbooks*; protein content according to Watt and Merrill (1963); plant-to-animal protein conversion factors according to Sims and Johnson (1972).

products (various brans and hulls with a total of 250,000 t N), and about 140 million t of feed grains (mainly corn) and other concentrate feeds (an equivalent of 2.2 million t N): as for roughages, some 130 million t of hay were consumed annually, about two-thirds of it alfalfa (with 2.5 million t N), the rest mostly in clovers (total of 1 million t N). The total of field concentrates and roughages thus comes to roughly 7.3 million t N.

To this must be added grasses grazed by animals on 40 million ha of pastures and almost 340 million ha of the currently used rangeland in 11 Western and 6 Plains states. According to Cook (1979), these rangelands provide a year-long forage supply for 17 million animal units which, using the USDA's standard conversions, translates roughly into an equivalent of 40 million t of corn grain containing some 600,000 t N. As for pastures, I shall assume average yields of 3 t/ha and mean nitrogen content of 1.1% to get a total of over 1.3 million t N/year. This brings the feeding grand total to 9.2 million t N.

Checking the totals from the other end—by looking at the output of animal wastes—is another possibility. There are several estimates of United States animal waste production which are in surprisingly close agreement (all values in million dry t): White-Stevens's (1977) 170, Benemann's (1978) 180, USDA's (1978) 157.7, and mine (Smil, 1983) 171.2. Selecting average nitrogen contents for different animal manures can be done only very approximately so that the total nitrogen in 170 million t of dry wastes can be put as low as 5 million t, using Van Dyne and Gilbertson's (1978) averages, or as high as 7.5 million t (USDA, 1978). For comparison, the NAS (1972) estimate for 1970 was 4.2 million t N, and for 1975, 5.3 million t N (NAS, 1978a). As about 75% of all nitrogen in animal feeds is excreted in manure (USDA, 1978), multiplication of waste totals by 1.33 results in annual plant nitrogen consumption totals between 6.7 and 10 million t.

And to check the validity of these multiplications, it is easy to add the nitrogen incorporated in (edible and inedible) animal tissues to the waste nitrogen total. Starting with roughly 800,000 t of nitrogen in edible animal foods and extending this total by 40% to account for inedible parts, and then by another 25% to cover maintenance of sire and dam animals and losses owing to disease and spoilage, results in some 1.4 million t N incorporated annually in animal tissues—and in plant nitrogen totals of 6.4–8.9 million t.

These calculations are certainly sufficient to establish a reliable consensus: definitely no less than 6.5 million and possibly as much as 10 million t of plant nitrogen is needed to produce meat, eggs, and daily foods containing 800,000 t N (the contribution of 300,000 t of nitrogen in feeds of animal origin, largely fish, is easily disregarded at this level of approximation). With 9 million t of plant feed nitrogen used as the best single value, the overall average efficiency of plant-to-animal protein conversion would be just below 9%, i.e., more than 11 units of the nutrient in phytomass must be fed for each nitrogen unit in animal foods.

This loss of more than nine-tenths of the fed phytomass nitrogen is clearly the key conversion inefficiency in United States farming—but a waste which is not to be eliminated, or even reduced, easily or rapidly. Leaving aside the obvious fact that concentrated feeds such as grain corn, sorghum, or soybeans cannot be used as food and hays such as alfalfa or clovers cannot be directly consumed by humans, any advocacy of skipping the animal link and eating "closer to the sun" in order to raise food chain conversion efficiency ignores two essential realities: different, and mostly superior, qualities of animal protein and the existence of common taste preferences and nutritive desires.

In spite of the far from settled recommendations for the optimum intake and composition of dietary protein (WHO, 1973; Rao, 1976; Hegsted, 1978), there is enough scientific evidence to argue that even drastically lower animal protein consumption levels than those currently enjoyed by the American population would be perfectly acceptable to maintain, even to improve, health and longevity. But deep-rooted taste preferences characterizing American food culture point in exactly opposite directions and so the channeling of almost 10 million t of nitrogen through animal stomachs shall continue.

In fact, even more of the United States-grown plant protein goes that way owing to the country's large grain exports most of which end up again as animal feeds: led by about 60 million t of corn, over 40 million t of wheat, and more than

20 million t of soybeans, a surprisingly large share of nitrogen harvested in American field crops—roughly 3 million t, or over one-third of the total—has been shipped abroad annually in recent years.

With almost 4 million t of crop nitrogen fed to animals, roughly 0.5 million t directly consumed as food, and 3 million t of nitrogen exported, one should have nearly 8 million t of nitrogen in annual field crop harvests (excluding forages). Indeed, the sum for the country's seven top field crops comes to 7.5 million t N and the addition of 10% to account for all lesser crops brings the sum to just over 8 million t N. About 4 million t of nitrogen is incorporated in above-ground crop residues (above all in corn stover and soybean straws). Consequently, the grand total is some 12 million t N incorporated annually in United States field crops in recent years. For comparison, Legg and Meisinger (1982) calculated (for the 1977 harvest) 10.63 million t N and Power (1981) 11.9 million t N, a very good consensus.

According to the USDA (1978), about 70% of all above-ground residues are left on the fields so that only about 1.2 million t N is removed annually in straws and stalks, largely for bedding. The USDA's (1978) estimates are perhaps also most reliable with regard to animal waste recycling: of the 7 million t of nitrogen voided by cattle, hogs, poultry, and sheep, only 39% is produced in confinement and of this total, 73% ends up on the fields. Consequently, automatic recycling by grazing animals amounts to some 4.2 million t N/year while nitrogen initially available in field manures adds up to no more than 2 million t N. However, the storage and application losses are estimated by the USDA (1978) to cut the total actually available to crops by 63%, to less than 750,000 t N.

Three more transfers can be established with satisfactory accuracy. Chemical fertilizer applications can, of course, be given with higher certainty than any other input or loss value: in the late 1970s they were 9.5 million t N/year and at the beginning of this decade they surpassed 10.5 million t. In comparison, nitrogen losses in eroded soil and inputs from atmospheric deposition can be fixed much less precisely but differences between the extreme estimates are almost certainly less than twofold. In the first case I shall use the total of 1.7 million t N (in terms of available nutrient) carefully established by Larson et al. (1983).

The range for atmospheric deposition is easily estimated by assuming an average minimum of 5 and an average maximum of 10 kg N/ha of fields and pastures (total of 530 million ha): 2.7–5.3 million t N would then be deposited annually. A better approach is to consider a variety of measured NO_x, NO_3^-, and NH_3/NH_4^+ precipitation concentrations and deposition values (see Sections 3.1.2, 3.3.1, and 3.3.2) and typical precipitation patterns (1000 mm over most of the country's arable land, 300 mm over most of the rangeland) to use a mean of 8 kg N/ha for farmland and 3 kg N/ha for grazing land for a total of 2.5 million t N annually; however, the NAS (1978a) puts this total much higher, at 5.4 million t N/year.

Having previously discussed the difficulties, if not impossibilities, of coming up with representative averages for biotic fixation (Section 3.2.1) and nitrogen field losses through volatilization, leaching, and denitrification (Section 3.4.1), it should be obvious that the following values will be just the best guesstimates I can offer. For fixation I shall stay on the conservative side by using just 1 kg N/ha for nonsymbiotic fixation in nonleguminous crop fields, 100 kg N/ha in soybean fields, 250 kg N/ha for alfalfa, 50 kg N/ha in other hay fields and in managed pasturelands, and 1 kg N/ha in all other grazing land.

Multiplying these values by appropriate areas (with fallow land included in nonleguminous crop total) gives a total fixation of about 8.2 million t N. For comparison, Power's (1981) differently derived estimate is 7.2 million t N, not a great difference—but both of these values have a large margin of uncertainty. But even the most conservative assumptions would not yield less than 5 million t of biotically fixed nitrogen, while a total between 10 and 12 million t would be a credible high-range limit.

The easiest, and relatively most reliable, volatilization estimate is for NH_3 losses from animal wastes: staying with the USDA's (1978) average of 63% results in some 4.4 million t N released from pastures, feedlots, sties, stables, and poultry lots. Using Lauer's (1975) higher average of 85% would raise the total to 6 million t N/year. In any case, there is a fairly narrow range for the extremes between 4 and 6 million t N.

Evidence presented for volatilization losses following chemical fertilizer applications showed a very wide range, most commonly 5–20% but with recorded extremes at virtually 0 and close to 80%. With more widespread proper injection of ammonia, I shall stay on the conservative side and assume just an average 10% loss for a total of 1 million t N; double this rate would almost certainly be too high.

Determining NH_3 losses from plant tops is the most difficult part of volatilization estimates. Even a conservative assumption of average losses at only 100 g N/ha per day during 60 days of peak vegetative period results in over 800,000 t N just for the field crops. Slightly more liberal assumptions and inclusion of other nitrogenous compounds volatilized from the tops may bring the total easily to 2 million t N/year. But I shall stay with just 1 million for a total volatilization loss from animal wastes, fertilizers, and plant tops at 6.5 million t N.

Denitrification presents no smaller estimation pitfalls. I will use an average of 0.9 kg N/ha per year for all grazing land and idle cropland (a total of just 300,000 t N) plus a mean 15% loss of the applied fertilizer for a sum of 2 million t N annually. Doubling this total would not appear excessive for a high-range estimate. Finally, leaching total is composed of the following parts: rates of 15 kg N/ha for heavily fertilized corn as well as for soybeans, 5 kg N/ha for other field crops west of the Mississippi, 10 kg N/ha east of the river, 2 kg N/ha in leguminous hay fields, and just 1 kg N/ha from all grazing lands. These assumptions add almost exactly to 2 million t N/year, an estimate fortuitously identical with value given by Stanford, et al. (1970); anywhere between 1.5 and 2.5 million t is the most likely total.

The only two important fluxes which remain to be estimated are immobilization and mineralization, both, again, hard-to-generalize processes. Assuming an average 8000 kg of mineralizable nitrogen in farmland soils made available at a rate of 0.5%/year, both being fairly conservative conditions, would release 25 kg N/ha per year for a total of 2.8 million t from all arable land used for annual crops. The NAS (1972) estimated the total for all United States land at 3.1 million t while Thomas and Gilliam (1977) showed typical annual rates to be 50 kg N/ha per year for corn, 15 kg for soybeans, and nearly 30 kg for wheat. I shall stay with roughly 3 million t N. Immobilization is primarily a function of the amount of recycled residues. Using an average of 10 kg N/ha for corn and 5 kg N/ha for other field crops (except for the leguminous ones as their higher nitrogen content is sufficient to support the decay organisms) adds up to about 0.5 million t N.

All the fluxes are listed in Table 3-10 and certainly the most notable outcome of this exercise is that without trying to balance the flows by adjusting the rates—but rather by simply using the best available data or most credible or acceptable estimates—the inputs and removals are within 7% of a perfect balance. Similarly, using equally unprejudiced rough estimates for plausible transfer minima and maxima results in input and removal totals which are again very close (14 and 5% difference, respectively, for the "low" and the "high" sets).

Unless the data, assumptions, and estimates used in this accounting are grossly in error, the only possible conclusions are that the United States agroecosystem is

TABLE 3-10
ANNUAL NITROGEN FLUXES
IN THE UNITED STATES AGROECOSYSTEM[a]

Flux	The single "best" estimate	The most likely range
Inputs	32.2	25–42
Fertilizers	9.5	9–10
Fixation	8.2	5–12
Animal wastes[b]	6.2	5–7
Mineralization	3.0	2–4
Crop residue recycling[b]	2.8	2–3
Atmospheric deposition	2.5	2–6
Removals	30.1	22–40
Harvested crops and residues	12.0	10–13
NH_3 volatilization	6.5	5–10
Hay and grazing	5.4	4–6
Denitrification	2.0	1–4
Leaching	2.0	1–3
Erosion	1.7	1–3
Immobilization	0.5	0–1

[a]All values are in million tonnes per year.
[b]Animal wastes and crop residues are just internal transfers rather than inputs and their totals are incorporated in the input category to show their importance. Double counting is avoided by listing theoretical totals for field and pastureland nitrogen removals. In reality, only 8.4 million t rather than 17.4 million t N is taken away in field crops and grassland phytomass converted to meat, the difference being the sum of animal waste and crop residue recycling.

currently either in balance with respect to its nitrogen flows or that there is a slight accumulation, most likely on the order of 5% of the annual inputs. Such an increment certainly does not appear unrealistic in comparison with the West German situation studied by Huppert (FAO, 1981b) and Dutch values reported by Frissel (1977b).

German gains of plant-available nitrogen appear to be about 5 kg N/ha annually while the transfer equivalent to 5% of the total current United States inputs (about 1.5 million t N) prorated over 190 million ha of cropland and improved pastures would come to less than 8 kg N/ha per year. Interestingly, the NAS's (1972) account of nitrogen inputs and returns for the United States in 1970, derived in a different manner, also ended up with net annual retention of 1.5 million t N in soils and waters. Considering the country's unusually large area under leguminous forages and soybeans, the gains in the two agroecosystems appear comparable and the argument for the United States nitrogen increments receives another bit of support.

However, all of the "evidence" in my claculations or in the cited German and Dutch trends is merely circumstantial: as stressed before, most of the figures are just the best estimates and even the conclusion of steady-state nitrogen fluxes in the United States agroecosystem may be wrong. As Kohl et al. (1978) rightly recall in their critique of the well-publicized study of nitrogen fluxes in the San Joaquin Valley (Miller and Wolfe, 1978), "retrospective results from long term experimental plots show, under most agricultural conditions, a gradual decrease in total soil N with time."

But Figure 3-11, which illustrates this process in the famous Morrow plots at the University of Illinois (Urbana–Champaign), also clearly indicates that a new steady level may be reached with proper management—and that there is nothing inevitable about continued decline. Nor, I would strongly argue, about perfect balance. All of the available nitrogen models for smaller United States regions—for the upper Santa Ana River Basin (Ayers and Branson, 1973), San Joaquin Valley (Miller and Wolfe, 1978), Wisconsin (NAS, 1978a), and peninsular Florida (NAS, 1978a; Messer and Brezonik, 1983)—are perfectly balanced by assuming *a priori* equilibrium and forcing some fluxes into its confines.

Unfortunately, as Viets (1978) reminds us, whole nitrogen models cannot as yet be verified or experimentally tested so that the "conclusions deduced must be compared with what appears to be reasonable and consistent with data from subsystems that are more susceptible" and "emphasis should be placed on the concentrated sources of nitrogen: human and animal wastes and the misuse of N fertilizers." Of course, the "reasonability test" is highly dependent on the reigning scientific consensus and its conclusions may appear dubious in just a few years.

Concentrating on the managed part of the cycle, whose inputs and some removals can be calculated with satisfactory accuracy, several conclusions seem to be fairly incontestable. First, chemical fertilizers are either already the single largest nitrogen input into the United States agroecosystem or are just fractionally smaller than biotic fixation (most likely between no less than two-thirds and nine-tenths).

Second, their share in total nitrogen inputs appears to be somewhere between one-quarter and one-third of the total. Third, when animal wastes are added, anthropogenic nitrogen inputs into the agroecosystem are at least two-fifths and most likely about one-half of the total in flow.

And as symbiotic fixation in soybean, alfalfa, and clover fields accounts for a large share of biogenic nitrogen, the total human-managed inputs is on the order of two-thirds of all additions. Clearly, human actions are now decisive in running the system. Of course, by counting crop residues this share would rise further but recycling of stover, straws, and stalks is a follow-up to removal operations (crop harvests) rather than managed *prima facie* input.

Fourth, nitrogen contained in animal wastes is a relatively large part of the total input (on the order of 20%) but with most of the nutrient voided in unconfined places, only a small part (roughly one-quarter) reaches crop fields where some two-thirds of it is volatilized before it can be taken up by crops. Here is a large potential source of manageable nitrogen—but one whose greater utilization with higher efficiency will be both difficult and costly.

Fifth, combined loss of manure and fertilizer nitrogen cuts the original nutrient availability by about one-half. Volatilization losses from animal manure applied as field fertilizers (1.2 million t N) and from chemical fertilizers (1 million t N) amount to 2.2 million t N, denitrification and leaching each account for about 1.75 million t N, and if three-quarters of nitrogen in the erosion losses is derived from the fertilizers, the total loss reaches 7 million t N, between two-fifths and one-half of the initial input. Primary conversion efficiency of the whole agroecosystem is raised only by about 8% (from 54% to 62%) when following Karlovsky's (1981) advice for calculating the value.

Sixth, the country's preference for animal protein requires large nitrogen transfers to produce its relatively high-nitrogen-content foodstuffs. About 12 million t of nitrogen removed from the agroecosystem for domestic consumption provides just over 1.2 million t nitrogen in foodstuffs, a loss of 90% largely caused by conversion of plant protein in animals and, to a much lesser extent, by processing losses and spoilage of plant foods. The overall efficiency of this transfer—a mere 10%—could be raised three to fivefold by switching to lactoovovegetarianism (assuming the retention of existing protein totals with two-fifths of all nitrogen coming from animal foods) and by nearly an order of magnitude by adopting a diet consisting exclusively of plant foods.

But Americans are not going to turn to yogurt, lentils, pickled cabbages, and bean curd—to emulate two of the world's greatest largely vegetarian food cultures—as long as the country's rich soils and rangelands keep providing the bountiful amounts of feed to turn out annually more meat per capita than the average consumer's body weight, and still to have a third of the total harvest available for export.

That these habits require managed nitrogen subsidies (in fertilizers, animal manures, leguminous forages) appreciably surpassing natural inputs and that they result in cycling of some 25 kg of nitrogen through the agroecosystem for each kilogram of the nutrient contained in the foodstuffs goes almost unnoticed—owing both to the ready, and still rather cheap, availability of the fertilizers, machines, and management infrastructures and to the relatively limited environmental effects of the practice.

Although the price is paid in local environmental degradation and some longer-term effects are even more worrisome (erosion, above all), nitrogen management in American farming has clearly not led to Commoner's dire prophecies. The practices are undoubtedly in need of adjustments and improvements but the country's overall nitrogen cycle does not appear to be either out of control or in a degradative slide requiring drastic emergency changes.

In spite of their problems and inefficiencies, temperate agroecosystems afford a better management milieu for nitrogen flows than tropical farming whose success will determine the livelihood of hundreds of millions of people in rapidly growing Latin American, African, and Asian nations but whose efficient operation faces some fundamental natural obstacles.

3.4.2.3. TROPICAL EXPERIENCE Although every generalization for a region with such a wide variation of climate, soils, and natural vegetation will abound with exceptions, it is quite obvious that nitrogen reservoirs, utilization, cycling, losses, and fertilization in the humid tropics differ substantially from those of temperate zones and that these differences must be well understood in order to manage properly the future required increases in tropical farm production.

To begin with, most of the climax vegetation of the region with whose clearing the process of cultivation generally starts—the still extensive expanse of tropical rain forest—is much unlike its temperate counterparts. Tropical rain forests usually have the largest standing biomass (around 450 t/ha appears to be the global mean), the highest share of green parts (around 10% of all biomass), the highest nutrient content of the phytomass as well as the fastest rate of primary production among all forest ecosystems (Whittaker and Likens, 1975; Marion, 1979).

Not surprisingly, the humid tropics also have the most rapid rates of mineralization with as much as 0.7–1.3% of all nitrogen in the fresh litter decomposed per day. Consequently, litter accumulation in tropical rain forests is small, merely a few tonnes per hectare per year. In contrast, in boreal forests where mineralization is very slow, litter can reach masses of 30–45 t/ha per year. But in terms of nutrient efficiency—phytomass addition per unit of nutrients—tropical forests rank last: while temperate coniferous forests, the best performers in that respect, can turn 1 kg of nitrogen into more than 400 kg of phytomass, the typical tropical rain forest will produce just 90 kg.

As a result, the efficiency of litter production per unit of nitrogen is lowest in tropical forests, with nitrogen in litter usually exceeding 100 kg/ha per year but with the dry mass/N ratio of that litterfall just between 60 and 90; in contrast, temperate coniferous forests shed nitrogen in litterfall at rates mostly below 50 kg/ha per year but the dry mass/N ratio of litterfall is as high as 200. Vitousek (1982) examined these relationships and concluded that nitrogen circulation in litterfall is a reliable indicator of nitrogen availability. In many tropical rain forests where potential symbiotic nitrogen-fixers are either dominant or codominant and have an essentially unlimited supply of available nitrogen (for which they have to pay, of course, in carbon and energy), nitrogen circulation is relatively high and efficiency of litter production is relatively low.

Clearing of tropical rain forests removes this intensive nutrient recycling link and rapid decomposition of the exposed litter—and rapid leaching by frequent, heavy rains degrades the site so speedily that sustained animal crop farming becomes impossible and hence it should not even be attempted. Jordan and Herrera (1981) reviewed this classical tenet of tropical ecology and found the well-known explanation to be valid for ecosystems on ancient highly leached soils on Precambrian bedrock or sands. However, not all tropical rain forests fit into this category; eutrophic ecosystems in the Andean region of South America or on volcanic rocks in Central America, Africa, and Southeast Asia are relatively rich in nutrients and forest clearance causes much less subsequent disturbance.

Hence, the answer to the question: "are nutrients really critical in the tropical rain forest?" must be yes for oligotrophic ecosystems which maintain their productivities through a variety of conserving mechanisms (the most important one being associated with the humus and root layer on top of the mineral soil) destroyed after clearing—and not really so much for eutrophic forests where, after cutting and burning, the mineral soil still has the capacity to absorb most of the released nutrients and the ecosystem can sustain much of its productivity.

The basic distinction between the behavior of eutrophic and oligotrophic ecosystems explains why, for example, Harcombe (1977) found in his clearance and recovery experiments in a tropical rain forest on nitrogen-rich soils in Costa Rica that the nutrients were not limiting to successional production after clearing—while most of the upland Amazonian soils appear unfit for continuous intensive farming (Sioli, 1973; Goodland, 1980). However, it turns out that this well-documented loss of soil fertility and complete failure of crops after just three or four harvests following the clearing of a primary or secondary tropical rain forest and planting of not only upland rice, or corn but also soybeans and peanuts can be prevented by careful management at least in some cases.

Experiments carried on in Yurimaguas (Peru) since 1971 have shown that it is possible to have continuous harvest in the typical Amazonian soils of high acidity (pH 4.0), high aluminum content, and serious nutrient deficiencies when appropriate rotations and adequate fertilization are used (Sanchez et al., 1982). For corn and rice crops, this meant applications of 80–100 kg N/ha as well as rectification of several micronutrient deficiencies. But this continuous cultivation method is clearly applicable only to level land spared of the erosion hazards and only at locations offering basic soil-testing and advisory services and having access to outside markets to justify the high fertilizer inputs.

But throughout most of the Amazon's Precambrian shield, the relief is sharply to moderately undulating with unforested land open to massive erosion (up to 100 t of topsoil per ha can be washed from 15° slopes planted to annual crops), soil-testing services are nonexistent and poor road access precludes efficient marketing of even the existing small harvests (N. Smith 1981). The prices of fertilizer are, hardly surprisingly, exorbitantly high and the management skills of the poor subsistence farmers are very meager. Consequently, the likelihood of opening up appreciable areas of oligotrophic tropical rain forest remains minuscule and Goodland's (1980) advise to concentrate on the development of ecosystemically much more robust *cerrado* appears to be the best solution.

But large and increasing areas of the tropical rain forest are being converted to cropland through the ancient practice of slash and burn farming. No accurate figures can be given but globally no less than 200 million people live off the fields established on cleared climax forest and, as the forest's areas are shrinking and population pressures increasing, more and more frequently on cleared secondary ground. The method is sustainable indefinitely as long as the cleared areas are relatively small and rotations fairly long; clearing of larger adjacent areas and insufficient fallow periods spell massive fertility declines and crippling erosion.

Total nitrogen reserves in the climax forest are considerable (600–2000 kg N/ha) and net accumulation rates of nitrogen in the fallow vegetation are at least 100–125 kg/ha per year during the first 2–3 years and they can be much higher; even in dry northern Nigeria the fallow buildup of nitrogen is at least 25 and up to about 80 kg N/ha per year (Bartholomew, 1977). Although most of the aboveground nitrogen is volatilized in the fire, there may be some gain of the nutrient owing to the removal of woody residues whose high C/N ratio otherwise leads to considerable immobilization of nitrogen during early phases of decomposition. The soil nitrogen that builds up during the fallow is then mineralized to support at least one good harvest but rapid leaching may cut the nutrient's availability afterwards.

Leaching rates in the humid tropics are generally more rapid than in the temperate zone, owing not only to high rainfalls but also to frequently poor soils. Data summarized by Bartholomew (1977) show that with rainfall between 500 and 1000 mm, each millimeter of rain produces roughly 1 mm of downward nitrate transfer while higher precipitation (around 1500 mm) can result in movements exceeding 4000 m. Naturally, wide local variations reflect rainfall intensities and distributions, soil and crop properties but it is not unreasonable to assume that most of the nitrogen mineralized after the fallow period can be removed beyond the root zone after a single cropping season. Part of this nitrogen transferred into subsoil layers can be used during dry seasons by deep-rooting fallow species, perhaps even by some arable crops, so that subsoil nitrogen is hypothesized to be an important part of the nitrogen cycle in tropical farming.

Even this very brief review of some critical problems concerning nitrogen in tropical agroecosystems should have been sufficient in conveying the inappropriateness of temperate zone experience for managing tropical farming—and considerable difficulty in arriving at sustainable practices based on local conditions. Undoubtedly, this is an area where further substantial research and practical advances are sorely needed. In contrast, our appreciation of the effects which nitrogen compounds have on human health is sufficient to set safe limits on ambient concentrations of nitrogenous air and water pollutants and to avoid hazardous exposures.

3.4.3. Nitrogenous Compounds and Human Health

While humans are totally dependent on the consumption of nitrogen containing amino acids synthesized by plants and animals for their survival, ingestion or inhalation of several nitrogenous compounds increasingly introduced into the environment through fertilization and combustion may carry mild to severe health risks; another health risk, greater incidence of skin cancer owing to the reduction of stratospheric ozone content aided by higher N_2O emissions, was already discussed in Section 3.4.1.3. The following three sections will summarize the current state of our knowledge regarding the effects of both acute and long-term exposure to nitrogen oxides and possible toxic and carcinogenic consequences of ingestion of nitrates and nitrites.

3.4.3.1. Nitrogen Oxides As with all other major components of polluted air, the health effects of nitrogen oxides have been investigated extensively through laboratory exposures of animals and human volunteers as well as through epidemiological appraisals. Some 80 such studies from the 1950s and 1960s are considered in the USEPA's (1971a) air quality criteria volume and the later evidence is reviewed in even greater detail in the NAS's (1977a) publication on nitrogen oxides, in a review by Gardner et al. (1979) which lists over 120 studies, and in a symposium volume edited by Lee (1980).

Starting with the extremes, the fatal concentration of NO_2 is nearly 300 mg/cm^3, or some three orders of magnitude above the levels normally encountered in polluted urban air. Exposure to 50–140 mg/cm^3 will cause pneumonia and bronchiolitis, both reversible and both apparently without any long-term effects. On the other end, the lowest detectable concentrations of NO_2 are by smell at 0.23 mg/cm^3 and through changes in dark adaptation at 0.14–0.50 mg/cm^3.

Most studies of animal toxicology have been repetitive, short term exposures of high levels of NO_2 to rats, guinea pigs, and mice, although there have been some interesting long-term trials. Perhaps the most interesting finding in animal studies is that, in spite of different sensitivities, the response to short-term high levels of NO_2 has a uniform course characterized by delay between exposure and effect

This means that intermittent short-term exposure, if repeated at intervals shorter than the time needed for the onset of the effect, its course, and the subsequent recovery, may eventually become equivalent to continuous long-term exposure. The lowest concentrations inducing either biochemical changes or reduced resistance to infectivity in experimental animals are 940 μg of NO_2 per cm^3 for 3 weeks to 3 months and the recovery period from a single short-term exposure appears to be at least 17 hr if the cumulative effect is to be avoided.

As always, direct translations of animal injury models to people are extremely uncertain but it is not unreasonable to assume, given the existence of the same cell types in lungs and the need for their renewal, that the human response may be similar. But the temporal sequence of NO_2 toxicity in humans is unknown; nor can we say anything on the chances of adaptation as most of our knowledge of NO_2 effects on humans has come from short-term volunteer exposures (as short as 15 min, as long as half a day, mostly 1 or 2 hr) under carefully controlled conditions. With healthy subjects, there have been no observed effects on pulmonary functions with concentrations up to 1 mg/m^3 or even more; increased airway resistance was definitely encountered at concentrations above 3 mg/m^3, and decreased pulmonary diffusing capacity set in above 7.5 mg/m^3. The few studies done with asthmatics and bronchitics found a similar lack of response with concentrations up to 1 mg/m^3 and exposures up to 2 hr.

In contrast to the multitude of epidemiological studies of SO_2 effects, community assessments of nitrogen oxides are scarce. No differences in pulmonary function were found for: Boston policemen (exposed to the annual 24-hr average of 100 μg NO_2/m^3) compared to their suburban counterparts; office workers in Los Angeles (annual mean of 130 μg NO_2 and 137 μg O_3/m^3) and in San Francisco (65 and 39 μg/m^3 for the two gases); nonsmokers in Los Angeles and San Diego (roughly twofold difference in average annual NO_2 level). And differences of only borderline significance emerged from a Chattanooga study of 4000 children repeatedly exposed to hourly maxima of between 1.4 and 2.8 μg NO_2/m^3.

On the other hand, there have been statistically significant findings of higher incidence of upper respiratory illnesses and more frequent visits to doctors for both children and adults living near large NO_2 sources. Unfortunately, in the most impressive cases of this effect, SO_2, H_2SO_4, and/or particulate matter were also present at high concentrations, making it impossible to assign the effects to a single pollutant. Consequently, there is no clear-cut cause–effect relationship between chronic exposures to elevated NO_2 concentrations and excess respiratory disease, and the available epidemiological studies offer even less guidance than their SO_2-targeted counterparts (see Section 4.4.4).

With the ambient NO_2 concentrations being so considerably lower than any demonstrable biochemical changes, with the existence of several potentially harmful compounds in polluted atmospheres greatly complicating any epidemiological studies, and with undoubtedly more worrisome effects of SO_2, it is rather unlikely that potential health effects of nitrogen oxides will rise to figure prominently in any deliberations about NO_x controls.

Of course, NO_2 is the precursor of nitrates which, together with sulfates, are now being transported over long distances from large fossil-fueled sources exposing many more people to airborne pollutants previously confined to urban and industrial areas. Atmospheric nitrates may be respiratory irritants, as are the sulfates. While in the case of sulfate we appreciate considerable differences in irritation potency of different compounds (see Section 4.4.4), we do not have such a rating for nitrates but as they are mostly NH_4NO_3 they should not be too bothersome. Still, there are reports of nitrates exacerbating asthmatic conditions and correlating with increased asthmatic attacks (Knelson and Lee, 1977). Also, if there were significant conversions of nitrates to airborne nitrosamines, higher carcinogenic risks would have to be considered.

Assessing the possible effects of nitrate aerosols would thus seem to be more important than delving further into NO_2 toxicology, but comparisons of the levels and diffusion of airborne and waterborne nitrates must conclude, especially in the light of the already well-documented health hazards of the latter, that nitrates in soils, waters, and food are of much greater concern.

3.4.3.2. NITRATES Concern about nitrates in food and water does not arise from their toxicity—in fact, acute poisoning with nitrates is possible only if one swallows several teaspoonfuls of the compound at once—but rather from the fact that enzymatic reactions will convert them readily to nitrites whose acute toxicity can cause methemoglobinemia and whose long-term effects have been linked to increased incidence of cancer. With the exception of cured meats, our food and water contain little nitrite but considerably more nitrate (e.g., the differences in nitrate versus nitrite content of vegetables range from two to four orders of magnitude!) so before discussing nitrite loadings and health effects in the next section, a discussion of nitrate intakes is a must.

Establishing typical nitrate intakes for large populations is not easy owing to wide ranges in per capita consumption of vegetables. To compensate for these differences, the NAS Committee on Nitrite (NAS, 1981b) used average intake estimates together with values for vegetarians (having four times the average intake of vegetables) and persons with high intakes of cured meat (also four times the mean). Average American intake is about 75 mg NO_3^-/day with 87% (65 mg) coming from vegetables, 6% from fruits and juices, 2.6% from water, and 1.6% each from cured meats, baked goods, and cereals. Vegetarians would ingest roughly 3.5 times as much (268 mg, 97% in vegetables), while the difference for persons with high intakes of cured meat is minimal (78 mg total).

Except for the relatively small areas with high nitrate levels in drinking water, consumption of fresh and processed vegetables is the dominant source of nitrate intake and, consequently, a great deal of nitrate-related research done during the 1970s concentrated on the processes of nitrate accumulation in vegetables and on the factors promoting its occurrence. Lorenz (1978) offers an outstanding example of this thorough research—with over 25 important vegetables grown at low to high nitrogen fertilization levels for subsequent nitrate determinations in leaves, stems, roots, fruits, and floral parts—while Maynard et al. (1976) and NAS (1981b) present exhaustive summaries of vegetable nitrate concentrations and research efforts.

These summaries include historic reviews which are interesting because the nitrate concentrations recorded in the first decade of this century, the late 1940s, the 1960s, and numerous data from the 1970s enable comparisons between the period when nearly all fertilizers were organic and the period of rapidly increasing inorganic fertilizing. Unfortunately, neither the comparison between organic and inorganic practices (1907 versus 1949), nor the one for low and high inorganic applications (1949 versus 1970s), allows any meaningful generalizations: nitrate-N concentrations are higher for some vegetables, unchanged or lower for others.

Moreover, repeated sampling of supermarket vegetables over several months showed as high as eightfold differences in nitrate levels for a given species, reflecting genetic and nutritional factors and environmental conditions before and at harvest (most notably, light is required to activate nitrate reductase for nitrate assimilation in shoots so that NO_3^- concentrations tend to be lower in the afternoons and on sunny days), and in the absence of detailed knowledge of these variables it would be impossible to rely on historic comparisons even if they did display clear trends.

A more fruitful approach, then, is to determine through carefully controlled trials whether increasing levels of nitrogen application increase nitrate concentrations in vegetables. The answer is usually yes and, not surprisingly, nitrate sources lead to higher NO_3^- accumulation than the ammoniacal ones. In his extensive experiments, Lorenz (1978) found higher NO_3^- accumulation with fertilization rates rising from 0 to 560 kg N/ha for all tested species with the most dramatic increases in beet and kale petioles, the lowest in onion bulbs, carrot and beet roots, and broccoli heads.

This, indeed, is a general rule confirmed in many analyses: petioles and stems are the top accumulators, leaves are next, roots are generally much lower, and fruits and floral parts store very little nitrate. As for the species, beets, kale, radishes, spinach, broccoli, lettuce, and celery are the leaders, with more than 0.5% and not infrequently more than 1% of dry weight as NO_3^--N, while cucumbers, melons, peppers, and onions almost always have very low (less than 0.2% and often well below 0.1% of dry weight) nitrate presence. These varying storages are basically natural and necessary—without some nitrate accumulation, certain species will not yield satisfactorily—but there is no shortage of options to reduce the nitrate intake from such vegetables (Maynard et al., 1976).

In some species (e.g., lettuce), differences among cultivars are small, but in others (e.g., spinach), choice of cultivars may bring nitrate reductions of up to 75%. As already noted, ammoniacal sources are preferable to nitrate fertilizers and further reductions can be achieved by application of nitrification inhibitors (for more details see Section 3.5.2.2): reductions on the order of 50–70% are possible in this way. Afternoon (instead of morning) harvest and sunny (instead of cloudy) day picking can make a 25–60% difference and similar reductions could be gained by discarding kale and spinach petioles or lettuce or cabbage wrapper leaves.

Although nitrate intake in water is usually low (United States average is a mere 1.3 mg NO_3^-/liter of drinking water), people living in areas with nitrate-rich water could ingest about as much as do vegetarians: for a population consuming 1.6 liters of water containing on the average 100 mg NO_3^-/liter, the daily total would be 233 mg NO_3^- with 68% (158 mg) coming from drinking water. However, as noted in Section 3.4.1.4, concentrations of 100 mg NO_3^-/liter are hardly common even in heavily polluted farming regions. Drinking water with 10 mg NO_3^-/liter (the U.S. Public Health Service standard) would result in ingestion of 16 mg NO_3^-/day per capita, still only about a quarter of the average intake from vegetables.

There is no reliable way to assess the intake from polluted air because there are no data on the actual conversion of NO_2 to NO_3^- in humans (NAS, 1981b). But even if all of the inhaled NO_2 were converted to NO_3^-, the average urban concentration of about 60 $\mu g/m^3$ and total daily adult inhalation of 20 m^3 of air would translate into intake of roughly 1.5 mg/day, and double that amount in heavily polluted Los Angeles—both values again a small fraction of vegetable NO_3^- ingestion. On the other hand, NO_3^- from cigarettes and cigars may add 10 or 20 mg/day, a significant contribution.

Clear conclusions made on the basis of this brief review are that except for localities where well or surface water is heavily contaminated with nitrates, vegetables are the only major source of their dietary intake and several preventive and corrective measures can be undertaken to reduce this ingestion by very large margins. After ingestion, most of the nitrate is rapidly absorbed and excreted by adults without any known cases of acute toxicity. However, there may be acute toxic effects in infants after NO_3^- conversion to NO_2^- and further reactions of nitrites are now suspect as producers of carcinogenic substances.

3.4.3.3. NITRITES AND NITROSAMINES There are two very different worries here. The first effect—acute toxicity, cyanosis, hypoxia and anoxia of methemoglobinemia—has been well known for a long time: it is rapid and dramatically life-threatening but it can be readily treated, it is avoidable with simple precautions, and it has been generally declining in the rich world. The second one—possible cancer-inducing effects arising from NO_2^- reaction with nitrosatable substrates such as amines to produce potentially carcinogenic *N*-nitroso compounds (nitrosamines)—arose only since the 1960s, peaking in intensive research, review, and administrative activities of the early and mid-1970s and continuing since then at somewhat lower rates, and it is diffuse, impossible to quantify, and, as will be seen later in this section, difficult to deal with through simple precautions.

Acute nitrite toxicity has been limited almost entirely to infants younger than 3 months owing above all to the fact that at birth the gastrointestinal tract lacks acidity which would prevent its colonization by bacteria reducing NO_3^- to NO_2^-; high nitrate intake, mainly in water and also in pureed vegetables, can thus lead to toxic levels of nitrites absorbed into the bloodstream where they change ferrous iron, the carrier and transmitter of oxygen, into ferric iron which cannot perform these vital functions. This process is compounded by the facts that infants' hemoglobin is converted more rapidly to methemoglobin than in adults, that their red blood cells are less able to convert back to hemoglobin, and, of course, that their fluid intake is much higher per unit of body weight than in adults, resulting in relatively much higher nitrate ingestion.

Normal infant methemoglobin levels in blood are 2.7–4.0%. The first symptoms of cyanosis appear at between 5 and 10%; in hypoxia survivable with treatment, methemoglobin is at 30–40%, and death is likely at 50%, although survivals have been recorded with levels of up to 80%. Depending on the severity of attack, treatment consists of either oral or intravenous administration of ascorbic acid or methylene blue; recovery is rapid, complete, and lasting without any side effects. Even without treatment, methemoglobin level drops rapidly after the intake of contaminated water or food had been stopped.

Vitamin C deficiencies may cause greater susceptibility to nitrites and as such deficiencies are fairly common among infants and children—for example, even in the United States perhaps up to 30% of children in low-income groups receive less than the recommended daily intake of vitamin C—combination of this effect and natural sensitivity to nitrite among infants may create a situation of hypersusceptibility (Babich and Davis, 1981).

Excessive nitrates in drinking well water on farms were first identified as the cause of methemoglobinemia in Iowa in 1944, and 181 cases of infant cyanosis were recorded in the four Midwestern states during the 1940s, 32 of them resulting in death; 30 cases, including 7 deaths, were reported from 13 other states in the same period (Lee, 1970). About 2000 cases of methemoglobinemia were reported in North America and Europe between 1945 and 1971, while in West Germany alone 745 cases were noted between 1950 and 1965, over 97% of them caused by drinking water from private wells containing close to or more than 100 mg NO_3^-/liter.

Because there is no requirement to report methemoglobinemia, the true incidence of toxic episodes in North America is not known but its frequency must now be very low indeed. Urban populations receiving drinking water from public waterworks are well protected by the quality standard limiting NO_3^- level to 10 mg/liter; actual NO_3^- levels in drinking water drawn from protected reservoirs are just a fraction of the standard, and even in the cases of those city water supplies where nitrate concentrations reach or slightly surpass the standard during spring months (e.g., Decatur and Danville in Illinois in the midst of the heavily fertilized corn region), no problems have been identified (Aldrich, 1980).

In fact, there is considerable evidence that normal, healthy children receiving enough vitamin C will be protected even with nitrate levels well above the standard (in most cases up to 20 mg NO_3^-/liter). Moreover, farmers in heavily fertilized areas, especially in the Corn Belt, are now aware of the problem and families wishing to be on the safe side simply use bottled water for infants during the period when nitrates in the wells might be close or above the standard.

The situation is obviously different in many poor countries where in some regions, especially those with rice double cropping, synthetic fertilizers are now used in fairly large quantities and drinking water is often drawn from shallow wells or from lakes, ponds, and streams which may frequently contain elevated nitrate levels. Nitrates can also seep into wells improperly located too close to animal sheds or compost heaps. While there are no statistics to indicate the frequency of these problems, there can be little doubt that especially in the rice-growing areas of Asia, relatively large numbers of infants must have elevated methemoglobin and the risks of acute cyanosis are greater there than anywhere else.

Concern over nitrosamines is an affair more recent than methemoglobinemia risks: it first arose in the late 1960s when analyses found the compounds in cured meats, fish, and cheese but almost at the same time it was realized that nitrosamines could be formed endogenously when nitrites react with amines in the stomach. Soon the pressures developed to ban or to reduce drastically the amount of NO_2^- used in the curing of foods and since then the controversies have continued along with extensive research and new regulations outlawing the addition of nitrates to most products and reducing the use of nitrites. In the late 1970s, nitrite *per se* became suspect as a carcinogenic agent.

This is not the place to go into extensive discussions of these controversies so well reviewed and referenced in publications cited in this, and in the preceding, section. I will outline just the basic facts and assumptions about intake and formation of nitrites and nitrosamines before discussing the broader questions of related uncertainties and risks.

The nitrite content of drinking water is usually negligible; the levels in fresh and processed vegetables are in most cases much lower than 1 mg/kg and the differences among species are not large: only beets, corn, and spinach have levels of a few milligrams per kilogram. Similarly, natural nitrite levels in fresh meats are quite low (1 mg/kg may be a good average) so that the largest dietary source of nitrite is cured meats. Nitrite added to cured meats together with other components in the curing salt mix inhibits the growth of spores of putrefactive and pathogenic staphylococci, bacilli, and clostridia, most notably *Clostridium botulinum*, providing protection against the risk of lethal botulism; it also fixes the distinctive pink color of the cured muscle tissues and imparts specific flavor to some products, especially hams (Cassens et al., 1978; NAS, 1981b).

A large number of nitrate and nitrite analyses of cured meats show the former ranging between less than 1 and 1400 mg/kg, the latter between a trace and 640 mg/kg with average United States values of, respectively, about 40 and 10 mg/kg (NAS, 1981b). Per capita daily intake of nitrite in North America is thus a small fraction of nitrate ingestion, just 0.77 mg with some 40% coming from cured meats, 35% from baked goods and cereals, and 16% from vegetables. Even those with high intake of cured meats (four times the average consumption) would ingest no more than 1.7 mg/day. These data are, however, only useful indicators of relative exposure rather than reliable totals of ingestion: they do not reflect the changes during food preparation and hence do not show the levels actually ingested.

In any case, directly ingested nitrite appears to be dwarfed by endogenous generation of the compound. Ingested nitrate is absorbed from the gastrointestinal tract and most of it is excreted in urine; however, about 25% of the ingested NO_3^- is secreted by salivary glands and roughly one-fifth of this amount, i.e. some 5% of all ingested nitrate, is reduced to nitrite by bacteria residing in the mouth. Normal NO_2^- content in saliva is 6–10 mg/liter and the concentrations appear to be directly proportional to the amount of nitrate ingested. The total daily nitrite exposure in the upper gastrointestinal tract of an average adult with normal gastric acidity is then the sum of dietary NO_2^- intake, about 0.7 mg, and salivary NO_2^-, about 3.5 mg, for a total of 4.2 mg.

So far the best technique for estimating the extent of *in vivo* formation of nitrosamine is by assaying urine samples for nitrosoproline. This conversion also appears to be proportional to the amount of NO_3^- ingested but any estimates of average amount of endogenously produced nitrosoproline are very uncertain as the actual intake of nitrosatable amines is unknown, nor can we assume that dietary amines are nitrosated at the same rate as proline—or that the stomach is the primary site for these reactions. Nevertheless, making these and other assumptions, the NAS (1981b) estimates that with average daily exogenous intakes of 75 mg NO_3^- and 0.77 mg NO_2^-, 2.2 mg of nitrosoproline is produced endogenously if there is no ascorbic acid available to react with nitrite.

This, of course, is a most unlikely assumption as most NO_3^- comes from vegetables which are substantial carriers of vitamin C, a compound known unequivocally to prevent the formation of nitrosamines and slowing the conversion of NO_3^- to NO_2^-. Coingested ascorbic acid will thus prevent most of the nitrosation and only 0.32 mg of nitrosoproline will be produced assuming one-half of the vegetables and all fruit are eaten raw. As for the exogenous nitrosamine sources, the NAS (1981b) estimated the average daily per capita intakes from cured meat (bacon) at 0.17 μg, a value much smaller than ingestion from beer (0.34–0.97 μg). Dermal penetration from cosmetics was put at 0.41 μg, inhalation in car interiors as 0.50 μg the latter value being two orders of magnitude lower than nitrosamine inhalation from a pack of cigarettes (17 μg). Some occupational exposures are also orders of magnitude above the dietary intake. Consequently, exposure to nitrosamines which might be traced back to nitrate from vegetables and nitrite added to cured meats appears to be well hidden within other daily exposure—and it is made negligible in the case of smokers.

What brief conclusions are fitting to respect the complexities involved here? On the most general level there are the fundamental problems of research involving hazardous substances: experiments will much more easily prove that something is unsafe than safe. In fact, Preussmann et al. (1977) believe that while it is possible to establish a no-effect level for a known carcinogen, it is impossible to define a "safe" dose.

There are also several intriguing specific complications. As already noted, we do not know the actual nitrite intake for large populations as the compound's levels decline in storage and cooking, and only direct meal assays could capture these changes. Moreover, it is certain that typically some two-thirds to four-fifths of the nitrite entering the stomach comes from saliva and that even with diets low in nitrate, nitrite from saliva dominates the total input of nitrite into the stomach, where (another interesting twist) ascorbic acid ingested with nitrate-bearing vegetables has a clearly protective effect in amine-nitrite toxicity.

Very low direct dietary intakes of nitrites as compared to natural generation in the mouth, number of exogenous nondietary nitrosamine exposures (dominated by smoking), counteraction by ascorbic acid, absence of direct data on complex gastrointestinal fate of NO_3^-, NO_2^-, and amines in man, and questions of synergistic effects, irreversibility, and safe maxima combine to make the nitrate–nitrite–nitrosamine problem one without simple and categoric answers—now and for a long time to come. As the NAS (1981b) concludes, evidence implicating these three kinds of compounds in the development of human cancer is circumstantial: in none of the studies hypothesizing the link was there a direct attempt to investigate actual ingestion of the compounds in individuals who developed cancer; moreover, several other plausible causative agents could be considered.

Extensive animal testing clearly proved the carcinogenicity of most nitrosamines: they induce cancer in every species and in every vital tissue tested. Yet, as so often with such tests, their value in prediction of risks to people is unknown as are dose–response relationships. On the other hand, protection against botulism would certainly decrease if the essential uses of nitrite were eliminated without substitution by an effective and safer preservative, And the available numbers, though they should be trusted only in relative fashion, show persuasively that society would be much more ahead in terms of resources spent on additional research and, above all, in monies saved from reduced health hazard, by achieving even marginal reductions in smoking rather than introducing restrictions on vegetable fertilizing or bans on traditional meat curing, steps of dubious validity but with widespread economic impact.

If nitrate accumulation in vegetables and nitrite addition in meat curing are problems, the risks arising from them must be kept in perspective: when they are already very low in comparison with other environmental exposure and when they can be further lowered very substantially by such simple actions as removing spinach petioles, consuming bacon moderately or eating it with citrus fruits to inhibit nitrosation, then there is no reason left to consider these human interventions in the flow of nitrogenous compounds as disturbingly hazardous and worthy of fundamental modifications.

There is, of course, a need for reducing nitrogen leaching during fertilization as it not only pollutes underground and surface waters but also represents a clear economic loss. Similarly, nitrogen oxides may in general be just a marginal health hazard but in those urban and industrial areas where they participate in frequent photochemical reactions leading to heavy smog episodes, controls are certainly needed. And as NO_x from large power plants are the most important source of nitrates contributing to acid deposition problems after long-range transport, these sources, too, may need controls.

3.5. CONTROLS AND MANAGEMENT CHOICES

As the sources of anthropogenic nitrogen are so diverse—mobile, small, and relatively low temperature from millions of cars versus stationary, very large, and high temperature from fossil-fueled electricity generation plants, nitrogen oxides from combustion versus nitrates, ammonia, and oxides from applications of synthetic fertilizers—their controls are quite varied, ranging from modifications of traditional emission-generating processes aimed at reducing the release rates, to removal of nitrogen compounds from the effluents, to improved management practices designed to minimize the changes of unwanted nitrogen losses.

The basic split, leaving aside relatively insignificant emissions from chemical processes (production of nitric acid, ammonia, and nitrogenous fertilizers whose emissions are relatively easily controllable by standard methods), is clearly between controlling atmospheric emissions of nitrogen oxides resulting from combustion and reducing nitrogenous compound losses following fertilization.

3.5.1. Controlling Nitrogen Oxides

Considerable advances have already been made in reducing NO_x emissions from two of the largest sources—gasoline-fueled passenger cars and large thermal electricity generation stations—in the United States, Japan, and some European nations and these achievements either slowed down or even resersed the long-term trend of rising NO_x releases. But a larger share of coals in future electricity generation, further increases in the number of diesel-fueled cars and off-road vehicles with their inherently higher NO_x releases, or a tightening of existing emission standards would call for much improved controls: as much as a doubling of the current reductions might be called for.

3.5.1.1. Controls of Mobile Sources Efforts to reduce nitrogen oxide emissions from car exhausts are relatively recent and came first as a result of previous controls of carbon monoxide and hydrocarbons. Increased air/fuel ratio assuring better oxidation of these pollutants led to a substantial rise of NO_x emissions, and in 1971 California imposed a limit of 2.5 g NO_x/km (4 g/mile) for all new cars sold in the state to return NO_x releases to their pre-1966 levels when there were no exhaust emission limitations. Federal standards came 2 years later and since then progressively stricter controls became universal and research into NO_x reduction techniques advanced rapidly not only in North America but also in Japan and western Europe.

Successful control techniques are basically of two types: various modifications of the combustion process or treatment of exhaust gases. The complexity of internal combustion engines offers many effective adjustments. Although the high temperatures inside the combustion chambers are mainly responsible for the formation of nitrogen oxides, other factors—including the air/fuel ratio, the compression ratio, spark timing, intake-manifold vacuum, coolant temperature, deposits in the combustion chamber, and distribution of the air–fuel charge to the cylinders—also determine the overall emission rates and these variables can be manipulated to lower NO_x formation (Crouse and Anglin, 1977).

However, toying with some variables of the internal combustion process to lower NOx emissions may worsen both engine performance and fuel economy: reduction of compression ratios is a perfect example of one such unwelcome trade-off. A relatively simple combustion modification technique which has been favored since the earliest years of NO_x control and which has achieved widespread applications is exhaust gas recirculation (EGR).

A part of the exhaust gas is returned to the intake manifold and thus increases the air/fuel ratio (an undesirable shift for NO_x control), but the noncombustible gas acts as a heat sink, lowers the temperature of the air–fuel mixture, and reduces NO_x generation. Both internal EGR, via valve overlap, and an external arrangement are possible but the latter technique is preferable in meeting the most stringent NO_x standards.

Japanese carmakers, who have been very active in design and development of emission controls, have relied on combustion modification to achieve compliance with even strict standards. An excellent example is the thermal and thermodynamic exhaust emission control (SEEC-T) developed by Fuji Heavy Industries for their Subarus. This system combines several familiar techniques—including lean carburetion, secondary air injection, EGR, and retarded timing—to meet worldwide regulations of CO, NO_x, and hydrocarbon emissions (Fukushima et al., 1977). Use of external EGR enabled Fuji to meet the final Japanese strict emission goal of 0.25 g NO_x/km.

Toyota's turbulence-generating pot (TGP) added to the company's lean-burn engine is charged from the homogeneous mixture in the main chamber during the compression stroke and it reduces NO_x emissions owing to a decrease of the temperature gradient of the burned gases and to higher cooling loss to the chamber wall (Konishi et al., 1979). Honda's in-cylinder EGR cuts NO_x emissions from the CVCC stratified-charge engine by up to 20%: this branched conduit system permits all gases in the main combustion chamber to recirculate through the torch passage.

Nissan researchers were able to demonstrate that short combustion together with heavy EGR not only reduces NO_x flux but also that it actually improves combustion stability and fuel economy (Kuroda et al., 1978). Their EGR alone could meet the Japanese standard of 0.25 g NO_x/km and in combination with fast burn the emissions are merely 0.13–0.16 g NO_x/km. Although American researchers also came to the conclusion that combustion modification with technology available in the late 1970s will permit achievement of 0.25 g NO_x/km emissions in a 2000-kg vehicle (i.e., in a large American car) with virtually flawless driveability (see, e.g., Parker, 1979), Detroit chose to augment these processes by catalytic converters.

The first converters, installed in the 1975 model year, reduced only carbon monoxide and hydrocarbon emissions. Three-way catalysis in American-made vehicles entered commerically in 1981. During the process, NO and NO_2 are reduced to N_2, and also to some NH_3, in reactions with hydrocarbons and hydrogen while hydrocarbons and CO are oxidized. Conversions are virtually complete within the wide range of 200–600°C and the catalysts are generally selected on the basis of their selectivity toward NO_x reduction. By far the most common catalytic combination for three-way reactions is that of platinum and rhodium in an Rh-rich proportion.

The main factors governing the design of three-way converters are static and dynamic air/fuel ratios, combustion temperature, thermal aging of the catalyst and its poisoning, either by covering of active sites with solid depositions or by doping of precious metals with lead, manganese, phosphorus, sulfur, and other impurities. As with combustion modification techniques, three-way catalysis has been a subject of intensive research not only in the United States but also in West Germany and Sweden.

Review of the United States NO_x emission reduction progress shows best how the combination of combustion modification techniques and three-way catalytic converstion reduced automotive NO_x releases by a large margin in a relatively short period of time. Compared to the precontrol level of 2.5 g NO_x/km, emission controls enabled compliance with federal standards of 1.9 g/km in force between

1972 and 1976 and representing a 24% reduction of NO_x emissions, then 1.25 g/km between 1977 and 1980, and 0.63 g NO_x/km in the early 1980s (76% reduction). While in 1972 only 0.8% of cars in America had some NO_x controls, that percentage rose to 54% by 1978 and to about 77% by 1982 (MVMA, 1983).

Reductions of NO_x emissions from light-duty trucks have not been as impressive as passenger car controls: United States standards today are still 1.45 g NO_x/km, or only a 36% reduction of the precontrol levels. And the NO_x controls in diesel-powered vehicles, especially in heavy-duty trucks, are certainly most difficult as excess oxygen in the combustion chamber causes greater NO_x production than in gasoline-fueled cars.

Several techniques will have to be combined to reduce NO_x emissions in heavy-duty diesel engines, the foremost ones being electronic engine controls, EGR, redesign of combustion chamber, retarded injection timing, and high-pressure fuel injection. The total hardware cost of the optimum package was estimated by the USEPA at $(1980) 700 per truck compared to just about $270 for controls in heavy-duty gasoline-fueled engines.

The USEPA's control targets for 1986 are emissions of 1.7 g NO_x per brake horsepower-hour while the actual emissions of heavy-duty diesels operating in California in the early 1980s were 4.75 g and elsewhere in the country the levels were above 6 g/BHP-hr. As heavy-duty diesel engine trucks account for about 30% of all highway NO_x releases, the still large emissions from this source have kept the overall contribution of mobile sources from declining even faster than it did during the 1970s. While in 1970 highway vehicles produced 40% of all NO_x, by 1981 this share fell to 34% (USEPA, 1982)—and it is not expected to change much during the remainder of this century.

Air transport is, in the aggregate, a small NO_x source compared to automobile traffic but controls of aircraft ground operations at major airports can bring substantial reduction of local air pollutant emissions. Reducing the extent of taxiing, queuing, and terminal time, during which about one-third of all on-the-ground NO_x are emitted, could thus bring significant emission decreases.

In practice, this would require more than just the optimum flow management: aircraft towing and shutdown of one engine during taxi operations would be the two most significant measures (Gelinas and Fan, 1979); there are obvious safety complications with the first option so the second one is the most viable choice as it also adds fuel savings without any apparent safety problems.

Controlling aircraft NO_x emissions in flight is certainly more difficult. Advanced combustor technologies for jet planes have projected costs two to ten times more per tonne of removed gases than controls for other NO_x sources, an expenditure hard to justify in view of the relatively very low contribution of air traffic to total NO_x emissions (Naugle and Fox, 1981). The outlook is even more problematic as the more efficient engines will use higher pressure ratios and inlet temperatures resulting in higher NO_x emissions. Financial gains from improved combustion might then largely be negated if the airlines were forced to develop currently unavailable control technologies.

3.5.1.2. Controlling Stationary Combustion If NO_x emissions are to be significantly reduced in densely populated regions of the Northern Hemisphere, then successes in complying with even strict automotive standards are not enough as in a typical industrialized area with large fossil-fueled power plants, heat and steam boilers, and chemical factories, stationary sources will account for roughly two-thirds of all emitted nitrogen oxides. These compounds, as already noted, originate from oxidation of atomic nitrogen—which is chemically bound with the fuel structure and in coals is contained in both volatile and char fractions, with the former usually responsible for about three-quarters of fuel NO_x emissions—and to a much larger extent from fixation of dinitrogen in the combustion air, a process exponentially dependent on flame temperature and negligible below about 1500°C.

Unlike with sulfur oxides, where the primary control thrust has been to clean the flue gases or to react sulfur oxides inside the combustion chamber and remove them in the form of disposable or saleable compounds, controlling of nitrogen oxides in large stationary sources relies on a variety of boiler and firing system designs.

Key designable variables are burner zone liberation rate, firing geometry, and fuel and burner characteristics (Vatsky, 1981). Net heat input to the burner zone divided by the effective projected surface defines the burner zone liberation rate which determines the amount of thermal NO_x generation and a boiler designer can, by enlarging the surface, lower the liberation rate and reduce NO_x flux. As for firing geometry, opposed-firing always has lower NO_x emissions than the single-wall arrangement and tangential firing has lower NO_x flux than wall-fired boilers, yet a boiler's firing configuration must be chosen above all with regard to specified plant capacity and economic considerations.

Total nitrogen content and heating value of the fuel will determine the fuel nitrogen emissions which must be taken into account in assessing the effect of various control methods. Rapid mixing and short intense flames of high turbulent burners promote high NO_x emission and hence reductions of burner air in staged combustion and controlled mixing of air and fuel at the burner will moderate the NO_x releases.

The four most common means to lower flame temperature, quantity of excess air, and residence time of the fuel and air in the furnace chamber are tangential firing, two-stage combustion (off-stoichiometric firing), low excess-air firing, and flue gas recirculation (Cheremisinoff and Eaton, 1976). During the late 1950s it was thought possible that any anticipated NO_x regulations could be met by installation of tangentially fired furnaces (Blakeslee and Burbach, 1972). The first of these furnaces was designed in 1927 and by the mid-1970s they accounted for more than half of the installed fossil-fuel-fired electricity-generating capacity in the United States.

In these furnaces, fuel is introduced at the corners of the combustion chamber through alternate compartments and both the fuel and air streams from each corner aim tangentially at the circumference of a circle in the furnace's center, creating a large flame swirl. Fuel-rich or air-rich streams are thus extensively blended and this vigorous internal recirculation, together with slower fuel and air mixing, results in inherently low NO_x emissions with all fuels. For example, older horizontal-fired oil-burning units used to produce 500–700 ppm (average 560) NO_x and their gas-fired counterparts emitted 300–500 ppm; switching to tangential firing cut the oil-fired emission to an average of 293 ppm, a 52% reduction with a similar (47%) reduction for natural gas (Sensenbaugh, 1966).

Two-stage combustion works quite well, especially for oil- and natural-gas-fired units: in the primary combustion, less air is supplied to the burners than is necessary for complete combustion so that this fuel-rich zone has lower temperatures and less NO formation than under standard procedure; fuel combustion is completed in a subsequent lean zone at even lower temperature. The efficiency of this staged process goes up as the primary stoichiometry is reduced toward 75% of theoretical air, but in coal-fired furnaces reductions cannot go below 95% in order to avoid serious slagging and corrosion. Still, the procedure can result in a 30–50% decline of NO_x emissions under everyday operating conditions and it is perhaps the most cost-effective way to comply with moderately strict standards for new power sources regardless of their nitrogen content.

For example, the nitrogen content of United States coals ranges from about 1% for the best anthracites to nearly 2% for medium-volatility bituminous varieties and these values translate more or less proportionately into NO_x emissions: while the anthracites will produce only 400–500 ppm in the premixed-fuel lean-combustion mode (5% O_2 in stack), medium-volatility bituminous coals will evolve 1000–1150 ppm (Martin and Bowen, 1979). In staged combustion, the effects of differences in the nitrogen content of coals nearly disappear as the emissions are 300–400 ppm almost regardless of initial fuel N content.

Low excess-air firing requires no major modification of the steam generator but combustion control instrumentation must be excellent to prevent repeated releases of unburned combustibles: furnace flue gases are continuously monitored and as soon as unburned combustibles are detected, fuel flow is reduced to a preset minimum to create a high excess-air situation. Otherwise, excess-air reductions below 5% (to as low as 2%) are standard and in both natural-gas and oil-fired units, NO_x emissions can be reduced by at least 15 and as much as 35% with performance in the 20s quite common.

Flue gas recirculation (either gas-to-furnace or gas-to-precombustion air systems) restricts NO_x formation by producing lower flame temperatures and reducing the need for higher excess-air flows. Again, the method works much better with hydrocarbon-fired units than with coal furnaces but it requires increased mechanical complexity to recirculate the gas and hence a higher capital cost. Lowering of the flame temperature can also be achieved by water injection to either the air or fuel stream, a preferred method for NO_x control in gas turbine engines. Overall NO_x reductions achievable with air recirculation or water addition are up to 70–90% for natural gas furnaces, 20–50% for heavy oil, and no more than 10–30% for coal.

A large variety of improvements, modifications, and innovations were introduced during the 1970s both to cope with anticipated stricter NO_x emission standards and to ease environmental complications of the widely preceived need for massively increased combustion of coal. Most of the United States patent literature on NO_x controls published during this period of accelerated interest was collected by Yaverbaum (1979), and the advances in the two countries most aggressively pursuing NO_x reductions, the United States and Japan, are reviewed in two volumes of a symposium chaired by Cichanowicz and Hall (1981). Only some notable innovations will be mentioned here.

Combustion Engineering, one of the world's largest boiler producers, developed a low-NO_x concentric firing system whose large-scale testing started in the early 1980s. Mitsubishi Heavy Industries, another world leader in boiler technology, introduced innovations such as the widely praised premix (PM) burner fuel (based on rocket design theory) and separate gas recirculation (SGR) burner (Kawamura and Frey, 1981). In the PM burner, combustion of gaseous fuels proceeds in two different zones, in the fuel-rich and fuel-lean mixture, with the same total excess air ratio as in conventional burners.

With liquid fuels, burning is done under diffused flame conditions in the fuel-rich mixture to maintain a stable flame; in tangential firing, stacked fuel nozzles produce an inner conical spray for fuel-rich combustion and an outer envelope for fuel-lean burning. Control of oxygen in the fuel-rich zone is achieved by separate gas recirculation through nozzles on the sides of the oil compartment.

The new distributed-mixing burner developed by the Energy and Environmental Research Corporation (Irvine, California) was tested at firing rates of up to 29 MWt and under optimum conditions, NO_x releases were below 65 mg/J (Folsom et al., 1981). Foster Wheeler Energy Corporation's two low-NO_x burners, the controlled flow and the controlled flow/split-flame designs, differ significantly in the coal nozzle design and the systems are now commerically available for large units (Vatsky, 1981).

In Babcock's dual register burner, NO_x generation is lowered by delaying combustion rather than by staging. Controlling the mixing of coal and air initiates combustion process at the burner throat and makes it possible to vary the zone of completion in the furnace chamber. The peak furnace temperatures are thus lowered (minimizing thermal NO_x conversion), a larger portion of the furnace water-cooled surface is utilized to further depress the peak flame temperatures, and minimized oxygen availability reduces fuel nitrogen formation (Barsin, 1981).

Advantages of this method as compared to two-stage burning result from the maintenance of an oxidizing environment in the furnace which minimizes slagging and reduces potential for furnace wall corrosion with high-S coals, better carbon utilization, and lower oxygen levels, and NO_x emissions can be reliably reduced by 40–60% from uncontrolled levels.

The second generation of combustion modification systems now under intensive development is expected to bring reductions of up to 85% and the most advanced of these techniques is the Low NO_x Combustor (LNCS) funded jointly by Babcock and Wilcox and the EPRI; it employs a deep staging approach (to avoid serious corrosion accompanying combustion of high-iron and sulfur coals in two-stage setting) with stoichiometry held at 65–75% in the flame zone (Barsin, 1981). Cyclone furnaces also show good promise for low-NO_x burning of lignites.

New techniques will be much needed to comply with stricter emissions standards. Current United States new source performance standards for steam generators with more than 73 MWt prescribe no more than 0.6 lb $NO_x/10^6$ Btu (i.e., 257 ng NO_x/J or 450 ppm in the flue gas) for bituminous coal-fired units, 0.5 lb $NO_x/10^6$ Btu (214 ng/J or 375 ppm) for subbituminous fuel, 0.3 lb/10^6 Btu (129 ng/J or 230 ppm) for oil, and 0.2 lb/10^6 Btu (86 ng/J or 161 ppm) for natural gas-fired boilers.

At the end of the 1970s, even as late as 1981, it was expected that the regulations for coal-fired units would be made much stricter soon, perhaps as low as 0.2 lb/10^6 Btu, or an 80–85% reduction in comparison with uncontrolled releases (Rittenhouse, 1981). This arose largely from the concern about NO_3^- contribution to acid rain and it fitted into the USEPA's general intentions to toughen all performance standards. However, as of 1983 the standards for new stationary sources remained at the indicated level (Pahl, 1983) as the regulatory process eased during the first years of the new administration.

In any case, reductions of 80–85% (i.e., emissions around 150 ppm NO_x) can be met by combustion controls in use or under development (and available in the late 1980s) but performances below 150 ppm, especially below 100 ppm, would require expensive flue gas treatment. There appears to be no shortage of proposed flue gas treatment methods with the most promising one being the ammonia-based selective catalytic reduction (Matsuda et al., 1978). In this process, ammonia is injected into the flue gas after the economizer where it mixes with nitrogen oxides and then reacts with them downstream in a fixed catalyst chamber preceding the air heater to produce molecular nitrogen.

Tests in the presence of fairly large amounts of sulfur oxides showed NO_x removal efficiencies exceeding 90%. The Japanese are also investigating techniques of high-efficiency joint SO_x–NO_x removal but the usual developmental problems and high capital and operation costs will preclude diffusion of these techniques during the 1980s. No such obstacles are in the way of numerous practices which are already at our disposal for minimizing the losses of nitrogen in agroecosystems.

3.5.2. Efficient Fertilizing

The basic goal of more efficient fertilizing is to maintain the yields with smaller nitrogen applications by making more of the nutrient available at the times of highest need. As so many variables affect the yield response—Keeney (1982) offers a nearly exhaustive list—there are almost as many opportunities for management intervention and resulting savings. The three most critical variables determining the outcome are the rate of application, its timing, and the placement of nitrogen, and agronomic literature abounds with experiments testing their variations with many different crops.

3.5.2.1. Minimizing the Losses Many experiments were undertaken to compare the relative fertilizing efficiency of different nitrogen carriers in temperate as well as subtropical and tropical conditions. There are some obvious advantages in selection, most notably in applying ammonia compounds in rice cultivation: isotope tagging tests in 15 rice-growing countries showed that less than 5% of the plant's nitrogen was derived from $NaNO_3$ incorporated at a 5-cm depth at transplanting, while ammonium sulfate and urea similarly incorporated at transplanting contributed about 20% of all nitrogen (FAO, 1980a; see also Section 3.4.2.1).

However, for dry land crops, numerous tests with wheat, corn, sorghum, and millet showed no significant difference between ammonia and nitrate compounds. The FAO (1980a) document on fertilizer efficiency sums this well: "Little if any difference in chemical carriers is likely to exist so long as each is used in accordance with its own limitations." Small or no yield differences owing to the source of applied nitrogen, clearly suggest that nitrogen purchases should be based on the cost per unit mass and not on the source; costs of handling and application must also be considered.

Two conditions required for efficient use of fertilizer by plants are at the same time the major causes of its losses under current application procedures. First, the initial nitrogen level must be high for sufficient uptake rates and, second, enough of the element must be present above this level to supply the crop with adequate nutrition when the potential leaching and volatilization losses are taken into account. To satisfy these conditions, most of the current practice consists of dumping large quantities of nitrogen in just a few applications, the heaviest one being in periods of relatively low uptake capacity (during, or shortly after planting). Logical remedies are to ascertain as closely as possible what the optimum application rate is and then to delay most of the fertilizing until such time when the crop can absorb the nutrient rapidly.

The first approach requires soil testing or plant tissue analysis, undoubtedly the most practical, most precise, and yet relatively cheap methods to determine optimum fertilizer requirements and thus to help in maximizing yields and profits (Walsh and Beaton, 1973; FAO, 1980b; Cottenie, 1980). Soil testing involves a variety of biological or, more often, chemical tests to be used as an index of nutrient availability and a guide to fertilizer application and is usually done before planting. Plant or foliar analysis is possible only after general relationships have been established between the nutrient content of the whole plant or a selected leaf or a petiole and the need for additional fertilizer.

Although these tests are now readily available through private or public laboratories in all rich nations, their use is surprisingly infrequent even in the richest farming regions of North America, where farmers do not test their soils and plants regularly and systematically and where even some extension advisors do not promote proper sampling and testing. For example, Robertson et al. (1980) report that in Michigan, 60,000 soil samples are tested annually, representing an estimated 365,000 ha or just 15% of all of the state's land on which major field crops are grown.

Playing the averages is inadvisable because average soil fertilities are not known for most areas or most kinds of soils—and even more so owing to large variability with soil types reflecting both natural differences between soils as well as the past cultural practices. Yet soil testing for nitrogen has not been without complications and failures. Viets (1980) calls the methods for establishing nitrogen availability "among the most baffling of all soil tests" as they range from estimates of organic matter and calculations of total nitrogen via C/N ratios to determinations of nitrate released after aerobic incubation.

Complications arise for several reasons (Smith, 1980). As previously described (Section 3.4.2.1), much of soil nitrogen is in organic compounds mostly unavailable to crops in adequate amounts: testing for the total nitrogen is thus not very revealing. Nitrogen fertilizer recommendations based on tests for inorganic nitrogen (NO_3^- and NH_4^+) have been most successful for grains, sugar beets, and potatoes in semiarid (400–500 mm annual precipitation) areas but much less so in regions of higher precipitation where NO_3^- is much more mobile. Soil sample depth may greatly influence the results: in general it is best to sample to the maximum depth at which the roots will be removing water and nutrients, which may mean going as far as 180 cm although 60 cm will usually suffice.

To represent a field correctly, a composite sample of a few dozen cores must be taken and as nitrates are not uniformly distributed in soils after rill or furrow irrigation, special sampling procedures are necessary in those cases to avoid biased results. Then the analytical errors may be surprisingly large with studies from the southern and western United States showing a threefold range in values using the same soil samples and supposedly identical methods (Viets, 1980). And when the results are in, there is no optimum way to evaluate the data for fertilizer recommendations as the nitrogen availability and effectiveness in the coming growing season will be so closely linked to a host of other variables, above all available soil water, precipitation, and temperature.

Consequently, Smith (1980) believes that frequent lack of success with nitrogen soil testing stems from the users' "unwillingness to try combinations of data from various factors affecting crop yield." There is definitely place for both complex equations and simpler indices as well as for use of analyses of organic nitrogen. In wet fields, satisfactory determinations are usually much harder, but their management can be brought to even a high level of perfection as illustrated by the case of California rice farming, one of the most mechanized and most efficient cropping practices in the world, where most of the nitrogen is introduced before planting in dry or liquid form (urea, ammonium sulfate, or aqua ammonia) at a rate of 135 kg/ha 5–10 cm deep into the soil, below the oxidation zone at the water–soil interface, so that it remains reduced and continuously available to the roots, and both plant tissue and soil tests are carried out subsequently to detect possible nitrogen shortages and only the required amount is added by aerial spraying (Rutger and Brandon, 1981). But this is an exception rather than a rule among fertilizer management practices.

The best way to close this brief discussion of soil testing complications is to describe Daigger's (1974) experiment in which five different laboratories (four private plus the University of Nebraska Soil Testing Service) analyzed the same soil samples and made their recommendations which were followed on plots replicated four times in the same field of irrigated beets. These recommendations ranged from 162 kg N/ha to just 72 kg N/ha with a 4.5-fold difference in the total fertilizer cost (including other macro- and micronutrients)—yet the maximum yield (achieved with the least amount of nitrogen and the cheapest overall treatment) was just 11% above the minimum one and the highest sugar content was a mere 3% above the lowest value. So it is hard to disagree with Viets (1980) when he concludes that the inherent complexity of the soil–water–plant system will probably always keep soil (and plant) testing in an empirical realm: fertilizing will for long remain as much an art as a science.

Decisions on the timing of application and on the placement of fertilizer are not easy to generalize either. Controversies regarding the merits of fall fertilizer application thus cannot be resolved unequivocally. Spreading out the labor, equipment, and fertilizer demands are obvious advantages for both the fertilizer industry and the farmers but yield response experiments comparing fall and spring application continue to range from no significant differences to large gains with spring fertilizing, perhaps an inevitable outcome caused by climatic fluctuations and variations of numerous other abiotic and biotic factors determining the harvest.

However, there is sufficient consensus to favor spring fertilizing with ammonia or ammonia-producing materials: field evidence shows that even when these materials are applied when soil temperature drops below 10°C, nitrification at reduced rates continues until freezing and will restart with spring warming. Combination of warm and rainy spring weather may then bring substantial denitrification *and* leaching losses.

There appear to be only two unexceptional rules to follow, regardless of the fertilizer used or crop grown, in a great majority of cases with clear benefits: delay the major application until some time after planting and incorporate the fertilizer into the soil. Anhydrous ammonia losses during and immediately after application can be practically eliminated by injection to appropriate depth (deeper in coarse-textured soils), proper spacing of injector knives, and timing to coincide with soil moisture content and physical state guaranteeing fast and effective closure of the injection sites (Nelson, 1982), Volatilization losses after applications of dry or liquid ammoniacal fertilizers are minimized by avoiding surface spreading of alkaline varieties (urea, ammonium phosphate), using ammonia sulfate or nitrate on neutral and acidic soils and nitrate only on calcareous soils, and applying to moist soils prior to rainfalls and at lower temperatures (Nelson, 1982). In wet fields, application just before flooding is much preferable to later broadcasting on the water surface.

Split applications can be adjusted to match crop needs at different growth stages but each pass over a field increases operation costs and unless the fertilizer is distributed in irrigation water, use of standard field equipment limits spreading to early crop stages. In any case, profitable recommendations must be crop or soil specific or both. Just two examples, for rice and corn will illustrate these specifics.

In an international study involving 16 countries, rice crops were fertilized with 60 kg N/ha at seven different times: all at transplanting, all halfway between transplanting and 2 weeks before primordial initiation, all at 2 weeks before initiation and with split doses (halves, thirds) at various intervals; sampling 3 weeks after the last application showed the highest share of fertilizer-derived nitrogen after the single dose of ammonium sulfate 2 weeks before primordial initiation and no advantage in split applications.

This study also confirmed clear advantages of placing fertilizers at some depth in the soil rather than broadcasting them on the surface or placing them on the topsoil in rows: incorporation of $(NH_4)_2SO_4$ 5–15 cm into the soil raises the nitrogen utilization almost twofold in comparison with surface applications, although there was not much difference in uptake efficiency between the three placement depths.

Use of slow-release fertilizers, the best known being sulfur-coated urea, is not suited to high-yielding field crops such as corn which need large amounts of nitrogen in short periods of rapid growth and whose response would have to be exceptional to justify the much higher (25–50%) cost of delayed-action compounds. High nutritional needs also make foliar applications of ammonium and urea to most major field crops impractical. To prevent leaf burns, maximum nitrogen concentrations in the spray must be kept to no more than 1–3% which means that only 10–20 kg N/ha can be applied at one time and as corn requires 2–3 kg N/ha each day during the period of fastest growth, it would have to be sprayed once or twice a week, obviously a choice too costly and impractical.

Besides optimum rates, timings, and placements, many additional measures can be taken to increase the efficiency of fertilizer utilization—and to have other beneficial effects as well. All possible efforts to reduce the rates of soil erosion would make a large difference on a global scale and in virtually every important farming region in both temperate and tropical zones as the amounts of nutrients removed in the eroded soil may easily surpass combined volatilization and leaching losses. Reduced tillage or no-till practices resulting in a less oxidative environment which may not only conserve both soil and fertilizer nitrogen owing to lowered losses but which may also enhance the efficiency of nutrient use are thus especially appealing.

Naturally, antierosion measures should, above all, ensure the sustainable existence of field farming over the long run but even where the soil losses are no immediate threat to cropping *per se,* their deleterious effects on soil fertility would be sufficient reason for action. On the other hand, while a variety of commendable soil conservation practices (contouring, strip cropping, terracing, crop residue recycling) will cut down erosion and limit the movement of soil nitrogen, these measures will not necessarily control other nitrogen losses and may actually increase the rates of leaching or denitrification (Keeney, 1982).

Proper rotations everywhere and intercropping in countries with inexpensive labor can go a long way toward economizing the use of fertilizers. Various cropping sequences where deep-rooted legumes scavenge nitrates leached to deeper levels after grain crops (besides, naturally, providing more agroecosystemic stability in other ways) are a must for proper long-term management. For example, some

Illinois calculations show that on corn fields fertilized with 224 kg N/ha, about two-fifths of all nitrogen removed by subsequent soybeans came from the residual fertilizer (Johnson et al., 1975).

And in many poor countries, where fertilizers are both expensive and improperly applied, use of the synthetic fertilizer can often be avoided by following and expanding traditional methods of intercropping of grains with legumes. Both ancient know-how and modern controlled experiments confirm the supremacy of this approach to monocropping (Ahmed and Gunasena, 1979). Intercropping will increase total yield with the more efficient utilization of nitrogen playing a large part; in cereal–legume combinations, net protein utilization value of the systems appears to also be enchanced, a much-needed benefit for subsistence farmers of the poor world whose diets would receive the greatest benefit from this practice.

Yet another traditional way of reducing nitrogen (leaching) losses is to incorporate relatively large amounts of crop residues whose high carbon content (C/N ratio typically around 50) leads to considerable nitrogen immobilization. In practice, this approach is limited to returning the *in situ* biomass, which may not bind enough nitrogen to make a substantial difference.

And then there is a large realm of basic agronomical choices and practices whose management will add up to a considerable difference in the ultimate efficiency of nitrogen fertilizer, thus potentiating or greatly negating the careful choice, timing, and placement of the nutrient (Aldrich, 1980). G. S. Sekhon's estimates of potential reductions in fertilizer efficiency caused by these omissions and neglect show how far-reaching their cumulative effect can be (Table 3-11).

Consequently, positive factors must be maximized: the best available, the most responsive, crop variety or hybrid should be chosen, its planting should be timely and at an appropriate rate: for example, modern corn hybrids have the highest yields

TABLE 3-11
RANGES OF REDUCTION IN FERTILIZER EFFICIENCY OWING TO VARIOUS AGRONOMIC FACTORS[a]

Possible sources of reduction in fertilizer efficiency	Reduction in fertilizer efficiency (%)
Poor seedbed preparation	10–25
Inappropriate crop variety	20–40
Delay in sowing	20–40
Improper seeding	5–20
Inadequate plant population	10–25
Inadequate irrigation	10–20
Weed infestation	15–50
Insect attack	5–50

[a]Source: FAO (1980a).

only near the optimum, dense populations. Irrigation should be carefully scheduled because in irrigated farming, increases in efficiency of water use will obviously reduce nitrate leaching: the traditional methods of flood and furrow irrigation are much more wasteful than the more expensive but fine- turnable central-pivot sprinklers (suitable for most field crops) and drip irrigation (for specialized high-value crops). On the other hand, negative influences must be minimized: weeds, competing for the nutrient, must be controlled as well as pests, timely harvesting will cut field losses, and proper drying and storage will eliminate postharvest waste.

The latest additions to the effort of boosting fertilizer use efficiency are slow-release fertilizers and nitrification inhibitors. The traditional application of organic wastes is, of course, a way of slow-release fertilizing—but one not well suited to any crop demanding high N take in a short time. By far the most popular innovation in this area was the development of sulfur-coated urea (by the Tennessee Valley Authority) which works best with crops on coarse-textured soils with severe leaching problems: only there would the benefit outweigh higher cost (25–50% above those for normal urea). Rice, high-value crops, and turfs are the best choices for sulfur-coated urea treatments (Nelson, 1982).

Nirtification inhibition is an approach so fundamentally different from all those outlined above—one aiming at the key link between soil and plant nitrogen and hence at the source of the greatest potential losses—that it deserves a separate section.

3.5.2.2. Inhibiting Nitrification Nitrification, carried out by *Nitrosomonas* (NH_4^+-N to NO_2^--N) and *Nitrobacter* (NO_2^--N to NO_3^--N), is a common oxidation process (see Section 3.1.2.1) which may be slowed down by root exudates or decomposing phytomass in many forest, orchard, and pasture soils but which is stimulated in soils under many field crops and with manure applications, and which is essential in making nitrate available to plants after fertilizing with ammonia, an increasingly preferred practice in United States crop farming. But this oxidation replaces relatively immobile NH_4^+ by freely mobile NO_3^-, hence opening the way for leaching and denitrification losses. Inhibiting the rates of nitrification would thus lead to higher efficiency of NH_3 fertilizer use and to lower nitrogen losses.

There are many nonspecific inhibitors affecting the activities of soil microorganisms, mostly among agricultural chemicals. Hauck (1980) lists six commerical insecticides, three herbicides, three fungicides, three fumigants, and one wood preservative whose toxic effects were reported to inhibit either NH_4^+ or NO_2^- oxidation or, in several cases, both of these processes. But there are substantial differences among these compounds' mode of action. Neither insecticides nor herbicides have inhibitory properties at normal field application rates; multiple concentration increases are needed to start the action which, in any case, may be quite short-lived (Prasad et al., 1971). On the other hand, fungicides and fumigants are very effective even at normal rate and some of these compounds suppress nitrification for 4–8 weeks.

Since the late 1950s hundreds of chemicals have been tested for their ability to serve as commerical nitrification inhibitors. Compounds for common use should be specific blockers of ammonia but not nitrite oxidation, should be nontoxic to animals and humans at effective application levels which also should not interfere with other beneficial soil organisms, and should retain their effectiveness for weeks following fertilizer application.

One dozen of such chemicals meet at least some of these requirements and have been, as of the early 1980s, either in commercial production or patented for specific use as nitrification inhibitors (Hauck, 1980). With two exceptions, all of these compounds were developed in Japan by such major chemical companies as Sumitomo (inhibitor known under DCS symbol), Mitsui Toatsu (AM compound), Mitsubishi (MAST), and Showa Denko (DNDN). Thiourea is also used as a nitrification inhibitor in Japanese farming.

But it is an American product which has emerged as perhaps the most important inhibitor: nitrapyrin, chemically 2-chloro-6-(trichloromethyl) pyridine, the active ingredient in Dow Chemical's N-Serve. This chemical inhibits *Nitrosomonas* activity for up to 6 weeks following the application of anhydrous NH_3 and its main metabolic breakdown product (6-chlorpicolinic acid) is nontoxic. The compound is also nonphytotoxic at effective rates, is not found as residue in plant tissues, is compatible with a variety of fertilizers, and has only a short half-life in soil (Huber et al., 1977). Details on the compound's biochemical actions as well as on the effects of other patented inhibitors can be found in Hauck (1980).

Although the first nitrification inhibitor was ready in 1962, it was not marketed until the early 1970s when the interest in efficient nitrogen utilization was spurred first by the concerns about nitrate in waters, then by the rapid rise of feedstock prices, and finally by worries about the effect of N_2O releases on the stratosphere. Since then, much testing has been done with different crops in all important farming regions in the United States and these experiences are well summarized in separate regional chapters in the American Society of Agronomy's review of nitrification inhibitors (Stelly, 1980).

The greatest benefits would obviously be derived in the Corn Belt where, under certain conditions, reduction of nitrogen losses by nitrification inhibitors was demonstrated to increase corn yields, decrease incidence of disease and NO_3^- water loadings, and improve grain gravity (Nelson and Huber, 1980). But the results have been far from uniform. In Indiana, crop yield increased followed nitrapyrin use in more than 70% of the trial additions conducted over half a dozen years—while in Wisconsin and Illinois, nitrification inhibitors added to fall- and spring-applied ammonia did not increase corn yields. And in the western part of the Corn Belt and in large areas of the Great Plains where losses from denitrification and leaching are minimal, addition of nitrapyrin may bring hardly any improvement to dryland crops (Hergert and Wiese, 1980).

Studies of inhibitor effects on crop yields were done with many crops—ranging from the most obvious choice, grain corn, to spinach and cotton—with increases attributable to inhibited nitrification ranging from just 3–4% to as much as 25–35% with grain crops and up to 200% with spinach (Huber et al., 1977). For

Corn Belt grain crops, the most common range appears to be 4–10% (Nelson and Huber, 1980).

Besides higher yields, there are two other important beneficial effects of nitrification inhibitor use (Huber et al., 1977). The first is reduced concentration of nitrate nitrogen in plant tissues and increases in amide, amino, and protein nitrogen. This is, of course, a very desirable effect: as discussed in Section 3.4.3.2, several rapidly maturing vegetables (radishes, spinach) can store large quantities of nitrate and vegetables are the principal source of the compound's sometimes undesirable dietary intake. Nitrate levels in plants fertilized with ammonia followed by nitrapyrin applications were only 30–40% of the levels found in plants fertilized with ammonia or nitrate nitrogen without a nitrification retarder.

Nitrapyrin applications can thus result not only in higher yields but also can lead to higher protein content of grain crops, a most advantageous "side effect" of the inhibiting process. The other benefit is the reduction of incidence and severity of some plant diseases which are influenced by the form of nitrogen taken up by the crops.

Given the variety of soil environments and fertilizer treatments, it is obvious that the inhibitors may act quite differently under seemingly very similar conditions. Keeney's (1980b) review of soil and environmental factors affecting the persistence and bioactivity of nitrification inhibitors (especially nitrapyrin) noted frequent conflicting conclusions in the research literature and supported the following generalizations. First, nitrapyrin bioactivity can be lowered by sorption on organic matter, by chemical hydrolysis, or by volatilization. Second, the compound's limited soil movement often lessens its effectiveness in band fertilizer applications. Third, in neutral and alkaline soils, more rapid recovery of nitrifiers will limit the compound's effectiveness. Finally, the inhibiting action is more effective at low temperatures or under other circumstances favoring low nitrification rates.

Clearly, this clever, environmentally innocuous way of improving fertilizing efficiency deserves the widest possible commercialization and extension—but it is equally certain that even when more research does clear many remaining uncertainties, application of nitrification inhibitors will long remain a special technique with a limited effect, rather than a common method, to cut nitrogen losses in the world's arable farming.

Even these brief reviews of various methods available to minimize escapes of nitrogen oxides and losses of fertilizer nitrogen in agroecosystems should have made it clear that the prospects for highly effective controls of nitrogenous pollution are excellent. Most of the techniques and practices are not only well tested and readily available but also are not so expensive as to hinder more widespread future adoption. Especially in comparison with sulfurous pollution, whose examination is next, there is little doubt that human interference in the nitrogen cycle, although both intensive and extensive, is manageable with relative ease and without excessive cost.

4 SULFUR

Bring me sulphur, which cleanses all pollution,
and fetch fire also that I may burn it, and
purify the cloisters.

—Odysseus to Euryclea in Homer's
The Odyssey, Book XXII (S. Buttler translation)

. . . cycling of sulfur in nature is no less
relevant to carboxylation than the cycling of
carbon and nitrogen.

—E. S. DEEVEY, JR. (1970)

Odysseus's orders to his old nurse cited above sound strange taken out of context. With the burnt element's acrid odor, the returned traveler wanted to cleanse his rooms polluted by the presence of his wife's suitors. In more ordinary circumstances, burning of sulfur—usually not directly but as a part of fossil fuels—does anything but "cleanse all pollution": combustion of coal and oil products, and also processing of hydrocarbons and smelting of nonferrous metals, now appear to be the single largest terrestrial input into the element's complex global cycle, the inflow resulting in worrisome regional environmental degradation simplifyingly labeled as "acid rain."

On the other hand, although it is quite rare in comparison with carbon and nitrogen, sulfur is one of the three doubly mobile elements whose grand cycling is essential to sustaining life. Masswise, it is only the tenth most abundant element in the universe and in the biosphere it occupies the same rank—the same order of magnitude as phosphorus or aluminum, an order below nitrogen and three orders below carbon. Such a simple quantitative comparison tells nothing of sulfur's critical role in life on this planet—bonding together segments of complex molecules which make up proteins, enzymes, antibodies. Although sulfur atoms of higher oxidation states can also be joined by a sulfur–sulfur link, a disulfide bridge is far the most frequent, an essential glue of living matter binding together cysteine residues either on the same polypeptide chain or on two neighboring strands.

Sulfur's key role as the fortifier of proteins—as Deevey (1970) puts it graphically, without it they would "coil randomly, like a carelessly dropped rope"—is also the explanation for its relative biospheric rarity: dominant plant tissues, cellulose and hemicellulose, have very low protein contents. Naturally, animal and human bodies contain larger shares of protein and hence more sulfur. Among the earth's three envelopes permeated by the biosphere, normal, unpolluted, atmosphere holds only trace amounts of sulfur compounds (although, as will shortly be seen, the total mass of sulfur cycled through it annually is quite large), in the earth's crust it is only the 13th most abundant element (however, in sediments it is eighth or ninth), while in the ocean sulfur ranks sixth, at just about half of magnesium's mass. As in the two preceding cases, I shall examine the magnitude of these reservoirs and the transfers among them before concentrating on the ways and scales of anthropogenic interference in the sulfur cycle.

4.1. THE SULFUR CYCLE

Cycling of sulfur is no less intricate than the flows of carbon and nitrogen—and its study is no less beset by approximations and uncertainties. In fact, the sulfur cycle has been studied less extensively than its two other companions analyzed in the preceding chapters. Still, there is no shortage of more or less comprehensive accounts: these began with Junge's (1960, 1963) and Eriksson's (1963) pioneering work, were followed a decade later by Robinson and Robbins's (1972), Kellogg and colleagues' (1972), and Friend's (1973) appraisals and were then, in quick succession, drastically revised by Bolin and Charlson (1976), Hallberg (1976), Granat (1976), Garland (1977), Davey (1978), and Ivanov (1981).

Consequently, there is no difficulty in identifying major reservoirs and most likely there is also no important flux missing from the global sulfur cycle model assembled in Figure 4-1 which identifies the reservoirs and exchanges using the same graphical conventions as for carbon and nitrogen flows. The following account, though a comprehensive one, will stress the fluxes of atmospheric sulfur, that part of the cycle where human interference is highly prominent and growing.

4.1.1. Major Reservoirs

As with carbon, the lithosphere dwarfs the earth's other reservoirs of sulfur: crustal rocks contain about 11 times more sulfur than the ocean and, in turn, fresh waters carry less than a third of the marine total. But unlike with carbon which, in the form of CO_2, is present in the atmosphere in a mass approaching 800 billion t, only less than 4 million t of sulfur resides—in a continuous and rather rapid flux—in the air, a difference of five orders of magnitude. Similarly, the sulfur content of biota is orders of magnitude below that for carbon. So while most of the later sections of this chapter will deal with sulfur in the atmosphere and its eventual effects on life and materials, only the two largest natural reservoirs of sulfur—the lithosphere and hydrosphere and the exchanges between them—will be introduced here.

4.1.1.1. Lithosphere Holser and Kaplan's (1966) geochemical inventory of sulfur in the earth's crust results in an aggregate of 14,600 ± 3400 trillion t roughly split between igneous and metamorphic rocks and various sediments. Evaporites, though of relatively very small mass, have by far the highest sulfur content and are thus the single largest reservoir of terrestrial sulfur, about a third of the total, and

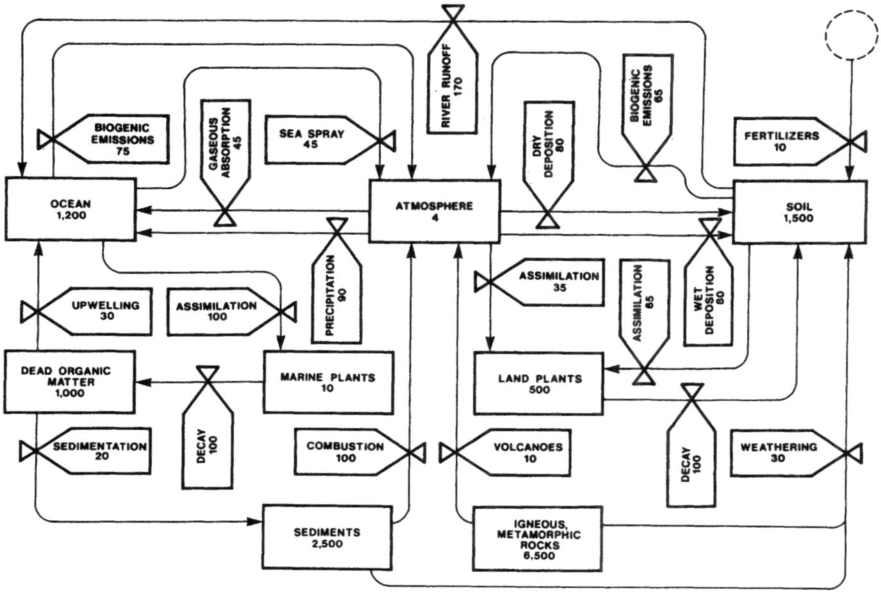

FIGURE 4-1. Global biogeochemical sulfur cycle. Values for all major reservoirs (rectangles) and annual fluxes (valves) are estimated in billion tonnes of sulfur. Sources: Tables 4-1 and 4-2, Eriksson (1963), Kaplan (1972), and Lasaga (1981).

virtually all of it in sulfates. In mafic, metamorphic, or volcanic rocks, nearly all sulfur present is in sulfides, which are often concentrated by hydrothermal processes in major mineral ore deposits. Elemental sulfur is frequently encountered in bituminous shales and may also be trapped in salt domes.

The world's verified recoverable coal reserves contain only a negligible fraction of the crustal sulfur (of organic origin and as pyrites)—no more than about 13 billion t—but the extensive mining and combustion of bituminous and lignitic coals is now among the leading inputs of sulfur into the atmosphere and, subsequently, into soils and fresh waters (see Sections 4.1.3, 4.2, and 4.3). Coal and salt dome sulfur enter the natural cycle only if recovered by mining, whereas sulfates and sulfides flow into the streams, lakes, and oceans as a result of rock weathering (see Section 4.1.1.3).

4.1.1.2. HYDROSPHERE Sulfur storage in this reservoir is dominated by the abundance of sulfate in seawater where it is, after chloride, the second most important anion: each kilogram of ocean water contains on the average 19.35 g NaCl and 2.712 g SO_4^{2-}. As the total ocean mass can be put with sufficient accuracy at 1.35×10^{18} t, the total amount of sulfur in the oceans is 1.22×10^{15} t. Sulfates are also the second most abundant compounds dissolved in river waters. Their average global loading is given at 10.6–11.2 mg/kg compared to 55.9–58.4 mg for bicarbonate and 7.8–8.1 mg for chloride (Livingstone, 1963; Gibbs, 1972). However, while the sulfate content of the open seawater does not fluctuate substantially among the various oceans, sulfate concentration in individual rivers, determined largely both by the types of rocks present (weathering effect) and by the runoff in a river's basin, are not uncommonly an order of magnitude lower or higher than the cited global mean.

Also, the sulfate loading in rivers can change rapidly with seasons and can be greatly altered in the course of geological history. In contrast, studies of evaporite deposits indicate that oceanic sulfur mass has not varied by more than an order of magnitude during the past several hundred million years (Lasaga and Holland, 1976). Residence time of the marine sulfate can be calculated from the river input and the total sulfur content of the oceans as 7.9 million years, ten times shorter than that of chloride but 100 times longer than the residence time of bicarbonate, the key link in sequestering of carbon.

Ocean sulfur is removed from water by formation of both reduced (mostly pyritic) and oxidized (sulfate) sediments. The initiation of the first process can be commonly seen in black FeS sediments coloring tidal flats and coastal muds, whereas the other deposition occurs above all in basins with restricted access to the open ocean; the Persian Gulf evaporites are perhaps the best example. The ultimate sink for oceanic sulfates may be their removal in the water cycling through the midocean ridges: even when assuming an annual water flow of just 1×10^{11} t and removal of only 1 g of sulfate per kg of water, some 33 million t of sulfur would be sequestered each year in this way (Lasaga, 1981). However, the process of formation of evaporites and their subsequent weathering appear to provide an even stronger link between the hydrosphere and the lithosphere reservoirs of sulfur.

4.1.1.3. EVAPORATION AND WEATHERING During seawater's evaporation in enclosed areas or in shallow, restricted-flow basins, carbonates are precipitated first and only when one-fifth of the initial water volume remains do sulfates follow. Globally, this process is capable of removing huge quantities of sulfur in what is geologically an instant: when a shallow (1 m deep) basin of seawater evaporates in 1 year, 0.27 g of sulfate will be left behind for each square centimeter, and at such a rate only about 0.05% of the planet's ocean surface has to behave like an evaporite basin to completely remove all annual sulfate inputs from rivers and from earlier evaporites (Holland, 1978). The subsequent fate of evaporite sulfur is not well known and so the convenient and logical way to balance the cycle is to equate the input of sulfur into evaporites to that amount removed by weathering in such formations (Lasaga, 1981).

Both gypsum ($CaSO_4 \cdot 2H_2O$) and anhydrite ($CaSO_4$), the two dominant evaporite sulfates, release their sulfur by dissolving, without the participation of other compounds: in surface and ground waters, their solubility is 2–3 g/kg H_2O (Blount and Dickson, 1973) and so even in relatively dry regions with annual runoff of just 30 cm, outcrops of these minerals are denuded about ten times faster than the normal erosion rates. On the other hand, sulfur from pyrites enters the cycle almost always through oxidation to sulfate, a process greatly accelerated in acid environments by microbial mediation of thiobacilli (Goldhaber and Kaplan, 1974; Holland, 1978).

Total weathering flux can be calculated from the average sulfur content of exposed rocks and from the aggregate suspended load of river sediment carried annually into the ocean. The latter figure is about 20 billion t and the former one is around 0.38% when assuming that about three-quarters of exposed rocks are sediments containing 0.49% S (owing largely to evaporites with 7% and shales with 0.27% S) and the rest are sulfur-poor igneous and metamorphic formations averaging 0.032% S. Consequently, weathering of all rocks transfers annually about 77 million t of sulfur into the rivers: by rock type, 2 million t from igneous and metamorphic rocks, 25 million t from sulfide weathering, and 50 million t from the breakdown of evaporites. Clearly, any changes in the rate of evaporite formation must have a profound effect on the whole sulfur cycle (Lasaga, 1981).

4.1.2. Atmospheric Fluxes

Sulfur moves into the atmosphere by three natural routes. Volcanic activity appears to be the smallest contributor, sea spray is an incessantly large input, and recent measurements of biogenic emissions from anoxic soils and from the ocean have confirmed their importance suspected previously on the basis of theoretical calculations and indirect evidence. However, none of these fluxes is known with any great accuracy as attested by a comparison of all major published global sulfur cycle models (Table 4-1). With the exception of sea spray emissions, there are wide disagreements, with up to 30-fold differences in the most controversial cases. The following sections will show that new measurements put our estimates of atmospheric sulfur inflows on a safer, though still far from satisfactory, ground.

4.1.2.1. Volcanic Emissions and Sea Spray Regular volcanic plumes and intermittent major explosions are certainly the most spectacular natural sources of atmospheric sulfur but they remained unmeasured until the 1970s so that all the older calculations of total annual production were based on indirect multiple approximations. Thus, Kellogg et al. (1972) started with an assumption of 3.9×10^{17} cm^3 of lava extruded between 1500 and 1914, applied an average density of 3 g/cm^3, assigned total gases as 0.5% of the lava mass and SO_2 as one-tenth of the total gas weight to end up with approximately 1.5 million t of SO_2/year. However, if one assumes a higher gas content of typical lavas—for example, Cadle (1975) uses 2.5%—then the SO_2 annual total rises to nearly 8 million t. All such exercises are made very uncertain by the highly error-prone estimates of past laval extrusions as well as by highly variable lava gas contents and their SO_2 shares.

Here the remote-sensing correlation spectrometry of different volcanic plumes offers a superior approach and since 1971 such measurements have been available for volcanoes in Japan, Central America, Hawaii, and Italy. Data for the Guatemalan, Nicaraguan, and El Salvadorean volcanoes are most extensive: 125 plume transects yielded values between 20 and 420 t SO_2/day (Stoiber and Jepsen, 1973). Considering the standard measurement errors and deviations (up to 36%), these

TABLE 4-1
ESTIMATES OF ATMOSPHERIC SULFUR FLUXES[a]

	Volcanoes	Sea spray	Biogenic flux		Anthropogenic emissions	Precipitation		Dry deposition[b]		Total sulfur involved in atmospheric cycling
			Oceans	Continents		Land	Ocean	Land	Ocean	
Eriksson (1960, 1963)	ND[c]	45	170	110	40	65	100	100	100	365
Junge (1963)	ND	45	160	70	40	140		75		315
Robinson and Robbins (1972)	ND	44	30	68	70	70	ND	20	ND	212
Kellogg et al. (1972)	1	44	18	71	50	86	72	25	ND	183
Friend (1973)	2	44	48	58	65	86	ND	35	ND	217
Bolin and Charlson (1976)	3	44	28	3	65	ND	ND	ND	ND	143
Hallberg (1976)	3	44	34	3	65	ND	ND	ND	ND	149
Granat (1976)	3	44	27	5	65	43	ND	28	ND	144
Garland (1977)	ND	44	46	106	70	112		154		266
Davey (1978)	10	44	26	60	60	123		77		200
Ivanov (1981)	29	60	19	23	70	ND	ND	103	ND	201

[a]All values are in million tonnes per year.
[b]Dry deposition values include absorption by vegetation.
[c]ND, not done.

Central American data mean that a typical quiescent volcano with a noticeable vapor plume releases about 100 t SO_2/day, as much as a coal-fueled power plant of 700 MW burning fuel (22 MJ/kg) with a sulfur content of 2% (35% efficiency and 65% load factor).

Extension of these Central American measurements to the world's 100 or so active volcanoes (half of them with prominent plumes) adds up to only about 7 million t SO_2/year. Addition of the minimum possible amount (8 million t) of SO_2 produced annually on the surfaces of ejecta from violently erupting volcanoes boosts the total to 15 million t, an order or magnitude above Kellogg and colleagues' (1972) indirect calcuation. Even higher values would be obtained by extrapolating from Mount Etna measurements whose daily averages were no less than 1100 t SO_2/day during the quiescent period after a major explosion in May 1977 and between 2700 and 4800 t SO_2/day during June 1975, a time of normal activity with a slow lava effusion on the northern flank (Zettwoog and Haulet, 1978).

Extending just the 1100 t/day rate to 50 major active volcanoes would result in some 20 million t SO_2/year and a few large eruptions could push this total close to 30 million t SO_2. For example, eruptions of El Chichón on March 28 and April 3 and 4, 1982, put large volcano gas clouds in the atmosphere and a preliminary estimate, based on spectral reflectonic mapping by the Nimbus 7 satellite, suggests that the sulfur dioxide deposited in the stratosphere by the two early April explosions amounted to 3.3 (+1.0, −0.2) million t (Krueger, 1983).

Even when taking 15 million t of SO_2 derived from more representative Central American plumes and eruptions, and adding underwater volcanism and degassing from a few major eruptions would make easily 20 million t of SO_2 annually, i.e., 10 million t of sulfur. I will use this value as the medium estimate in the coming summation of annual emissions.

In contrast to the spectacle of volcanic plumes and eruptions, sea spray is an inconspicuous source of sulfur's ingress into the atmosphere and yet the total annual flux is almost certainly greater than that originating from geothermal emissions. As noted in Section 4.1.1.2, sodium sulfate is the second most abundant compound dissolved in seawater, and incessant collapsing of myriads of bubbles at the ocean's surface propels tiny jet drops into the air at speeds of up to 10 m/sec. These minute (1–20 μm in diameter) drops are the principal source of airborne marine sulfate. Eriksson (1960) estimated this flux at 44 million t S/year and this value, or his later (Eriksson, 1963) rounding to 45 million t, has been accepted by all subsequent modelers of the cycle except for Ivanov (1981) who uses 60 million t. Also generally accepted is Eriksson's estimate that nine-tenths of the sulfate-carrying spray is returned to the ocean by gravitation or preciptation and that only 10% is deposited over the coastal regions of continents. This would mean a temporary landward flux of only 4.5 million t S/year.

4.1.2.2. Biogenic Emissions from the Ocean Balancing of the global sulfur cycle requires considerable oceanic releases ranging from Eriksson's (1963) 170 million t S to Davey's (1978) 26 million t (Table 4-1). Marine sulfates are reduced above all by two genera of Schizomycetes (belonging to Eubacteriales)—

Desulfovibrio is the dominant one, *Desulfotomaculum* appears to be much less frequent—which are largely heterotrophic and thrive at about neutral pH and in the absence of oxygen although they can be found over a wide range of pH and salinities (Goldhaber and Kaplan, 1974). H_2S as well as reduced organic sulfur compounds are mitted by biogenic sulfur reduction; in the late 1970s, the first background concentration measurements became available from open oceans and although far from conclusive they enable first generalizations based on wider empirical backgrounds instead on just abstract balancing of fluxes.

Nguyen et al. (1978) measured dimethyl sulfide (DMS) concentrations in the surface waters of the Atlantic and Indian oceans and in the Mediterranean Sea and concluded that its mean oceanic conent is on the order of 30×10^{-9} g/liter. Calculations based on this value resulted in an annual flux of at least 20 million t of sulfur in the form of DMS alone. Barnard et al. (1982) conducted the most thorough investigation of marine DMS to date and found convincing evidence against its postulated high [40–200 ng S (DMS)/m^3] concentration in the remote background atmosphere; their measurements during a transatlantic cruise from Hamburg to Montevideo averaged only 6.1 ng S (DMS)/m^3.

Model calculations based on this value and on DMS levels in seawater (sampled frequently by Barnard et al.) suggest an annual global flux of 34–56 million t of DMS sulfur from the oceans to the atmosphere, a rather good agreement with the value of Nguyen et al. (1978). Similarly, Andreae et al. (1983) sampled seawater from various ecological zones off the shore of Florida and found surface ocean concentrations averaging 50 ng S (DMS)/liter, corresponding to an annual flux of 36 million t S.

Slatt et al. (1978) were the first to measure H_2S concentrations in remote marine atmospheres of the tropics and subtropics at several locations on both the eastern and western sides of the Atlantic. The levels found ranged from 5 to 50 pptv, or 4 to 40 times lower than the previously accepted values (e.g., Robinson and Robbins, 1972), and concluded that Friend's (1973) estimate of marine biogenic sulfur emissions, 48 million t/year, is substantially correct. Yet that total completely ignored DMS concentrations and the combination of the two estimates based on actual concentration measurements puts the biogenic marine emissions at no less than 75 million t S, considerably higher than in any of the theoretical cycle models of the 1970s (Table 4-1).

Kritz's (1982) steady-state analysis of sources, sinks, and exchanges in the remote marine atmosphere suggests that about one-third of the background concentrations of SO_2 and excess sulfate in such locations mixes in from the free troposphere and the balance of about 100 μg S/m^2 per day comes from the local oxidation of reduced sulfur-containing precursors in the marine boundary layer.

4.1.2.3. BIOGENIC EMISSIONS FROM THE CONTINENTS For two decades this was the source of the greatest uncertainty in the global sulfur cycle with published estimates ranging over two orders of magnitude and, as with the ocean emissions, with most of these totals calculated simply as residuals to balance the atmospheric sulfur cycle. Actual measurements available until the early 1980s were rare and unsystematic, although most students of the sulfur cycle had little doubt of the biogenic land flux's importance.

For example, Grey and Jensen (1972) noted the importance of bacteriogenic sulfur in and near Salt Lake City by isotopic measurements of precipitation and air samples; Brinkmann and Santos (1974) reported on biogenic H_2S from Amazonian floodplain lakes; Aneja et al. (1979) measured direct H_2S, COS, and $(CH_3)_2S$ emission rates (0.2–0.66 g S/m^2 per year) from salt marshes on the coast of North Carolina; Delmas et al. (1980) found 0.19–0.24 g S/m^2 per year for three inland soils in France; Hitchock (1976) measured sulfate levels in New England and Middle Atlantic states and concluded that bacteriogenic production in fine-grained anoxic muds contributes most of the sulfate observed in the cities during summer and fall, with even an annual share up to 50%; and Hansen et al. (1978) found some high H_2S fluxes, up to 440 g S/m^2 per year, in Danish coastal lagoons covered with decomposing grass.

Basing global, or even regional, flux estimates on such a sparse data base is, naturally, a quite unreliable and frustrating exercise. Fundamental change came only with the publication of a reliable and satisfactorily widespread group of direct measurements of land sulfur emissions by Adams et al. (1981). Inevitably, these values must be used to appraise all of the past theoretical estimates and assumptions. Over a period of 4 years, Adams et al. took field measurements of biogenic sulfur emissions from soils, and also from some water and vegetated surfaces, at 35 sites in the eastern and southeastern United States during a well-designed study, encompassing different soil orders (from coastal, saline marshes to dry mineral soils) and adjusted for the annual mean temperature of each site; the results were conveniently averaged for 80 × 80-km blocks of the grid system used in the EPRI's Sulfate Regional Experiment study. Consequently, these measurements offer the best available information on a disputed phenomenon based on extensive coverage: nearly 4 million km^2, or 4% of the Northern Hemisphere's continental area.

Regressions for ambient air temperatures versus sulfur flux of all soil orders resulted in a common intercept of 0.160 g S/m^2 per year at 5°C. Recorded extremes ranged from 0.004 g in Texan alfisol to 650.9 g S/m^2 per year in a Sanibel Island saline marsh. High fluxes of coastal emissions were not unexpected—7% of the study's areas covered by tidal flats and intertidal marshes produced 41% of all recorded biogenic emissions—but the study's most significant finding is that the inland soils (93% of all area) contributed nearly 59% of measured biogenic sulfur. Clearly, dry inland mineral soils covering most of the Northern Hemisphere's surface are a considerable source of biogenic sulfur.

The other notable conclusion is the existence of an exponential southward increase of the measured biogenic sulfur flux, a phenomenon caused by a combination of increasing biomass, higher temperature and humidity, and more extensive wetlands. While near 50°N the flux is only about 0.015 g, it rises to around 0.03 g at 35°N and to 0.3 g at 25°N; extrapolations all the way to equatorial regions [confirmed by Ivory Coast measurements by Delmas et al. (1980)] give values up to 1.0 g at around 5°N and as much as 2.3 g near the equator.

Using these values for global calcuations results in an aggregate biogenic sulfur flux of about 65 million t/year (or nearly 0.43 g S/m^2 per year), the value falling almost perfectly in the middle of the extreme theoretical balance calculations of 110 and 3 million t S (Table 4-1). Until more extensive long-range measurements (especially from tropical areas) become available, the work of Adams et al. (1981) and the careful conclusions drawn from it will remain the best adjudication of the long-standing puzzle of biogenic sulfur from continental sources. Another question settled perhaps more definitively by the measurements of Adams et al. is the speciation of the biogenic land flux.

Until the early 1970s, SO_2, sulfates, and H_2S were assumed to account for virtually all atmospheric sulfur. Lovelock et al. (1972) were the first to suggest the relative importance of DMS. Shortly, Rasmussen (1974) added dimethyl disulfide (DMDS) as another nonnegligible compound. Of the six compounds analyzed by flame photometric gas chromatography in Adams et al. (1981), H_2S was almost always highly dominant: up to over 90% in the Everglades and in Sanibel Island marshes, 50–70% in most dry inland soils, though occasionally it was not detected at all. Methyl mercaptan was the second most frequent compound in the marshes, DMS the third.

While SO_2 and sulfate measurements have been performed with satisfactory accuracy for several decades, reliable measurements of atmospheric H_2S date only since 1972 when the use of a fluorescence method brought sensitivities of 0.002 $\mu g/m^3$. Since then, measurements in clean atmospheres (in locations such as Colorado, Sylt, or Ivory Coast) found H_2S concentrations mostly between 0.04 and 1.0 $\mu g/m^3$ but maxima an order higher; in polluted atmospheres, levels on the order of 1 mg/m^3 appear to be common (Aneja et al., 1982).

In spite of many new measurements of various sulfur species, the latest summary of our knowledge of biogenic sulfur fluxes shows continuing major uncertainties: the mean flux of all sulfur species is estimated at about 10.7 g/m^2 year but the mean maximum at 373 g S/m^2 per year, implying a total biogenic sulfur contribution to the troposphere as low as 4 million t/year or as high as 140 million t, and a biogenic contribution of COS and CS_2 to stratospheric sulfate accounting for just 1% but possibly for more than 100% of the needed total (Aneja et al., 1982). Emissions of only 1.8×10^{-4} g S (as CS_2 and COS)/m^2 are needed to account for all stratospheric sulfate aerosol content (Crutzen, 1976b). These huge uncertainties should be kept in mind when trying to account for anthropogenic sulfur emissions with errors of no more than a few million tonnes per year.

4.1.3. Anthropogenic Emissions

Burning of fossil fuels, above all coals, is by far the largest source of sulfur liberated by industrial activities. Smelting of nonferrous ores, led by copper production, is much less important, and other industrial contributions—synthesis of sulfuric acid, recovery of elemental sulfur, conversion of pulp into paper, incineration of refuse, and flaring of natural gas—probably total less than the error associated with the estimates of global SO_2 emissions from coal combustion.

There are at least two other widespread human activities releasing considerable amounts of sulfur into the environment but as they involve no emissions of sulfur compounds into the atmosphere, they will not be treated here in any detail—just mentioned before starting a systematic discussion of atmospheric emissions. Acid mine drainage is a complex and still only partly understood process which creates serious local pollution problems and complicates extraction of higher-sulfur coals around the world (Rose et al., 1983). Abundance of pyrite and other iron sulfides is the principal factor in determining acid generation, but accurate predictions of future acid outputs have proved difficult and no reliable global quantifications of the process are available.

Sulfur inputs to soils in fertilizers are, of course, much more widespread than those to waters by mine drainage. Sulfur is a part of common fertilizers, most notably $(NH_4)_2SO_4$, it is included as a micronutrient in compound materials, and most recently it has been used increasingly as a coating for urea. Worldwide annual inputs are put by most authors at about 10 million t, except for Friend (1973) who set the total at 26 million t. When in the form of a sulfate, these applications contribute to soil acidity, and they will be mentioned later during discussion of soil acidification processes (Section 4.4.3.2).

4.1.3.1. Coal Combustion Calculation of the sulfur emission from coal combustion is made easy by the availability of sufficiently accurate statistics on the output of bituminous, lignitic, and anthracitic coals—but far from precise because of the very large variety of coals (see Section 2.2.1.2). Consequently, it is not surprising that ultimate analysis reveals sulfur contents ranging from a low fraction of 1% to nearly 8% of the total mass (Averitt, 1974; World Energy Conference, 1974). National figures for the world's top two coal producers show similar ranges: Soviet bituminous coals contain 0.8–7.5% S, the American ones 0.3–7.7% (World Energy Conference, 1974). Detailed statistics available for the remaining reserves of American coals illustrate substantial differences among major coal varieties (DeCarlo et al., 1966; Agnew, 1977; Hibbard, 1979): at less than 1% the total mean is agreeably low, owing to large subbituminous and lignite resources west of the Mississippi, but half of all bituminous coal (which accounts for some 45% of all coal resources by mass and for about nearly two-thirds of energy content and which forms the bulk of eastern production) contains more than 2.5% S.

A finer disaggregation just for the bituminous coal reserves shows well the pitfalls of using average values in calculations: the mean of 2.5% is actually the least frequent value while the modes lie at the extremes. Averages are elusive even for an individual coal basin, even for a single mine, where strata of different ages at various depths may contain substantially different quantities of sulfur. In any case, data on sulfur in the coal actually mined rather than on reserves are needed to calculate the gaseous emissions—and such information is very scarce.

Most of the inorganic sulfur in coal is usually in sulfides (pyrite and marcasite), the rest in hydrous ferrous sulfate ($FeS_4 \cdot 7H_2O$, a product of pyrite weathering) or in gypsum. Pyritic sulfur content can be lowered by various washing and cleaning procedures (Section 4.5.3.1) so that only about 30% of the original inorganic sulfur remains in the fuel: depending on the kind of coal, this can reduce the total sulfur content by as much as half or hardly at all because the mechanical cleaning does not remove organic sulfur present within the matrix of coal-forming vegetal matter.

No global data are available on the extent of pyrite sulfur removal but statistics on the sulfur content of shipped coal systematically available for the United States only since the latter half of the 1970s show that the cleaning of coals as well as the selection of low-sulfur varieties for metallurgical coke and electricity generation to meet the USEPA's emission standards have resulted in a mean sulfur value of 2% for all coal consumed in the late 1970s and that the mean for the utilities, which consume three-quarters of American coal, has been just short of 2% (Agnew, 1977).

Even these data do not allow a calculation of the precise averages as the shipments according to sulfur content are given in interval categories. Still, the figure of 2% appears to be a very good approximation for the United States, and a vareity of information on sulfur content of coals from the leading bituminous and lignite basins in the USSR (where the range is similar to American values and cleaning is also extensive), China (where cleaning is not at all widespread but where most coals have less sulfur, 0.3–1.5%) and Europe (where some lignites extracted for power generation contain well over 3% S) suggests that using 2% for global calculations may also be the best approximation. This translates into an emission factor of 40 kg of SO_2 per t of coal; Robinson and Robbins (1972) used 48.2 kg, as did Cullis and Hirschler (1980). An even more difficult question to answer concerns the share of the evolved sulfur oxides that will be retained in ashes and fly ashes.

A considerable amount of sulfur may be retained not only in copious ashes from inefficient domestic small stoves but, a fact generally much less appreciated, also in fly ash electrostatically removed after the combustion of pulverized coal in large modern power plants, now the leading consumers of coal worldwide. In the most detailed study to date examining 183 sets of data from actual operation of various power plants, Davis and Fiedler (1982) found that sulfur retention in fly ash from coal-fired boilers is highly correlated with the total alkaline metal oxide content of the ash. For pulverized coal-fired boilers, this correlation was as high as 95.8% and the graphical analysis of the data in Figure 4-2 shows that the percentage of SO_2 removal owing to fly ash retention may range from negligible shares for coals with low alkalinity to more than 40% for a coal with 10% ash, 0.5% S, and 35% total alkaline metal oxide content in the ash.

With the absence of any data on average fly ash composition even in the United States, and also with Davis and Fiedler's (1982) reminder that in spite of excellent correlations they found the range of observed values of sulfur retention for any given fly ash content to be greater than a factor of two in most cases, it is clearly impossible to offer a single global generalization. For the basic calculations, I will

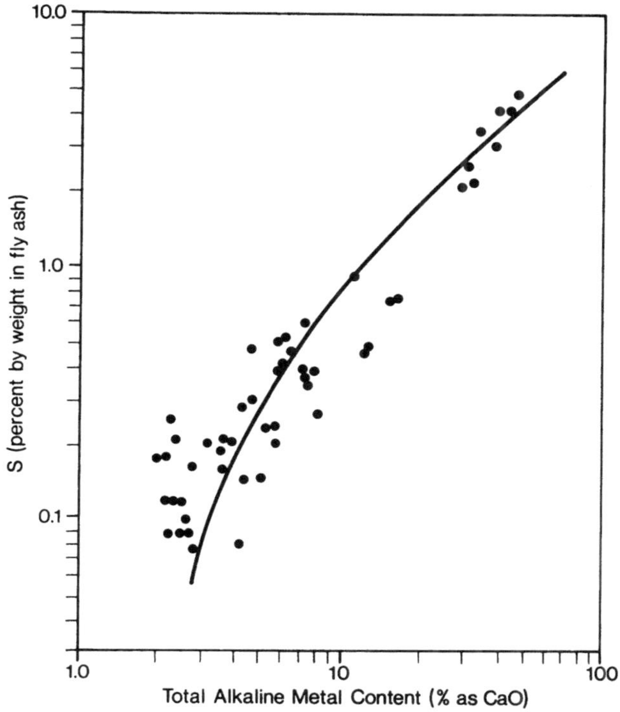

FIGURE 4-2. Sulfur retention as a function of the total alkaline content (as CaO) in fly ash (Davis and Fiedler, 1982).

thus neglect this possibly important effect but I will include it in an alternative estimate.

Coking, a destructive distillation now overwhelmingly done with the by-product process rather than in beehive ovens, starts with low-sulfur coal charges (usually 0.8–0.9% by weight) and only about one-third of all sulfur present is transferred to coke-oven gases which are used in the underfiring or elsewhere in steel production. Depending on the efficiency of the production, overall SO_2 emissions may thus be as low as 1.5 kg/t coke, or as high as 6 kg. The USEPA (1973a) uses a factor of 2 kg. Rohrman and Ludwig (1965) worked with 5.4. Using 2 kg for the OECD nations (producing about 170 million t) and 4 kg for the rest of the world (annual output of about 150 million t) gives about 470,000 t of sulfur annually released into the atmosphere.

Sulfur releases from coal used in production of portland cement appear to be only slightly larger. Limestone, the essential raw input into the kilns, reacts vigorously with sulfur oxides (hence it figures high on the list of ingredients used to desulfurize flue gases) and the available measurements of actual sulfur uptake by portland cement clinker and collected dust show that at least 86 but more frequently

95–99.8% of sulfur oxides never enter the atmosphere (Gagan, 1974). In industrialized countries, modern cement works use about 230–280 kg coal/t cement (Kreijger, 1976; IEA, 1982) and of the original 10 kg SO_2 evolved (assuming 2% S content), only about 1 kg would be released with typical retention and with dust collection.

A breakdown of fuel consumption in the cement industry is available only for the OECD nations: in 1980, coal was used to fire 40% of the organization's 380 million t cement)OECD, 1982). Multiplying 155 million t by an emission factor of 1 kg SO_2/t results in a total of a mere 80,000 t S. For the rest of the world production—another 350 million t—I shall assume that 75% of it is coal-fired and that the mean emission factor is 3 kg SO_2/t cement, a value reflecting not only the frequent absence of controls but also higher energy use per tonne of clinker. These assumptions result in annual emissions of some 520,000 t S and a global total of roughly 600,000 t, a contribution easily lost in the error inevitably committed in coal combustion accounting.

Subtracting the coal used for coking (globally about 500 million t, equally split between the OECD nations and the rest of the world) and cement production (globally 100 million t, in OECD 30 million t) from the global consumption total of 3.7 billion leaves 3.1 billion t for conbustion in households, factories, and electricity generation plants. With the average sulfur content of 2% and no assumed ash or fly ash retention, this yields 62 million t S into the atmosphere and this total could be raised by another 1 million t with addition of coking and cement industry contributions. Whatever the actual number, there is no doubt that coal combustion represents a major flux of sulfur into the atmosphere, a flow comparable to total biogenic terrestrial flux.

4.1.3.2. HYDROCARBON PROCESSING AND COMBUSTION Although ultimate analysis of crude oils shows much more uniformity than in the case of coals, sulfur is the only one of oils' minor constituents which can range as widely as it does in coal—from hundredths of a percent up to 8% by weight (King et al., 1973). Heavy crude oils (property measured in °API gravity where the higher the number, the lighter the specific density of the oil) invariably have more sulfur than those with a larger share of more volatile fractions (Nelson, 1972).

Among the oils dominating the world market, Saudi Arabian heavy (28.2 °API) oil contains 2.84% S, Saudi light 1.80%, Iranian oils 1.4–1.6%, Kuwaiti 2.5%, and Nigerian and Algerian lights only 0.13–0.14%; most North American oils contain less than 0.5% and Soviet oils (the country is now the world's top producer) usually 1.5–2% (Aalund, 1976, 1983; Semb, 1978). The fundamental difference, of course, arises through the processing of crude oils: less than 100 million t, or only some 3–4% of global production, is burned unrefined.

Modern crude oil refineries comprise many separate operations ranging from simple distillation to catalytic cracking and chemical treating. The main sources of sulfur emissions are, besides the boilers and process heaters fueling the operations, fluid catalytic cracking units: emissions from these sources range from 142 to 238 kg (average 224 kg) of SO_x per 1000 barrels of fresh feed (USEPA, 1973a). Using the global average of 7.4 barrels/t, this translates to 1.65 kg SO_2/t processed crude oil. For older, less efficient units outside the industrialized nations, the rates are higher. Cullis and Hirschler (1980) used 2 kg SO_2/t crude oil and even this may be too low for many older, smaller refineries. I will assume 1.6 kg SO_x/t for the OECD refineries and 2.5 kg elsewhere; for 1980 these assumptions give emissions of 3 million t S.

Refining yields a wide variety of liquid fuels of which gasolines, distillate fuel oils, and residual fuel oils are by far the most important with jet fuels and kerosene far behind. Reasonably accurate worldwide statistics are available for all of these products in the annual UNO publications; multiplications of these totals by average sulfur content of the fuels—0.04% for gasoline, 0.25% for distillate fuel oil, 2% for residual oils, 0.05% for jet fuel, and 0.3% for kerosene (Considine, 1977)—sum up to 18 million t S in 1980. At least another 1 million t should be added for combustion of the unrefined crude and for consumption unaccounted for in the UNO statistics: 19 million t S is thus the best approximation I can offer for the combustion of liquid fuels, about 22 million t for combustion and refining. This is considerably less than the 27.15 million t calculated by Cullis and Hirschler (1980) for 1976: although they used very similar sulfur content values, they made an inexplicable error of starting with 1976 crude oil output 30% higher than the actual production.

As for natural gas, one of the main reasons why its consumption has been expanding in all parts of the world has been the clean combustion of the fuel which usually contains only a trace of sulfur or from which, in cases of somewhat higher contamination, the sulfur is removed relatively easily. Consequently, neither Robinson and Robbins (1972) nor Cullis and Hirschler (1980) considered SO_2 emissions from natural gas in their detailed appraisal of man-made emissions. Indeed, multiplying the 1980 natural gas consumption by the fuel's average sulfur content of 4.6 kg/10^6 m^3 gives less than 7000 t S/year, a completly inconsequential amount.

4.1.3.3. Nonferrous Metallurgy Production of nonferrous metals, above all copper with lesser contributions by lead, zinc, and nickel, is the third largest source of antropogenic sulfur and often a very large one and highly concentrated: tall stacks of some large smelters are point sources with very high SO_2 fluxes (tens of kilograms per second) and devegetated and subsequently eroded neighboring downwind areas are perhaps the starkest examples of severe damage caused by excessive sulfur burden in forest ecosystems (for details see Section 4.4.3.4).

To produce copper, the usually low-grade sulfide ores are concentrated, then roasted, smelted, and concerted at high temperatures. Considerable quantities of sulfur oxides are liberated at each of the last three stages. The USEPA (1973a) puts emission factors for primary copper smelters without controls at 30 kg for roasting, 160 kg for smelting in reverberatory furnace, and 435 kg for converting for a total of 625 kg SO_x per t of concentrated ore produced. As about four unit weights of concentrate are needed to produce a unit of metal, uncontrolled emissions would total 2500 kg SO_x/t copper.

Although copper smelter SO_2 emissions can be effectively controlled—combination of sulfuric acid plants (using the smelter's SO_2 as raw material) and lime slurry scrubbing can reduce the sulfur oxides to just a tenth of their initial mass and many large smelters have been producing H_2SO_4 for decades—usually only a fraction of the gas flow is so treated. Moreover, half of the world's copper originates in nations outside of the OECD and one-quarter comes from a handful of poor Latin American, African, and Asian countries where controls are usually minimal or nonexistent.

Consequently, I will assume an average value of 1700 kg SO_2/t cooper for the OECD nations (one-third of emissions removed) and 2000 kg elsewhere. For comparison, Robinson and Robbins (1972) did their often-cited calculations of global SO_2 emissions with a single factor of 2000 kg (Stanford Research Institute data), a value also adopted by Cullis and Hirschler (1980) for their global estimates. For the 1980 primary copper production of 8 million t, this works out to 7.5 million t S.

Lead production follows a route similar to copper smelting and overall uncontrolled plant emissions are given by the USEPA (1973a) as 660 kg SO_x/t lead; the same source gives losses of 1100 kg SO_2 from roasting and sintering of zinc ores. The actual factors implied in the agency's national air quality estimates are 650 kg for the two metals combined. Robinson and Robbins (1972) used 500 kg for lead and 300 kg for zinc; Cullis and Hirschler (1980) applied, respectively, 470 and 200 kg for the years 1974–1976. I shall assume averages of 400 kg (Pb) and 600 kg (Zn) for the OECD countries and 500 kg (Pb) and 800 kg (Zn) for the others. For the 1980 global output of 3 million t of smelter lead and 5.3 million t of zinc, these assumptions result in totals of, respectively, 650,000 and 1.75 million t S.

Nickel smelting contributed in 1980 just a bit more than lead production: only 750,000 t of nickel was produced that year and with emission factors very close to those of copper smelting, the global output of sulfur from this source was no more than 0.75 million t annually. In the aggregate, nonferrous metallurgy contributed almost 11 million t S in 1980, or about 13% of the total (85 million t) released by combustion and processing of fossil fuels.

4.1.3.4. MINOR CONTRIBUTIONS Among the minor industrial sources of sulfur compounds, emissions from the production of sulfuric acid are by far the most important and widespread because H_2SO_4 is a basic raw material required for a large number of industrial processes. The acid's synthesis in contact process plants produces two kinds of sulfurous pollutants in the exit gases, SO_2 and acid mist. SO_2 emissions are inversely proportional to the rate of SO_2–SO_3 conversion. Most of the well-designed single absorption plants operate with conversion of 95–98%, releasing 13–35 kg SO_2/t 100% H_2SO_4 (USEPA, 1973a).

In contrast, the USEPA performance standard for new and modified plants is only 2 kg SO_2/t H_2SO_4, an equivalent of 99.7% conversion efficiency in an uncontrolled plant. The quantity and particle size distribution of acid mist are largely determined by the feedstock and the strength of the synthesized acid, with oleum plants burning spent acid producing up to about 5 kg of small-particle (hence difficult to control) mist/t H_2SO_4 while 98% H_2SO_4 plants charging elemental sulfur have emission rates below 2 kg/t (USEPA, 1973a).

Very good controls have been available for years for both gaseous and mist emissions (USEPA, 1971b). Dual absorption process raises the SO_2–SO_3 conversions to an average of 99.7% or higher, and where it proves impractical, sodium sulfate–bisulfate scrubbing can remove SO_2 from the absorber exist gases so effectively that, again, less than 2 kg SO_2/t H_2SO_4 is emitted. Acid mist can be reduced by electrostatic precipitators to 0.06 kg/t H_2SO_4 and fiber mist eliminators may leave a mere 0.01 kg/t H_2SO_4, a tiny fraction of SO_2 controlled emissions. Obviously, such excellent controls are not applied even to most plants in the OECD nations, which produced two-thirds of all H_2SO_4 in 1980. I shall assume that the OECD emissions average 15 kg SO_2/t (one-third of plants uncontrolled with 95% conversion rate, one-third controlled to 99% rate, and the last third controlled to 99.5%), while for the rest of the world the mean is 25 kg SO_2/t H_2SO_4. These assumptions lead to a global total of 1.6 million t S in 1980. In comparison, emissions from production of elemental sulfur by the modified Claus process are very small: using emission factors of 2 kg/t pure sulfur in the OECD nations and 5 kg elsehwere adds only some 80,000 t of the element into the atmosphere.

Another widespread source of minor sulfur emissions, though locally often most unpleasant, is wood pulping, the production of cellulose from wood by dissolving the lignin. SO_2 emissions originate from the oxidation of reduced sulfur compounds, and the frequently objectionable smell in the vicinity of kraft plants derives from a variable mixture of hydrogen sulfide, dimethyl disulfide, and mercaptans. Typical SO_2 emissions are low, 2.5 kg/t air-dried unbleached pulp and are usually vented uncontrolled; H_2S emissions are just about 0.5 kg/t and the quantities of disulfides and mercaptans are even smaller. Air pollutant control of the kraft process has been rather difficult (USEPA, 1973b) and so I will assume averages of 2 kg SO_2 for the OECD nations and 2.5 kg SO_2/t pulp for all other producers. The global total then comes to about 100,000 t of sulfur as SO_2, and another 50,000 t of H_2S, disulfide, and mercaptan sulfur must be added to this total.

A new source of sulfur has emerged with flue gas desulfurization: emissions from stored desulfurization sludges. Adams and Farwell (1981) measured their strength at 13 sites and found total sulfur emissions ranging from less than 0.25 to nearly 0.75 kg S/day for an equivalent of 100 ha of sludge impoundment surface. This represents only a minor source, with maximum daily emissions amounting to a mere 0.0001% of the sulfur discharged to ponds.

Two contributions very hard to quantify are burning of biomass and incineration of urban and industrial refuse. The latter process emits at least 1 kg SO_2 per t of the waste—the USEPA (1973a) lists 1.25 kg for uncontrolled emissions—and open burning of municipal refuse releases about 0.5 kg SO_x per t of material. As noted before (Section 2.2.2), there are no global statistics on the total amount of solid wastes and even when approximations are made, the uncertain assumptions about the shares incinerated and burned in the open make the whole exercise questionable. Even if the total annual output of about 1 billion t of combustible solid waste were burned (half of it in the open, half incinerated), the total annual contribution would not surpass 500,000 t S. In reality, only a fraction is so disposed so that actual emissions are on the order of 10^3–10^4 t S/year, comparatively negligible on a global or continental scale.

Fuelwood usually contains only a trace of sulfur and hence its contribution can be disregarded. Straws and stalks of cereals and other crops contain 0.04–0.2% S (Staniforth, 1979; NAS, 1982b). Using 0.1% as the average and multiplying by the best available global estimate of nearly 500 million t of crop residues used as a fuel results in emissions of 500,000 t S annually.

Sulfur compounds other than SO_2 are also released during biomass burning but their aggregate mass is not large at all. By relating emission quantities of various trace gases to those of CO_2 in fire plumes, Crutzen et al. (1979) estimated that worldwide burning of biomass releases annually about 100,000 t of sulfur as carbonyl sulfide, i.e., about 60,000 t S, again a negligible total compared to many other uncertainties.

4.1.4. Uncertainties and Balances

With all important and some minor natural and man-made contributions of sulfur into the atmosphere accounted for, it is possible to assess the degree of human interference in the cycle not only in global but also in hemispheric and in selected national terms. As all of these assessments suffer by far-from-accurate knowledge of several major fluxes, a brief review of outstanding uncertainties and their effects on the quantitative conclusions will precede quantitative comparisons of anthropogenic and natural flows and a brief discussion on the magnitude and direction of net sulfur flow between the continents and the oceans, one of many outstanding problems in understanding the cycle.

4.1.4.1. Major Uncertainties In spite of at least two decades of lively interest in the global sulfur cycle, none of the major natural fluxes into the atmosphere is known with a satisfactory degree of accuracy and differences on the order of ±50% cannot be excluded for any of the estimates presented in Section 4.1.2.

Although my estimate of volcanic sulfur is much higher than all but two published values (see Table 4-1), it still may be too low. Tremendous progress in volcanology has not brought, unfortunately, great advances in our understanding of quantities and composition of gases evolved from extruded magma: such measurements remain very difficult and some of the best available values are now more than half a century old (MacDonald, 1972). Accounting of volcanic emissions is complicated by the variety and irregularity of the sources as gases of widely different composition evolve not only from extruded magmas but also are emitted from craters, fissures, fumaroles, and solfataras of active and dormant volcanoes and, of course, escape in huge volumes during major eruptions.

Lamb's (1970) data on nearly 500 years of major volcanic eruptions indicate that since 1500 there have been about 60 such events per century, or one every 20 months, but only most recently have such eruptions been investigated from the point of atmospheric chemistry and the conclusions do not afford any satisfactory generalizations. Previously cited measurements of Etna's plume recorded up to 12,400 t SO_2/day during a major (Punta Lucia) explosion in January 1976 (Zettwood and

Haulet, 1978)—while air samples collected within 24 hr of the largest explosion of the 1979 eruption of Soufrière, on the Caribbean Island of Saint Vincent, showed only 0.8–3.3 ppb COS and 0.2–1.2 ppb CS_2, and comparable concentrations (0.5–1.5 ppb v) of sulfate (Cronn and Nutmagul, 1982).

Similarly, during the probably best investigated volcanic phenomenon of recent years—Mount Saint Helens eruptions in spring and summer 1980—multiple traverses of the plumes and simultaneous measurements of horizontal winds established peak fluxes of up to 100 kg H_2S and 20 kg SO_2/sec (Hobbs et al., 1981) but the means for 5 months of activity were only about 1–2 kg/sec, translating to the aggregate flux of only several tens of thousand tonnes of sulfur between April and August 1980. In view of these disparities, I will assume that annual volcanic emissions may be less than 5 million t or as high as 30 million t S.

As for the biogenic emissions of sulfur compounds, Delmas et al. (1980) are undoubtedly correct in stating that many more measurements in different zones of various ecosystems will be needed before arriving at a satisfactory total. Measurements in the paddy fields are especially needed as these wet formations with anoxic soils are known to be the strongest biogenic source of methane. Uncertainties arising from the paucity of existing measurements in the world's warmest regions where the emissions of sulfur compound are most abundant can easily be responsible for differences amounting to a few tens of millions of tonnes of sulfur per year.

While Adams et al. (1981) presented the best set of land emission measurements (see Section 4.1.2.3), in their global estimate they had to extrapolate for the tropical regions (their southernmost monitoring was in southern Florida) and used values of up to 2.3 g S/m^2 per year near the equator. However, Delmas et al. (1980) concluded from 90 measurements in six locations of the Ivory Coast tropical rain forest that the mean H_2S emissions vary appreciably with moisture averaging only 35 μg/m^2 per hour during the dry season and going up to 300 μg/m^2 per hour during the rainy season. The latter value would give 2.62 g/m^2 per year, an excellent agreement with Adams and colleagues' model—but as it lasts for only about 3 months the mean annual H_2S flux appears to be somewhere between 0.3 and 0.9 g/m^2 per year, far from below the value Adams et al. (1981) used to calculate tropical sulfur emissions and their speciation of the emitted sulfur makes it highly improbable that other compounds could make up the difference.

More data are also needed on the fluxes and subsequent behavior of carbonyl sulfide and carbon disulfide, two substances which until recently were not even mentioned in the sulfur cycle studies but which are now believed by some researchers to be major forms of atmospheric sulfur on a global scale: Sze and Ko (1980) estimate that oxidation of COS and CS_2, initiated by reaction with OH^-, could provide an annual flux of 8–12 million t S. They also estimate, on the basis of photochemical considerations and by comparing calculated profiles and burdens of SO_2 and SO_4 with observational data, that the combined global fluxes of H_2S and DMS are about 20 million t/year and argue that the high fluxes proposed by Eriksson (1963) and Junge (1960) are unlikely, while the total biogenic fluxes around 100 million t S/year should be ruled out.

Obviously, selecting the most likely ranges remains in the realm of feelings and preferences still only imperfectly guided by hard data. I shall assume that the continental biogenic flux may be only as little as one-half of the medium value offered in Section 4.1.2.3 (65 million t) but that the maximum is no more than 25% above the standard. Ocean emissions will be treated analogically and the total natural flux would then range between 115 and 265 million t, with 195 million t S/year as the most likely medium value.

Uncertainties surrounding anthropogenic emissions are much smaller but the final aggregate differences can still add up to tens of millions of tonnes. In the absence of worldwide data on the extent and degree of control of gaseous emissions from production of copper, nickel, lead, zinc, or sulfuric acid, errors of ±10–15% must be commonly anticipated and they may be further compounded by using the USEPA6's (1973a) uncontrolled emission factors as the basis of estimates because backward technologies in many poor countries produce even greater uncontrolled quantities of pollutants. However, it is the accounting of sulfur from fossil fuel combustion where by far the greatest errors are made inadvertently.

As stressed before, averaging of coal's sulfur content is very difficult even on a small scale and in global terms a multiplier just one-tenth of a percent higher or lower (i.e., 1.9 or 2.1 instead of the used 2% average) adds or subtracts 3 million t S/year, i.e., more than the most likely total from all minor contributions treated in Section 4.1.3.4. Another important fact to notice is that the major expansions of coal mining initiated recently in many countries around the world have brought some shifts to lower-sulfur coals—as well as some substantial increases in extraction of coals with very high sulfur content. Thus, for example, the average sulfur content of low-energy (around 12 MJ/kg) lignites mined in the North Bohemian Basin (annual output is now about 70 million t) rose rapidly from 1.47% in 1973 to 2.35% in 1980 and will continue to increase. Once again, on these grounds alone, a 10% margin of uncertainty must be seen as quite realistic—and it must also be extended to the combustion of liquid fuels. Retention of sulfur compounds in coal ashes and fly ashes could have an effect at least as large as it is well within the range of measured possibilities to lower the total emissions by 10% (see Section 4.1.3.1).

Combinations of these uncertainties will result in the smallest likely anthropogenic emissions of just over 80 million t S/year, a medium estimate of almost exactly 100 million t, and a possible maximum of a bit over 110 million t in 1980. Comparisons of these values with natural fluxes (Table 4-2) raises many interesting points and many more challenging unresolved puzzles.

4.1.4.2. Balancing the Cycle Regardless of what the values of atmospheric influxes actually are, these inputs are removed in a relatively short time from the atmosphere and transferred back to the soils, vegetation, fresh waters, and oceans. Of the general validity of this grand balance there can be little doubt. Concentrations of sulfur compounds measured in remote places are invariably very low and show no accumulation trends, nor does the atmospheric content of these gases and solids rise continuously in even the most polluted areas: temporary increases in ambient air concentrations during periods of limited mixing are soon, in a matter of hours or at worst days, reduced to lower levels.

TABLE 4-2
COMPARISONS OF THE BEST ESTIMATES OF NATURAL AND ANTHROPOGENIC ATMOSPHERIC SULFUR FLUXES[a]

	Estimate		
	Minimum	Medium	Maximum
Natural flux			
Volcanoes	5	10	30
Sea spray	40	45	60
Biogenic S—oceans	40	75	95
Biogenic S—continents	30	65	80
Total	115	195	265
Man-made emissions			
Coal combustion	52	63	70
Liquid fuels	20	22	25
Nonferrous metallurgy	8	11	12
Minor sources	2	3	5
Total	82	98	112

[a]All values are in million tonnes per year.

A variety of processes—dry and wet ones, physical and chemical—remove sulfur from the atmosphere and more will be said about them when dealing in detail with SO_2 and SO_4^{2-} (Section 4.2). Here is the appropriate place to stress the fact that, although unexceptionable as a general proposition, balancing of the global sulfur cycle by postulating relatively rapid atmospheric removals presents another set of major uncertainties. First, the estimates of the total mass of the element to be removed range from 143 to 365 million t, a 2.5-fold difference (Table 4-1). In the medium version of atmospheric inputs presented in the preceding sections, this total was 294 million t and the extremes were put at 197 and 377 million t S. This alone shows our inadequate understanding of the removal processes as the different models implicitly endow them with very different rates.

A second uncertainty concerns the partitioning among the removal processes. As seen in Table 4-1, of those models with complete estimates one attributes merely 14% of the removal to dry deposition while the rest puts this share between roughly 40 and 60%. And the third critical uncertainty has to do with the spatial division of the removal: although clearly not long-lived, sulfur compounds reside in the atmosphere long enough to be carried tens and hundreds of kilometers before final deposition or precipitation. Consequently, depending on the location of strong sources and the course of prevailing air flows, there may be significant net transfers between the continents and the ocean—or exchanges of sulfur compounds of different origin (i.e., fossil fuel emission carried over the ocean, while sea spray and biogenic sulfur from coastal areas are deposited on land) with no important shift in any direction.

The latter appears to be the favored treatment in global sulfur models although their review by Cullis and Hirschler (1980) indicates that some authors suggest massive net transfers from sea to land (Junge, 1963; Eriksson, 1963) while others favor large shifts from land to sea (Robinson and Robbins, 1972; Garland, 1977). This is a completely mistaken impression created by erroneous accounting use by Cullis and Hirschler (1980) to calculate what they call net transfer from land to sea by adding continental biogenic emissions, volcanic flux, and man-made sources and subtracting from this total biogenic emissions from the ocean and sea spray sulfur. This difference tells simply how much larger or smaller is one group of emissions—but nothing at all about the places of their deposition. A closer look at the published cycle models will reveal, for example, that instead of the claimed transfer of 64 million t S from land to sea, Robinson and Robbins (1972) clearly state that this net flow is only 22 million t; or that the 65 million t shift from sea to land ascribed to Eriksson (1963) is merely a 15 million t transfer.

In a way, all these attempts, assumptions, or errors have little importance: unlike with carbon dioxide where gradual increase in background concentrations is measurable all over the planet, the overwhelming majority of man-made sulfur emissions originate on the continents of the Northern Hemisphere and as the residence times of common sulfur compounds are not long enough to permit any significant interhemisphere exchange (see Section 4.2.3.1 for details), it is more profitable to look at balances, or rather imbalances, of the sulfur cycle at smaller scales where both the dominance and the effects of anthropogenic sulfur flux stand out quite unmistakably.

4.1.4.3. Comparing the Sources To return for a few paragraphs to global balances, the importance of human contributions to atmospheric sulfur flux is, of course, far from insignificant even on a global scale. The obvious combinations of totals offered in Table 4-2 show that in the medium estimate, man-made sources are almost exactly half of the natural global flux, i.e., one-third of all sulfur emitted into the atmosphere; in the minimum estimate they account for 42% and in the maximum one, 30% of the total; finally, crossing the minima and maxima yields as little as 24 and as much as 49% of atmospheric sulfur originating from human activities. The extremes roughly (25–50%) circumscribe well the extent of our uncertainties while the most likely share based on a review of all atmospheric fluxes is between one-third and two-fifths of the global total. For comparison, the published sulfur cycle accounts (Table 4-1) imply anthropogenic shares as low as 13% (Eriksson, 1963) and as high as 45% (Bolin and Charlson, 1976).

Even in the lowest presented case this would be a considerable interference and the most plausible values signify an uncommonly large "enrichment" of the global account. However, as stressed in the closing of Section 4.1.4.2, one must look at the Northern Hemisphere, and more specifically at its heavily industrialized and urbanized regions in Europe, North America, and the USSR (between 30° and 55°N), to appreciate the magnitude and the gravity of man-made intervention in the sulfur cycle. Hemispheric comparisons of natural and man-made sulfur emissions are hardly easier than the global ones: the key problem is what shares of natural emissions should be allocated to the hemispheres.

The Northern Hemisphere represents 70% of all land and 42% of the oceans, and so dividing the appropriate biogenic emissions and sea spray accordingly and halving the volcanic contribution, would give, in the standard case, some 100 million t of natural sulfur flux annually (extremes of 57 and 136 million t), just over half of the global flux. In contrast, with the single exception of copper metallurgy where the smelters of Chile, Peru, Zaire, and Zambia account for about one-third of the industry's sulfur emissions, the Southern Hemisphere's contributions to anthropogenic sulfur flux are well below 10% of the global total: 6% for coal, 5% for refined liquid fuels. Using 94:6 division for all emissions means that Northern Hemisphere countries put into the atmosphere at least 77, mostly likely about 93, and as much as 105 million t S each year. Comparing the hemispheric estimates raises the man-made share to 48% (the values for matched minima and maxima are 57 and 44%, for cross comparisons 36 and 65%). Fifty percent is certainly a good approximation.

Naturally, the shares of man-made emission will rise further when comparisons are made for major industrialized and urbanized regions. Here the choice of the area is critical. In small regions with a high number of large fossil-fueled power plants, refineries, mines, smelters, and other industries, the anthropogenic sulfur flux will completely dominate the atmospheric inflow with emission totaling hundreds of tonnes of sulfur per square kilometer. An outstanding example of such an extreme concentration is the North Bohemian Brown Coal Region, a relatively small (just 1000 km^2) coal mining basin located in a narrow (10 km) intermountain valley (its floor is 400–700 m lower than the tops of main ridges flanking it in a SW-NE direction) with relatively very low precipitation (rain shadow effect of the Ore Mountains).

These natural conditions produce limited ventilation, frequent temperature inversions and air mass stagnation, and slower removal of any compounds emitted into the atmosphere. This small region now generates sulfur emissions at the astonishing rate of 500 t/km^2—and because of the limited atmospheric mixing in the valley, most of this burden will be, after causing ambient air concentrations greatly surpassing all standards, deposited back onto the area and on the forests of the surrounding mountains. At the same time, biogenic sulfur flux in the area (50°N) can be up to 20 kg/km^2, a difference of four orders of magnitude!

In contrast, there are other areas with high densities of sulfur emissions but with relatively low immissions, and vice versa. Perhaps the best such examples are the English Midlands where large coal-fired power plants are the main sources of fluxes averaging up to 3 t S/km^2 over areas of several thousands of square kilometers—but where the prevailing winds carry the emissions released from tall stacks for hundreds of kilometers to be finally deposited in southern Norway, an area whose local man-made sulfur flux is as low as 60 kg/km^2 annually, just three or four times the level of biogenic emissions.

On a national scale, Czechoslovakia has the greatest density of SO_2 emissions with about 9 t S released per km^2 per year, closely followed by Belgium with approximately 8 t; the United Kingdom averages nearly 6 t and the Netherlands about 5 t S/km^2 (Semb, 1978). In North America, national averages are obviously much lower owing to huge unsettled areas but local and regional anthropogenic sulfur fluxes are no less staggering.

In Canada, two-thirds of the total annual SO_2 emissions of about 5 million t comes from just two provinces, Ontario and Quebec, and one-quarter of this originates in two huge point sources, INCO's nickel and copper smelter at Copper Cliff near Sudbury, Ontario, and Noranda Mines copper smelter at Noranda, Quebec (Sub-committee on Acid Rain, 1981). The Sudbury discharges from a 381-m-high "superstack" are perhaps the world's largest point source, and the current rate of about 2270 t SO_2/day (the maximum permitted by Ontario legislation) followed appreciable reduction in smelter output rather than improved controls. However, the flue gas SO_2 removal had previously reduced the emissions from as much as 5500 t/day in 1969 (i.e., nearly 2 million t SO_2/year) to 3300 t/day by 1978. The goal is to reduce the Copper Cliff emissions to just 750 t/day.

While Canadian SO_2 emissions originate overwhelmingly (about 70%) from industrial processes led by nonferrous metallurgy, the United States anthropogenic sulfur flux is heavily dominated by thermal electricity generation. In 1940 this largely coal-fired power generation released just 13% of all SO_2 but this fraction rose rapidly to 23% in 1950, 43% in 1960, 56% in 1970, and currently stands at almost exactly two-thirds of the total SO_2 flux (Table 4-3). About 75% of United States SO_2 emissions are released east of the Mississippi and Clark's (1980) detailed mapping by 80-km grid squares reveals the dominance of large power plant concentrations in the Ohio Valley (Figure 4-3).

In the early 1980s there were three coal-fired power plants releasing in excess of 300,000 t SO_2/year: Paradise (Kentucky) at 370,000 t and Muskingum and Gavin (both Ohio) each with 340,000 t. In Ohio, the state with by far the largest SO_2 emissions in the United States, the annual flux of SO_2 in the early 1980s surpassed 2.4 million t (U.S./Canada Work Group, 1982). Even when prorated over all of the

TABLE 4-3
UNITED STATES ANTHROPOGENIC SO_2 EMISSIONS, 1940–1980[a,b]

	Total	Stationary combustion	Thermal electricity generation
1940	19.4	15.1	2.5
1950	21.4	16.5	4.9
1960	21.0	15.8	9.1
1970	29.8	22.6	16.3
1975	26.1	20.8	16.7
1980	26.9	24.3	17.7

[a]Sources: Cavender et al. (1973), Hunt (1976), USEPA (1980a), and U.S./Canada Work Group (1982).
[b]All values are in million tonnes per year.

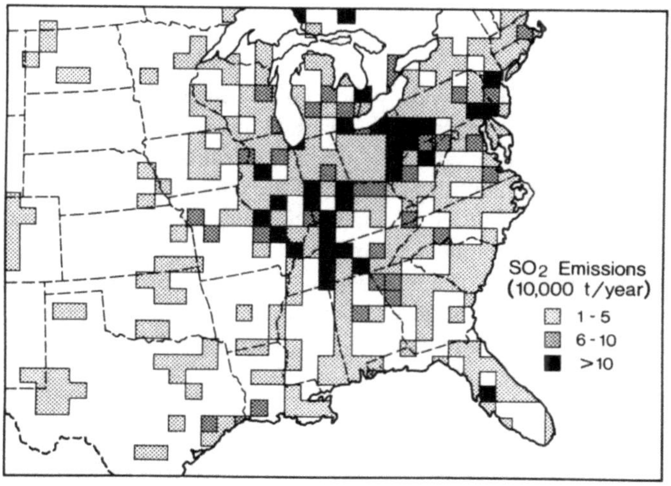

FIGURE 4-3. Annual point source emissions of sulfur dioxide in the eastern United States and Canada mapped in 80 × 80-km grid (Clark, 1980).

state's territory (116,000 km^2, the size of an average European country), this averages to a bit over 10 t S/km^2. In contrast, biogenic emissions from soils and waters would, according to Adams et al. (1981), average at this latitude (40°N) no more than 30 kg/km^2, or a mere 1/400th of the man-made flux. For the eastern United States as a whole, SO_2 emissions prorate to about 3.8 t S/km^2 while the annual biogenic flux in the same area appears to be no more than 100,000 t S (Adams et al., 1981), or just 0.03 t/km^2, about a 125-fold difference.

A change of scale is thus imperative: careful assembly and intricate balancing of the global sulfur cycle may be challenging and intriguing but to understand the causes and effects and to suggest some remedies to human interference in natural sulfur flow, one must leave the megascale of uncertain planetary flows to descend at least to a continental or regional macroscale where the detailed data are easier to get—and not infrequently all the way to a local microscale, to a single lake or to a group of kindergarten children in a polluted area. But before dealing with effects and management choices, more must be said of sulfur in the atmosphere.

4.2. SULFUR IN THE ATMOSPHERE

A more precise title would be "anthropogenic sulfur in the atmosphere" as I am not going to say anything about several sulfur compounds which can be detected with sensitive instruments in very low concentrations (usually in parts per trillion) over the open oceans and vegetated surfaces of the planet: hydrogen, carbonyl, and dimethyl sulfides, dimethyl disulfides, and methyl mercaptan. During the 1970s we came to realize how important in the aggregate these compounds are in global sulfur balance and how intricate their atmospheric behavior is, of which we still know far from enough.

However, it is sulfur dioxide which holds center stage in man-made sulfur emissions—and various sulfates into which it is, sooner or later, oxidized. Of course, not only SO_2 is released during fossil fuel combustion, the source of at least 85% of all anthropogenic sulfur: up to 5% of the total sulfur present in fuel may be converted to the higher oxide, SO_3, and the concentration of this gas increases with higher combustion temperatures and oxygen concentrations, as well as in the presence of metallic catalysts (V, Fe, Ni) whose oxides are commonly found in fly ash. Still, SO_3 concentrations in the flux gas are usually only 5–15 ppm, rarely exceeding 35 ppm. The focus thus remains on SO_2 and on sulfates arising from it.

4.2.1. Sulfur Dioxide

Owing both to its ubiquity in polluted atmospheres of rich fossil-fuel-burning nations of the Northern Hemisphere and to its obvious harmful effects on materials and on biota, sulfur dioxide has been among the most studied air pollutants. Yet there are huge gaps in our knowledge, foremost among them the surprising fact that after hundreds of laboratory experiments and plume transect measurements, the aqueous-phase chemistry of SO_2 oxidation is very poorly known. And as will be seen later (Section 4.3), there are more than a few disagreements about SO_2's damaging effects.

4.2.1.1. Physical and Chemical Properties Recapitulating the essentials is certainly not redundant: in recent decades when fashionable sicentific problems have attracted the transitory attention of many instant experts, one often came across unfortunate examples of discussing involved concepts without the appreciation of basic underlying facts. For this reason the essential physical properties of SO_2 will be given. It is a colorless, nonexplosive, nonflammable gas with high solubility in water. Owing to this property makes it is easier to precipitate the gas from the atmosphere as well as to ease its ingress into plant and human tissues.

Chemically, the gas is avidly reactive and can be either a reducing or an oxidizing agent: it can oxidize H_2S at ambient temperature (the Claus reaction) and while it needs 400°C to yield SO_3 this process can proceed at ordinary temperatures in the presence of several catalysts—platinum, charcoal, vanadium pentoxide, or chromic oxide—but many other substances commonly present in polluted air will aid the conversion. Among them are nitrogen oxides, ferrous sulfate which catalyzes the process directly to sulfuric acid, and several metal oxides (MgO, Fe_2O_3, ZnO, Mn_2O_3, CuO) which can carry the reactions all the way to sulfates.

An interesting property of such an aggressive gas is its relatively high sensory threshold. Although most people can detect it by taste at concentrations of 0.86–2.86 mg/m^3, more than 8 mg/m^3 may be needed to perceive its pungent, irritating odor. However, there appears to be a wide range of individual sensitivities with some people being able to smell the gas even at levels below 3 mg/m^3 (Smil and

Karfik, 1968). In any case, the latter are much above the generally agreed maximum allowable concentrations not only for 24 hr exposure but even for 1 hr limits (Table 4-4). In sum, SO_2 is not only a ubiquitous gas of polluted atmospheres of the Northern Hemisphere but also an invisible, exceedingly water-soluble and highly reactive one—besides being hard to detect by smell at usual concentrations.

4.2.1.2. MEASUREMENTS AND CONCENTRATIONS Assessment of human interference in the sulfur cycle is critically dependent on the availability of reasonably accurate data on atmospheric concentrations of the emitted compounds and their subsequent reactants. For SO_2, scores of continuously monitoring commercial instruments are available, employing spectrophotometry, conductivity, nondispersive infrared coulometry, electrochemistry, flame emission spectrometry, and condensation nuclei principles, as well as a variety of remote-sensing techniques (Forrest and Newman, 1973; Hodgeson et al., 1973; WHO, 1976; Katz, 1980).

Lack of sensitivity has been the most important limitation of commercial instruments whose lowest realistic detection limits are usually about 30 $\mu g/m^3$. This value is also the lower detection limit of the West–Gaeke procedure of SO_2 sampling and analysis which was designated as the reference method by the USEPA and in which the gas is absorbed in potassium tetrachloromercurate to make a stable complex whose color formed by reaction with formaldehyde and acid-bleached pararosaniline is measured photometrically. Most of the published ground-level SO_2 measurements of the past two decades have been done with the West–Gaeke

TABLE 4-4
AMBIENT AIR QUALITY STANDARDS FOR SO_2 IN VARIOUS COUNTRIES[a,b]

	Year	Day	Hour	30 min
Bulgaria	—	50	—	500
Canada[c]	60	300	900	—
Council of European Communities	80–120[d]	—	—	—
Czechoslovakia	—	150	—	500
Federal Republic of Germany	—	150	500	—
Israel	—	260	—	750
Japan	—	100	260	—
People's Republic of China	20–100[e]	50–250	—	150–700
Sweden		250	625	—
United States	80	365[f]	—	—

[a]Sources: Martin and Stern (1974), Council on Environmental Quality (1980), and Siddiqi and Zhang (1982).
[b]All values are in micrograms per cubic meter.
[c]Maximum acceptable values; maximum desirable levels were set, respectively, at 30, 150, and 450 $\mu g/m^3$.
[d]Lower value when suspended particulates are above 40 $\mu g/m^3$; the higher mean when particulates are below that level.
[e]The extremes refer to standards in class I and class III areas (cleanest and most polluted).
[f]Not to be exceeded more than once a year.

technique which is highly specific and accurate within the given limits (0.01–0.4 ppm, or 30–1100 μg/m^3).

For measurements aloft, remote-sensing techniques are clearly preferable but of the many methods available, only two are sufficiently sensitive to detect trace amounts of SO_2, correlation spectrometry and differential lidar (Hamilton et al., 1978). The latter technique has an outstanding sensitivity of a few parts per billion to ranges in excess of 1 km with resolutions of 100 m and 10 sec.

Another notable recent innovation has been the development of a long-term SO_2 monitor based on permeation of gaseous SO_2 through a dimethylsilicone polymer membrane into a manganese salt solution and subsequent turbidimetric analysis of the resulting sulfate (McDermott et al., 1979). Detection limit is 10 μg/m^3 for a 7-day period and exposures as long as 3 months are possible, enabling one to assemble—relatively cheaply, without electricity or complicated peripheral equipment—integrated values for ambient SO_2 concentrations which can be readily related to air quality standards.

Undoubtedly, measurement techniques of sufficient accuracy are now available to provide satisfactory assessment of both natural background concentrations and man-made emissions. However, the second necessary requirement for such appraisals is in place only in a few regions of the world: long-term monitoring networks measuring SO_2 and other gaseous particulate pollutant concentrations with both frequency and density sufficient to make generalizations and to discern trends and relationships. While there are some extensive, long-term arrays of SO_2 measurements in some polluted urban and industrial atmospheres, data on spatial and temporal distribution of the gas in continental regions remote from human settlements and over the oceans are not at all abundant. Yet such information is needed to calculate the residence time of atmospheric sulfur, to determine removal rates, and to appraise the intensity and effects of anthropogenic emissions. Consequently, I will first review the available natural background values.

In a review of SO_2 concentrations in remote continental and oceanic areas, Mészáros (1978) pointed out that sufficient data are available only for Europe, North America, and the Atlantic Ocean. In fact, the measurements of background SO_2 in Latin America and Africa are so meager that they do not allow any generalizations, totaling just short-term studies in four nations. In Panama, Lodge and Pate (1966) sampled at three sites during three 2-week periods and established values from less than 1 to 5 ppb with no clear pattern. Later measurements in Panama and in Brazil found concentrations of 0.3–1.0 ppb under tropical rain forest canopies, 0.3–0.9 ppb above them, and 0.3 ppb along interior rivers (Lodge et al., 1974). Ivory Coast measurements by Delmas et al. (1978) averaged 6.1 μg/m^3. From a relatively large number of measurements in rural areas in Europe and North America, it is possible to generalize that the background SO_2 concentrations are typically 5–20 μg/m^3, although in northern Scandinavia values below 1 μg/m^3 were recorded (de Bary and Junge, 1963; Eliassen and Saltbones, 1975; van Dop et al., 1980).

Interesting Dutch measurements on the 213-m-high tower of Cabauw, a rural site in the central part of the country, confirmed some expected variations of SO_2 concentrations and enabled some surprising generalizations (van Dop et al., 1980). There is an annual cycle at all levels (measurements taken at 3,100, and 200 m), with May to July minima (mostly below 10 $\mu g/m^3$ at 3 m but often above 20 $\mu g/m^3$ at 200 m), and maxima in October, December, and March (above 20 or 30 $\mu g/m^3$ at 3 m, up to more than 70 $\mu g/m^3$ at 200 m). General concentration increases with height appear to continue sometimes up to 1400 m, a clear confirmation of long-distance SO_2 transport at elevated levels. Typical diurnal variations predictably reflect the atmospheric stability: during daytime mixing, surface concentrations reach the average boundary layer values, whereas the nighttime stable stratification prevents the surface levels from rising until the onset of vertical mixing (fumigation processes were often observed).

For the oceans most remote from any anthropogenic sulfur emissions, we have a series of measurements made aboard oceanographic ships during Antarctic and South Pacific expeditions (Nguyen et al., 1974). They ranged between 0.05 and 0.5 $\mu g/m^3$ for the air above Antarctic areas (60°–70°S) and 0.04–0.9 $\mu g/m^3$ for subantarctic (40°–60°S) ocean air. Their averages, 0.13 and 0.18 $\mu g/m^3$, are representative of natural SO_2 oceanic background levels, unlike the measurements over the Mediterranean or in the Northern Atlantic which are undoubtedly affected by man-made sources. Although the concentrations over much of the Mediterranean Sea may be as low as 0.2–1 $\mu g/m^3$, between Greece and Turkey up to 9 $\mu g/m^3$ was recorded and Nguyen et al. (1974) offer 2.27 $\mu g/m^3$ as the average of their measurements.

A relatively large number of values are available for the Atlantic Ocean [see Mészáros (1978) for their review] and they indicate that without the anthropogenic influence the background levels across a wide latitudinal span (60°S–60°N) would be fairly constant and rather low at about 0.1–0.2 $\mu g/m^3$. However, between 30° and 50°N the recorded SO_2 concentrations have repeatedly exceeded 1 $\mu g/m^3$, almost certainly owing to man-made emissions from North America and Europe. Another notable observation is that SO_2 levels over the Atlantic appear to be vertically constant, at least in the lower half of the troposphere (Georgii, 1978).

While the clean background levels of SO_2 in areas affected only by natural sulfur fluxes will rarely rise above 10 $\mu g/m^3$, such concentrations are very common in urbanized and industrialized regions from which we have large numbers of measurements. Literally thousands upon thousands of various SO_2 values can be found in the air pollution literature of all industrialized nations together with frequent summaries of secular trends and with interesting statistical analyses of frequency distributions [for good summaries of the last topic, see Larsen (1971) and Pollack (1975)]. Here I will just list some typical low, mean, and high values for a few selected places around the world.

Cities using hydrocarbons for residential heating and power generation have never had very high SO_2 concentrations: not only warm San Francisco or Los Angeles, where the annual averages were no higher than 25–50 $\mu g/m^3$ even before the promulgation of air quality standards, but also much cooler Denver, where the

1962–1967 mean was 50 μg/m^3 and 1973 values averaged just 15 μg/m^3 (Ferman et al., 1977), or cold Winnipeg where, according to annual Environment Canada reports, the annual means in the 1970s were uniformly below 25 μg/m^3. Large cities in colder climates with heavy concentrations of coal-fired power plants and chemical industries have, naturally, fared much worse. Chicago in the mid-1960s had annual averages around 300 μg/m^3 and Philadelphia about 200 μg/m^3, but these conditions changed soon.

A notable development during the 1970s was a gradual decline of average annual SO_2 concentrations registered by national monitoring networks of several industrialized countries. Data for 258 selected "trend sites" in the United States showed an appreciable improvement for the 90th percentile for both annual arithmetic mean (from close to 100 μg/m^3 to just over 50 μg/m^3) and 24-hr maximum concentration (from 240 μg/m^3 to 140 μg/m^3) during the years 1970–1974 (NAS, 1978b); afterwards, the concentrations leveled off, while the 50th and 10th percentile remained stable throughout (annual averages at, respectively, around 20 and 5 μg/m^3). Consequently, by 1975 a mere 3% of United States counties were in violation of USEPA SO_2 standards which are 80 μg/m^3 for the annual average (USEPA, 1980a).

In Canada, SO_2 average of the annual means of National Air Pollution Surveillance (NAPS) stations fell from 45 μg/m^3 to 35 μg/m^3 between 1974 and 1979, a 25% decline from an already low level (Environment Canada, 1981). As a result, only 4% of NAPS stations exceeded acceptable annual air quality objectives in 1979 compared to 18% in 1974. And just one more example, this time one of Europe's most polluted regions in the Netherlands: van Dop and Kruizinga (1976) analyzed SO_2 concentrations in the vicinity of Rotterdam for 13 consecutive winters between 1961 and 1974 (values considered representative for an area of about 75 km^2) and found an average decrease of 11 μg/m^3 per year, from 247 μg/m^3 in 1961–1962 to 111 μg/m^3 in 1973–1974.

This phenomenon is not difficult to understand: most monitoring sites are in urban areas—for example, over 90% of the United States National Air Surveillance Network are so located while all of Canada's 161 sampling stations of the NAPS Network are in cities—where the domestic burning of high-sulfur coals or oils has been almost completely eliminated and where, unlike before 1960, new coal or oil power plants are no longer built. Reduced coal emissions, relocation of major new sources, and also the increased emission height (tall stacks of mine-mouth power plants) combined to bring about these considerable urban SO_2 improvements.

However, there are still many places with very high SO_2 averages, annual or seasonal. In some cities of the North Bohemian Brown Coal Region, the maximum permissible concentrations of 500 μg/m^3 are exceeded on 30–50% of days during the winter heating period and the recorded short-time maxima go up to 1500 μg/m^3. Similarly, in Chinese cities, so heavily dependent on the combustion of coal for household as well as industrial uses, winter means are several hundred micrograms per cubic meter; in 1980, Chongqing in Sichuan had an annual average of 520

$\mu g/m^3$ while maxima recorded in some northern cities surpassed 2000 $\mu g/m^3$ (Kinzelbach, 1983).

Even higher short-term concentrations were recorded during high air pollution episodes which became classical examples of acute effects in the epidemiological literature. In Donora, Pennsylvania, no measurements were taken during a 5-day severe air pollution episode in late October 1948, but judging by the excessive deaths and morbidity the SO_2 concentrations peaked almost certainly well above 2 mg/m^3. In London's famous smog of December 4–9, 1952, SO_2 concentrations rose from normal daily averages around 290 $\mu g/m^3$ to 2100 $\mu g/m^3$ on December 8 and the highest observed value reached 3720 $\mu g/m^3$ (Wilkins, 1954). And during the last of the notably high air pollution episodes in New York (November 22–26, 1966), SO_2 concentrations peaked at 2.6 mg/m^3 and for 84 hr they did not slip below 400 $\mu g/m^3$ (Fensterstock and Fankhauser, 1968).

To generalize, annual averages of SO_2 concentrations in urbanized and industrialized regions range typically from a few tens to a few hundreds of micrograms per cubic meter, while the recorded daily maxima in heavily polluted locations surpass 1 mg/m^3 and the short-term peaks are still two or three times higher. High as these maxima appear to be, they are actually quite low in comparison with values which would be reached if SO_2 removal from the atmosphere did not proceed so rapidly.

For example, during the London smog of December 1952 cited above, a stagnating air mass limited the mixing volume to about 200 km^3 and into this air an estimated 2000 t SO_2 was added per day. If all of this increment had stayed for just 1 day, the ambient SO_2 concentrations would have grown to 10 mg/m^3 and if the accumulation would have continued for the four peak days of the episode, Londoners would have had to (or rather could not) breathe air with SO_2 concentrations approaching 50 mg/m^3.

Yet such large increases were never observed: the highest daily average was only about 20% of the theoretical accumulation and the maximum recorded peak was less than 10% of the calculated 4-day aggregate. A simple calculation will show that when the observed daily maximum was 2.1 mg while the increment was 10 mg SO_2/m^3, the average residence time of the gas in the atmosphere had to be only 5 hr (and 4 min, to be exact). An understanding of the removal processes and rates is thus of critical importance for evaluating the effects on major sinks of the gas—soils, waters, and vegetation.

4.2.1.3. Reactions and Removals Sulfur dioxide can be removed from the atmosphere by either dry or wet deposition and before this happens the compound may be involved in complex chemical transformations. Both the transformation and deposition processes were studied with increased intensity throughout the 1970s but both remain the areas with quite a few fundamental uncertainties and hence with considerable need of further research to understand the actual behavior of the atmosphere as opposed to theoretical expectations derived from laboratory measurements or from limited open air observations.

Whereas Urone and Schroeder (1969) remarked about the surprisingly small number of studies measuring the chemical reactions of SO_2 in polluted atmosphere, there is now considerable laboratory and field evidence to consider: detailed reviews by Möller (1980) and NAS (1983a) list 60 different pathways and their rates embracing direct photooxidation, reactions of triplet SO_2 with O_2, reactions of SO_2 with radicals (HO, HO_2, CH_3O_2), molecules (e.g., NO_2, NO_3, N_2O_5), and hydrocarbons, as well as a variety of liquid-phase oxidations and surface reactions of SO_2 with particles.

Photochemical gas-phase oxidation is now recognized as a very important route, especially the reactions with free radicals which have a mean reaction rate constant of 1.2×10^{-6} sec, considerably faster than direct photooxidation at 10^{-7} sec. Gas-phase oxidation produces sulfates throughout the year in lower latitudes (below 35°N) but in higher latitudes (45°–55°N) it appears to be significant only during the summer. The most important intermediate for summer gas-phase oxidation of SO_2 is undoubtedly the hydroxyl radical: high HO concentrations in polluted sunny skies (the radical is formed largely by photochemical processes) result in SO_2-to-H_2SO_4 conversion rates averaging 0.8%/hr (NAS, 1983a).

Even during winter the process is dominated by reaction of SO_2 with HO, while HO_2 and CH_3O_2 contribute in a minor way (Altshuller, 1979). In typical urban atmospheres, SO_2 photooxidation will proceed on sunny summer days at rates of about 2–4%/hr (Calvert et al., 1978), occasionally even higher, and this process was shown to be important even in western European environments where it was considered unlikely owing to less sunlight, lower temperatures, and smaller amounts of free radicals (Atkins et al., 1972).

Aqueous-phase reactions proceed much faster than photooxidation with free radicals: mean reaction rate constants are 10^{-5}–10^{-4}/sec in water droplets with pH of 4–5 (Möller, 1980) so that all of the gas present may be oxidized in just 1 hr. Although there remain many disagreements about the reaction rates, especially as far as uncatalyzed transformations are concerned, it is most likely that liquid-phase oxidation is the dominant worldwide source of atmospheric sulfates produced from SO_2 (Hegg and Hobbs, 1978). Catalytic oxidation of SO_2 in stack and urban plumes, where transition metals (above all Fe, Mn, and Cu) are present in relatively high concentrations, appears to be of great importance in heavily polluted atmospheres, while the role of NH_3 in aqueous SO_2 oxidation is perhaps not as important as claimed in earlier stages of SO_2 reaction research.

Besides catalytic aqueous oxidation, another transformation process appears to be important in stack plumes: absorption and oxidation of SO_2 on particles. Both particles just released from the stacks as well as those already residing in the atmosphere absorb SO_2 and the subsequent rates of conversion to sulfates are largely determined by the presence of alkaline compounds (Liberti et al., 1978). Coffer et al. (1980, 1981) also found enhanced SO_2 oxidation on carbon particles in the presence of NO_2, a gas always present in stack plumes, as well as N_2O and O_3. No reliable rate measurements, however, are available for these surface reactions.

What actually happens inside stack plumes, which are now the largest carriers of anthropogenic SO_2, is quite variable. Disagreements still prevail regarding the dominant path for the oxidation—perhaps the best current appraisal is that the basis for choice is not definitive and that both homogeneous and heterogeneous mechanisms are at times operative (Newman, 1981)—but approximate quantitative evaluation of the reaction rate is now possible. Pioneering measurements of SO_2 oxidation in power plant plumes identified relative humidity as the key factor: at humidities of up to 70% oxidation was sluggish while with higher moisture there was initial rapid conversion (as much as 22% after just 12 min) and the highest observed totals (up to 55% in 108 min) were produced in mist (Gartrell et al., 1963). Since then, theoretical calculations confirmed that the rate of SO_2 oxidation is very sensitive to the ambient humidity and that it is highest in high relative humidities where droplet acid concentrations will be restricted to low values (Foster, 1969)—but numerous new measurements have shown that even in aged plumes, the SO_2/SO_4^{2-} molar ratio is as large as 5 and that oxidation cannot usually proceed at such fast rates as indicated by the early studies.

Newman (1981) believes that limitations in collection and measurement techniques were responsible for the clearly flawed results and in his review of recent power plant and smaller plume studies he demonstrates that the average rate of oxidation of SO_2 in plumes moving through and mixing with clean air is typically less than 1%/hr, although in polluted air the conversion rate can be at least double this; in terms of diurnal variation, the transformation rate is often negligible during the night and reaches approximately 3% during midday hours.

Brief reviews of a few interesting studies will show better the actual recorded means and ranges. For example, instrumented aircraft measurements within the dispersing plume of the coal-fired Cumberland power station in Tennessee (2600 MW, 305-m-high stack, coal with 4.1 ± 0.2% S) confirmed that most of the oxidation occurs in the immediate vicinity of the plant, the conversion rates beyond 10 km being a mere 0.2%/hr (Meagher et al., 1978). On the average, only 1.1% (minimum 4.3%) of the plume sulfur was present as sulfate. Forrest and Newman (1977) measured particulate S/total S concentration ratios in various humidities (32–85%), temperatures (10–25°C) and for distances up to 70 km and durations up to 200 min and established that in four coal-fired power plant plumes, SO_2 oxidation rarely exceeded 5%/hr, and that virtually all of it occurred within the first few kilometers after emission.

Cooling tower plumes should provide an environment favorable for SO_4^{2-} formation owing to dissolved oxidants which can accelerate the liquid-phase reactions, and this theoretical expectation was confirmed by plume measurements at the Paradise plant in Kentucky (Meagher et al., 1982). Similarly, air chemistry measurements at the Bowen power plant in Georgia found a high correlation between SO_2 conversion rates and plume water vapor content (Liebsch and de Pena, 1982).

Chan et al. (1980) used a helicopter to trace the oxidation over a wide range of conditions (temperatures of −11–18°C, humidities of 34–87%, plume ages up to 5 hr) in plumes from INCO's 381-m stack in Sudbury, Ontario, and found that particulate sulfur appeared predominantly as H_2SO_4 droplets and that at distances up to 100 km from the source it accounted for a mere 4% of the total plume sulfur, with oxidation rates typically less than 0.5%/hr. In earlier measurements of plumes from the same source, Lusis and Wiebe (1976) found somewhat higher values—up to 10% of sulfur as SO_4^{2-} within the first 100 km of the plume and oxidation rates generally less than 3%/hr with 1% as the average—but even these figures show clearly the persistence of unreacted SO_2 for many hours and long distances. Obviously, no accurate quantitative generalizations are possible but it appears imperative to assume very low rates of conversion in most circumstances, especially in low humidities. This unequivocal conclusion from stack plume studies is further supported by the available data on urban plume conversions as well as by calculations for larger regions.

For example, Cox (1974) put the SO_2–SO_4^{2-} transformation rate in an urban plume with photooxidation in the presence of NO_2 and hydrocarbons, or thermal oxidation with ozone and olefins, at 1–10%/hr, while Alkezweeny and Powell (1977) estimated from tetroon and aircraft data that the conversion rates for the St. Louis urban plume are 14 ± 4%/hr for the first day and 10± 2% for the second. Eliassen and Saltbones (1975) used SO_2 emission data for Europe and measurements from 11 OECD ground sampling stations to establish the transformation rates giving the best agreement between observed and computed SO_2 and SO_4^{2-} concentrations; they ranged from 0.28% (for Waldhof, West Germany) to 1.73%/hr (for Wageningen, the Netherlands). Cass (1981) calculated the rates of SO_2 oxidation in Los Angeles for each month from 1972 to 1974: the 3-year mean was lowest for December, a mere 0.8%/hr, and highest for May and June (6.3%) with monthly maxima ranging from 0.5 to 8.0%.

Clearly, oxidation of SO_2, formation of sulfates and their subsequent settling or washout from the atmosphere (for details see Section 4.2.2) is only part of the complex process of SO_2 removal which proceeds relatively fast owing to the existence of several (often aggregate) decay mechanisms among which dry deposition of the gas may account for removal of about half of all SO_2 emitted to the atmosphere (Garland, 1978). There is no shortage of dry deposition velocity measurements which are usually taken 1–2 m above the surface.

In a comprehensive review covering a wide range of publications from the 1930s to spring 1978, McMahon and Denison (1979) listed ten laboratory measurements of 18 different surfaces and 31 field investigations for nearly 50 different surfaces ranging from alfalfa to wheat and from the Atlantic Ocean to snow. Sehmel's (1980) review of published SO_2 deposition velocities lists over 40 values ranging from 5–8 cm/sec for St. Louis in 1975 to 0.04 cm/sec for asphalt with most of the values for different grasses, soils, waters, and artificial surfaces between 1 and 4 cm/sec. Clearly, SO_2 deposition velocities increase with increasing vegetation height, ranging from usually less than 1 cm/sec for short grasses to several centimeters per second for forests. However, there are many exceptions with extreme measured values as high as 3.5 cm/sec for just 5-cm-high grass and as low as 0.3 cm/sec for Scotch pine forest.

Two notable SO_2 deposition velocity studies published after McMahon and

Denison's exhaustive review are those for snow in northeastern Alberta by Barrie and Walmsley (1978), which established an average value of 0.25 ± 0.2 cm/sec from simultaneous measurement of deposition and ambient concentration; and measurements with stirred chambers in arid regions of northern Australia by Milne et al. (1979). This latter work showed velocities much lower than those found typical in Europe and North America, no doubt an effect of the area's low humidity; of 17 measurements over grasses (15–50 cm high), nine were below 0.10 cm/sec and of the remainder only one surpassed 0.2 cm/sec; two values for bare soil were over 0.04 cm/sec and four measurements for trees ranged merely from 0.08 to 0.09 cm/sec. This should be contrasted with values offered as typical by Husar et al. (1978) in their summary of day deposition velocities (all values are in cm/sec): 0.5 for grasses, 0.8 for calcareous (pH ≥ 7) soils, and 0.4–0.6 for acid (pH 4) soils.

In general, good sinks for SO_2 are water surfaces (except when already highly acid), wet snow, taller vegetation in at least moderately humid climates, and basic soils. Acid soils, fresh dry snow, most short grasses everywhere, and all kinds of vegetation in arid regions are at best only modest, more often very poor, sinks for the gas. Regional or national generalizations cannot be very precise but do convey the intensities of the process. Fowler (1978) estimates that the mean SO_2 deposition velocity for all of Britain's agricultural areas is 0.6 cm/sec (72 kg SO_2/ha), while Owers and Powell (1974) put the average at 0.8 cm/sec over the whole British Isles, a value also used in the OECD study of long-range sulfur transport (OECD, 1977). However, for much drier areas these values may shrink even below 0.1 cm/sec.

As for the built-up urban environments, it is difficult to extrapolate the known, laboratory-measured and relatively high SO_2 dry deposition velocities over cement and stucco (Judeikis and Stewart, 1976) to a larger area. Similarly, it is difficult to offer an average value for the ocean. There is little doubt that, contrary to earlier suggestions (see, e.g., Kellogg et al., 1972), the ocean—owing to its high pH and good carbonate buffering action—is a good sink of SO_2; in fact, it may never be a direct source (Beilke and Lamb, 1974). However, the actual deposition measurements are too few to offer a meaningful average.

Wet deposition of sulfur occurs predominantly in the form of sulfates (see Section 4.2.2.3 for details) but dissolved sulfur dioxide is also an important constituent of rains, fogs, and dews in polluted areas. In spite of this far from negligible rate, only a few measurements of SO_2 in precipitation are available, largely owing to differences accompanying the sampling and presentation of the solutions (Hales and Dana, 1979).

Beilke and Georgii (1968) were probably the first to demonstrate conclusively that the gas phase should not be ignored when considering the incorporation of atmospheric sulfur into rainwater. They found that while rainout of solids contributed ten times more to sulfur concentrations within clouds than did SO_2, this ratio was reversed for weather near the ground owing to high SO_2 concentrations in polluted atmospheres. Since then, Brimblecombe's (1978) calculations have shown that transfer of SO_2 into leaf wetness (dew) is relatively rapid but that the volume seems to be insufficient, unless neutralizing alkaline compounds are present, to absorb more than 0.3% of all British SO_2 emissions, although two field analyses indicated that the actual share may be as high as 2.4%. In any case, dew is not a major SO_2 sink even in such a humid climate though it is very important in controlling the pH of leaf surfaces and, in turn, in the loss of cations from leaves in polluted atmospheres.

More significantly, during the Large Power Plant Effluent Study in 1969 and 1970 when rain samples were analyzed for SO_2 as well as for sulfates—owing to the absence of network measurements of precipitation-borne SO_2, these are the best values available for the purpose—share of SO_2 sulfur in precipitation that did not fall through the studied power plant plumes ranged from 6 to 27% with the average around 15% (Hales and Dana, 1979). The same authors also made rough estimates of fractions of sulfur existing as dissolved SO_2 in rain in the northeastern United States; for 0°C these values ranged from 17% in Pennsylvania to 49% in Maine, and for 25°C the range for the same locations was 4–18%. And even higher ratios—14–82%—were reported by Davies (1976) in precipitation collected in an industrialized area in England.

Recently, Fisher (1982) attempted to describe the removal of SO_2 in a rain system with a general set of dynamical equations and his conclusions are that the overall fraction removed is not very sensitive to meteorological parameters (i.e., depth of clouds, updraft velocities) but is extremely dependent on the values of SO_2 oxidation in cloud and rainwater and on their air acidities; 40–100% of SO_2 can thus be removed in the rain but there appears to be self-regulation on how much the acidity of the precipitation can increase: in less acid rains, SO_2 scavenging can contribute considerably, and for already highly acid precipitation the effect may be small. Davies (1983) tested predictions of SO_2 washout ratios by hourly measurements in a heavily industrialized area (Sheffield's Don Valley with its iron and steel complexes) and found good agreements, although some recorded values were also lower than the predicted ones.

4.2.2. Sulfur Aerosols

Oxidation of SO_2 supplies the atmosphere with rather substantial quantities of sulfates which have been increasingly credited with environmental damage formerly attributed to the gas (Rowe et al., 1978). There are also primary sulfate emissions and natural flows but most of the sulfur aerosols are secondary pollutants (Larson, 1980). They may originate within plumes immediately after their release from stacks or may arise from sluggish oxidation hundreds of kilometers from their source; their small size makes them transportable over long distances and their occurrence in complex solid or liquid mixtures with other ions, trace metals, and organics makes their accurate and disaggregated monitoring fairly difficult; also, the relationships between SO_2 and SO_4^{2-} in the ambient air are far from understood.

Consequently, much more effort is necessary before measurements of H_2SO_4 and acid aerosol will be as routine as SO_2 monitoring and before well-founded air quality standards can be adopted for sulfates and sulfate-related substances (Rowe et al., 1978). Although the following sections will attest to a rapid increase in our knowledge of sulfates in polluted air, especially since the mid-1970s (for a good review, see Kneip and Kioy 1980), they will also reflect the gaps and uncertainties which, as I will argue, will not be overcome easily or soon.

4.2.2.1. KINDS AND PROPERTIES The recent use of X-ray powder diffraction led to identification of several sulfates previously not reported in ambient air and made possible a better classification of the compounds based on their origin (Biggins and Harrison, 1979). Primary sulfates are those emitted directly to the air or those formed there from nonsulfate compounds; secondary ones arise from reactions of primary sulfates with other compounds; and tertiary sulfates are formed by reactions of secondary sulfates with nonsulfate compounds.

Among the primary sulfates, $MgSO_4$ is present in sea spray aerosol (by weight, sea salt has a magnesium sulfate content of 5–7%), $CaSO_4 \cdot 2H_2O$ (gypsum) is found near quarries (as in Na_2SO_4), and it is also released in cities by decay of plaster, while $PbSO_4$ and $PbO \cdot PbSO_4$ are emitted in stack plumes of lead smelters. H_2SO_4 is mostly a secondary air pollutant but as it normally originates from oxidation of SO_2, which is a nonsulfate compound, Biggins and Harrison (1979) classified it as a primary sulfate; naturally, it is one of the dominant sulfates and its neutralization by ammonia gives rise to HSO_4 and $(NH_4)_2SO_4$, two abundant secondary compounds. Each of these three sulfates is hygroscopic and is encountered in the air commonly as liquid droplets. Of the tertiary compounds, zinc ammonium sulfate was reported as the major component of aerosol during an air pollution episode in Donora, Pennsylvania.

The physical size distribution of atmospheric sulfur aerosols is characterized by the trimodal model (Whitby, 1978). Nuclei mode has geometric mean size by volume from 0.015 to 0.04 μm and volumes from 0.0005 over the remote oceans to 9 $\mu m^3/cm^3$ on an urban freeway; accumulation mode (mean size 0.15–0.5 μm) has volume concentrations as low as 1 in clean areas and as high as 300 $\mu m^3/cm^3$ in polluted cities; finally, coarse particles (5–30 μm) have modal volume concentrations from 2 to 1000 $\mu m^3/cm^3$. The average geometric mean diameter of sulfates was calculated at 0.48 ± 0.1 μm but formation, transformation, and removal of fine particles (less than 2 μm in diameter) are basically independent from the processes affecting coarse sulfates.

Field measurements show repeatedly that in polluted atmospheres, sulfates are the leading aerosols and that they are especially dominant in the fine size category. For example, Altshuller's (1982) measurements at ten sites around St. Louis established ammonium sulfate as by far the most important constituent of fine aerosol mass with up to 70% of the total at nonurban sites during the summer months and 40–60% of the annual average for all sites. Sulfur was thus the leading fine aerosol element, followed by carbon, and accounted for 25–35% of all particulate mass measured below 20 μm.

Similarly, Alpert and Hopke (1981) analyzed a large number of air samples from St. Louis (Missouri) to determine the makeup and sources of airborne particles, and found that in 1 week beginning July 31, 1976, sulfate accounted for 84%

of all fine-fraction (less than 2 μm in diameter) particles (steel industry emissions were a distant second with 7%) and for 11% of the coarse aerosols (limestone and soil aerosols led this group by, respectively, 39 and 31%).

Chemical properties are, obviously, different for different sulfates. Four communalities are that SO_4^{2-} is chemically virtually inert at ambient temperatures, though the associated cation may be highly reactive; that the sulfate ion is practically nontoxic, although some sulfate compounds are definitely toxic; that sulfate aerosols in polluted atmospheres are mixed, internally rather than externally, with many impurities including metals, inorganic ions (above all nitrate compounds), and a wide variety of trace organics (from alkanes to phenyl acids); and that the sulfate compounds, together with other ions and impurities, cause the small particles to accumulate rather large volumes of water (Charlson et al., 1978).

Yet it is only rarely that information is available on individual compounds—primary, secondary, or tertiary—making up the sulfate fraction of the suspended particulates: our measurements so far have been overwhelmingly nonselective and for many sources and locations we simply have no breakdown at all (i.e., at least into sulfate, nitrate, metal compounds), just the values for total aerosols.

4.2.2.2. Concentrations and Removals As in the case of SO_2, our knowledge of background sulfate levels is much inferior to our records in urban and industrialized areas. Unlike with SO_2, substantial advances in monitoring and analytical methods appear to be essential before rational air quality standards could be set for suspended sulfate particles. This need arises from the characteristics just described: as the sulfates are a complex mixture of mostly secondary pollutants of small particle size, and hence transportable over long distances, sensitive, accurate, and discriminatory detection techniques are required for their satisfactory monitoring.

Unfortunately, the most widely used analytical method based on water-soluble sulfates extracted from total suspended particulate samples does not meet these criteria. Moreover, it does not distinguish between H_2SO_4 aerosols, neutral, metallic, and acid sulfates and sulfites, compounds which vary considerably in their health and other effects. Also, some filters used for sampling (including the most common glass fiber ones) can convert SO_2 to titratable acid and thus influence the measured totals (Forrest and Newman, 1973).

Coutant (1977) tested the performances of various filter media used in collection of ambient sulfates and found that appreciable errors can arise owing to the absorption and subsequent oxidation of ambient SO_2 in the presence of basic components (Na, K) of the filter medium. With the use of common glass fiber filters under normal high-volume sampling conditions, loading errors of 0.3–3 μg/m^3 can be expected, quite significant increases when measuring nonurban sulfates where the concentrations are almost always below 10 and very often below 5 μg/cm^3. A detailed discussion of difficulties to be overcome can be found in Rowe et al. (1978).

Starting with background concentrations over the remote oceans and in the clean continental air far from anthropogenic emissions, the recorded SO_4^{2-} concentrations are usually below 5 μg/m^3. Seasonal averages in higher altitudes in the northern midlatitudes are typically 0.5–1.5 ppb (by mass) but sulfate concentrations in the Mount Saint Helens stratospheric eruption plume (measured at 13.1–20.1 km) in May–June 1980 were mostly above 5 ppb and rose up to a maximum of 263 ppb (Gandrud and Lazrus, 1981).

In Antarctic and subantarctic oceans the values are as low as 0.5–1.6 μg/cm^3 (Nguyen et al., 1974) and comparably low levels can be found in the Colorado Rockies, in deserts of the southwestern United States, and in northern Scandinavia (Rodhe, 1976; Georgii, 1978; Hoffer et al., 1979). In contrast, sulfate levels over the North Atlantic (8°–50°N) are in excess of 2 μg/cm^3, over the Mediterranean Sea more than 8 μg/cm^3, values comparable to nonurban sites in the eastern United States and western and central Europe (Mészáros, 1978).

A notable property of ocean sea spray is that its ratio of sulfate to sodium exceeds that in seawater by 10–30% (Garland, 1981); this sulfate enrichment is sufficient to explain some measurements of excess sulfate levels in clean midocean precipitation—but it is still too low to account for large SO_4^{2-} excesses, approximately equivalent to the marine sulfate, brought by the westerlies from the Atlantic to Scandinavia.

For the industrialized parts of the United States, many urban and regional sulfate studies became available during the 1970s and have been contrasted with older measurements of the national monitoring network. This network's biweekly sampling at 96 stations in 1964–1965 resulted in an arithmetic mean of 10.6 μg/cm^3 and a maximum of 107.2 μg/cm^3 (Morgan et al., 1970), while the 5-year average from 1964 to 1968 was 11.4 μg/cm^3, with pronounced differences between eastern and western locations (Altshuller, 1973).

In eastern urban sites the mean was 13.5 μg/cm^3 with 7% of all values above 20 μg/cm^3; in the West the mean was 6.4 μg/cm^3 with 88% of all values below 10 and none over 20 μg/cm^3. With substantial increase in sulfur emissions from the northeastern coal-fired power plants in the 1970s, this regional difference became even more pronounced and, unlike with SO_2 whose levels declined even in urban areas, sulfate concentrations in cities have not decreased and in eastern rural areas they have actually increased (Altshuller, 1976; Pierson et al., 1980).

Thus, in 1974 in all but five states (Florida, Georgia, the Carolinas, Maine) east of the Mississippi, even rural values were above 10 μg/cm^3 with the Ohio Valley above 15 μg/cm^3 (USEPA, 1975b). Subsequently, it was confirmed by Hidy et al. (1978) that the 24-hr average sulfate concentrations have clear summer maxima prevalent over a distance of some 1600 km. Their studies showed that the high SO_4^{2-} levels of the winter tended to be localized and lasted only a day or two, in contrast to summer concentrations which persisted up to 5 days, covered areas extending over hundreds of kilometers, and often surpassed 20 μg/cm^3 in the most affected area between Indiana and New York. Canadian measurements in coastal Nova Scotia showed a range from nondetectable amounts to 15 μg/cm^3 with a mean of 3.8 μg SO_4^{2-}/cm^3 and with clearly higher levels when the summer air came from regions of high-sulfur emissions (Shaw, 1979).

These summer sulfate maxima could not be explained either by seasonal variation of anthropogenic emissions or by biogenic contributions but appeared to be most influenced by large-scale air mass movements—the highest concentrations closely corresponding to the influx of maritime tropical air masses channeling warm, humid air over the region of the highest emission density and thus promoting faster SO_2 conversion. Period of low sulfate levels resumed with penetration of relatively dry continental polar air masses from Canada (Hidy et al., 1978).

Relatively high sulfate levels may also be encountered in otherwise clean locations near major highways. Since 1975, North American cars have been equipped with oxidation catalysts to reduce hydrocarbon and CO_2 emissions but these devices also oxidize some exhaust SO_2 to SO_3 which is, in turn, rapidly hydrolyzed to H_2SO_4. During a massive experiment conducted by General Motors to simulate a freeway situation, maximum measured increases over the background level were as high as 15 $\mu g/cm^3$ next to the roadway and they fell off rapidly with height and distance; sulfate increases in vehicles ranged from 1 to 20 μg, averaging 4 $\mu g/cm^3$ (Cadle et al., 1977). Home air is also far from sulfate-free. Measurements of respirable particles in 68 homes in six American cities, taken over a period of at least 1 year, showed the mean infiltration rate to be about 70% and that the increased indoor levels of fine sulfates are associated with smoking and also with gas stoves (Dockery and Spengler, 1981).

Our understanding of sulfate levels, sources, correlations, transport, and removal was greatly advanced by the Electric Power Research Institute's Sulfate Regional Experiment (SURE) which lasted 7 years and entailed continuous air quality measurements at nine stations (in Massachusetts, Pennsylvania, Delaware, Ohio, Indiana, North Carolina, and West Virginia) between August 1977 and June 1979 and intermittent, seasonally intensive, sampling at 45 sites between the Mississippi and the Atlantic coast during 1977 and 1978. Results of this massive study, are available in 20 separate EPRI reports, with the summary presentation (EPRI, 1981) surveying all the key results and implications.

All of the sulfate concentrations were above the measurement threshold of 0.5 $\mu g/cm^3$, the median daily value for all sites during the 18 months of monitoring was 6.8 $\mu g/cm^3$, approximately one-half to three-quarters of the urban sulfate levels. Frequency distributions show that 90% of all samples were below 20 $\mu g/cm^3$ and 99% were below 30 $\mu g/cm^3$, but episodes of elevated SO_4^{2-} concentrations (above 40 $\mu g/cm^3$) were observed and these were clearly associated with major SO_2 source areas (whose influence was generally limited to 200–300 km, rarely exceeding 500 km) and with stagnation and ducting, meteorological conditions well known to be conducive to high air pollution levels.

Concentration differences attributable to weather are impressively illustrated by a comparison of 24-hr sulfate level averages for two summer days (Figure 4-4). While August 5, 1977 was the last day of a maritime tropical air mass-channeling situation when vigorous southwesterly winds distributed sulfate accumulated during the previous 2 days across the Northeast, resulting in relatively low uniform areal concentrations, July 21, 1978 came near the end of an extended spell of very hot

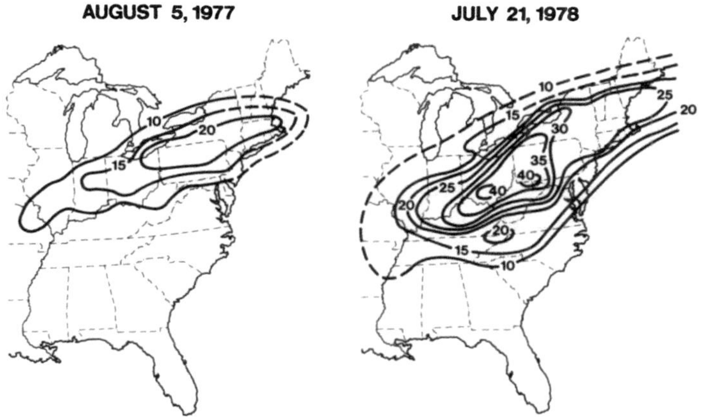

FIGURE 4-4. Sulfate concentrations (24-hr averages in μg/m³) during two summer days in the northeastern United States as observed by the SURE project network (EPRI, 1981).

and humid weather when northeastward channeling was much weaker and localized concentrations reached unusually high levels.

Of the total suspended particulate matter, sulfates accounted mostly for 20–30%, with the highest proportions occurring in summer in the areas with the highest SO_2 emission densities. As expected, most of the sulfate—82% in summer and 66% in winter—was in fine (less than 2.5-μm aerodynamic diameter) particles, accounting for 30–50% of the total fine particulate matter during high sulfate events, again with summer peaks. Multivariate statistical analyses showed small but significant seasonal dependence of sulfate levels with a summer peak, their strong association with SO_2 concentrations (hence also emissions), and their significant dependence on high-pressure stagnation episodes.

One of the rather surprising findings of the SURE study was that on the average, only 5–10% of the SO_2 emitted in the Northeast is converted to dry (nonprecipitation) sulfate particles with the balance of the sulfur either transported outside the studied area or deposited. Long-range transport will be discussed in a separate section (4.2.3.2); here a brief discussion of deposition mechanisms is in order. Deposition of sulfates, as any particles, is very much a function of their size. Most of the sulfates usually have diameters below 1 μm and their overall deposition velocities are very low as both gravitational settling and turbulence are rather inefficient in this size class. Consequently, such particles have the longest atmospheric residence times.

Garland (1978) calculated the mean speed at 0.025 cm/sec for all particles below 1 μm and 0.56 cm/sec for all larger sizes (see also McMahon and Denison, 1979). Larger particles would thus be removed rather swiftly, in a matter of 2–4 days, but the submicrometer fraction would linger for weeks: even from just 1 km above the ground (and hot plumes from large power plants rise much higher) the settling of submicrometer aerosols would take almost 7 weeks. Processes other than gravitational settling then become important as the tiny particles are deposited through impacts and interception on rough surfaces. This is the main reason for the inadequacy of our field measurements: a plastic bucket, commonly used for sampling, is clearly a poor substitute for a tree in determining the total dry decomposition rates.

There are no regional data available to map dry sulfate deposition and yet there is no doubt of its importance in acidification: from dozens of reported values of sulfate-sulfur deposition rates for throughfall and incident precipitation in world forests, it is clear that on the order of 50% of net throughfall sulfate comes from the dissolution of dry deposition (Parker et al., 1980).

However, for the smallest sizes, rain and fog are the principal determinants of residence time. Owing to their ubiquity, small size, and hygroscopicity (they become wet at a relative humidity of 80%), sulfate particles act as essential cloud condensation nuclei: practically all droplets in continental clouds are formed on sulfate particles. Aerosol sizes between 0.02 and 0.5 μm in radius, which are most abundant at cloud-forming altitudes, could form droplets by coagulation or, if the temperature is below freezing, by distillation onto ice crystals; complexities of these rainout processes are well reviewed and referenced by Garland (1978). However, of all these mechanisms, only the rainout of condensation nuclei seems to be able to produce the sulfate concentrations of several milligrams per liter encountered in water samples.

Washout of larger particles, which make up only a small portion of all sulfates, is also an important contribution but one which ceases rapidly after the first few millimeters of rain. On the basis of measurements of particulate tracers with a size distribution similar to sulfate, Garland (1978) concluded that the rainout of sulfate aerosols results in washout ratios (concentration per unit mass of rain/concentration per unit mass of air at the ground level) between 300 and 1000 and that the compounds are removed by rain with a time constant on the order of 10^{-4}/sec.

As with SO_2, sulfate removal rates are thus often slow enough to enable considerable long-range transfers but before focusing on this phenomenon I shall look in some detail at the SO_2–SO_4^{2-} relationship. Although it may seem that the strong correlation between SO_2 concentrations (and emissions) and SO_4^{2-} levels cited as one of the key findings of SURE's statistical analysis is trivially obvious, there are more than a few studies which did not find such a link at all, or only barely. These contradictions clearly indicate that the whole SO_2–SO_4^{2-} relationship is much more complex than might be expected.

4.2.2.3. SULFATES AND SO_2 Measurements from remote areas indicate that in the absence of man-made emissions, the SO_2/SO_4^{2-} ratio is fairly steady around 0.1 above the oceans and ranges between 0.2 and 0.5 in clean continental sites, while over polluted urban and industrial areas it rises one and even up to two orders of magnitude. For example, in the SURE study the average values of SO_2 (around 10 μg/cm^3) and SO_4^{2-} (around 7–8 μg/cm^3) yield a ratio between 1.25 and 1.40.

Whatever the actual ratios, it might be expected, especially in urban areas and in regions of high sulfur emissions, that the particulate sulfate levels in a given location will be closely correlated with SO_2 concentrations. Yet this is not the case in a surprisingly large number of instances. Altshuller's (1973) large-scale analysis of SO_2 and SO_4^{2-} concentrations in urban sites in the United States was perhaps the first one to demonstrate convincingly that the relationship of SO_2 to SO_4^{2-} is nonlinear over the range of city concentrations encountered and that as a result, reduction of SO_2 in a statistically average site by 90% would result in only a 53% reduction of sulfate.

A basic explanation is not difficult: in any industrial–urban area, primary sulfates will be emitted together with SO_2 from a variety of sources but considerable concentrations of sulfate (and other) aerosols may be brought to the city from upwind emissions. Local emissions, the area's location with respect to other large SO_2 sources, and the prevailing wind flows should thus explain most of the discrepancy—though their importance will vary with specific sites.

Of the several recent studies examining this lack of clear relationship, I shall summarize the main findings of just three projects notable for their weak SO_2–SO_4^{2-} correlations. Cooke and Wadden's (1981) analysis for Chicago found that the only two conditions of statistically significant correlation were those when relative humidity was less than 66% and when the minimum temperature was below 7.2°C, both conditions occurring primarily on days when the sulfate levels were below 13 $\mu g/cm^3$. However, the correlation coefficients were low, 0.58 and 0.49, respectively, explaining a mere 33 and 24% of the variance. For high-sulfate days, there was no significant correlation and the authors estimate that during such periods, only about 40% of SO_4^{2-} levels may be attributed to local sulfur emissions.

Factor analysis of particulate sulfate, meteorological and air quality data performed by Henry and Hidy (1979) for Los Angeles and New York revealed that in southern California, neither ambient SO_2 levels nor dispersion factors are important in explaining sulfate variability: photochemical activity variables and atmospheric moisture content are the main predictors. While New York has a similar photochemical component, in its case dispersion and SO_2 concentrations are also important in explaining sulfate levels.

Statistical evidence for a weak SO_2–SO_4^{2-} relationship comes mainly from United States studies but a similar phenomenon must also occur in Europe, of which the Dutch developments are perhaps the best available example. As noted in Section 4.2.1.2, Dutch urban ground-level concentrations of SO_2 decreased precipitously during the first half of the 1970s owing to the large-scale introduction of natural gas—but measurements in Arnhem and Rotterdam (Rijnmond) areas showed a much smaller decline in sulfate levels; this, together with the uniform sulfate distribution over the Netherlands, clearly indicates that the recorded levels are caused mainly by long-range transport (Elshout et al., 1978).

Two more studies deserve to be introduced here. Perhaps the best illustration of the complexities of the SO_2–SO_4^{2-} relationship is the fine report by Pierson et al. (1980) designed to find various links among sulfates, rates of SO_2-to-SO_4^{2-} conversion, air flow characteristics, rainfall, visibility, SO_2, aerosol mass, and so on. Ambient sulfate measurements were made in July and August 1977 on the summit of Allegheny Mountain (elevation 830 m) in southwestern Pennsylvania, a location in a clean wooded environment but right on the path of westerly flows carrying sulfur pollution from the Ohio Valley.

The three most notable findings of this experiment were the establishment of H_2SO_4 dominance in the aerosol, the key role of SO_4^{2-} in atmospheric turbidity, and the failure to prove a genetic relationship between atmospheric SO_2 and aerosol sulfates. The first finding is unusual as at high sulfate levels, most of the filter-sample SO_4^{2-} was H_2SO_4 and ammonium concentrations were low but somewhat similar evidence was previously found in Sweden (Brosset, 1978). The second finding is very much as expected (see Section 4.4.1 for more details) but the third one is most surprising, intriguing, and perplexing.

Pierson et al. (1980) found no SO_2–SO_4^{2-} correlation and no effective relationship between SO_2 conversion to sulfate and relative humidity, wind speeds, non-SO_4^{2-} aerosol mass, presence of transition metals or alkaline compounds, or aerosol NH_4^+/H^+; also the conversion showed no diurnal variation. The inevitable conclusion that SO_4^{2-} does not originate from SO_2 is unacceptable and the authors attempt plausible reconciliations. For example, if the SO_4^{2-} is too acidic to have formed in solution, then no correlation of the conversion rate and relative humidity can be expected.

The question of sources is most troublesome: SO_4^{2-} was generally associated with transport from the West but SO_2/Pb and SO_4^{2-}/Pb ratios show that cities were the origin of less than 20% of atmospheric sulfur at the site—and, most surprisingly, power plants may not be the major source either. Ambient SO_4^{2-} concentrations attributable to power generation in the Ohio Valley have, by coincidence, their maxima near the monitoring site (Sheih, 1977)—but the observed concentrations were four times higher! Whatever the sources were, they also, curiously, contributed very little to the aerosol except SO_4^{2-} and perhaps H^+ and NH_4^+.

Analysis of sulfate concentrations in relation to directional air trajectories in the Tennessee Valley region yields findings almost as surprising as the puzzles of the Allegheny Mountain: Reisinger and Crawford (1980) found that the greatest concentration maxima occur during northeasterly flow, a fact explainable by the region's own emissions and by the transport of modified continental polar air from the Ohio River Valley. However, the highest mass transport comes with the southwesterly maxima tropical flow in spite of the fact that the anthropogenic emissions north of the Tennessee Valley are more than eight times those originating south of it! Trajectory trace shows southern Louisiana and Mississippi as the favored source areas, pointing to likely biogenic origins of these sulfates. This southwestern sulfate flux surpasses 66 $\mu g/m^2$ per sec on 30% of all days, almost twice the relative occurrence of such high fluxes with wind flow from other sectors.

In contrast to the nonlinearities found at Allegheny Mountain and the Tennessee Valley the NAS (1983a) review concludes that "direct evidence of a strongly nonlinear relationship between net deposition of sulfate and SO_2 emissions is limited to extensive historical data in Europe" and that the long-term records at

Hubbard Brook provide no evidence of a strong nonlinearity between annual deposition and annual emissions in the northeastern United States.

This conclusion implies that if all else is more or less constant, ambient concentrations of SO_2 and SO_4^{2-} will be largely determined by patterns of SO_2 emissions—but it cannot be held with much certainty as it is based on reliable long-term records for a single station and as the uncertainties characterizing meteorological processes and physical and chemical removals and reactions make any simplifications difficult. There is enough equivocal information and associated uncertainty to foresee that the questions concerning the SO_2–SO_4^{2-} relationship will not be settled for years to come.

I devoted considerable space to the SO_2–SO_4^{2-} relationship because it illustrates perfectly the multiple uncertainties we are facing in understanding the source, reactions, removals, and transfers of atmospheric sulfur of anthropogenic origin. I would like the reader to keep these puzzling examples in mind as I trace the effects of sulfur compounds after they are deposited, absorbed onto, or absorbed into animate and inanimate surfaces and tissues.

4.2.3. Residence Times and Transport

Even when relatively rapid, the transformation of SO_2 to sulfates and dry and wet removal processes of the gas and the aerosols are not accomplished in minutes; at least many hours and more likely several days are needed, with the actual residence times depending on a host of variables of which the nature of the source (large tall point sources such as power plant or smelter stacks versus area sources emitting near the ground), mixing depth (determined by temperature stratification and wind regime), cloud volumes and chemistry, precipitation frequency and amount, duration and intensity of sunshine, and the kind of land cover are the most important.

After a brief review of typical residence times on various scales, I will give more space to examples and appraisals of long-range transports of anthropogenic sulfur which can transfer the emissions from major sources over distances exceeding 1000 km, thus creating virtually continentwide impacts in the European case and very large regional effects in the eastern half of North America.

4.2.3.1. Residence Times Having discussed the variety of processes which remove SO_2 from the air and having reviewed many conversion and deposition rates measured, calculated, or estimated mostly during the 1970s, it will be appreciated that the residence times for atmospheric sulfur compounds can easily span an order of magnitude. For example, one can assume on the basis of published reports that in an arid, sparsely vegetated area, SO_2 oxidation will proceed at a rate of only 1%/hr, SO_2 dry deposition at no more than 0.4 cm/sec, and sulfate settling at 0.1 cm/sec; with a mixing depth of at least 1 km (i.e., release from a tall power plant or smelter stack), about 3 days could be needed for conversion of all SO_2 to SO_4^{2-} or for dry deposition of the gas and it could take nearly another 6 days to deposit all the sulfate.

The only field measurements from such an area are those of Williams et al.

(1981) for northern Australia: studying the plumes from Mount Isa smelter, they found that dry deposition and atmospheric oxidation contributed about equally to SO_2 removal from the air and that the process was not completed in less than approximately 14 days. In contrast, SO_2 released from low sources over surfaces with higher deposition velocities (such as calcareous soil or perhaps a city) and in humid weather may be removed in just a matter of a few hours as it was, indeed, during the London 1952 smog (see Section 4.2.1.2). These extremes fix the residence range from a handful of hours to several weeks.

Still, averaging generalizations are needed and have been offered by many researchers over the years and just a few of them will quickly establish that most likely range. Meetham's (1952) classical value, derived from many years of measuring SO_2 concentrations and deposition rates, was 10 hr for all anthropogenic sulfur. Rodhe returned to the problem several times. In and around Uppsala during the winter of 1969, he found the lower limit for the atmospheric residence of sulfur to be 3 hr (Rodhe, 1970); in the mid-1970s he estimated the average residence times of anthropogenic sulfur in the European midlatitudes as follows (all values in hours): overall residence for SO_2 25, for SO_4^{2-} 80, and for total sulfur (SO_2 + SO_4^{2-}) 50; with less certainty for SO_2 dry deposition 60, SO_2 wet deposition 100, and transformation to SO_4^{2-} 80, which would imply that about 30% of SO_2 is converted to sulfate before being deposited (Rodhe, 1976, 1978). Baker et al. (1981), who analyzed precipitation chemistry measurements for the first year of the Multistate Atmospheric Power Production Pollution Study (MAP3S), found the residence time to be about 100 hr for all sulfur species.

Garland (1978) used average dry deposition velocities of 0.8 cm/sec for SO_2 and 0.1 cm/sec for sulfate and a time constant of about 10^{-4}/sec for SO_4^{2-} removal in rain (ignoring SO_2 washout and rainout) as inputs into a simple model resulting in a mean residence time of approximately 2 days for SO_2 before dry deposition or oxidation to sulfate (roughly equal shares are removed by each), 5 days for all emitted sulfur. All of these values are considerably higher than Rodhe's estimates with the overall residence time being about twice as long. Garland (1978) felt these values to be adequate on a continental or large national scale but not for smaller areas where a variety of detailed information is needed before an accurate appraisal can be made.

On the other hand, one can easily make calculations on continental and global scales just to establish the most probable ranges. Assuming that the atmosphere contains 1.25 million t of sulfur (Mészáros, 1978) while the natural and anthropogenic emissions total roughly 300 million t (see Section 4.1.4.2), the average global residence time comes to about 36 hr; assumptions of higher or lower fluxes (see Table 4-2) give residence times of just 29 or as much as 56 hr. Taking just the sulfur burden over the continents—roughly 800,000 t—and the terrestrial natural and man-made fluxes between 117 and 222 million t (average 174 million t), and disregarding any land–ocean transfers, results in just slightly different totals of between 32 and 60 hr (mean of 40 hr) for all atmospheric sulfur.

But over relatively rainy Europe the residence time is almost certainly much shorter. According to Mészáros (1978) the air column above the continent contains 1.9 mg SO_2-S/m^2 and 3.3mg SO_4^{2-}-S/m^2 or a total of 55,000 t of atmospheric sulfur; divided by annual emissions of 25 million t, this gives an average residence time of only 19 hr.

So perhaps the most realistic approximation fitting the largest number of actual situations would be around 40 hr for the complete removal of sulfur by several concurrent or consecutive processes. This is clearly too short a time to make anthropogenic, or natural sulfur compounds global pollutants (except perhaps for the small mass of sulfates thrown into the stratosphere by volcanic explosions) but sufficient to move them over considerable distances.

4.2.3.2. LONG-RANGE TRANSPORT As the large number of measurements and estimates reviewed in the preceding section have shown, the processes removing SO_2 from the atmosphere, converting the gas to sulfates and then depositing SO_4^{2-}, operate usually with rates slow enough to enable transport of sulfur from point sources or urban plumes easily over tens, frequently over hundreds of kilometers, and up to 1000–2000 km. During the 1970s, much research effort was devoted to documentation and modeling of long-range transfers of sulfur compounds but considerable improvements are needed to achieve consistent and reliable predictions (Pack et al., 1978).

A Swedish study of air pollution across national boundaries, prepared for the UNO Conference on the Human Environment, was the first comprehensive effort of this kind (Royal Ministry for Foreign Affairs, 1971). The report concluded, based on the average atmospheric residence of sulfur compounds, "that the deposition in any one place will be dependent on the emissions within a surrounding area with a radius of one or two thousand kilometers. The problem is thus an international one."

The study suggested the extent of possible long-range transport by charting end points of trajectories calculated each day for a period of 1 year for parcels of air originating at a location in northern Poland and traveling at a height of 1.5 km above the ground, i.e., at the typical altitude in which air pollution is carried. After just 24 hr, more than half of all parcels have moved beyond a 700-km radius and after 60 hr, 50% of the trajectory end points are outside a 1000-km radius. Moreover, in this particular case the spatial distribution is relatively symmetrical and hence during a course of 1 year a portion of central European pollution could be carried virtually all over Europe (with the exception of the Iberian peninsula, Ireland, northernmost Scandinavia, and the easternmost part of European Russia) in just 2½ days.

Since 1971, many continental, regional, and national studies have illuminated the European long-range transport of sulfur compounds with considerable detail and many researchers concentrated on the development of transfer models which would enable satisfactorily accurate predictions of sulfur deposition originating from remote sources. Here a pioneering study was Fisher's (1975) account of the British SO_2 contributions to European sulfur pollution. Its results showed that on the average about 50% of daily SO_2 emissions from low-level sources leave the United Kingdom and that this share is about 70% for elevated sources (mainly power plants). Dry and wet deposition of this British sulfur predicted by Fisher's (1975) model was highest, as must be expected, in England, Wales, and Scotland (virtually all of these territories are within an isoline of 5 g SO_2/m^2 per year), while northern France and central parts of Belgium and Holland receive 1–2 g SO_2/m^2 per year. The British share of Swedish deposition was put at just 10% of the total, or one-half of the Swedish emissions. The often-heard simplification about British SO_2 "exports" to southern Scandinavia is thus put into a different perspective: the rest of Europe, as well as natural sulfur deposition, contributes more than the United Kingdom.

Inputs of Fisher's (1975) work came from the broadly based OECD cooperative technical program for the study of long-range transport of air pollution (Ottar, 1976, 1978; OECD, 1977). This program used emission figures for SO_2 (estimated by grid squares of 0.5° of lattitude by 1° of longitude) and meteorological data, together with atmospheric dispersion models and with observations of SO_2 and SO_4^{2-} in a network of nearly 70 ground stations, to calculate the concentration field for man-made sulfur compounds. Aircraft measurements showed that the anthropogenic sulfur emissions were generally contained and carried within the lowest 1–2 km of the atmosphere, and the large number of ground sampling stations in southern Norway, separated by emission-free zones of the Baltic and the North Sea from the major sources, was instrumental for detailed documenting of the long-range sulfur transfers from the southerly quadrant.

Although the observed airborne concentrations were not high in comparison with source area levels, they were about an order of magnitude above the mean local backgrounds and could influence the precipitation chemistry to an astonishing degree, especially in southernmost Norway where orographic rains in southerly flows prevailing for a few days and over small areas cause an increase in sulfate deposition rates comparable to sulfur emission rates in the continent's major source regions. A Lagrangian model developed by Eliassen and Saltbones (1975) was used to calculate dry deposition of SO_2 and wet deposition of sulfate for the period between December 1973 and March 1975. Both deposition patterns are shown in Figure 4-5: dry removal is generally much more important over Europe than is wet deposition, the only exceptions being small areas of high orographic precipitation (southern Norway, parts of the Alps and Scotland).

Weather, naturally, is a key determinant of the frequency and intensity of the transfer and several situations were found especially favorable for long-range transport of sulfur (as well as other pollutants) in Europe (Barnes, 1979). Stationary anticyclones are not frequent either in northwestern Europe or over heavily polluted central Europe but when they do occur, very high concentrations of pollutants accumulated during a few days in the limited mixing layer are eventually channeled elsewhere bringing high pollution levels to areas remote from combustion sources. In contrast, cold strong breezes at anticyclonic margins have little rain but enough humidity to sustain oxidation to sulfates and enough speed to move sulfur over long distances in a short time.

In Europe, this means that when the sulfur sources are west of the recipient

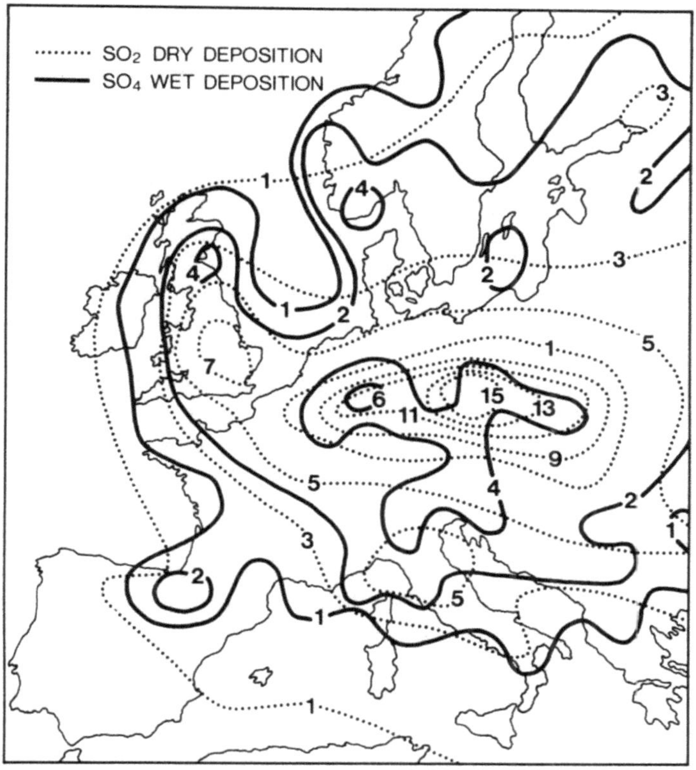

FIGURE 4-5. Estimated dry (SO_2 in g/m^2) and wet (SO_4^{2-} in g/m^2) deposition patterns for Europe in the mid-1970s (Ottar, 1976).

areas, concentrations will be relatively low but aggregate depositions will be high. In southern Scandinavia, the region most susceptible to acidification, higher concentrations are usually associated with southerly or southeasterly flows but these flows may not result in particularly high annual deposition totals (OECD, 1977). Southern Scandinavia is usually singled out owing to the well-monitored effects ascribed to long-distance transfer of sulfur but, owing to the location of major sources and to the deposition of pressure centers most frequently influencing the air flows over Europe, it is far from being the biggest recipient of transboundary sulfur flows.

The variability of transport winds, removal rates, and source strengths makes it extremely difficult to calculate reliable average export shares for each European country on an annual basis but a good appreciation of the relative magnitude of international transfers can be gained from daily calculations of transboundary exchanges with the EURMAP-2 model published by Bhumralkar et al. (1981). This model, as its OECD (1977) counterpart, is in far from satisfactory accord with actual SO_2 and SO_4^{2-} measurements taken during the investigated episodes as it ignores effects of complex terrain and dry deposition processes and has other simplifying features. In fact, I would argue that there is little advantage in constructing these relatively elaborate models which show little or no improvement over simpler ones when compared with the real world.

Regardless of the actual values (they do change daily anyway), there is no doubt that at any given time a large share of sulfur emissions are deposited outside the countries of their origin: the locations of European industries and urban areas and the prevailing winds leave no other possibility. EURMAP-2 calculations show that 45% of all sulfur emitted in Europe was transferred from the source countries to their neighbors and beyond, a reality making any national sulfur emission efforts inevitably ineffective.

Working with another simple dispersion model, van Egmond et al. (1979) estimated that in 1977, only about 30% of SO_2 levels in the Netherlands were contributed by Dutch sources, with the rest coming from the Ruhr (15%) other neighboring areas in West Germany and Belgium (20%) and other European countries; however, the LRTAP model (OECD, 1977) estimated the Dutch domestic contribution at 60% in 1974.

The North American experience with long-range transport of pollutants is not dissimilar. The distance between central Europe and southern Sweden or between England and southern Norway is just about the same as the distance between North America's leading sulfur source areas in southern Ohio and the Adirondacks in New York: about 1000 km away from the power plant-rich Ohio Valley, the soils and lakes of the Adirondack mountains are susceptible to acidification in much the same way as their Scandinavian counterparts, and most of this unwelcome change has again been attributed to long-range sulfur transfer.

Just a few typical examples will be given. Two-year monitoring of precipitation in the New York region established a significant variation of pH depending on the sector from which an air parcel approached the area: the pH of eastern quadrant rains averaged 4.83, that of northern flows 4.76, and that of precipitation coming from the west and southwest 4.01 and 3.74, respectively (Wolff et al., 1979). Similarly, Rao and Sistla's (1982) examination of the relationship between urban and rural sulfate levels in the Niagara–Buffalo region of New York revealed that high concentrations at both kinds of sites occurred under persistent southwesterlies, the lowest values under persistent northerlies, apparently a clear indication of long-distance transfer from the Ohio Valley.

Measurements at Whiteface Mountain in the Adirondacks show that 62% of all sulfate (and 65% of all nitrate) deposited annually comes with precipitation associated with air parcels arriving from the Ohio River Valley and the Midwest (Wilson et al., 1982). But as 56% of all precipitation comes with that southwesterly air, a high share of the Northeast's and deposition originating in the Midwest may be simply a natural consequence of precipitation's cyclonic origins. Six of eight principal cyclone tracks crossing the North American continent converge on New England; four major cyclonic systems—Alberta, North Pacific, North Rocky Mountains, and Colorado lows—pass over the areas of the highest SO_2 emissions in Ontario, Michigan, Illinois, Indiana, and Ohio as they converge on northern New York, Vermont, New Hampshire, and Maine.

Another very significant transport mechanism bringing sulfur compounds to the Northeast is the southerly clockwise flow of air on the backside of high-pressure systems frequently entering over the southeastern United States during late summer (Bermuda highs). A detailed interregional transport case study, including airborne and ground-based SO_2 and SO_4^{2-} measurements, trajectory trace, emission inventory and long-range transport modeling, indicates very well the complex makeup of an air parcel's sulfur content after its transcontinental journey (Reisinger and Crawford, 1982). A graph of a schematic development of the parcel's total sulfur budget during its transit from northern Florida across Alabama, Tennessee, Kentucky, Indiana, Ohio, and Lake Erie to northern New York is, regardless of the inevitable errors, a good replication of the still far from well-charted complexities (Figure 4-6).

Concentration of the heaviest sulfur sources and the climatology of the eastern part of the North American continent thus combine to bring about frequent long-distance transport and deposition of sulfur compounds over heavily populated (and by themselves far from clean) areas of the eastern United States as well as in the otherwise relatively clean parts of northern New York and New England. However,

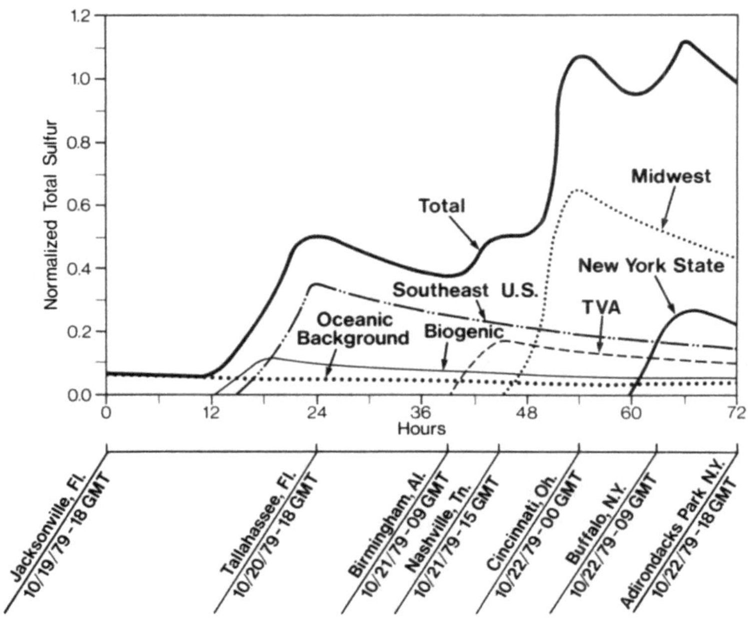

FIGURE 4-6. Schematic tracing of the total sulfur budget for the air parcel arriving at the Adirondacks on October 22, 1979 at 1800 GMT (Reisinger and Crawford, 1982).

complex interactions of atmospheric mixing and varying precipitation patterns tend to make acid deposition rates surprisingly uniform when considered on an annual basis. MAP3S measurements show a relatively gentle spatial gradient from the Ohio Valley's 35 mg H^+/m^2 to 27 mg at Pennsylvania State University in central Pennsylvania and at Ithaca to 18 mg at Whiteface Mountain (Wilson et al., 1982).

Modeling of these regional transfers and depositions has become a small growth industry in the early 1980s: the NAS (1983a) study reviews 13 such models by a convenient tabulation of their characteristics. As different as these models are in their treatment of trajectories, chemical transformations, and deposition processes, they all share fundamental weaknesses owing to many simplifying assumptions—and as there are no reliable data to validate or verify them, one must see them only as useful, qualitative research exercises. The NAS (1983a) sums up clearly: ". . . we do not regard currently available models as sufficiently developed to be used with confidence in predicting responses of the atmospheric system to alternative control strategies."

Uncertainties regarding long-range transport are also well illustrated by the lack of agreement concerning the ultimate fate of over 30 million t SO_2 emitted annually in the United States and Canada. While Galloway and Whelpdale (1980) concluded that nearly 40% of all sulfur inputs in the United States east of the Mississippi are deposited over the land and 25% are carried over the ocean, Shinn and Lynn (1979) disposed of 50% of inputs by dry and wet deposition and had almost 40% transported to the Atlantic. Similarly, there is no good agreement on annual transboundary SO_2 exchange between the United States and Canada with estimated totals being at least twice and possibly four times larger in the Canadian direction (USEPA, 1980b).

But it is a Canadian source which provides an unparalleled opportunity to study long-range SO_2 transport from a point source: INCO's 381-m-high stack at Sudbury, Ontario with fluxes around 35 kg SO_2/sec. Plumes from this source are frequently visible for many tens of kilometers, and Millan and Chung (1977) fortuitously detected a Sudbury plume as it passed over Toronto, 400 km to the SSE. The plume's SO_2 was recorded with a correlation spectrometer and trajectory analysis placed its origin at Sudbury about 18 hr earlier (average speed of 22 km/hr). Emissions at Sudbury were 37 kg ($\pm$10%) SO_2/sec while the calculated mass flux over Toronto was 28 kg ($\pm$25%) SO_2/sec.

Sulfur emissions from the eastern United States and Canada not deposited on the continent are transported over the Atlantic Ocean for considerable distances. Thus, Bermuda's marine atmosphere contains much higher concentrations of fine-particle sulfates than the air over remote islands in the Pacific, a clear indication of appreciable transport from the United States sources 1000 km or more away. Shipboard measurements between New York and Bermuda show that parcels of air which passed over the polluted Northeast 1–2 days earlier carry frequently high relative concentrations of sulfate aerosols with aerodynamic diameter less than 0.25 mm (Winchester, 1980).

Long-range transport of sulfur compounds is thus a well-documented fact—

although one we cannot model and predict with quantitative assurance. Given the right weather conditions (lofting between the ground-based and subsidence inversion), even a single, albeit a heavily SO_2-laden, plume can be detected several hundreds of kilometers from the source, and the compounds from combined industrial and urban emissions are traceable at concentrations well above the natural background in the recipient areas after distances in excess of 1000 km.

Yet, unlike with CO_2, there will always be some remote regions where anthropogenic sulfur compounds will not contaminate the air, the Antarctic above all. Analysis of well-dated snow samples from the years 1950–1975 collected in the Antarctic at the South Pole and Dome stations indicates that the main contribution to the snow's sulfate content is from the oceans, with added sporadic stratospheric injections from major volcanic explosions (such as Mount Agung in 1963), and gives no evidence of anthropogenic contributions (Delmas and Boutron, 1978).

This closes the sections on atmospheric sulfur and I will now look in some detail at acid deposition in general and acid rain in particular: as stressed in the preceding chapter on nitrogen, sulfur compounds are the dominant but far from the only contributors to these phenomena. Consequently, the following sections, although logically best fitting into this chapter, reach beyond the sulfur compounds to include all other contributions to acid deposition.

4.3. ACID DEPOSITION

Public awareness of this subject has come only recently: acid rain as a particularly insidious form of pollution, as an international environmental concern, as an object of political debates as well as intensive new research designs—all of these date only since the late 1960s in Scandinavia, since the mid-1970s (with increasing intensity) in North America. Yet, as mentioned in the introductory chapter, acidity of precipitation caused by anthropogenic emissions was the subject of a remarkably thorough investigation more than 100 years ago (Smith, 1872).

But Smith's work was an unusual exception and when one leaves Gorham's pioneering writings of the 1950s and 1960s aside [for their review see NAS (1981a)], it has only been since the early 1970s that acid rain publications have started to flood the journals. Flood is a most appropriate term as any researcher trying to survey and synthesize the accumulated knowledge is now overwhelmed by this surfeit of writings. Still, I hope I have not missed any essential references in the following sections which will start with discussions of basic concepts, move through a review of measurement difficulties, background acidities, and acid–base considerations to an assessment of anthropogenic contributions, and close with a reminder about the importance of dry deposition.

4.3.1. Precipitation and Acidity

The acidity of aqueous solutions, i.e., the concentration of hydrogen ions (H^+, more precisely hydronium ion H_3O^+), is measured on the familiar pH scale where values less than 7 are acid. But even in the absence of any acidifying emissions—natural or anthropogenic—the cleanest rains cannot be exactly neutral: as CO_2 dissolves in it, weak H_2CO_3 dissociates into bicarbonate (HCO_3^-) and hydrogen (H^+) ions, imparting, at normal CO_2 concentrations and pressures, slight acidity—pH 5.6—to all precipitation.

The standard account is then invariably simple: normal rain, fog, or snow is expected to have pH 5.6, i.e., the same as pure laboratory distilled water in equilibrium with the background atmospheric CO_2 concentration—and all precipitation with pH below this level is considered acid. This assumption can be found in scores of opening statements in acid rain publications written by the leading authorities in the field and by scientific news writers and printed in specialized scientific as well as popular sources (see, among many others, Likens et al., 1979; Glass et al., 1979; USEPA, 1980b; Likens and Butler, 1981; La Bastille, 1981).

As the pH scale is logarithmic (pH equals negative logarithm to the base ten of H^+ concentration), the numerical difference between solutions with pH 4.6 and 5.6 will be tenfold but the often-encountered claim that rain with pH 4.6 is ten times more acidic than "clean" precipitation is false as such a statement neglects the acidity of undissociated H_2CO_3 in normal rain, and the strong acid necessary to reduce precipitation pH from 5.6 to 4.6 actually represents only a twofold increase in total acidity (Stumm and Morgan, 1970).

4.3.1.1. Acid Rain Compared to the reference value of pH 5.6 (or 2.5 μEq/liter) for "clean" rain, precipitation over a large part of Europe and North America is now repeatedly recorded as considerably acid both during individual events and in terms of annual averages. There are extreme short-term values below pH 3.0—the lowest one recorded in Pitlochry, Scotland in a 1974 rainstorm at pH 2.4, as acid as lemon juice—but most of the affected areas in Europe and North America receive precipitation with an average pH between 4.0 and 5.5.

Historic reconstruction of rain acidities in the eastern half of the United States shows that in 1955–1956 the area with an annual mean pH below 5.6 (2.5 μEq/liter) was limited to the northeastern section (roughly east of 85°W and north of 35°N) whereas two decades later there were hardly any sites east of the Mississippi with a pH above 5.0 and the region most affected—extending from the Ohio River Valley to Vermont—had an annual pH below 4.3 (Figure 4-7; Likens and Butler, 1981).

Summer acidity values in the northeastern United States are typically about 200 μEq/liter higher (i.e., pH is about half a unit lower). For example, for the six-station network in the Washington, D.C. area, operated since April 1974 and characterizing pH values in the eastern megalopolis, there is a definite seasonal trend with summer pH minima (on several occasions they dipped just below 3.7) and winter acidity frequently above 4.5 (Bodhaine and Harris, 1982).

And the acidity appears to be spreading both westward and southward from its peak northeastern core. Collections of rainfall samples in the vicinity of the Kennedy Space Center in Florida between July 1977 and September 1979 yielded pH values between 3.4 and 5.1, the weighted average pH for the whole period being

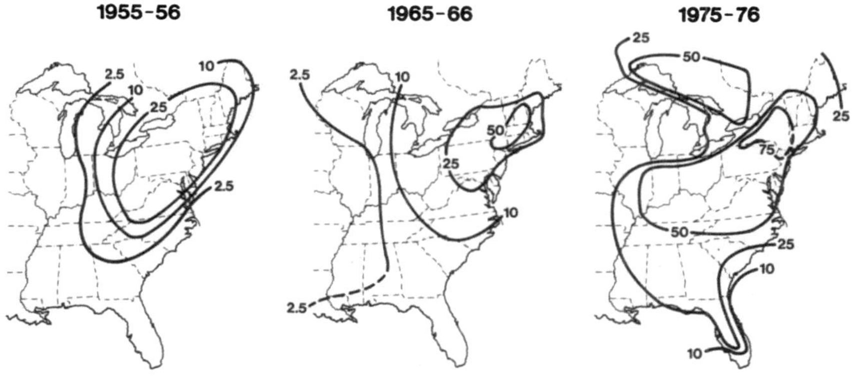

FIGURE 4-7. Distribution of acid precipitation in the eastern United States in the years 1955–1956, 1965–1966, and 1975–1976. Volume-weighted annual data shows H^+ concentrations in microequivalents per liter; for the first two periods these values were calculated from wet sampling or pH data, for 1975–1976 the values come mostly from bulk sampling (Likens and Butler, 1981).

4.55 (Madsen, 1981). H_2SO_4 and HNO_3 were shown to be the causative agents, with the ratio of SO_4^{2-}/NO_3^- typically greater than 1. West of the Mississippi, most monitoring sites still show values well above pH 5.0 but there are a growing number of pockets with pH between 4 and 5.

Continuous large-scale acid deposition measurements in North America are quite recent—the NOAA's network of ten widely spaced stations since 1972, the National Atmospheric Deposition Program (NADP) organized in 1976, and the Department of Energy's Multistate Atmospheric Power Production Pollution Study (MAP3S) begun in 1976—whereas in Europe, Egnér's systematic collection of precipitation for chemical analysis began in Uppsala in 1946 and in the mid-1950s he, Rossby, and Eriksson extended the monitoring from Sweden to Austria, Belgium, Finland, France, Holland, Iceland, Italy, Norway, the United Kingdom, and West Germany and this network later stabilized at about 50 stations (Granat, 1978).

Measurements by this extensive network show that whereas in the mid-1950s the region of high precipitation acidity (values below pH 5.0) covered only England, northernmost France, Benelux, and southernmost Norway, two decades later virtually all of Europe north of the Alps and the Loire Valley and east of the Vistula River, including most of Sweden, received rain and snow with such low pH while the industrialized core of northwestern Europe had precipitation with pH less than 4.3 (OECD, 1977). Even in southwestern Norway the mean pH (weighted for equal area) in the mid-1970s (4.35) was nearly identical to the average from the United Kingdom monitoring sites (4.24), and the frequency distribution was also very similar.

This quantitative information appears to be sufficient to conclude that since the 1950s both in Europe and in North America there have been appreciable increases in the peak acidities of precipitation as well as in the area affected by low-pH rain and snow. And this conclusion, vigorously publicized first by the Swedes in the early 1970s, later sanctioned by the OECD (1977) for Europe, and near the end of the decade taken up with zest in the United States and Canada, gave rise to a flood of fashionable scientific research, to healthy bureaucracies, to parades of catastrophic adjectives, and to much public concern.

Only the most recent publications have started to bring the necessary corrections, showing how much we do not know, pointing to hasty errors and simplistic assumptions which make some of the past conclusions untenable or misleading, and explaining better the great complexity of the phenomena unsuited for simple conclusions. Some of these key considerations will be introduced here, starting with the problems of acidity measurement.

4.3.1.2. Measurement Problems Assessments of regional acidification must be based on comparisons of current and historical data and the degree of uniformity, consistency, and comparability between any such two sets will obviously be critical for drawing reliable conclusions. And while it is not surprising that many past measurements cannot be classed as accurate, there is now considerable evidence to show the astonishing inaccuracies of many recent analyses as well. But first the pitfalls of temporal comparisons will be discussed.

As pointed out in the preceding section, the oldest systematically available data are relatively recent. The European precipitation chemistry network began its measurements in 1948, while Junge's project on chemical precipitation data at 24 sites in the eastern United States operated only during 1955–1956 (Emanuelsson et al., 1954; Junge and Werby, 1958). Thus, the longest span now available for comparisons is about three decades, which may not be enough to establish a firm trend—especially owing to the inaccuracies of past and present measurements.

Opportunities for considerable errors in this kind of monitoring range from inadvertent contamination to operators' skills. Kramer and Tessier (1982) demonstrated how tracing ionic changes over time can be significantly influenced simply by changing the containers used in rain sampling. The earlier use (generally before 1968) of soft-glass containers for water sampling makes comparisons of recent and historical acidity measurements dubious as the fresh glass surfaces could contribute 20–100 μEg of alkalinity per liter when samples are stored for less than 12 days, and around 140 ± 80 μEq/liter after 6 months of storage. Even with proper cleansing, with aging bottles and with rapid analysis, the increase of 20–60 μEq/liter must still be expected—and this covers precisely the range of claimed alkalinity changes in the eastern United States between the mid-1950s and the mid-1960s, the decade when the widespread use of glass containers was supplanted by much less reactive plastics.

And even more dramatic is the effect of rain sample collection in metal gauges, also a common procedure before the 1960s: Kramer and Tessier's (1982) experiments showed their rapid neutralizing effect and pH of almost 5.9 for a diluted H_2SO_4 sample of pH 4.39. The methyl orange alkalinity technique, used frequently before 1960, also imparted a significant error of an additional 80 μEq/liter beyond the end point of low-alkalinity samples. Interestingly enough, Kramer and Tessier (1982) believe that the colorimetric comparator pH method introduced in the early 1900s appears to be, with suitable corrections, the best choice for historical comparisons, offering results within 0.1 pH unit—but such measurements are not systematically available for the period before the 1960s.

And they, too, may not be so reliable as shown by Tyree (1981) who in his revealing paper performed an essential service only an experienced analytical chemist could: with a deep appreciation of the difficulties inherent in any sensitive microanalytical procedure, he looked at the actual applications of the three accepted methods of rainwater acidity measurement and found each to be deficient. For the first method, pH measurement, mean values reported by 25 laboratories of the World Meteorological Organization for four reference samples were much too low (i.e., too acid). Titrimetric acidity measurements tested in the same way showed even less accuracy and reproducibility with errors ranging in both directions.

The third method, Granat (1972) determination, requires properly combined accuracy and reproducibility of each of the values of the principal solute ions (eight of them usually), again a state not achieved in everyday measurements. Yet this was the method which, in the absence of pH measurements, was used by Cogbill and Likens (1974) to calculate the pH distribution for the eastern United States from 1955–1956 and 1965–1966 measurements. Before Tyree (1981), Liljestrand and Morgan (1979) pointed out some fundamental limitations of these calculations, concluding that "the uncertainty arising from applying the indirect methods to historical data is high enough that the date of transition from alkaline to acid precipitation is hard to determine."

Even more importantly, Tyree's (1981) critique did not stop at these revealing sample comparisons. He searched for a large number of actual rainwater samples for which acidity measurements by all three accepted methods could be compared, and found 315 such data sets. Comparisons of some of these measurements from Europe and the United States convey, in Tyree's accurate words, a gross inconsistency: differences of an order of magnitude are more the norm than the exception, and not infrequent are the cases when two methods fix the value at the same number—but with the opposite sign. Tyree's (1981) conclusion is that "if a credible assessment of the acid precipitation problem is to be made, the first step might be to assure competent chemical advice and supervision of the data acquisition."

The last of the critiques of American acidity measurements and comparisons to be reviewed here is also by far the most extensive and its conclusions confirm and strengthen the previously cited reservations. Hansen and Hidy's (1982) attention turned first to historical data, i.e., largely information gathered by short-term monitoring during the 1950s and 1960s when there were no consistent methods used to sample, preserve, store, analyze, and interpret the various observations.

The greatest disparities owing to sampling arise from comparisons of wet-only and bulk collection. As alkaline species are found in dust particles which are typically much larger than acid-bearing aerosols, the pH of bulk deposition samples will be higher than that of concurrently sampled precipitation. Junge's (1958) mid-1950s' values came from wet-only collection so that their comparison with the mid-1960s' bulk sampling is questionable: the secular shift in pH may be largely an artifact of improper comparisons.

Testing of variability of collocated samples shows that the uncertainty in the H^+ concentration owing to sampling and analysis is as large as 20% even today and the past values thus had certainly at least that much error. Other uncertainties introduced in sampling are siting of stations and failure to collect initial rainfall which tends to have higher alkalinity.

Analytical errors include exponential decay of H^+ concentrations in wet-only samples during the first 4–6 days, resulting in higher pH values for all samples collected in the 1950s and 1960s and analyzed many days after collection; and uncertainties in chemical analysis commonly introduce errors of ±10% for major alkaline and acid ions and 0.1–0.5 pH unit for acidity values. However, Hansen and Hidy (1982) believe that the interpretation of historical data is the most critical consideration among the arguments supporting acidification trends.

Where pH was calculated by Granat's (1972) method, the errors may be very large indeed: for example, pH calculations of the values found in Junge (1958) may have uncertainties of at least one pH unit. The impropriety of listing the pH values as averages for 1955–1956, 1965–1966, and 1975–1976, where a careful check of the original sources shows that the data are actually from up to 4 years preceding or following the compared years, is clear. Moreover, consecutive records show that temporal variability of pH data is such that annual values acquired at different sites and at different times cannot be offered as representations of regional acidity trends. Interpolations across distances of hundreds of kilometers is another grossly simplifying step, one which can result only in very crude approximations of regional patterns; the high degree of subjectivity in constructing isopleth maps from low-density samplings further magnifies this problem.

The only way to establish a trend is to analyze consistent and consecutive samplings—the USGS network of eight stations in New York and one in Pennsylvania provides the only set to be so treated. Hansen and Hidy (1982) looked at the decade 1967–1977 and found that by just comparing the initial and final values, five stations had a pH decline and four a pH increase in that period but, more meaningful, plotting the annual averages for 1965–1979 resulted in slopes whose magnitude was insignificant compared to the standard deviations—and hence no pattern should be ascribed to small ups or downs. Clearly, either there is no trend or 14 years of measurements has not been sufficient to reveal it.

Combined uncertainties of historical pH values are clearly summarized by Hansen and Hidy (1982) in a comprehensive table and put at as much as 0.25 ± 0.72 pH unit for the 1950s, at −0.6 ± 0.54 pH unit for the 1960s, and ± 0.63 pH unit for the early 1970s. Although it could be argued that the cumulative errors of the historical data are so large that they obscure any claimed secular changes, the authors prefer to interpret their findings as clear proofs of exaggeration of purported acidity shifts owing to all the uncertainties of the data acquisition and analysis. Another important and little appreciated point they stress is the relationship between climate and the total ionic composition of precipitation: once this is considered, one can easily appreciate the influence of the mid-1950s' droughts on elevated pH owing to much higher quantities of terrigenic alkalinity.

From all of the above considerations, it is clear that historical data are of insufficient quality and quantity to establish an indisputable secular trend toward higher acidity in the eastern United States, although there is no doubt that the region has precipitation that is more acidic than might be expected from natural conditions. But such low pH values were calculated (including the attendant uncertainties) from Connecticut precipitation chemistry data covering 19 years (1929–1948): Frink and Voigt's (1976) analysis of that period shows no secular trend and a mean pH of 4.2, virtually identical with Likens and colleagues' (1979) value representative of that region in 1975–1976. Similarly, pH monitoring for 17 years based on weekly bulk samples at the Hubbard Brook Experimental Forest (New Hampshire) shows no significant change (NAS, 1983a).

These measurements represent the longest continuous record of precipitation chemistry in the United States and fortuitously characterize the conditions in the Northeast. Comparisons of 1964–1965 and 1980–1981 values show that SO_4^{2-} deposition declined from 3.16 to 2.36 μg/liter (25% decrease), NO_3^- deposition rose from 0.70 to 1.66 mg/liter (a rise of 137%), and NH_4^+ levels remained unchanged at well below 0.5 mg/liter. These trends translated into no significant long-term changes in annual precipitation acidity with pH fluctuating between 4.03 and 4.20.

European data, as noted, have been continuously available for a much longer period and as they have been collected and analyzed by the same program their results should be less problematic. However, Kallend et al. (1983), who undertook a detailed statistical evaluation of these extensive precipitation acidity measurements gathered between 1956 and 1976, came up with results surprising in several ways. A simple regression analysis of weighted annual average precipitation acidity showed that only 29 of 120 sampling sites (24.2%) display a statistically significant trend of increasing acidity (and five sites a decrease), with the 1970s' concentrations typically three or four times higher than the 1950s' levels, and with more substantial increases for nitrate than for sulfate.

Detailed examination of the data for individual stations shows that in about one-third of the cases with significant correlations, the rise in acidity (from around 10 H^+ μEq/liter) came suddenly in the mid-1960s rather than gradually paralleling, as might be expected, the steady increase in sulfur emissions and that the higher annual levels of H^+ were not caused by a sustained high level of acidity but rather by an increased number of intermittent high monthly values as the post-1965 measurements show a much greater scatter with some months continuing to have low values similar to those prior to 1964. No clear anomaly in the data could be isolated to explain this sudden rise by changes in sampling or analytical techniques.

Even more intriguing are the results of statistical analysis for a group of adjacent sites (within 25 km of each other) near Uppsala for the decade 1966–1976: detectable common regional component in the variation in data accounts for no more than 20–30% of the variance with the bulk of the differences explainable only by measurement factors particular to each site. Clearly, many specific local factors must be considered in explaining the trends rather than simply referring to regional or continental trends and influences.

Although their work raises more questions than it answers, it provides an unassailable base for refutation of grand-scale generalizations displayed by isopleth maps of rising acidity. These unwarranted generalizations are misleading because of "failing adequately to show the pattern of time variation over the years, the large degree of uncertainty attaching to individual contours and assuming a geographical homogeneity which is not borne out by detailed calculations with data taken from adjacent sites" (Kallend et al., 1983).

Thus, a reliable monitoring program must utilize uniform collection minimizing the biasing errors, a variety of techniques to monitor the acidity, competent analytical work guaranteeing consistency and reproducibility of the results, a large enough network of stations, and a long period of data collection. The last point is critical: even if all other conditions are satisfied, great differences in composition of rain within and between events, seasonal and annual fluctuations tied to macroscale climatic patterns and mesoscale meteorological processes, emission characteristics, and atmosphere reactions present a complex phenomenon of such spatiotemporal dimensions that decades, rather than years, of reliable information might be needed before rain composition trends can be established with certainty (Jacobson, 1981).

Yet the longest operating United States network of precipitation chemistry has been collecting rain samples only since 1972. In 1980 this network became a part of the NADP which is now the most important cooperative monitoring network of 84 stations involving several governmental and private organizations. Begun in September 1977, the NADP samples precipitation weekly, instead of monthly as before, and uniform collection procedures and prompt analysis at the Illinois State Water Survey Laboratory ensure the quality and comparability of data. This program's Canadian counterpart, the Canadian Network for Sampling Precipitation (CANSAP) with 55 stations, began in April 1977 and the two networks now provide a fair coverage of the continent's most populated and industrialized regions.

At the end of the 1980s we will thus be able to look back at the decade's acidity trends with unprecedented confidence—but even then, as now, we cannot look at the phenomenon in isolation: we have to compare precipitation acidities prevailing in industrialized regions of the Northern Hemisphere with background levels and also to consider complex acid–base relationships.

4.3.1.3. Baseline Acidity Monitoring acidity changes, appraising the degradative shift, cannot be done without referring to the normal, in this case pH 5.65, pure water in equilibrium with atmospheric CO_2. As noted in Section 4.3.1.1, this is a standard assumption throughout the acid rain literature—yet it is a wrong starting point, a simplistic adoption of a reference value easily leading to erroneous scientific conclusions.

Only a rudimentary understanding of atmoshperic chemistry is needed to realize that there are no pure rains: unpolluted rainwater—to choose an extreme case, precipitation occurring millions of years before the appearance of the first humans—cannot be expected to have the properties of distilled water, and, of course, it does not. To begin with, natural aerosols making up the condensation nuclei in clouds may be neutral, basic, or acidic. Not surprisingly then, measurements of pH in clouds over England, Russia, Germany, and Australia show a wide range of recorded values from as low as 3.1 in subinversion stratiform clouds over Europe at 500 m up to 7.2 at 2000 m over England (Larson, 1980), whereas cloud water samples collected around Hawaii range between pH 4 and 5 regardless of sampling location (Parungo et al., 1982).

Also of neutral, basic, or acidic character are other solid and liquid aerosols and gases scavenged by "pure" rains. Consequently, Sequeira (1982) argues that only the pH of rainwaters containing nothing additional except dissolved sea salt may be expected to be close to 5.6. However, during the late 1970s it came as a surprise that even in remote ocean areas the actual pH values are substantially different.

The NOAA's measurements near the top of Mauna Loa (3500 m above sea level) on Hawaii (part of the Geophysical Monitoring for Climatic Change Program) yielded a predominantly acid pH; the range for 1973–1979 was pH 3.3–6.7, with median pH at 5.0 and volume-weighted mean pH at 4.9; even more interestingly, averages weighted for surface wind direction between June 1975 and December 1977 showed that rains coming from *every* quadrant were acid (means: Northwest 4.35, Northeast 4.18, Southeast 4.56, Southwest 4.63).

Measurements done at different sites on the island also indicate quite clearly that rain acidity increases with altitude: at Kapoho (sea level) the precipitation is least acidic (weekly values up to pH 5.6), whereas on Mauna Loa the pH can drop below 3.8. Similarly, at Cape Matatula on American Samoa, another of the NOAA's global climatic change monitoring sites selected for its extreme "cleanliness," the pH range in the late 1970s was 4.5–6.0, with median at 5.3.

To gather a more systematic body of information on rain pH from remote areas, the Global Precipitation Chemistry Project was begun in 1979 and its first results were published in 1982 (Galloway et al., 1982). The project's aim was to select sites very distant not only from any large industrial or urban area but also from local volcanic influences; annual precipitation in excess of 500 mm and the presence of trained personnel were the other key preconditions. The five stations selected are Katherine in northern Australia, San Carlos de Rio Negro in southern Venezuela, Poker Flat Research Range in central Alaska, St. George in Bermuda, and Amsterdam Island in the South Indian Ocean, halfway between Africa and Australia.

Precipitation collected at all five sites during the first year of the program was acidic: average volume-weighted pH values were 4.8 for St. George, San Carlos, and Katherine, 4.9 for Amsterdam Island, and 5.0 for Poker Flat. Even when the authors did some speculative adjustments, they had to put lower limits of the average precipitation pH in remote areas just at 5, considerably below the supposed baseline level of pH 5.65.

Clearly, the evidence is already rich enough to conclude that rains with pH 5, or perhaps even 4.5, should not be considered acid in terms of their departure from the ideal ''clean'' natural standard—because such standard cannot be recorded either in the mid-Pacific, or in Alaska, or in the South Indian Ocean. Two explanations are possible: either the acidity of rains in the remotest stations is of natural origin, or it is the result of considerable anthropogenic influences. As with most atmospheric phenomena, a much longer span of reliable measurements will be needed before giving satisfactory answers but several indications strongly militate against the universality of the second explanation.

Only in Bermuda could the acidity be explained largely by long-range transport of sulfate aerosols from the United States, while in Alaska anthropogenic emissions may be an important (seasonal) contribution. In other locations the man-made emissions could not have been decisive, especially in ''clean'' Southern Hemisphere locations such as northern Australia, American Samoa, and, most notably, Amsterdam Island. In the last two cases, anthropogenic sources can be excluded rather convincingly: a glance at a global map of pressure centers and prevailing winds will show that it is hardly possible to postulate major upwind pollution sources for Samoa or Amsterdam Island. Is oceanic production of dimethyl sulfide sufficient to cause this acidity?

For Mauna Loa there is a long-range transport explanation: eolian dust from deserts of northern China can be carried in less than 10 days 10,000 km eastward and detected at the observatory at concentrations in excess of 20 $\mu g/m^3$ with more than 8 $\mu g/m^3$ being additional sulfate; this phenomenon offers perhaps the best explanation of the relatively high acidity of rainwater encountered at higher altitudes in Hawaii (Darzi and Winchester, 1982).

But this process will be influenced little by anthropogenic sources: sulfurous pollution in China is released overwhelmingly near the ground together with heavy particulate loads and high concentration of soil dust which neutralize the emissions to a large extent and in Japan low-sulfur liquid fuels predominante; moreover, low-level East Asian wind flows—monsoonal regime with continentward flow in summer and a strong southerly component in winter—take the pollution in other directions. Finally, the known fate of SO_2 and SO_4^{2-} in the atmosphere makes it improbable that after the 5–8 days it takes for the high-level East Asian air to arrive over Hawaii, this flow would contain large amounts of sulfur compounds originating in Japan or China (see Sections 4.2.1.3, 4.2.2.3, and 4.2.3.1). Besides, the Mauna Loa rains are acid *regardless* of the direction from which they arrive.

Yet the long-range transport over distances of several thousand kilometers from both natural and anthropogenic sources can occur and although only very small masses of acid aerosols can then arrive at remote oceanic locations, they could cause low pH because in such areas there are few alkaline ions to neutralize even modest quantities of sulfates or nitrates. Clearly, attention must be given to overall acid–base balance (or imbalance), not just to the presence of acid aerosols alone.

4.3.1.4. ACID–BASE CONSIDERATIONS Far from being determined simply by the concentrations of sulfates and nitrates, acidity of solutions is a result of complex influences of many cations and anions and the often-ignored necessity to consider both the acidic and alkaline aerosols when evaluating acid deposition is perhaps best illustrated with the Chinese example. China, with 70% of its primary energy coming from raw coal burned with low efficiencies, has very high SO_2 releases and urban SO_2 concentrations frequently exceed 1 mg/cm^3 during winter months. Yet a series of measurements in Beijing during 1979–1980 showed its rainwater to be close to neutral, with most values between pH 6.0 and 7.0, i.e., *above* the value for "pure" rain (Zhao et al., 1980).

However, the same measurements showed, not surprisingly, heavy concentrations of SO_4^{2-} and NO_3^-, averaging 12 and 3.76 mg/liter, respectively (for comparison, natural background for these compounds would usually be just around 0.1 mg/liter). Zhao et al. concluded that the high concentration of suspended alkaline matter—dust blown from the deserts encroaching on northern China, and soil particles from the surrounding deforested mountains, dry fields, and the city's unpaved alleys—neutralizes the acids and keeps the pH up.

Most kinds of airborne particulates are sinks for H^+ ions, either by dissolution or by adsorption and ion exchange. Soluble particles have a much higher H^+ capacity than the active surface ones: $CaCO_3$ and FeOOH will exchange 0.02 and 0.03 μEq/μg, respectively, while most clay particles (amorphous silica, kaolinite, montmorillonite) have exchange capacities of only 30–1200 μEq/g (Kramer, 1978). Still, as these particles are commonly present in the dust, their combined effect is substantial.

Sequeira (1982) lists other measurements of continental rains in drier regions which show pH values much higher than the expected norm: 6.2–6.8 for 12 stations in Mexico, 5.8–8.9 for nine inland sites in India (medium pH 7.5), again the effect of large quantities of alkaline dust. In all such cases the anthropogenic emissions may, as in northern China, cause no or little change in pH owing to the significant buffering action of existing aqueous solutions, and anthropogenic acid deposition would then be better determined by a temporal lowering of alkalinity. This led Sequeira (1982) to argue that the focus on SO_4^{2-} or NO_3^- without consideration of the alkaline or neutralized components is an unsatisfactory way to provide quantitative information on strong mineral acid deposition.

He believes that the acid deposition problem in the northeastern United States and eastern Canada results both from the undeniably large anthropogenic sulfur outputs and from the disadvantageous location of the region with respect to terrigenous alkaline matter which neutralizes atmospheric acidity over large subtropical and temperate parts of the continent. This point is well illustrated by recent extensive mapping by Munger and Eisenreich (1983) who used numerous regional and local precipitation studies and the data from the NADP, CANSAP, and MAP3S/RAINE—altogether measurements for nearly 200 sites for various periods between 1971 and 1981—to prepare contour maps for all major cations and anions.

Munger (1982) determined major ions and trace metals in precipitation and snow cores from sites along a 600-km transect from the North Dakota prairie to the forests of northeastern Minnesota for the period between April 1978 and June 1979. Alkaline dust and gaseous NH_3 from the cultivated prairie kept the acidity there at one-quarter of the level at the eastern end of the transect and the Ca^{2+} and Mg^{2+} concentrations decreased two to three times going eastward.

Mapping of the ionic distribution in Munger and Eisenreich's (1983) study is worth a detailed look. Concentrations of Ca^{2+} and Mg^{2+} have pronounced regional maxima over the cultivated Canadian Prairies and the northern area of the Great Plains (Figure 4-8) and the close ion–ion association is even visually clear; the correlation coefficient (significant at the 0.001 level) is 0.83, by far one of the highest among all precipitation ion pairs so investigated. Ammonium ions have a weaker, but still unmistakable maximum in the Corn Belt, naturally associated with animal husbandry and intensive crop fertilization (Figure 4-9).

Consequently, North America's heartland—roughly west of the Mississippi, north of the Ozarks, east of the Rockies, and south of the northernmost line of cultivation in the Canadian Prairies—has the highest combination of NH_4^+ and basic cations to neutralize acidic precipitation. In some coastal areas, especially in California and New England, contributions of Na^+ are also significant (Figure 4-9).

In contrast, acid anions have maxima in the industrialized eastern part of the continent although in the case of NO_3^- the distribution is surprisingly uniform (a phenomenon explainable by the combination of great dispersion of sources and relatively long atmospheric residence times), while SO_4^{2-} displays much clearer peaks between the Mississippi and the Atlantic coast (Figure 4-10).

Cation/anion balances are thus regionally quite distinct, with the highest portions of total acid anions neutralized by NH_4^+ and major cations in the western half of the continent (75–96%, with most of the effect owing to alkaline elements). In contrast, in the northeastern United States, 52% of all acid anions are not neutralized and must be balanced by H^+ to satisfy electroneutrality. Consequently, H^+ in precipitation peaks in the northeastern United States with concentrations ten times above the minima on the plains and prairies of the North (Figure 4-11).

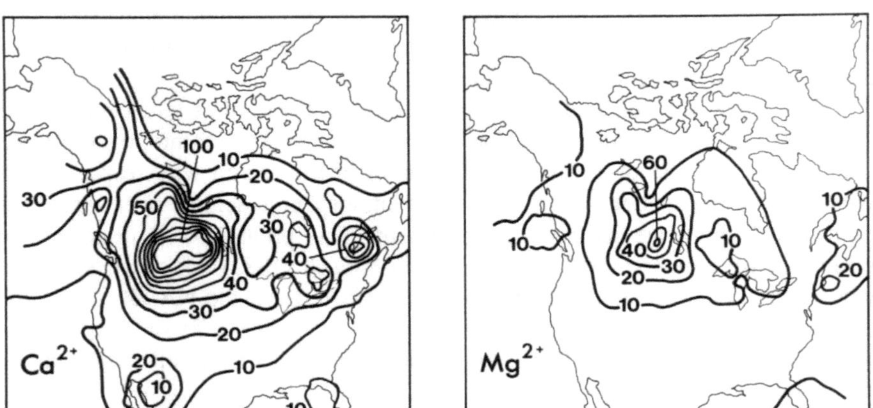

FIGURE 4-8. Average concentrations of calcium and magnesium cations in North American precipitation in microequivalents per liter (Munger and Eisenreich, 1983).

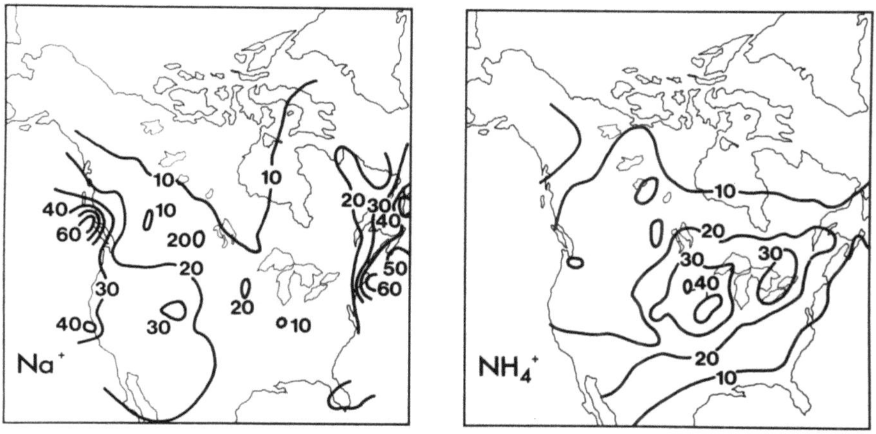

FIGURE 4-9. Average concentrations of sodium and ammonium cations in North American precipitation in microequivalents per liter (Munger and Eisenreich, 1983).

These acid–base relationships also have a most intriguing temporal dimension. Sequeira (1982) points out an interesting fact that while the National Air Sampling Network measurements for the years 1959–1968 showed median Ca^{2+} rainwater concentrations at 3.9 mg/liter for Chicago and 2.4 mg/liter for Cincinnati, the currently recorded NADP and MAP3S values are merely around 0.5 mg Ca^{2+}/liter: from this, could it not be that the considerably lower levels of alkaline dust in today's precipitation are the cause of the lower pH levels recorded throughout the Northeast?

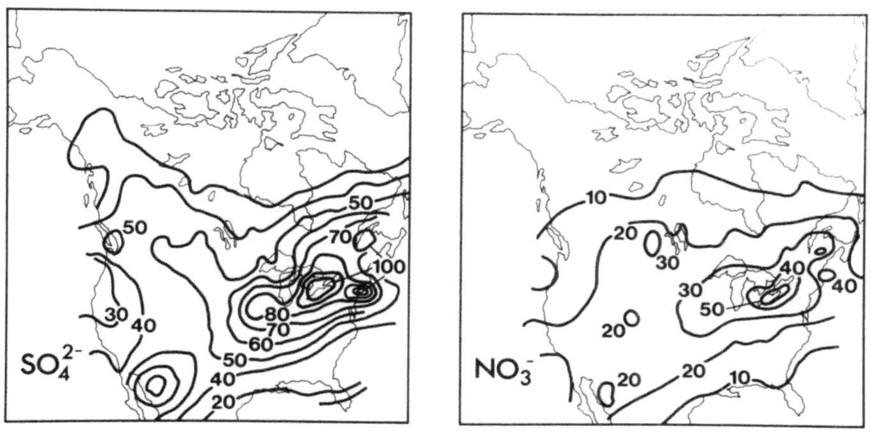

FIGURE 4-10. Average concentrations of sulfate and nitrate anions in North American precipitation in microequivalents per liter (Munger and Eisenreich, 1983).

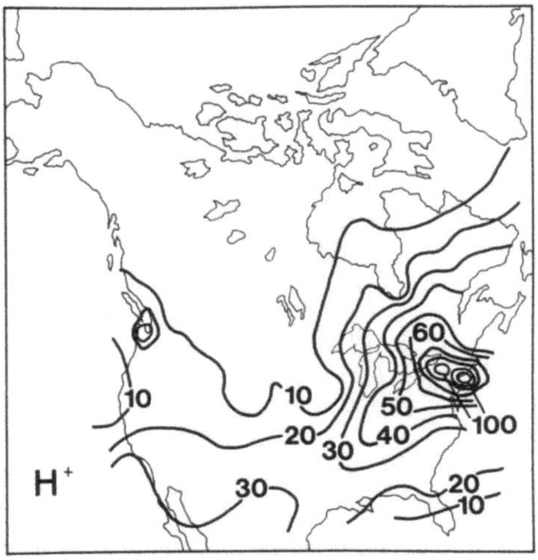

FIGURE 4-11. Average concentrations of hydrogen ions in North American precipitation in microequivalents per liter (Munger and Eisenreich, 1983).

I have investigated this suggestion and reached some surprising conclusions. Between the mid-1950s, the years for which the first satisfactory network measurements of precipitation's ionic content are available, and the mid-1970s, when the acidification trend was established for the northeastern United States, there was indeed a great loss of airborne alkaline material owing to the large-scale replacement of coal as household, transportation, and industrial boiler fuel by hydrocarbons as well as by the nearly uniform and highly efficient removal of fly ash from large power plants, now the top consumers of coal.

While it is impossible to quantify this loss with a high degree of accuracy, the following calculations will show that its effect was undoubtedly quite important. In 1955, utilities burned 140 million t of coal, generation of industrial steam and heat consumed some 90 million t, household combustion amounted to 59 million t, and transportation used 15 million t. Uncontrolled household and transportation combustion released at least 650,000 t of particulate matter. About 60% of large stationary sources were equipped with particulate controls averaging 80% efficiency (Walker and Brown, 1971) so that about half of all emissions from these sources—some 8.6 million t/year—were released, for a grand total of about 9.25 million t.

However, by 1975, in spite of the fact that large stationary combustion rose to 480 million t/year (420 million t in power plants, the rest in industrial boilers), only 3.6 million t of particulates was released from these sources as 95% of them had particulate controls averaging 95% efficiency; coal was no longer used in transportation, and households, burning just 10 million t/year, released only some 100,000 t. Consequently, the mid-1950s' total of about 9 million t shrank to some 4 million t by 1975, a decline of 5 million t of particulate matter.

Alkaline content of coal ashes varies rather widely: I have quoted Davis and

Fiedler (1982) in Section 4.1.3.1 and other investigations—Davison et al. (1974), Hecht and Duvall (1975), and Bechtel National (1981)—offer similar ranges. Consequently, assuming that alkaline constituents (Ca, Mg, Na, and K) comprise on the average 10% of the particulate matter released by coal combustion does not appear exceedingly liberal. If so, then coal burned in 1975 contributed to the atmosphere 500,000 fewer tonnes of alkaline elements than in 1955. How large this amount is can be, again approximately, demonstrated by assuming that about four-fifths of it—say 400,000 t of Ca equivalent—were released in the Northeast (an area of roughly 1.5 million km^2) where the annual precipitation averages 100 cm. If uniformly distributed in rains and snows over the whole area, alkaline elements would impart an average concentration equivalent to 0.25 mg Ca^{2+}/liter precipitation.

Consulting Junge and Werby's (1958) map of calcium ions in United States precipitation between July 1955 and June 1956 will show that at that time the Northeast was receiving between 0.25 and 1.0 mg Ca^{2+}/liter, with the 0.5 mg isoline running through its SW–NE axis. Taking 0.5 mg Ca^{2+}/liter as a good average, the comparison shows that the area experienced a large loss of anthropogenic alkaline material between 1955 and 1975, the period when its sulfur emissions rose by at least 50%.

Naturally, this comparison ignores the fact that a significant part of the emitted alkaline particulate matter would be deposited before being precipitated and that its atmospheric residence time would be most often shorter than that of SO_2 or SO_4^{2-}; on the other hand, these settled aerosols would nonetheless contribute to neutralization of the total acid deposition. Moreover, the total "loss" of anthropogenic alkaline matter over the Northeast was certainly much higher than indicated by calculations for stationary coal combustion: controls in iron and steel and cement industries also advanced appreciably between 1955 and 1975, and the areas of barren land shrank with advancing settlements, paving, grassing, and considerable regrowth of forests.

Whatever the actual value of the decline might be, there is little doubt that successful particulate controls have kept from the atmosphere over the northeastern United States a substantial part of the alkaline aerosol the region used to receive from coal combustion, and because in this wet and vegetated region terrigenic alkaline aerosols are much rarer (3–20 times) than in the drier and less vegetated areas of the continent, this decline of alkaline aerosols certainly contributed to the Northeast's higher precipitation acidity. And as this acidity arises not only from sulfur emissions but also from anthropogenic nitrogen releases, the relative contributions of the two ions must be examined in some detail, including their different fates after deposition.

4.3.1.5. Anthropogenic Contributions Although acidic or potentially acidifying substances include many compounds of sulfur (SO_2, SO_3^{2-}, SO_4^{2-}, and H_2SO_4) and nitrogen (NO, NO_2, NO_2^-, NO_3^-, and HNO_3) and also hydrochloric acid, organic acids (above all H_2CO_3), and Brønsted acids (dissolved Fe, NH_4^-), precipitation acidity is primarily caused by the presence of the two strong mineral acids, H_2SO_4 and HNO_3.

As anthropogenic releases of HCl are relatively small, the consideration of man-made acidity narrows just to sulfates and nitrates. Not surprisingly, no simple generalizations can be made regarding the shares contributed by sulfates and nitrates to precipitation acidity. The chemical composition of precipitation in both polluted and unpolluted (remote) areas shows correlations (nearly always statistically significant, often very strong ones) between the SO_4^{2-}, NO_3^-, and NH_4^+ ions so it appears that there are chains of chemical reactions involving all of these ions, but while the role of NH_4^+ is straightforward acid neutralization, the relationship between NO_3^- and SO_4^{2-} is more difficult to define (Marsh, 1978).

On the basis of a simple photochemical model comparing sulfate and nitrate measurements in northwestern European precipitation, Rodhe et al. (1981) offered several tentative conclusions. First, owing to the relatively rapid reaction of NO_2 with OH, HNO_3 is generated faster than is H_2SO_4 and if their precipitation removal is not greatly different, it must be expected—and it has been confirmed by the authors' observations—that the nitrate in rain will have a more local origin than the sulfate and that the latter will be transported over considerably greater distances. Second, production of both acids will be enhanced by higher emissions of hydrocarbons. Third, changes in NO_x emissions may affect not only HNO_3 formation but also, through their effects on oxidant concentrations, the H_2SO_4 levels: higher NO_x releases will use up more OH in forming HNO_3, thus depressing H_2SO_4 generation rates; doubling of NO_x emission rate is expected to lower H_2SO_4 formation by 45%.

Rodhe et al. (1981) illustrated their conclusions by mapping molar ratios of sulfate to nitrate for different areas in northwestern Europe. While in the region of major emissions in West Germany, Benelux, and England the ratio is around 1.5, in southern Sweden it increases to over 2.0 and in northern Scandinavia it surpasses 5.0. However, the actual atmospheric relationships of the two pollutants mainly responsible for precipitation acidity are far from understood and so, for example, Cleveland and Graedel (1979) claim, in contradiction to the conclusions just cited, that according to their chemical kinetic model, reduction of NO, a precursor for nitrates in precipitation, brings a reduction in sulfate levels. Nor is the role of increased NO_x emission well understood: Rodhe et al. concluded that the rise of NO_x/SO_2 emission rate may not be directly responsible for higher HNO_3 precipitation contents, while the growing evidence from the United States would support a rather clear link between higher emissions of NO_x and higher shares of HNO_3 in monitored precipitation—but not the conclusion of higher sulfate/nitrate molar ratios with increasing distance from major emission sources.

The best long-term evidence supporting the relationship between higher NO_x emissions and greater importance of nitrate in precipitation comes from the Hubbard Brook Experimental Forest records analyzed by Galloway and Likens (1981) and reviewed in NAS (1983a). Average minimum contribution of HNO_3 to acid precipitation during the years 1964–1979 was about 25% in summer and 45–55% during winter; for H_2SO_4, the summer maxima were 80–90% and the winter peaks up to 70–80%, largely an expected breakdown. However, linear regressions showed a steady secular increase in HNO_3 contributions which rose from maximum contributions of 20% in summer 1964 and 44% in winter 1964–1965 to 39% in summer 1979 and 61% in the winter months of 1978–1979, about a 50% gain relative to H_2SO_4.

Of the three possible explanations for increased formation of HNO_3—higher fossil fuel combustion, greater fertilizer applications, higher frequency of lightning—only the first one appears to be decisive. Indeed, a plot of the equivalents of NO_x relative to SO_2 from fossil-fueled combustion between 1940 and 1980 shows the ratio increasing from about 0.25 in 1940 to nearly 0.6 in the late 1970s, and Section 3.3 provides details on the rise of NO_x emissions which may, or may not, continue depending above all on the expansion of fossil-fueled power generation and implementation of both NO_x and SO_2 controls. In large cities with their heavy traffic, the NO_x/SO_x ratios are much higher; for example, in Chicago they range from about 5 in winter to 10 in summer on a volumetric basis and there is no doubt that acidity of urban rain is primarily caused by nitrogen oxide emissions (Seliber, 1981).

However, American data do not offer any evidence of rising SO_4^{2-}/NO_3^- molar ratio with long-range atmospheric transport. A detailed appraisal of these ratios for the northeastern United States is now possible as the first 4 years of the MAP3S/RAINE network measurements—embracing nine stations in New York, Pennsylvania, Virginia, Illinois, Delaware, and Ohio have been reviewed (The MAP3S/RAINE Research Community, 1982).

While the latest published Hubbard Brook sulfate/nitrate ratio is 1.42 for the years 1980–1981 (NAS, 1983a), the annual average for the eight stations was just 1.15 and seasonal disaggregation shows the ratio uniformly higher in summer (about 1.4 compared to 0.86 in winter), most likely a result of less efficient photochemical conversion of SO_2 during the cooler months as SO_2 emissions are relatively constant throughout the year. The seasonal difference is thus clear but the spatial pattern is surprisingly uniform, particularly in summer when the ratios are either identical or only insignificantly different, with stations in Illinois and Ohio (source areas) having the same molar ratios as stations in northern New York (recipient region). Vigorous advective atmospheric mixing is seen as the principal factor causing this uniformity.

Moreover, annually averaged precipitation sulfate/nitrate molar ratios correspond rather closely to molar ratios of SO_2/NO_x emissions in eastern North America which lead the NAS (1983a) to conclude that in this continent, there is no evidence for Rodhe's hypothesis of ratios changing with distance and that the net oxidation of SO_2 resulting in precipitation SO_4^{2-} is as efficient as the analogical transformation of nitrogen oxides.

Thus, it might appear that a conclusion about the validity of the growing importance of nitrogen compounds in acid precipitation in eastern North America is warranted, with the addition that seasonally (during winter months) this contribution becomes easily dominant. Moreover, NO_x emissions are expected to keep increasing faster than SO_x releases. These conclusions would, obviously, have essential implication for acid precipitation control strategies as efforts to reduce much more diffuse NO_x flux are both less advanced and less manageable than desulfurizing the few scores of large power plants and smelters which are responsible for most regionally deposited sulfates. But, as a last demonstration of the uncertainties surrounding the relative importance of ionic makeup of acid rain, McLean (1981) argues that the contributions of the two compounds to the acidification of ecosystems are not equivalent and that, with a possible exception during the spring thaw, HNO_3 is responsible for much less acidification than H_2SO_4.

His main argument is based on a common-sense nutritional balance. Unlike SO_4^{2-} in precipitation, which may react in the watershed but whose outputs usually about equal the inputs (i.e., the net effect equivalent to no reaction), NO_3^- output from monitored ecosystems is much less than the precipitation input, usually less than 25%, as the nitrogen in one of its reactive forms is extensively utilized by the plants: in assimilation, which can be represented by

$$HNO_3 + 2CH_2O \rightarrow NH_3 + 2CO_2$$

1 mole of hydrogen is consumed per mole of nitrate. Moreover, the same exchange accompanies denitrification if the reductant is a carbohydrate:

$$4HNO_3 + 5CH_20 \rightarrow 2N_2 + 5CO_2 + 7H_2O$$

Consequently, HNO_3 in acid precipitation is effectively decomposed by biota and although the NO_3^- reduction reactions proceed better under anaerobic and neutral conditions, they can also occur, though slower, aerobically at pH 4.0 or less so that nitrates can be reduced not only in organic soils but even in acidified waters. Additions of HNO_3 to an ecosystem affected by acid rain will thus not cause acidification if the reduction of NO_3^- proceeds at a sufficient rate.

Additions of HNO_3 can become detrimental only after extensive acidification of waters (when the rate of NO_3^- reduction is greatly lowered)—or during spring thaw when the acid accumulated in snow is suddenly released into waters and when it may not be used up rapidly enough. However, even in this special case HNO_3 appears to contribute slightly less than the H_2SO_4 to the overall excess acidity, but detailed studies of HNO_3 storage in snow, the subsequent release and effects on aquatic life are needed to clarify these processes. That the process of NO_3^- reduction is more complex than outlined by McLean (1981) is clear—see, for example, Harkov's (1981) reaction to McLean's paper—but of the basic difference between ecosystemic disposal of NO_3^- and SO_4^{2-} there can be little doubt and hence the deposition of sulfur compounds should be considered the main cause of anthropogenic acid precipitation.

4.3.2. Wet and Dry Sulfur Deposition

Discussions of sulfur dioxide and sulfate removal processes (Sections 4.2.1.3 and 4.2.2.2) made clear the importance of dry deposition: any appraisal of sulfur immissions limited to acid precipitation (wet deposition) is bound to be a serious underestimate for all highly affected areas of the Northern Hemisphere. Consequently, in this section I will review the best available measurements and computations of both wet and dry sulfur deposition.

For wet deposition it might seem that the best way (and, of course, the easiest) is to start with the values measured at the stations monitoring precipitation chemistry. However, as Granat (1978) points out, a better approach is to first construct a concentration field and then to convert it to deposition using the precipitation data from the much denser network of normal rain gauges. This method is especially desirable in regions where precipitation has a considerable spatial variability.

Conversely, spatial variability of wet deposition is a function of total precipitation, and normalization of deposition values per 1 cm of precipitation results in much lower differences than the total would indicate: a perfect illustration is provided by comparing total and normalized SO_4^{2-} and NO_3^- values for MAP3S stations for 1977–1979 (Wilson et al., 1982). While, for example, the Penn State site had three times as much sulfate and nearly four times as much nitrate than Urbana (Illinois), their normalized depositions are virtually identical.

The northwestern European deposition field for excess sulfate in precipitation derived by this method shows maxima surpassing 3 g SO_4^{2-}-S/m^2 annually in a small part of southern Norway (owing to the high orographic precipitation), more than 1.6 g over most of Belgium and the Netherlands, in mountainous areas of Wales, northern England, and Scotland, with the 1 g SO_4^{2-}-S/m^2 per year (i.e., 10 kg S/ha or 33 kg SO_4^{2-}/ha) isoline embracing northern France, virtually the whole British Isles, southern Scandinavia, and most of central Europe.

In eastern North America, wet deposition of sulfate is higher than 0.3 g S/m^2 per year (10 kg SO_4^{2-}/ha) everywhere south of a line running from southeastern Manitoba across the southernmost part of the Hudson Bay to central Labrador and in excess of 0.6 g (20 kg SO_4^{2-}/ha) throughout most of southern Ontario, southern Quebec, and the Maritimes in Canada, and in virtually all of the United States east of the Mississippi. The spatial distribution of mean annual wet deposition of sulfate weighted by 1980 precipitation is shown in Figure 4-12. The average value for the eastern part of the continent was in that year less than 30 kg SO_4^{2-}/ha (below 1 g S/m^2) while Galloway and Whelpdale (1980) put the average wet deposition in the eastern United States at about 1.2 g SO_4^{2-}-S/m^2 per year for an annual total of 2.5 million t of sulfur for the whole region.

Regional values for dry deposition are calculated, in the absence of direct network measurements, by using representative air concentrations and typical deposition velocities (Sheih et al., 1979). Galloway and Whelpdale (1980) used mean SO_2 levels of 3.5 μg SO_2-S/m^3 and 2 μg SO_4^{2-}-S/m^3 for southeastern Canada and 7 and 2 μg for the eastern United States and average deposition velocities of 0.6 (SO_2) and 0.4 cm/sec (SO_4^{2-}) to calculate average annual dry deposition of about 1.6 g S/m^2 (16 kg/ha) in the United States east of the Mississippi and 0.92 g S/m^2 in Canada south of 49°N. The NAS (1983a) calculated average annual rates of 30–60 kg S/ha within an ellipse embracing Indiana, Ohio, Pennsylvania, and parts of adjacent states where dry inputs dominate the total acid deposition.

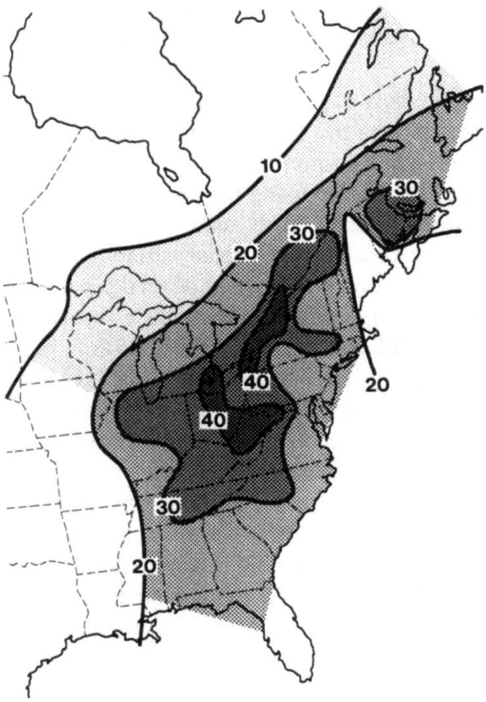

FIGURE 4.12. Annual wet sulfate deposition in eastern North America. Values in millimoles per square meter (1 mmole/m^2 = 0.961 kg/ha) are weighted by the amount of precipitation in 1980 (U.S./Canada Work Group, 1982).

There is considerable uncertainty in all of these values and Galloway and Whelpdale (1980) estimate that their annual dry deposition totals of 1.2 million t for eastern Canada and 3.3 million t S for the eastern United States may be off, respectively, by as much as ± 102 and $\pm 87\%$. Keeping these major uncertainties in mind, their mean values for wet and dry sulfur deposition result in wet/dry ratios of 1.22 for all of eastern North America (0.75 for the eastern United States and 2.5 for eastern Canada), and in an average total deposition of 1.6 g S/m^2 per year (2.7 g in the eastern United States, 1.1 g in eastern Canada).

This means that of eastern North America's total annual anthropogenic emissions of about 16 million t S (a value used by Galloway and Whelpdale), some 10 million t (63%) appears to be deposited over the land and a calculation of outflow to the Atlantic indicates a loss of another 4.3 million t. For the eastern United States, this loss, 3.9 million t, would be larger than either dry (3.3 million t) or wet (2.5 million t S) deposition. Still, this leaves a major gap between the inputs and outputs to the sulfur budget of the region and Galloway and Whelpdale (1980) believe that both deposition fluxes may be overestimated (wet more than the dry) while the eastward loss is underestimated. In any case, oceanward flux is significant and thanks to it at least one-quarter of anthropogenic emissions are not deposited on the eastern part of the continent.

Similar deposition rates, breakdowns of fluxes, and estimation uncertainties are found in Europe. Bolin and Persson's (1975) regional dispersion model for central and northwestern Europe shows wet deposition rates as high as 4 g S/m^2 annually, with most of the continent lying within a 0.5 g S/m^2 isoline; for dry deposition there are localized peaks in excess of 8 g S/m^2 and, once again, most of the region (except for northern Scandinavia) lies within a 0.5 g S/m^2 isoline. Figure 4.5 shows the estimated patterns for Europe but as the values are given, respectively, in grams of SO_2 and in grams of SO_4^{2-} for dry and wet deposition, they must first be multiplied by 0.5 and 0.33 to obtain the readings in grams of sulfur.

Combined anthropogenic deposition annually is thus between 4 and 12 g S/m^2 in parts of Germany, Czechoslovakia, Poland, and England and 1–2 g/m^2 nearly everywhere in the region. Subsequent calculations by Rodhe (1976) show total average dry and wet deposits over northwestern Europe (area of about 4 million km^2) at 2.4 g S/m^2 per year, by Garland (1977) at 2.25 g, while the OECD's (1977) figure for a 9 million km^2 area (virtually all of Europe) averages to 1.84 g S/m^2 annually. Fisher's (1978) improved deposition model for Europe shows the largest total annual values around 5 g S/m^2 near the main emission areas in central Europe with a gradual concentric decline; the isoline of 0.5 g S/m^2 per year embraces virtually all of continental Europe except for northernmost Scandinavia and the Iberian peninsula.

Comparison of three atmospheric sulfur budgets for Europe shows that only in the case of dry deposition is there a substantial agreement: Rodhe (1976) assigns 22–45% of all anthropogenic sulfur removal to it, Garland's (1977) figure is 45%, and Mészáros et al. (1978) estimate 43% of sulfur is so removed. For wet deposition the three values differ appreciably at, respectively, 32, 19, and 48%, leaving the wind loss fraction widely disparate, ranging from just 9% to as much as 46%.

Several generalizations are possible in spite of the surprisingly large errors inherent in the calculation of regional imissions. Typical total sulfur deposition rates in the most polluted industrialized regions of the Northern Hemisphere (areas of several hundred thousand square kilometers) are well in excess of 5 g S/m^2 per year (50 kg/ha) and several smaller areas (on the order of a few thousand square kilometers) receive considerably more than 10 g/m^2 per year. Large parts of both European and North American continents are affected by long-range transport of sulfur compounds and the total sulfur deposition thus averages 1–3 g/m^2 per year (10–30 kg/ha) for areas covering several million square kilometers. Numerous effects of this environmental interference will now be considered in detail in separate sections.

4.4. EFFECTS

Even a nonspecialist reading beyond the headlines of scientifically fashionable topics of the 1970s could not miss the recurring litanies of dire consequences brought by acid rain—fish are being "suffocated" in the "deadly" waters of southern Scandinavia's lakes or in "previously pristine" Adirondack streams; "breath of the machine age" has cut down visibilities over vast areas of North America, reducing crop yields in the Ohio Valley, wearing away European gothic cathedrals and the Parthenon (the Taj Mahal's turn is soon to come); as it moves across thousands of kilometers of oceans and continents, it acidifies precipitation in some of the world's remotest places, subverts ecosystem balances, accumulates in soils, and, not unexpectedly, endangers human health.

Enough evidence has been introduced in the preceding chapters to demonstrate that both closer looks at the facts as well as the numerous and far from negligible uncertainties besetting our understanding of the atmospheric part of the sulfur cycle and of the complexities of acid deposition make it imperative to write with less assurance, to use less dramatic adjectives, and to avoid the prophecies of doom ("Will the 21st century be silent?") which the 1970s discredited in other environmental affairs.

Of course, many effects of higher concentrations of anthropogenic sulfur in the atmosphere and of significant sulfur deposition on soils, waters, and ecosystems susceptible to acidification are undeniably deleterious and well documented; however, in many other cases, overdramatized speculation has taken the place of scientifically proven facts, and there are also numerous instances where higher sulfur inputs are thoroughly benign, even beneficial. All of these cases will be discussed in the following sections before moving to reviews and critiques of management options.

4.4.1. Atmospheric Effects

In most industrialized urban areas of the Western world, visible black smoke from factory boilers and household stoves has been eliminated by the combination of particulate emission controls and a switch to liquid and gaseous fuels, and the modern coal-fired power plants have, if properly run, highly efficient electric precipitators removing virtually all visible particulates. And yet the visibilities in many of these areas were reduced even further: the smaller black carbon particles and gross soot are gone but high SO_2 emissions remain to be oxidized to sulfates which are, beyond any doubt, very important in reducing visibility through light scattering and absorption, thus imparting the haziness of enclosed horizons not only to cities but seasonally also to large parts of rural countrysides in most industrialized countries—and, possibly, influencing large-scale radiation balances.

Atmospheric visibility is determined by the mass of fine (0.1–1.0 μm) suspended particulates and as the sulfates (and nitrates) are dominant contributors to these fine aerosols in both urban and nonurban air, it must be expected that if not always the majority then at least the largest share of the observed light scattering and extinction will be caused by SO_4^{2-}. Moreover, condensation nuclei associated with polluted atmospheres laden with large quantities of SO_4^{2-}, NO_3^-, and NH_4^+ provide more favorable conditions for generation of dense advection fog than condensation nuclei in clean atmospheres (Hung and Liaw, 1981)—and these fogs are, not surprisingly, highly acidic.

For example, measurements in Southern California repeatedly recorded fogs 100 times more acidic than the average rain in the area (Waldman et al., 1982): while in 1981 the volume-weighted means for precipitation at sites in Los Angeles showed ranges (all values are in μEq/liter) of 7–56 for SO_4^{2-}, 11–44 for NO_3^-, and 4–39 for H^+, the values recorded in the city's fogs were 62–5000 for SO_4^{2-}, 130–12,000 for NO_3^-, and 5.6–3020 for H^+. Precipitation pH was 4.4–5.4, whereas in fogs it dropped as low as 2.52.

An interesting difference between rain and fog is their inverted NO_3^-/SO_4^{2-} ratio: in fog water this measure is about 2.5 : 1, reflecting the relative NO_x–SO_x emissions in the region, but in rainwater the ratio is close to or less than 1. An obvious explanation is that fog water forms around the secondary aerosol from recent emissions. Clearly, the previously held belief that Southern California's fogs are relatively clean—an opinion understandable when relying only on measurements of gaseous air pollution—is untenable after these findings.

Fog with acidity ten times higher than rainwater has also been measured on Whiteface Mountain in the Adirondacks and similar differences would most likely be encountered in just about any location in the Northeast where such acidic fogs may be implicated in the decline of the red spruce which grows only at elevations frequently receiving fog and rime (condensed fog frozen onto the surface; see Section 4.4.3.4).

But considerable visibility reduction occurs in the presence of sulfate aerosols even in drier or very dry atmospheres. Pierson et al. (1980) concluded that 87% of the dry aerosol scattering coefficient variability at Allegheny Mountain was explainable by sulfate levels—but as their monitoring site was about 50 km downwind from the country's largest concentration of anthropogenic sulfur emissions, their reported SO_4^{2-} share might be seen as atypically high. However, the authors' conclusion was basically confirmed by thorough measurements on the western slope of the Blue Ridge Mountains (Shenandoah Valley, Virginia) conducted by Ferman et al. (1981).

During the study, no pollutant levels exceeded the national air quality standards (SO_2, NO_x, and hydrocarbons were all near background concentrations) but visibility was noticeably reduced with 92% of the total light extinction attributable to scattering by particles; sulfates were by far the most important, accounting for 55% of the fine aerosol mass and, together with associated water, for 78% of the light extinction (the rest was largely caused by crustal materials). Air mass trajectories showed that the upwind anthropogenic emissions were responsible for some four-fifths of the total light extinction.

Similarly, in drier regions, such as in California, sulfates are also the leading cause of reduced visibility (Tang et al., 1981). According to Trijonis's (1982) analysis of visibility–aerosol relationships based on 3 to 8 years of data collected at 12 California locations, sulfates account for about 40% of total extinction, with the rest equally split between nitrates and the remainder of the total suspended particles. Sulfates are even more important at southern sites (Los Angeles, San Diego), whereas nitrates lead in the San Joaquin Valley and the other aerosols are the greatest contributors in the Sacramento Valley. In Los Angeles, sulfates were found to be much more effective light scatterers per unit mass than other particulates and are estimated to be responsible for over half of all light scattering in the downtown area (Cass, 1979). Other confirmations of sulfates' key role in visibility reduction, based on strong correlations between ambient sulfate concentrations and visibility or light scattering measurements, are available for the United States Northwest, Midwest, and South as well as for western and northern Europe (Cass, 1979; Leaderer and Stolwijk, 1981).

Temporal changes in the visibility are even more difficult to pinpoint. To begin with, there is a choice of four basic spatiotemporal scales: diurnal mesoscale, synoptic, seasonal, and secular (Husar et al., 1981). The first level, changes during 1 day with air transfers up to 500 km, is too detailed and too variable (frequent intense haziness in the humid early morning hours) to evaluate any larger trends. On the synoptic scale, large parcels of hazy air accumulated over polluted regions and then transported elsewhere are investigated during 1 week and over distances up to 2000 km, and 2 PM extinction coefficients will be used to characterize a day's visibility. Again, a considerable variability must be expected. A recognizable pattern arrives with the shift to seasonal scale, perhaps the most meaningful frame for studying visibility as all the determining factors (emission strengths, atmospheric reactions, removal rates, transport, and properties of the aerosols) have marked seasonal patterns.

Comparisons of 1961–1966 and 1972–1975 turbidity measurements in the eastern United States show clear warm season haze maxima, with the lowest values between November and February and a pronounced peak in July and August (Husar et al., 1981). The most significant finding of this study, however, was that of a large increase of summer turbidity in a belt extending from Arkansas to Delaware. This deterioration was further confirmed by the secular trend of the mean extinction coefficient for 70 stations in the East in the 27 years between 1948 and 1974: there were drastic increases of this value during summer in the Carolinas, Ohio River Valley, Tennessee, and Kentucky with this region's visibilities declining from 24 km in the early 1950s to 13 km two decades later.

This, of course, is the region with the largest coal combustion increases during that period and although there are still nontrivial problems in relating the sulfur sources to ambient SO_2 and SO_4^{2-} levels and transports (see Sections 4.2.2.3 and 4.2.3.2), the coincidence of concentrated coal-fired power generation and reduced summer visibilities is telling. Or, as Sloane (1982) put it in a recent study of Mideastern visibility, the parallels are striking, but the correspondence is not complete as many fluctuations in the trends are undoubtedly caused by meteorological events. Sloane's (1982) analysis for the years 1948–1978 also reveals a bit more complex pattern than some previous findings: it identifies the years 1953–1955 and 1970–1972 as, respectively, the region's highest and lowest visibility periods and it clearly shows the summertime decline at all sites in the 1960s—but it also points to a leveling off of the warm season's haziness in the early 1970s, to an absence of persistent decline at all sites in the area, and to a significant improvement in the fall and winter quarter visibilities at the moderate-growth sites (both metropolitan and rural) in the 1970s.

These improvements were investigated in detail by Leaderer and Stolwijk's (1981) analysis for the New York region, based on 1948–1977 surface weather data at La Guardia Airport. Here the turnaround in annual medium visibility can be dated since 1968, the year when the strict emission control law was introduced and when the decline of average citywide SO_2 and SO_4^{2-} concentrations began. However, this annual visibility extension (from about 16 km in 1967 to some 24 km by 1977) was mainly the result of improved winter conditions; the summer values show hardly any change in spite of reduced levels of atmospheric sulfur, and the authors postulate aerosol transport from distant sources as the main explanation of this discrepancy. In contrast to the eastern United States, visibility in the Southwest remains high as demonstrated by a USEPA-sponsored study by Trijonis and Yuan (1978). Indeed, some of the region's remote locations offer visibilities approaching the value of blue-sky (Rayleigh) scattering—145–185 km.

Although natural baseline visibilities cannot be reasonably determined from the historical trends—thus making it impossible to assess any absolute secular declines—regional deterioration and improvements documented since the 1950s, together with theoretical expectation of SO_4^{2-} effects, make it clear that the reduction and elimination of sulfur emissions would be by far the most important ingredient in efforts to improve or to preserve visibilities. These efforts were made national goals by the Clean Air Act of 1977 (section 169S) in federally designated Class I areas, national parks and wilderness sites where no new major source should be located unless it can ensure that no significant deterioration of visibility would occur.

Yet, as stressed in the Sulfate Regional Experiment's conclusions (EPRI, 1981), reduction of only SO_2 emissions may not be sufficient: it is conceivable that the total conversion of SO_2 to sulfates is determined by the oxidant concentrations produced through photochemical reactions (rather than by the total SO_2 present) and because these processes are not dependent on SO_2 levels, reduction of SO_2 emissions alone may be ineffective in reducing sulfate levels. This remains, again, to be proved or refuted by future research. Another interesting aspect in need of further clarification is the fact that in the immediate vicinity of large sulfur pollution sources, the effects of sulfates (as well as nitrates) on visibility may be quite small. Hobbs and Hegg (1982) measured the mass distribution of SO_4^{2-} and NO_3^- in three power plant plumes out to distances of nearly 60 km (in the arid Southwest) and found that most of the aerosol mass was not contained in the optically-critical size range of 0.3–1.5 μm associated with the first scattering peak.

The last case of visibility reduction ascribable to sulfate to be described in this section is the most unexpected one and it has been systematically investigated only since the mid-1970s—the Arctic haze which blankets the huge circumpolar area during winter months, especially in the later winter. The first airborne measurements of the haze were taken in April and May 1976 over Point Barrow, Alaska and since then a network of ground sampling stations (in Alaska, Canadian Arctic, Greenland, Spitzbergen, and Scandinavia) and numerous aircraft traverses (including the North Pole overflight) have provided enough evidence to conclude that the haze is almost certainly of anthropogenic origin (Hileman, 1983).

The haze can be impressively thick, brownish or orange, extending as high as 10 km above all latitudes (continuous layer up to about 3 km), covering an area almost as large as North America and restricting horizontal visibility to 3–8 km (Rahn, 1981). The anthropogenic origin of a large part of the haze has been established by the presence of graphitic carbon particles (soot) whose particular structure can arise only during combustion, by enrichment of aerosol crust by vanadium (which originates mainly from combustion of residual fuel oils), and by sulfate concentrations too high to be accounted for solely by the sea salt or soil dust particles. The haze is rather aged as a long time is required to disperse it throughout the Arctic (distances involved are up to 5000 km) and as there is little precipitation during the Arctic winters to remove it from the air (dry deposition is also limited owing to strong temperature inversions).

Sources in central Russia, and, later during winter, in western Europe appear to be mostly responsible for the seasonal buildup which can, in terms of sulfate, reach concentrations up to an order of magnitude higher than elsewhere in remote regions of the Northern Hemisphere; however, more typical levels of Arctic haze sulfate derived from pollution are around 2 $\mu g/m^3$.

Not surprisingly, black soot in the haze absorbs appreciable amounts of radiation (at least 10%) but the climatic effects of this absorption remain uncertain: should the increased absorption lead to higher surface temperature, the effect would be indistinguishable from postulated CO_2-induced warming nor would the consequences of higher sulfate levels be easier to discern. As there are few alkaline ions to buffer the acid deposition, spring thaw may deliver relatively high acidity to the tundras—but with other stresses abounding in this extreme environment, determination of any effects on biota would be difficult.

The final note before leaving the subject of atmospheric effects concerns the possibility of sulfates' interference with large-scale radiation balances, a matter of uncertain calculations to be outlined here on the basis of Bolin and Charlson's (1976) paper. The authors start from the well-known facts concerning the effects of noncloud aerosols on incoming visibile radiation and from the correlations between sulfates and the total extinction coefficient and turbidity and then proceed to exclude clouds, infrared radiation, and all other aerosols to limit their calculations to cloud-free situations, visible wavelengths, and sulfates only. This choice removes any resemblance to reality and the authors admit that their quantitative estimates may be off by an order of magnitude.

This is easily seen from their final results which indicate that anthropogenic sulfate haze over Europe and the eastern United States would, without horizontal transport of energy in and out of these regions, cut their temperature through backscattering by 3–6 °C. Recognizing the impossible enormity of this value, they guess that the "actual" value would be somewhere between 0.1 and a few degrees Celsius if sulfate were the dominant substance controlling the backscattering.

Yet the SURE data (Section 4.2.2.2) show that sulfates in the eastern United States do not account for more than 30–50% of all fine particles and that 40–60% of the fine aerosol mass during increased haze periods in both summer and winter and in all parts of the studied region consisted of materials not associated with either sulfates or nitrates (EPRI, 1981). And, as the authors admit in closing, these other substances, as well as infrared radiation, could change their figures appreciably. Their intent—demonstration of the need to investigate the role of an individual class of substances in the earth's shortwave radiative climate—is commendable but their final conclusions may be misleading to casual readers who, unlike the experienced authors, do not realize the limits of their inquiry. After appreciating the deep uncertainties surrounding any surface temperature change calculations connected with CO_2 increase and discussed in Chapter 2, one has a strong urge to forego any quantifications of this kind! Anthropogenic sulfates in the troposphere should thus not be regarded as a looming cause of major climatic changes. Except for reduced visibilities, the effect of sulfur compounds on the biosphere is felt on the ground rather than in the air.

4.4.2. Weathering and Corrosion

Unlike most other degradative effects of air pollution, corrosion and weathering rates in polluted and clean atmospheres can be easily tested by comparative exposures of metals and building materials or by investigating the frequency and kind of maintenance required to protect or renew exposed surfaces. Often the differences between corrosion and weathering rates in polluted urban and industrial areas and in cleaner locations are immediately obvious: especially numerous are the cases of ancient statues or facades having deteriorated more during the past 50–100 years than during many previous centuries or even millenia.

Sulfur compounds damage buildings by reacting directly and irreversibly with their surfaces or by accelerating corrosion of iron and steel used as the buildings' structural supports. Three environmental factors aid these processes above all: high moisture, high temperature, and the cycles of freezing and thawing which cause cracks and spalling on stones. Concern about the effects of air pollutants on buildings has been greater in Europe with its riches of priceless old stone structures of which the marble ones are, obviously, most susceptible.

Their crumbling and black crusts are not caused by sulfur compounds alone—nitrates and organic particles contribute their share—but $CaCO_3$ reactions with SO_2 and SO_4^{2-} are responsible for most of the damage. A small fraction of the resulting gypsum ($CaSO_4 \cdot 2H_2O$) penetrates into the intergranular spaces (more porous marbles, such as the famous Carrara stone from northern Italy, thus have thicker weathering zones, up to 4 mm), while most of the gypsum is either deposited (in crusts in protected areas underneath cornices and domes where continued weathering behind the encrustations causes more serious damage than in open areas) or washed away.

And, unfortunately, this attack can continue without reaching any new chemical equilibrium: in interesting experiments, Braun and Wilson (1970) showed that weathered limestone surfaces not only from Marlow Bridge (built 1831) but also from a 15th century building of Eton College could remove as much SO_2 as fresh cut surfaces of the same block. Clearly, some limestones, if washed sufficiently by rain removing the $CaSO_4$ in solution, can continue to absorb sulfur from the air indefinitely—while others may become temporarily saturated rather rapidly.

While thousands of susceptible marble or marble-clad buildings in polluted cities around the world were deteriorating with little notice, the damage done by air pollutants to ancient monuments became a subject of international concern. And no case is more famous than the accelerating deterioration of the Acropolis, one of the most memorable sites of the Western world. Under normal circumstances, its Pentelic marble—dense, almost pure $CaCO_3$ with low porosity and fine structure—weathers slowly and little damage of this kind was done to the stones until the last century. In fact, most of the damage occurred very recently, since 1945 with rapid industrialization of the area and shift to high-sulfur fuel oils.

Most of the household heating in this city of 3.5 million is done with fuel oil containing up to 1% sulfur and until outlawed in the late 1970s, heavy residual oil with as much as 3.5% sulfur was also commonly burned; combined with substantial SO_2 releases from industrial zones near Piraeus harbor and at Eleusis Bay and aggravated by the city's basin location, these emissions have been leaving increasingly destructive marks on the Acropolis (Yocom, 1979). Physical damage occurs as surface changes from the marble's reactions with high levels of sulfur compounds, and as cracking of stones caused by corrosion and reinforcing steel pins and beams installed in the Parthenon, Erechtheum, and Propylaeum during the first decades of this century in order to slow down the structural deterioration (as with $CaSO_4{\cdot}2H_2O$, the resulting Fe_2O_3 occupies more space than the parent material and cracking ensues).

These now largely useless steel reinforcements can be removed and replaced by titanium pins but surface preservation remains an unsolved challenge, especially as the working group for the Acropolis is against the use of synthetic resin coatings until more is known about their effects and properties. The best form of protection is, of course, a stringent control of Athens's severe air pollution problem—but little progress can be seen as yet (Lalas et al., 1982).

As the prevention is immeasurably more effective than costly and still often unsuccessful protection and as our knowledge of marble decay is sufficient to make predictions of susceptibility, it is most unfortunate that one of the best preserved monuments of an Eastern civilization has been put into some jeopardy as were the numerous structures in Europe: the Taj Mahal is situated just 30 km from a new refinery in Agra whose emissions of 25–30 t/SO_2 per day are carried in the mausoleum's direction between October and March by the prevailing northwesterlies (Gauri and Holdren, 1981). As with the temples of the Acropolis, the Taj Mahal's surface deterioration will be accompanied by rusting of iron bars which were used copiously throughout the building to attach marble blocks to the structural framework and to join the adjacent blocks.

There are currently no proven methods for preserving stone structures containing $CaSO_4 \cdot 2H_2O$ in the weathering zone. For statues the best treatment is transfer to a museum's air-conditioned halls; for buildings, only a drastic lowering of sulfur compound concentrations in ambient air will do.

Investigation of damages done to economically important materials, above all metals and paints, has been a long-standing and frequent part of air pollution research. Consequently, there is no shortage of detailed analyses as well as broad surveys in this mature field and, as far as the specific effects of sulfur compounds are concerned, surveys by Kucera (1976), Sereda (1977), Nriagu (1978b), the OECD (1981), and the Economic Commission for Europe (1981) offer comprehensive reviews and numerous citations.

Most of the material damage caused by higher SO_2 and SO_4^{2-} atmospheric levels occurs in urban areas with their vast numbers of cars, metallic and painted surfaces and so it is not surprising that the secular decrease in SO_2 in North American and European cities (see Section 4.2.1.2) has reduced the corrosion and deterioration problems. What is surprising is the rapidity and magnitude of this decline exemplified by Gillette's (1975) estimate that in the United States between 1968 and 1972 the cost of SO_2-associated deterioration dropped from $(1975) 900 million annually to less than $100 million. And the OECD (1981) has estimated that a 37% reduction of the 1974 western European SO_2 emission total would save nearly $(1980) 1 billion in corrosion repair cost.

Needless to say, such estimates are based on a large number of aggregated assumptions and they should be seen, at best, as order of magnitude estimates. The tenuousness of all of these estimates can best be shown by their comparisons, although the figures are usually nationwide totals for damages caused to all materials by all air pollutants rather than values disaggregated according to the compounds responsible [for comprehensive U.S. accounts see, among others, Salmon (1970) and Waddell (1974)]. Whatever the actual cost, there is no doubt that the annual global total is in billions of dollars each year.

The most serious damage done by sulfur compounds is certainly in corroding many common metals, and numerous corrosion tests repeatedly made in polluted and relatively clean atmospheres in the United States, Canada, Japan, and many European countries have produced very similar results: unprotected carbon steel, galvanized steel, nickel-plated steel, nickel, and zinc all corrode faster in urban and industrial environments containing higher levels of SO_2 and sulfates than in cleaner rural areas. In contrast, copper is much less affected and aluminum and its alloys have high weathering resistance and, moreover, their corrosion rate declines with time.

The mechanisms by which SO_2 promotes atmospheric corrosion are complex and not fully understood but there is no doubt either about the accelerating effect the gas has on the corrosion rate or about the close correlation of steel and zinc deterioration with ambient SO_2 concentrations as several reports have concluded that SO_2 levels are the major factor controlling (linearly) the rate of zinc corrosion. Similarly, sulfates, including the most common $(NH_4)_2SO_4$, strongly stimulate metal corrosion. The corrosive action of sulfur compounds can be considerably potentiated by specific environmental conditions and no other among these variables is as important in determining the rate of deterioration as moisture.

For most common metals, the critical humidities above which the corrosion rates increase rapidly in the presence of SO_2 are between 55 and 75% and the total damage is best appraised by the extent of time during which these threshold humidities are exceeded. This measure is known as time of wetness; its precise definition is the interval during which the potential developed between platinum and zinc under normal atmospheric conditions exceeds 0.2 V (Sereda, 1977) but the value is applicable to other metals as well. Time of wetness varies greatly from year to year, typically ranging between 25 and 50% in the Northern Hemisphere's industrialized latitudes.

Effects of precipitation are equivocal—it may reduce corrosion rates by dilution and washout of corrosive material, or enhance them by continuous stripping of protective coatings. Acid precipitation, however, clearly promotes corrosion, especially in metals protected by a layer of basic carbonates, sulfates, or oxides. Temperature, in the range encountered in densely inhabited polluted areas, has a very small effect on corrosion rates.

Acid precipitation can also accelerate the already serious losses caused by underground corrosion: as a major ion in soil solutions, SO_4^{2-} is responsible for much damage done to buried concrete structures (0.1% of SO_4^{2-} in groundwater is enough to initiate the degradation) and to various metallic pipes; moreover, sulfate-reducing bacteria cause corrosion losses to iron pipes estimated globally to be on the order of billions of dollars per year.

Damages done by sulfur compounds to metals are rivaled only by those done to paints. These deteriorative effects have long been known: discoloration of lead-containing paints by H_2S, sulfur-induced deterioration of paint pigment, fungicides, and binders, increased drying time and loss of gloss caused by SO_2, crystalline sulfate blooms and complete destruction of paint layer on lime plaster surfaces have been common occurrences everywhere in polluted areas. In Europe, rapid deterioration of painted glass windows has also been quite common. Compared to metal corrosion and paint deterioration, other damages ascribable to sulfur compounds (degradation of fibers, elastomers, plastics) are of much smaller extent.

Economically important as they are, overall damages done by sulfur compounds to materials have become of relatively lesser concern with the decreased sulfur pollution levels in cities. In contrast, whereas acid deposition does not cause greatly accelerated material damage, its cumulative effects may eventually alter the structure and function of ecosystems, bringing changes far more difficult to remedy than replacing corroded pipes.

4.4.3. Acid Deposition and Ecosystems

While sulfates, once formed from SO_2, will lower visibility both near and (after long-range transport) far away from the emission areas, and while common materials such as marble or steel will be damaged by sulfur compounds in any location which experiences higher sulfur releases, there is no automatic link between acid deposition and observable environmental degradation. The substrate is the key: precipitation would have to be of sustained acidity much beyond the currently monitored levels to cause any unwelcome changes in waters, soils, and plants on calcareous ground.

Fortunately, large areas of North America and Europe have this critical property, calcareous, sulfur-deficient soils which neutralize acid inputs and make the ecosystems resilient even where very high deposition rates are encountered. As already noted (Section 4.3.1.4), the Great Plains area of the United States and the Canadian Prairies, as well as large areas of central Europe, also have an abundant presence of alkaline ions, making these extensive regions relatively immune to degradative acidification. In contrast, areas with bedrock highly resistant to chemical weathering produce acidic soils (with pH as low as 4 in the surface horizons, pH 5.5 deeper in the profile) and shelter water bodies which have very low alkalinities and meager calcium reserves. The natural neutralizing ability of such ecosystems is greatly limited and their low resilience is relatively easily broken by higher inputs of the dissociated strong acids. The resulting acidification of soils may then bring changes of primary productivity and the dropping pH of fresh waters can profoundly alter the makeup of aquatic biota.

Among these regions extremely susceptible to acid deposition are parts of southern and central Norway and Sweden where the phenomenon of "dying" lakes was first brought to worldwide attention and large parts of the Canadian Shield extending across southern Canada from the Manitoba–Ontario border to Newfoundland. In the United States the Adirondack State Park in northern New York is certainly the best known example of such a hypersensitive region although smaller areas of acidic bedrock are scattered throughout the eastern United States, especially in parts of Appalachia. I will start the survey of acid-induced ecosystemic degradation with the aquatic environment as its changes are usually the first to be noticed in susceptible areas.

4.4.3.1. Acidification of Waters Reliable long-term measurements are necessary to discern this trend—unless the changes have been relatively rapid as in the case of Swedish rivers where regular monthly examination of pH in 15 basins began in 1965 and a clear declining trend was evident a mere 5 years later—the pH of streams and lake outlets in southern Sweden had dropped by 8–23%/year. This, together with comparison of the pH values collected for some lakes over periods of 10 to 40 years and with testing of the pH in about 1000 locations in 1965 and again in 1970, constituted clear evidence of a widespread acidification of Swedish fresh waters, a key conclusion of the first comprehensive study on acid phenomena (Royal Ministry for Foreign Affairs, 1971).

Of Sweden's 85,000 lakes exceeding 1 ha, more than 18,000 now have pH values less than 4.5—and 4000 were acidified so seriously that their ecosystems suffered grave damage (Swedish Ministry of Agriculture, 1982). In southern Norway, lakes covering a total of more than 13,000 km^2 contain no fish and lakes with reduced fish stocks cover another 20,000 km^2. In North America some rivers in Nova Scotia have pH less than 5.0, too acidic to support salmon or trout reproduction; Ontario has nearly 50,000 lakes susceptible to acidification and hundreds of such lakes in Quebec and New York (Adirondack region) have shown acid stress in the form of diminished fish counts. Measurements in 1929–1937 showed that most of the 320 lakes investigated in the Adirondacks had pH values between 6.0 and 7.5, whereas in 1975 pH measurements of 217 lakes yielded a bimodal distribution with half of the lakes (51%) below pH 5.0 and the rest between pH 6.0 and 7.5; in the 1930s only 4% of all lakes had pH values below 5.0 (Schofield, 1976).

Not surprisingly, this rather easily observable damage to aquatic ecosystems has become perhaps the most intensively studied part of acid-induced environmental degradation not only in Sweden and Norway but also in the United States and Canada. In all of these countries, fishless lakes became the most publicized symbols of acid rain and the threat of "deadly waters" was taken up eagerly by the mass media. Still, the key to understanding the phenomenon is the realization that the acidity of fresh waters arises not simply from the acidity of precipitation but no less from the inability of some watersheds and lakes to neutralize the acid inputs.

The natural chemistry of affected waters determines the response to acid deposition. The alkalinity of each lake is supplied primarily by its watershed, and natural waters on calcareous substrate have alkalinities of 1–3 mEq/liter while those on poorly buffered rocks (such as most Canadian Shield lakes) have alkalinities of a mere 0.1 mEq/liter. In very well-buffered lakes, acids will be continuously neutralized; in well-buffered lakes, inputs of acids will lower the alkalinity but not to the point of rapid pH declines: for example, in 51 such lakes in southernmost Sweden, the pH decreased only 0.015 unit/year between 1935 and 1974 (Wright and Gjessing, 1976). As the alkalinity decreases to below 600 μEq/liter, the lakes become potentially susceptible to acidification and the danger becomes pronounced below 300 μEq/liter (Glass et al., 1979). Continuing acid deposits may then almost deplete bicarbonate buffer in the long run and shift the acidity below pH 5.5.

In soft and poorly buffered waters, the pH can drop rapidly below 5.0, first just periodically during spring when the acids accumulated in snow packs are rapidly released in meltwaters; later, when bicarbonate is essentially eliminated from the water, such low pH levels can become permanent. In Sweden the threshold of acidification for such sensitive lakes appears to be reached when the atmospheric acids depress average precipitation pH to 4.6 and below. The acidified waters have pH between 4.5 and 5.0 and sulfates in them replace bicarbonate as the major anion.

Consequently, sulfate loading of lake waters is a telling measure of acidification and Dickson's (1975) data show well the much more rapid response of very sensitive soft-water lakes where the sudden pH drop with the still relatively low sulfate loadings is especially notable. For a region with fairly constant SO_4^{2-}

loading, the pH response of lakes may thus be seen as a series of curves whose shape is controlled not only by bedrock composition but also by other environmental factors.

Foremost among these are the extent and volume of surface runoff: high precipitation and impermeable ground are obviously not conducive to deep penetration into the ground where contact with neutralizing minerals would lower the acidity of precipitation. The small sizes of many lakes and short turnover times of their water worsen this situation as there is little chance of diluting a sudden influx of acidic water, especially during spring meltdowns. Similarly, a small catchment area will offer less opportunity for precipitation to be neutralized as it travels through the soils, and thus will contribute in rapid acid surges. And if a lake is located at the water basin head where soils will be thinner and catchment travel very short, the risks are even more acute.

Consequently, lakes in the same region receiving the same acidity loading per unit of catchment and surface area will react in a very "individualistic" way and one can find heavily acidified small lakes surrounded by steep rocky slopes literally a proverbial stone's throw from much less affected larger water bodies in a shallower setting and with slower turnovers: no simple generalizations will apply.

Low pH and high sulfate concentrations are not the only negative changes in acidified waters. Once the pH reaches below 5.0, humus substances and aluminum start buffering against further acidification. While in moderately acid waters aluminum ions function as an acid and their reaction with water releases hydrogen ions

$$Al^{3+} + 3H_2O \rightarrow AL(OH)_3 + 3H^+$$

with continued acidification the aluminum hydroxide dissolves, neutralizes further pH declines—and puts large amounts of Al^{3+} into the water

$$Al(OH_3) + 3H^+ \rightarrow Al^{3+} + 3H_2O$$

Aluminum ions are strongly toxic to many aquatic organisms and Al^{3+} poisoning has been increasingly identified as the real cause of mass fish deaths in highly acid lakes. Acidified waters also have abnormally high concentrations of heavy metals (mercury, manganese, zinc, lead, copper, cadmium, nickel), very low phosphorus content (caused by the attachment of phosphate ions to the released aluminum), and often surprisingly high nitrate levels as nitrogen cannot be assimilated owing to phosphorus shortages. Comparisons of the chemical composition of unaffected and acidified soft-water lakes can be generalized by describing the first ones as calcium–magnesium–bicarbonate waters, the others as hydrogen–calcium–magnesium–sulfate waters with unusually high concentrations of several metals—and the numbers of the latter have undoubtedly been increasing in sensitive areas downwind from large anthropogenic sources.

In general, our knowledge of biotic degradation caused by aquatic acidification is both much more extensive and a great deal more specific than our understanding of terrestrial impacts. Fine reviews of the ecosystemic consequences of acidified waters are available, among others, in Hultberg and Grahn (1975), Muniz (1981), and NRC (1981) and the main effects will be described here in some detail.

Interference with nutrient cycling through changes of environment is a key degradative shift caused by aquatic acidification. Field studies and laboratory experiments clearly demonstrate slower rates of organic decay leading to abnormal accumulations of debris on lake bottoms (thus both preventing organically bound nutrients from being mineralized and sealing off the mineral sediments from the overlying waters), decreased bacterial activity, and inhibition of specific biochemical processes leading, among other changes, to smaller counts of ciliates and zooflagellates (Hendrey et al., 1976). Inevitably, reduced availability of nutrients affecting both microorganisms and invertebrates has, in turn, an effect on the lakes' amphibian, fish, and bird communities.

Phytoplankton species start disappearing relatively rapidly at pH below 6, with the greatest compositional changes in the range between pH 6.0 and 5.0. The total phytoplankton biomass in acid lakes with pH near 4.5 can shrink to as little as one-ninth compared to that at pH 6.5. As for the total lake phytomass, Canadian sampling through the euphotic zone showed that lowering of pH from 7.0 to 4.0 does not result in reductions of standing phytoplankton (NRC, 1981). Canadian phytoplankton productivity measurements also show no significant reduction in spite of the low concentrations of dissolved inorganic carbon. However, there is a clear alteration of community structure: while in nonacidic oligotrophic lakes of eastern Canada chrysophytes are the main genera, acidic lakes are usually dominated by dinoflagellates.

Experimental lake acidification and concurrent ecological appraisals carried on for several years by Schindler and co-workers [Schindler et al. (1980) is just one of many reports] in Ontario provide a wealth of information on a number of dramatic changes taking place in waters long before the degradation progresses to pH values that cause high fish mortality. In terms of macrophytic changes, oligotrophication of many Scandinavian lakes has been characterized by regression of original communities (dominated by *Lobelia* and *Isoetes*) and by massive invasion of *Sphagnum* and filamentous algae. Acidophilic and photophilic *Sphagnum* outgrows the flowering plants and its ability to withdraw calcium and other ions from solution reduces their availability to other biota. Besides these heavy moss carpets, epiphytic and epilithic algae (*Mougeotia* and *Batrachospermum* above all) invade the benthos not only in lakes but also in acidified streams. However, massive *Sphagnum* invasions were not found in acidified Canadian Shield lakes (NRC, 1981).

In soft-water streams, acidification increases standing crops of benthic algae and shifts their composition toward acid-tolerant filamentous species such as *Mougeotia* but productivity may remain unaffected (Burton et al., 1982). Effects on composition, health, and survivability of fish stocks have been certainly by far the most studied topic of aquatic acidification as the complete disappearance of many species is the most dramatic consequence of that ecosystemic degradation. Fish studies, as other acute manifestations of acidification, come overwhelmingly from

Scandinavia, Canada and Northeastern United States, number in hundreds and range from irritation of gills to acclimation of larvae, from changes in predation patterns of perch to recruitment failure in trout (NRC, 1981).

Changes preceding the eventual complete loss of individual species or total fish populations include poor recruitment, failure of females to reach spawning condition, very high mortalities of eggs and larvae, fin deformations, and shortening of gill covers. Total loss of fish species was first noted in southern Norway, where brown trout (*Salmo trutta*) and Atlantic salmon (*Salmo solar*) had been declining since the early decades of this century. In the southernmost region of the country (Sørlandet), two-thirds of all lakes now have pH below 5.0 and two-fifths of this number have no fish and the annual rate of loss continues to increase. In Sweden the western coast region has been affected the most, with extinctions of char (*Salvelinus alpinus*) and roach (*Leuciscus rutilus*). Smelter emissions at Sudbury (Canada) eliminated fish from 20% of surveyed lakes in the La Cloche Mountains in Ontario by 1973 and the 1980 total for Ontario was 140 fishless lakes; extinct species include lake trout (*Salvelinus namaycush*), smallmouth bass (*Micropterus dolomiec*), and lake whitefish (*Coregonus clupeaformis*).

Total extinction is preceded by gradually (sometimes rapidly) shrinking population size: for example, nearly 20,000 m of gill nets set overnight in an acidified lake near Sudbury yielded only 14 fish. Population structure is also altered drastically with fish younger than 1 year least often present in highly acidified lakes (pH 4.0–4.5), a clear sign of recruitment failure because in lakes with high pH (6.0–6.5) they would be most abundant. In some acidified lakes, older fish also succumb readily.

Negative effects on growth rates were noted among the earliest indicators of lake acidification in Sweden. However, some surviving species in acid waters can grow faster as intraspecific and interspecific competition for food declines. Changes in growth rates and maximum sizes are also determined by different interspecific tolerances which may be substantial and are often influenced by the developmental stage. The increased sensitivity of eggs and larval fish (the most susceptible life stage) is shown by greater egg and fry mortality and delayed hatching.

As for the physiological basis of fish kills in acidified waters, the earlier explanation (coagulation film anoxia) was discarded in favor of disruption of normal ionic and acid–base balance. Experiments show that the gill membranes are highly permeable to hydrogen ions and that low pH leads first to cessation of sodium influx, then to massive losses of plasma sodium and chloride, as well as to inhibition of RNA synthesis. Metal toxicity is no less important as low pH favors the formation of methylmercury and, above all, leads to mobilization of aluminum which is the prime cause of fish deaths in numerous lakes (Grahn, 1980). Aluminum ions irritate the gills, destroy protective mucus, erode gill filaments, and thus bring on eventual suffocation.

Amphibians can account for a significant share of an ecosystem's zoomass (more than either small mammals or birds) and hence their decline and loss must have widespread consequences throughout the food chain. As they often reproduce in temporary pools or in shallow waters which can be highly acidified during the spring snowmelts, eggs and tadpoles of many anurans and salamanders are especially at risk. There is a relatively rich literature documenting declines of frogs (*Rana temporaria*), toads (*Bufo bufo*), spring peepers (*Hyla crucifer*), and salamanders (*Ambystoma* spp.) in acid lakes and ponds of Scandinavia, the United States, and Canada (NRC, 1981).

Higher lake acidities bring lower diversity of rotifers and reduced biomasses of zooplankton; among invertebrates, gastropods and Ephemeroptera appear to be especially sensitive. The number of freshwater molluscs are also reduced but their abundance correlates better with calcium content or alkalinity rather than with pH. Among benthos organisms, higher acidity reduces mass and activity of both microbial and invertebrate decomposers and invertebrate community compositions shift toward air-breathing carnivores. Generalizations on acidification to about pH 4.5 by key benthic biota are as follows: bacterial growth rates inhibited, fungi replace bacteria as leading decomposers, crustacean survival equivocal, molluscs eliminated, insect mass and diversity down with Ephemeroptera and Plecoptera easiest eliminated while Coleoptera and Hemiptera are most tolerant (Grahn et al., 1974; Singer, 1982).

The effect on waterfowl is mixed: fish-eating birds will have to devote more energy to securing their food in acidifying lakes with declining fish variety or, in the case of fishless lakes, leave these habitats; on the other hand, benthic organisms become more abundant in fishless lakes and for ducks there are also more insects to eat in the absence of competition from fish. The most negative effect on water birds in the long run may be the accumulation of heavy metals.

Finally, acid deposition will change not only the pH of some lakes and streams but also the acidity of groundwaters in susceptible areas. In Sweden where much of the bedrock is hard and dense and moraines are almost impermeable, water cannot penetrate very deep and marked changes of hardness (doubling of Ca^{2-}) and alkalinity (halving of HCO_3^+) have been recorded in many localities (Swedish Ministry of Agriculture, 1982). But more worrisome than increasing hardness and higher sulfate loading is the dissolution of metals out of the ground and out of piping systems as the rates of metal releases rise sharply with acidity lowered below pH 4.

Of the greatest practical concern is the accelerated corrosion of copper pipes which raises the metal's content in drinking water up to 45 mg/liter when hot water is left overnight in the pipes—the WHO recommdendation's 1.5 mg Cu/liter as the maximum permissible level. Besides obvious health risks, tableware and clothes will after laundering become discolored by a greenish tint and so-called "erosion corrosion" phenomenon shortens the lifetime of the pipes. In municipal water supplies, water can be easily treated to maintain neutral pH so that the elevated metal content in drinking water is mainly a risk in small, private catchments.

Rather expensive alkaline filters can be installed in such circumstances but they do not remove aluminum easily. Being the third most abundant element in the crust (after oxygen and silicon with 8.13% of the total), aluminum is commonly found in soils, waters, and foodstuffs. There are no standards for its presence in drinking water; studies have shown daily ingestion ranging between 1 and 30 mg Al/day and other research indicates that the total amount of aluminum in the body is homeostatically controlled within the range of usual intake and the element should be seen as an essential metallic micronutrient (Sorenson, 1977).

The potential worry centers on the metal's possible role in Alzheimer's disease (neurofibrillary degeneration) affecting millions of old people worldwide. Studies of the molecular interaction of aluminum and the central nervous system (CNS) in animals have documented pathological changes similar to those in Alzheimer's disease. Yet these studies have been done with injections of aluminum compounds into the CNS, not a good model of actual human exposures; on the other hand, senile persons have markedly higher blood and brain levels of aluminum. This opens up a variety of interpretations, ranging from aluminum as the etiologic agent to a deficit of other essential metals in the brain which allows aluminum accumulation. Although uncertainties preclude making any clear links, Roberts (1982) stated in a recent review of therapies that all "known extra sources of entry of this metal [Al] should be eliminated insofar as is feasible."

Effects of aquatic acidification thus change from readily observable physical and biotic changes as the maximum depth of lake visibility rises as much as fourfold (up to 20 m) and various species disappear while a few thrive, to uncertain consequences of elevated metal levels in drinking water. The only practical control method in the presence of continuing acid deposition is lake and stream liming and its usefulness and prospects will be addressed in Section 4.5.5. And liming has been, of course, a time-honored practice to keep naturally acid soils near neutrality and will have to be used more frequently to reduce the rates of soil acidification in susceptible sites.

4.4.3.2. Acidification of Soils This is perhaps an even more complex phenomenon than the acidification of waters. Soils are complex, heterogeneous systems consisting of varied proportions of solid, liquid, gaseous, and colloidal components in which mineral weathering, organic decay, precipitation, leaching, and, in the case of crops, also fertilization and repeated tillage combine often in synergistic or antagonistic ways, to bring about constant minute changes in chemical and physical composition, processes whose greatly altered states may not be evident for decades, very often for centuries.

Soils can absorb SO_2 directly from the atmosphere and the rate of removal increases with higher moisture in the air or in the soil. Typically, dry soils absorb 1–15 mg SO_2/g while soils with 50% water-holding capacity will absorb up to 70 mg SO_2/g; a rough estimate gives absorption of SO_2 by soil (in kg S/ha) as 0.55 times the atmospheric concentration (in $\mu g/m^3$) (Prince and Ross, 1972). Abeles et al. (1971) put these absorption capacities much higher, calculating that some 40 billion t SO_2 can be absorbed annually by United States soils, a total three orders of magnitude higher than the country's overall SO_2 emissions.

But even in places far from sources of sulfur or nitrogen oxides or remote from recurrent air flows carrying relatively high loadings of sulfate and nitrates, there is no shortage of sources of hydrogen ions (Bache, 1978; Ulrich, 1978). To begin with, respiration of soil flora and fauna (tiny in individual size but large in total mass per unit of land) raises CO_2 concentration of soil air up to more than 1% (0.03% is, of course, the ambient mean) and this amount of gas dissolved in soil water gives a pH of 4.9 and total acidity on the order of 1 kEq/ha per year:

$$CO_2 + H_2O \rightarrow H^+ + HCO_3^- \rightarrow 2H^+ + CO_3^{2-}$$

Nitrification of ammonium ions released during natural biomass decay or added as fertilizers produces substantial quantities of hydrogen ions:

$$2NH_4^+ + 4O_2 \rightarrow 4H^+ + 2NO_3^- + 2H_2O$$

Each 100 kg of nitrogen added as ammonium salts contributes about 14 kEq H^+, whereas unfertilized soils nitrify 20–100 kg N/ha per year, or some 1.5–7 kEq. Organic acids from decomposing biomass may be a negligible source but they may add up to 5 kEq/ha, and oxidation of H_2S and FeS_2 in certain soils, and nitric acid deposited during rainstorms may be additional minor sources of acidity. Not surprisingly, any effects that the additions of smaller quantities of man-made sulfur compounds from atmospheric deposition may have will not be easy to separate and to quantify.

General changes of acidification, regardless of the source, are not difficult to predict with our knowledge of soil behavior. There will be an inevitable loss of base cations, above all Ca^{2+}, by leaching of free carbonates from the solution (which will act as an efficient buffer propping up pH above 6.5 until they are completely dissolved), and by displacement of Ca^{2+} from weak acid or from permanent-charge exchange sites and reduction in cation-exchange capacity; no less important, mobilization of aluminum ions will turn acid mineral soils into aluminum soils in which Al^{3+} levels may be orders of magnitude higher than the concentrations predicted for a soil solution dominated by carbonic acid and in equilibrium with amorphous aluminum and iron hydrous oxides.

Toxic Al^{3+} ions as well as liberated cations of heavy metals may have deleterious effects on many soil organisms. Acidification also dissolves and mobilizes fulvic acid, and hence leads to substantial losses of organic matter in those soils where the fulvic acid molecules are its major components, and slows various soil microbial processes (N fixation, nitrification, biomass decay). Detailed descriptions and analyses of these changes can be found in, among many others, Pearson and Adams (1967), Baes and Mesmer (1976), Bache (1978), Schnitzer (1978), Ulrich (1978), Wiklander (1978), Wainwright (1980), and Sposito et al. (1980).

Experiments examining the effect of acidified water treatment on soils showed, as expected, a clear increase in the leaching of metallic cations which increased considerably with decreasing pH of the solution, and field studies in Sweden and West Germany indicated a large throughfall enrichment of sulfur from the acid precipitation filtered through spruce canopies and increased calcium leaching from forest soils (Malmer, 1976; Ulrich, 1978). Inhibition of nitrification was confirmed both by laboratory experiments and by some field sampling, and mycorrhizal associations in Ore Mountain (Czechoslovakia) soils with heavy sulfur deposition were found to be abnormal. On the other hand, some studies of acidification's effect on soil microorganisms have been inconclusive (Wainwright, 1980).

An increasing number of studies have looked at the effects of acid precipitation on forest soils—Cronan (1980) gives a summary of more than a dozen completed by 1980—and found that with the shift of the historic leaching regime from organic and carbonic acids to H_2SO_4, nutrient cation losses were accelerated and aluminum mobilization and outputs of aluminum were almost always elevated.

But the response is far from uniform. In some Swedish studies, only a negligible effect was noted on aluminum leaching in organic-rich forest soils and Cronan (1980) found the same result in his treatments of intact hardwood and coniferous forest floors in New Hampshire. He subjected these to simulated throughfall of 3.5 cm at pH 5.7 and pH 4.0 every week during a 3-month period and after that sampled lysimeter percolates to measure changes in ionic outputs. As expected, acid precipitation depressed the percolate pH and raised significantly outputs of H^+ (by about half), Ca^{2+} (by 40–60%), Mg^{2+} (by 25–65%), and SO_4^{2-} (by 60–100%)—but there were no statistically significant increases in percolate outflows of K^+, Na^+, and Al^{3+}.

Some of the postulated soil chemistry changes were confirmed by a fairly realistic long-range experiment done in Corvallis, Oregon, with simulated acid rain applied to model forests containing either sugar maple (*Acer saccharum*) or red alder (*Alnus rubra*). Lee and Weber (1980) sprayed control plots with "pure" rain of pH 5.6 while on the test plots nozzles applied 2.8 mm/hr (for maple) or 3.7 mm/hr (alder) of acid rain (pH 3.0, 3.5, and 4.0) for 3 hr a day, 3 days a week throughout the year.

For the first 6 months, sulfate absorption by the soil prevented any differences between the control plots and the test patch receiving the most acid rain; afterwards, soil sulfate concentrations started to increase until, after 3 years, they were about equal to SO_4^{2-} levels in the rain. For pH 3.5 and 4.0 the trend was similar, except that the onset of soil acidification started 1 and 2 years after experimental spraying. Higher sulfate concentrations were accompanied by higher Ca and Mg levels and lowered pH throughout the top 20 cm of the soil (no changes were found at a depth of 1 m).

Great gaps in our understanding do exist and these may not soon be filled as long time spans are needed to observe the cumulative effects, to trace their origins or to exclude the processes masking the changes, and to consider properly all complexities involved in what should never be seen as a simple anthropogenic interference. The best, balanced evaluation of the complexities characterizing soil acidification is that of Krug and Frink (1983). Their main argument is that natural soil formation is often more important than acid deposition in determining the acidity of soils (and hence waters) and their analyses, falling into five broad groups, deserve to be summarized here in some detail.

First, they call attention to a commonly overlooked acidity of "normal" rain:

an annual precipitation of 1 m/ha with pH 5.6 can dissolve (depending on the value of thermodynamic constants) 400–500 kg $CaCO_3$/ha—while rain with pH 4.6 dissolves an additional 10–12 kg of leached or neutralized limestone. Such additional releases are small compared to common limestone applications which run between 2500 and 5000 kg $CaCO_3$/ha on acid farm soils of North America.

These conclusions are supported by Henderson et al. (1980) who used values for H^+ production following urea fertilization together with observed cation discharge to estimate potential increases in the release of cations from soils of experimental watersheds affected by higher precipitation acidity. Their calculations show potential Ca, Mg, K, and Na increments to be less than 0.5 kg/ha at pH 4.0 and less than 2.5 kg/ha at pH 3.5; such increases would be too small to be detectable among natural variation of cation leaching.

A second, and rather extensive, argument concerns soil formation processes. As soils mature, they become more acid than their parent material—regardless whether the bedrock was granite or limestone. Chemical oxidation accounts for a smaller part of this shift as the process is largely biotic and decidedly gradual (on the order of thousands of years) owing to the capacities of the parent material to neutralize the acids. Consequently, strong acidification of mineral soil is usually restricted to the upper profile and acid deposition, even should it last for a century, would not change this vertical extent.

A third group of corrections concerns acidification reactions and it opens by questioning the equivalence of increases in H^+ in acid rain (ΔH) and those in leached nutrient cations (ΔM). If, as commonly accepted, $\Delta M / \Delta H = 1$, then strongly acid soils would be rapidly acidified owing to their low nutrient content—while in reality, removal of nutrient cations by H^+ in soils more acid than pH 5 is sluggish; besides, the proportion of alkaline/acid ions in "acid rain" is commonly equal or greater than their share in strongly acid soils exposed to the rain!

Another misinterpretation results from neglecting the buffering mechanism of organic acids. Standard accounts hold that hydrogen ions in rain remain in solution as they move through the soil accompanied by sulfate and thus most of the H_2SO_4 in precipitation causes higher cation leaching proportional to rising sulfate flux. But Krug and Frink (1983) note that the strongly acid soils in the watersheds most sensitive to acid rain do not have a chemistry of simple ion exchange.

Humic acids richly present in these soils are made less soluble by added hydrogen ions and so the influx of H_2SO_4 promotes sulfate outflow but lowers the flux of organic anions and thus there is little or no measureable pH change. This explains the puzzling phenomenon as to why acid rain does not speed up cation leaching in very acid soils at rates matching those of wet deposition—and why in such "high-risk" areas as the northeastern United States, measurements fail to show any excessive chemical weathering attributable to acid rain.

A fourth new perspective deals with leaching of aluminum ions. As naturally acid soils have large amounts of soluble aluminum, there appears to be no reason to involve acid rain as the key mobilizing factor: any rain will do. Furthermore, the claim that Al^{3+} leached from organic-rich soils is dissolved by acid rain because it is complexed by organic anions has a limited validity as organic complexation, evident at soil pH between 4.5 and 5.0, drops drastically with further pH declines, as does the solubility and mobility of the complexed aluminum.

Finally, changing land use can have soil acidifying effects as strong as intensive acid deposition and Krug and Frink (1983) gathered many references to document these shifts. The best North American examples come from the mountainous areas of the northeastern United States where acid rain is commonly thought to be *the* cause of much described environmental degradation—yet repeated measurements have shown that it is the recovery of the region's forests from previous disturbances (burning, overcutting, overgrazing, severe erosion) which has resulted in increasingly acid surface soil horizons and in thickening and acidification of forest floors. Naturally, water moving through such soil will become more acid and leaching of Al^{3+} may also increase.

But the most impressive data on the acidifying influence of recovering forests come from Norway where the exchange acidity of surface humus under a 90-year-old Norwegian spruce forest growing on the site of an abandoned farm is equivalent to strong acids in about 1000 years of annual precipitation of 1 m/year with pH 4.3. These changes are the principal cause of acidifying runoff, and rapid declines of soil acidity in regrowing spruce–fir forests of northern England support this explanation.

Consequently, Krug and Frink's (1983) conclusions are unequivocal: additional acidity introduced by rain contaminated by anthropogenic emissions is a small fraction of normal rain's acidity; natural biotic soil formation processes are strongly acidifying but the effect is limited to the upper horizon; the standard hypothesis that increased deposition of acid and sulfate is a cause of equivalent leaching and acidification is unsupportable both on theoretical grounds and by field studies; Al^{3+} leaching is a result of much more complex, site-specific processes than the common generalizations imply; and ecosystems which were disturbed in the past are undergoing soil formation processes which result in the greatest increases in natural soil acidity.

Before closing this section on acidification of soil, I will take up one more key factor in greater detail: assessments of potential soil sensitivity to acid deposition, an essential piece of information for appraising the extent of possible future impacts. As the chemical and physical properties of soils vary so widely, it is only natural that their susceptibility to acidification will also. Both theoretical expectations and experimental studies lead to the conclusion that acid deposition should be most damaging in those soil types which are classified between brown earths and podzols (Malmer, 1976). Acidification of soil is the essential prerequisite of podzolization and while under natural conditions the time needed for development of a podzol may range from a few hundred to several thousand years, this period could be shortened by heavy anthropogenic acid deposition (Petersen, 1978). But the time spans involved are obviously such that no actual observations of this accelerated process have been possible during the short history of acidification research. Whatever the degradative rate might be, there is little doubt that noncalcareous sandy soils with pH above 5.0 are potentially most susceptible to acidification.

On the other hand, percolating water or smaller acid inputs should have much less pronounced leaching effects in soils already very acid, although even small nutrient losses might be detrimental in such cases. And the soils most resistant to acidification are naturally those with high pH and base saturation as their high calcium content lends them high buffer capacities. Consequently, it is not too difficult to come up with reasonably representative divisions of soils into acidification sensitivity categories and to single out the areas of the maximum expected impact as well as those where acid deposition is of little consequence, at least as far as soil properties are concerned.

Calcareous soils of central Europe, whose lime content is around 20% and whose thickness is up to several meters, are well buffered against the high acid deposition prevalent in the region, and so throughout most (by no means all) of Czechoslovakia, Hungary, Austria, and southern Germany, soil acidification will continue to be naturally neutralized. North of this resilient region, however, are the poor, sandy soils of northern Germany, Denmark, and Poland which are very susceptible to acidification, while the Scandinavian soils are a mixture of the sensitive and the resilient ones (Swedish Ministry of Agriculture, 1982).

But it is the North American situation that I will focus on to illustrate the surprisingly high degree of irresponsibility in disseminating clearly false information. One can frequently encounter simple outline maps of the continent or of the United States showing sensitivity of soils to acidification: Figure 4-13 is such an example with a large area in the eastern United States classified as "very sensitive."

Yet by the time these maps were published, relatively detailed studies appraising soil sensitivity to acid deposition were available for all states east of the Mis-

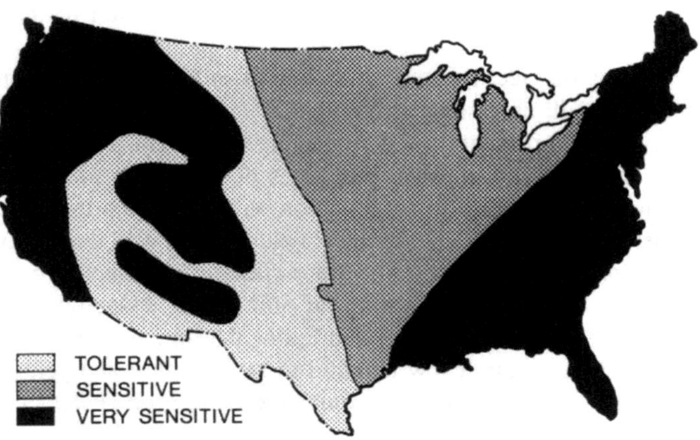

FIGURE 4-13. Susceptibility of soils to acidification in the United States (Sub-committee on Acid Rain, 1981).

sissippi and for Canada south of 52°N and east of 96°W and these convey very different images. One study of the United States (McFee, 1980) classifies sensitivity according to cation-exchange capacity and a 25-year span was chosen arbitrarily so that if the maximum probable annual acid input (1 m of precipitation with pH 3.7) during that period exceeded 25% of the cation-exchange capacity in the top 25 cm of soil (i.e., when the exchange capacity was less than 6.2 mEq/100 g), such soils were considered sensitive. Soils with cation-exchange capacity between 6.2 and 15.4 mEq/100 g were classified as slightly sensitive, while all soils having cation-exchange capacity in excess of 15.4 mEq/100 g, as well as all soils repeatedly flooded and all calcareous soils were classified as nonsensitive.

Figure 4-14 generalizes the study's findings. Wisconsin, Michigan, Illinois, Indiana, and Ohio have overwhelmingly well-buffered, nonsensitive soils; only in northern parts of Wisconsin and Michigan do there occur larger areas of coarse glacial deposits which might be slightly sensitive. The Mississippi Delta and virtually all of the central Southeast are also nonsensitive. The soils most susceptible to

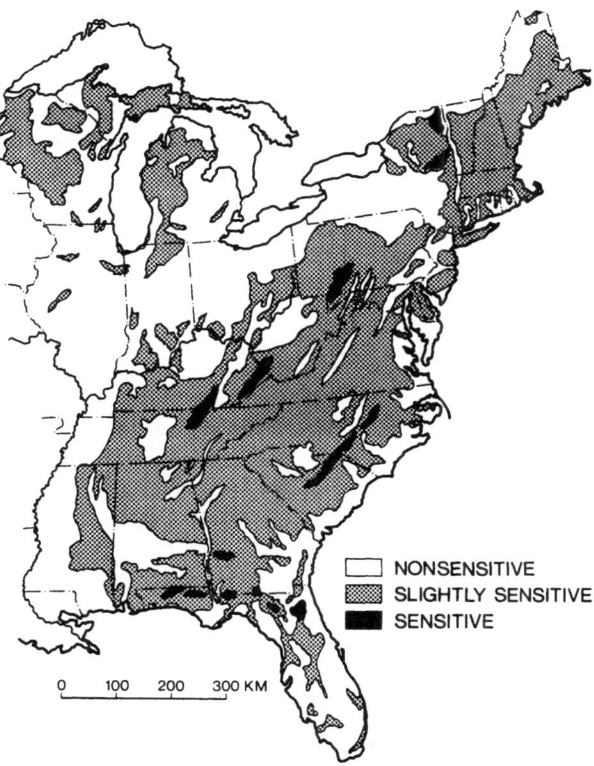

FIGURE 4-14. Susceptibility of soils to acidification in the eastern United States according to McFee's (1980) calculations.

long-term effects of acidification are some highly weathered Southeastern formations, shallow sloping soils of the Appalachian Highlands and the Adirondacks and coarse hills of New England.

The conclusion is clear: the total area of highly sensitive soils in the eastern United States is relatively small. Moreover, McFee (1980) offers 27 individual state maps which give detailed appraisal with the smallest mapped areas being on the order of 50 km^2, and an enlargement of the Adirondacks region, perhaps the most sensitive area of the eastern United States, shows that even there the pattern is complex and only rather small patches of the land are in the most sensitive category (Figure 4-15).

Similarly, in Ontario, where the conventional impression assigns most of the northwestern part of the province to the highly acidification-prone category owing to its granitic bedrock, comparison of bedrock pattern with that of calcareous glacial drift shows that most of the region is not in the extremely sensitive class as surficial sediments offer considerable buffering capacity. Simplified maps portraying a huge part of North America as supersensitive to acidification are thus dispelled as grossly misleading.

And yet another consideration weakens the claim of looming acidification damages to a large portion of these soils: even for much of the land shown as slightly sensitive in McFee's (1980) map (Figure 4-15), this label is only of theoretical interest, since, as would be shown by overlapping the sensitivity map with a land use map, in much of these soils under cultivation, and in nearly all such cases, effects of fertiliz-

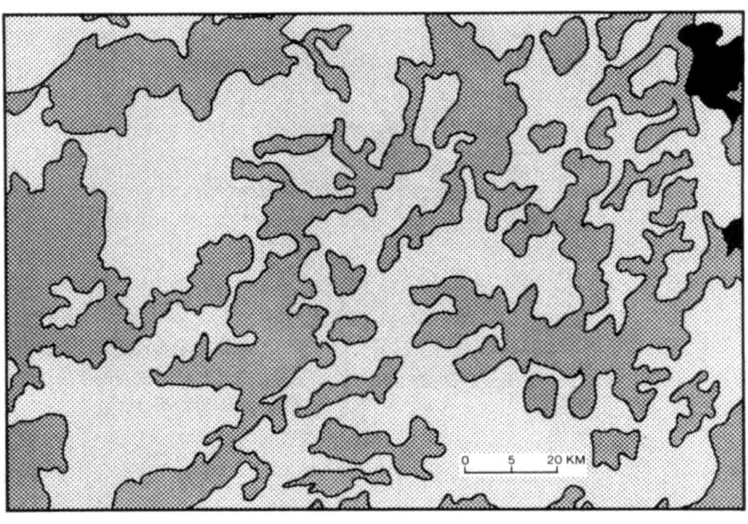

FIGURE 4-15. Susceptibility of soils to acidification in the Adirondacks region of New York State (McFee, 1980).

ing will greatly overshadow any possible consequences of acid deposition. As noted earlier in this section, 100 kg of nitrogen applied in ammonium compounds (just about average fertilizing rate per hectare throughout most of the intensively farmed regions of the Northern Hemisphere) would contribute more than 13 kEq H^+ annually—while the inputs by acid precipitation would rarely go above 1 kEq H^+/ha per year!

To offset this acidification accompanying high-rate application of nitrogen fertilizers, various liming materials (crushed limestone, burned lime, marl, hydrated lime) are regularly added to soils with low calcium content—and liming would thus be a simple and very cheap remedy to counteract any additional acidification caused by deposition of pollutants. Compared to acid contributions from nitrogen fertilization, almost any conceivable acid deposition on cultivated land is much smaller so that only fractions of today's limestone application would be needed to correct the possible deterioration: current annual applications of all liming materials in the United States average 200 kg per harvested ha of crops and on more acid soils in the Northeast they surpass 1000 kg/ha.

Standard practice has been to proceed with booster applications of liming materials after an interval of several years but increasing use of ammoniacal fertilizers may mean rapid pH declines in sensitive soils if liming materials are not added more frequently. Swedish experience illustrates quite well the extent of acidifying effects of nitrogenous fertilizers on a national level: up to the mid-1960s, fertilizer provided an alkaline increment worth up to 50,000 t of CaO annually, whereas during the 1970s they caused increased demand for lime now equaling 100,000 t CaO each year and forcing more frequent liming.

Comparisons with fertilizer-induced acidification in agricultural soils, widespread existence of adequate buffering capacities over large regions of the Northern Hemisphere (including many areas erroneously portrayed as very sensitive), possibility of maintaining the original pH through inexpensive liming (which, with greater expense, could be extended also to forests), general considerations of natural resilience of most soils to all but huge and rapid extraneous inputs, very long time spans needed to bring unwelcome changes, and dominance of natural acidification process during soil formation and as a part of vegetation succession—all of these considerations make it necessary to assess the problem of soil acidification as a relatively minor one.

Of course, it would be naive to maintain that no significant damage can be done to certain sensitive soils even within a short time (i.e., several decades), but then those effects will be fairly localized and can be counteracted. On any scale of threats and action priorities based on the best available scientific evidence, rather than on catastrophic speculations, acidification of soils ranks not only far behind the economic consequences caused by acidification of sensitive lakes but also behind the possible effects of acid deposition on crops and, above all, forests.

4.4.3.3. Effects on Crops That a strong acid will cause injury to living tissues is self-evident. For example, Wedding et al. (1979) produced severe lesions on 4- to 6-week-old soybean plants by spraying them with 10% sulfuric acid mist droplets or with heavy doses of a 1% solution. But such damage may not arise in other trying circumstances: when Wedding et al. exposed the plants to 95% H_2SO_4 aerosols (aerodynamic diameter 1.7 μm) for 2 weeks (loading up to 10^7 particles per leaf after 140 hr of exposure), no visible toxicity symptoms were found even after scanning the leaves by electron microscopy.

On the other hand, dry deposition of sulfur compounds and the direct effects of sulfur dioxide may seriously damage a wide range of plants as proven by numerous experiments and frequent field observations even during the early decades of the systematic scientific interest in air pollution-related phenomena. However, as stressed in the opening of this chapter, suflur is also among the essential structural elements of living matter and while its excess may be injurious or lethal, its severe deficiency or total lack can have effects no less profound.

Sulfur requirements and metabolism in plants are relatively well-researched topics, largely because of their close association with nitrogen's role in plant nutrition. The extremes for critical sulfur content (minimum concentration associated with maximum yield of dry matter) among the crops range from about 0.04% or less for sugarcane to 0.26% for white clover and this rather wide variability is almost solely determined by the amount of sulfur in protein (Thompson et al., 1970). As the incorporation of sulfur-containing amino acids (cysteine and methionine) into proteins is genetically controlled, N/S ratios of plant proteins are relatively constant.

Consequently, the N/S ratio in the phytomass is a better measure of sulfur requirements than the element's absolute level and the ratio has been established at approximately 15 : 1 for many different species grown under a wide variety of conditions (Dijkshoorn and van Wijk, 1967). There is some variation, from 17 : 1 for legumes to 11 : 1 for corn tops, but this spread is much smaller than the range of absolute sulfur content. This closeness of nitrogen and sulfur requirements can be further illustrated by experiments establishing that the nitrogen content of sulfur-deficient alfalfa is increased by sulfur fertilization or that nitrogen utilization in wheat is largely dependent on the sulfur sufficiency (Thompson et al., 1970).

Moreover, the symptoms of sulfur deficiency in plants closely resemble those of nitrogen deficiency. Thresholds of sulfur deficiency, however, vary widely: over 3000 analyses of plant foliage collected in southern Ontario during the 1970s showed that most evergreen trees and some deciduous species (e.g., birch) are low sulfur accumulators (0.13–0.15% S in the foliage), while most deciduous trees and shrubs contain medium amounts (0.16–0.26%) of leaf sulfur; only herbaceous crops (including corn, wheat, barley, tomato) and some deciduous trees and shrubs (poplars, dogwoods) accumulate over 0.26% S in their foliage (Linzon et al., 1979).

Thus, most annual farm crops will require sulfur inputs of 10–50 kg/ha per year. For example, corn, by far the most important crop in North America, will need 25–35 kg S for good yields of 5.5–6.5 t/ha, all *Brassica* species (from kohlrabi to cauliflower), sugar beets, and onions will need relatively more, while potatoes and cereals will require less, as will most of the tree crops or natural forests. Rough global calculation assuming annual net production of 100 billion t of dry phytomass containing, on the average, 0.2% S would result in incorporation of some 200 million t S in plant tissues, a total very similar to the global atmospheric flux of the element.

Indeed, many plants take a large share of their sulfur needs directly from the air: for example, Noggle (1980) used a radioactive tracer ($^{35}SO_4^{2-}$) to establish that cotton grows 3 and 16 km from a coal-fired power plant derived 44 and 26% of its total sulfur tissue content directly from the atmosphere. However, most of sulfur used by plants comes from soil sulfates, whose concentrations range widely in different types of soil and, in some areas, may be insufficient to support heavy cropping, a shortage often deepened by the increasing use of sulfur-free fertilizers (Coleman, 1966).

But fields, forests, or shrublands, especially those in more humid areas closer or adjacent to man-made sulfur emission sources, should not be expected to be sulfur deficient. In many of these cases the opposite is true as the plants are exposed either to occasional acute SO_2 fumigation episodes or to chronic lower SO_2 and SO_4^{2-} levels—or both. Research into the effects of gaseous pollutants on plants is a mature branch of air pollution studies with numerous fine summaries (Mudd and Kozlowski, 1975; Kozlowski, 1980; W. Smith, 1981).

The effects of SO_2 and the distinct acute plant injuries it induces are thus well known. The relative susceptibilities of common trees, shrubs, and crops differ appreciably but when injury occurs it follows the same pattern. In broad-leaved plants, the gas causes collapse of the spongy mesophyll and lower epidermis cells, then the distortion of chloroplasts and palisade cells and the collapse of the upper epidermis.

As the vascular tissues are injured least, the symptoms typically appear as bleached brownish, yellowish, or whitish spots, blotches, or infills of dead tissue between the veins or on the margins of leaves. For gymnosperms there is less of a distinct response but larger SO_2 doses cause an initial brown tip burn of needles which later spreads downward. There are well-known differences in species resistance: whereas alfalfa, barley, beans, clove, soybeans, and wheat among field crops and apples, birches, elms, larches, pears, and some firs, pines, and spruces among trees are sensitive to SO_2, celery, cherry and citrus trees, corn, grand fir, onions, potatoes, silver and sugar maple and white spruce are quite tolerant (Rennie and Halstead, 1977).

Chronic exposure of sensitive species causes advancing chlorosis and interferes with many metabolic processes, resulting in declines of nonstructural carbohydrates and proteins in most plant tissues, inhibition of leaf growth, and reduced primary production. Actual reductions in yields of important field crops chronically exposed to SO_2 may be locally significant but generally do not appear disturbing even in areas of rather high sulfur emissions. The best published case on this point is a study estimating crop yield losses from ambient air pollutants in the Ohio River Valley, the area with the highest concentration of sulfurous emissions in the eastern United States (Loucks and Armentano, 1982). The authors used a variety of experimental and field evidence to estimate typical loss coefficients for chronic exposure to SO_2 concentrations now prevailing in the valley: expressed as a percentage of yield reduction, these values ranged from 1 to 9% for soybeans, with 3–6% being most probable, 0–3% for corn, and 1–6% for wheat (the lowest values are for SO_2 concentrations up to 0.286 mg/m^3, the highest ones for readings above this level).

In comparison, ozone can reduce soybean yields by up to 26%, corn harvests by 15%, and wheat yields by 10% so that the potential yield increases from abatement of current oxidant levels in the valley might total nearly 7 million t for the three crops—whereas reduction of SO_2 levels would result in combined "clean air" yield gains of only less than 100,000 t for the three crops.

Our understanding of the effects of wet acid deposition on crop productivity is also increasing. In experiments with crops grown to harvest in field chambers and exposed to a weekly total of 30 mm of simulated H_2SO_4 rain with pH 3.0, 3.5, or 4.0 (and to a control rain of pH 5.6), yield production was lowered for five (radish, beet, carrot, mustard green, broccoli), stimulated for six (tomato, green pepper, strawberry, alfalfa, orchard grass, timothy hay), ambiguously affected potato, and was not altered in any statistically significant degree in 15 plant species (Glass et al., 1982). These experiments make possible the following generalizations: yield inhibition is limited to dicotyledons, legumes and fruits are more frequently stimulated by acid rain, and grain crops are the least sensitive, altogether a picture of surprising resistance (see also Lee, 1982).

The effects of chronic SO_2 pollution on natural grasslands may also be relatively minor. As the grasslands of the northern Great Plains overlie nearly two-fifths of America's minable coal reserves which might be converted to electricity in large mine-mouth power plants, Heitschmidt et al. (1978) performed a pioneering study of the possible consequences of such developments by continuously fumigating swards dominated by western wheatgrass (*Agropyron smithii*) at two Montana sites with SO_2 concentrations between 50 and 200 μg/m^3 for two growing seasons.

They found that SO_2 stimulated leaf growth by increasing the number of leaves per plant but did not significantly alter the net above-ground production, growth rate, and assimilation rate, and that leaf area ratios for both the western wheatgrass and the entire community remained unaffected. These findings are highly suggestive of sulfur deficiency in the studied grassland, a state which may be rather common to many drier ecosystems.

While annual crops and grasses may not be greatly affected, chronic exposure of sensitive fruit trees may be damaging as shown by studies in South Carolina orchards in sandy soils where pH, soils, and soil Al^{3+} are significantly correlated with each other and with Al concentration in peach tissue and the higher Al levels are linked to greater tree mortality (Jones and Suarez, 1980). Needless to say, degradative effects must also be expected in major forest species and such changes may lead eventually to profound alteration of whole ecosystems; however, determining cause–effect links in such cases is not easy.

4.4.3.4. EFFECTS ON FORESTS Severe injuries are, of course, unmistakable: their best examples are seen in studies done in the vicinities of intensive point sources of the gas, mainly ore smelters (Miller and McBride, 1975; Linzon et al., 1979). The common findings in surviving trees include generally much higher levels of foliar sulfur (up to more than four times than in trees growing in unpolluted air), moderate to severe leaf injury, and stunted growth; in the downwind areas most heavily polluted by SO_2, nearly all tree cover eventually disappears and is replaced by more resistant shrubs and herbs and these severe effects have been traced for up to several tens of kilometers from large point sources of SO_2.

Such severe damage—blots of destroyed vegetation downwind from large Canadian smelters could easily be picked up on LANDSAT images scanned from its altitude of 950 km—is not, however, a consequence of acid deposition from sources hundreds or thousands of kilometers distant from the recipient areas. In an overwhelming majority of these cases, no, or at worst very few, foliar injuries will be visible but as the plants will be accumulating sulfur in excess of quantities that would normally be taken up from the soil, their metabolism may eventually be affected and their primary productivity may decline.

Here I have to tackle perhaps the most controversial effect of long-range acid deposition: does it lower the productivity of common trees and will it lead to eventual destruction of some large forest areas on sensitive soils in heavily affected regions? No simple answers are possible as will be made clear by the following reviews of accumulated, and still unsatisfactory, evidence.

Scandinavian investigations begun in the late 1960s and extensive Norwegian experiments with artificial acid (H_2SO_4) rain found no decline in coniferous tree growth and the principal investigator in this research believes that "natural" acid precipitation, containing significant amounts of NH_4^+ and NO_3^-, will most likely have a positive, fertilizing effect on Scandinavian and other forests where nitrogen is the main limiting nutrient (Abrahamsen, 1980). Similar fertilizing effect would occur in forests on sulfur-deficient soils and only in those cases where magnesium and possibly calcium and potassium are limiting nutrients would the acid precipitation be a likely cause of reduced forest productivity.

Experimental studies on tree growth in acidified plots begun in Sweden in 1969 found no statistically significant differences with respect to nonacidified plots. In fact, a slight positive growth effect was noted for some tree seedlings, credited to nitrogen fertilization as HNO_3 was a constituent of the simulated acid rain (Tamm, 1976).

On the other hand, a Swedish study of the trend of forest growth based on examination of the annual rings in pine and spruce in southern Sweden suggested an annual growth decline of 0.3% during the years 1950–1965 within those areas which were seen as susceptible to acidification. A thorough update of this early 1970s' study compared again the years 1911–1950 with those of 1950–1965 and found the same declining growth rates (Jonsson and Svensson, 1982). However, addition of another 9 years to the more recent period (1950–1974) resulted in a positive growth trend during that period; and the phenomenon is more pronounced on better soils.

This exercise is a perfect illustration of how the length of the observation period influences the outcomes in any complex, naturally fluctuating process: at the end of the mid-1960s a clear downturn, a decade later an unmistakable rise and altogether a clear warning to all students of forest growth trends to hold their judgments for some decades before proclaiming any causal links.

The Scandinavian experience is best characterized by the summary of the Norwegian SNSF project: "decrease in forest growth due to acid deposits has not been demonstrated. The increased nitrogen supply often associated with acid precipitation may have a positive growth effect. This does not exclude, however, the possibility that adverse influences may be developing over time in the more susceptible forest ecosystems" (Drabløs and Tollan, 1980).

The North American experience is similarly inconclusive. Whereas Tomlinson (1983) concluded that there are growing indications that widespread dieback and decline of both European and North American forests are caused by airborne sulfurous pollutants, Johnson and Siccama (1983) maintained that the available evidence does not show any clear cause-and-effect relationship between acid deposition and forest decline and dieback in the United States. Their evaluation is worthy of a more detailed look as it points out well the complexities of the forest growth decline and tree dieback phenomenon.

That the appreciable dieback and large reductions in basal area and density in high-elevation red spruce (*Picea rubens*) forests of New York, Vermont, and New Hampshire have been ascribed to acid deposition is not surprising in view of the fact that the northeastern United States has become a textbook example of a principal "target" area of long-range acid deposition.

Comparisons between the mid-1960s and the early 1980s (1964–1982) for Whiteface Mountain (Adirondacks) show red spruce diameters declining by over 70% for both the largest and very small trees (over 10 and below 2 cm of breast height diameter) but diameters of younger firs increasing by up to 50% and those of birches dramatically expanding by some 250%. Consequently, long-term effects of these changes on total phytomass may be negligible although the species' composition will be altered appreciably. But are these shifts caused by acid deposition?

While it is certain that these montane forests do receive relatively high rates of acid deposition from rainfall as well as from clouds, and that the soils from which they grow are typically very acid (pH 3.5–4.5), looking for simple causes has proved unrewarding. A direct effect of gaseous SO_2 must be discounted as the concentrations in forested regions of the northeastern United States are an order of magnitude below the levels causing observable foliar symptoms with chronic exposure. The hypothesis of aluminum toxicity, suggested by Ulrich to explain dieback of spruce and beech in West Germany, does not fit American realities: declining spruces in Vermont and New Hampshire were found to have low Al concentrations in their fine roots whereas high levels were recorded in healthy root systems.

But acid deposition is far from being the only stress on these precariously placed ecosystems. Current decline can be a purely natural phenomenon initiated or sustained by drought alone: historical data document a wave of such diebacks during the last quarter of the 19th century. Or it can be a combination of drought and subsequent infection or drought, infection, and acid deposition-induced stresses.

Drought remains the best documented trigger of many historic forest declines and the tree ring records taken from red spruce cores in both the northern and southern Appalachians show the onset of a dramatic decrease in increment size in the mid-1960s, following an extraordinary dry spell at the beginning of that decade. As a delayed response to drought stress is well known, the triggering stress in this case appears to be well accounted for.

And so the best explanation Johnson and Siccama (1983) offer is that spruce dieback and decline are a multiple stress syndrome in which drought followed by infection by secondary organisms appears to be most responsible, with acid deposition enhancing the drought stress or vice versa. With both Scandinavian and North American evidence inconclusive, the only area for which there are now claims of unequivocal links between acid deposition and forest decline is central Europe and since the early 1980s West German studies describing this effect have received great attention at home and abroad.

As noted previously (Section 4.1.4.3), central Europe receives very high rates of acid deposition (commonly up to 60 kg S and 30 kg N/ha per year, in heavily industrialized regions twice these values), and it has been appreciated for several decades that the elevated areas exposed to winds carrying high concentrations of acid compounds from nearby sources—regions such as the Ore Mountains between Bohemia and Saxony, the Harz Mountains between East and West Germany, and many still extensively forested parts of West Germany from the Schwarzwald to the Westerwald—are at especially high risk.

Although damages were noted in many areas, only since the 1960s, when burning of brown coals for electricity has much increased in East and West Germany, Czechoslovakia, and Poland, has the areal extent of growth decline and frequency of dieback reached worrisome levels—and acid deposition was found to be the fundamental cause in the official West German report issued jointly by the Federal Ministry of Food, Agriculture and Forestry, the Federal Ministry of Interior, and the Committee for Protection from Emissions (Schütt et al., 1983).

The study's survey, conducted in summer 1981, included 60% of the country's forested area: its results were that 7.7% of West German forests are damaged, 75% of them slightly (loss of 10–25% of needles or leaves, some crown thinning), 19% moderately (25–50% needle/leaf loss, strong crown thinning), and only 6% or 35,400 ha severely (needle/leaf loss over 50%, greatly damaged or dead crowns). While the overall loss may seem low, species-specific damage is much worse, with some 60% of the country's silver fir (*Abies alba*) trees severely affected and 9% of all spruce (*Picea excelsa*) growth. Pine, beeches, and oaks are the other most afflicted species.

The report considered all possible causes such as drought, frost, pest infestations as well as poor silvicultural practices and concluded that while all these stresses have recently been present, none could account for damage of such extent and severity and that only heavy dry and wet acid deposition emerges as a common factor in all the widely distributed damaged forest areas. Besides high acid deposition rates in the affected forests, high sulfur content in needles and the fact that the heaviest damage occurs on western slopes or at other sites exposed to wind, rain, and clouds are cited as the telling signs of the injury process.

Differences between susceptibility of coniferous and deciduous species may be explained to a large extent by the much higher capacity of evergreens to filter the pollutants. Mayer and Ulrich's (1978) measurements in a beech and spruce forest near Göttingen show, respectively, ranges of 47–51 and 80–86 kg S/ha per year for bare soil deposition in the same area. Canopies of coniferous species can also retain much more acid deposition than seasonally bare broadleaves and hence there is a greater chance for erosion of cuticle wax and necrosis which may open sites for attack by insects and pathogens.

The whole syndrome described in the 1982 West German report is well familiar to me from my work in the North Bohemian Brown Coal Region during the 1960s when the needle loss, crown thinning, and eventual dieback among the firs and spruces—especially at the highest elevations of the Ore Mountains exposed to heavy winds, snow, and rain—were unmistakably advancing and were in a clear contrast to apparently unaffected or just slightly affected broadleaves.

Updated inventories published in 1983 indicated that as much as 34% of all West German forests may be afflicted by *Waldsterben* but in spite of naming acid deposition as the fundamental cause, the reality, I am certain, is more complex. The 1982 report itself admits that environmental factors play a contributory, if not causal, role with, not surprisingly, recurrent droughts of 1967, 1968, 1969, 1971, 1973, and 1976 and damaging frosts of 1978, 1979, and 1981 causing undoubtedly a good deal of reduced growth. Soil degradation from coniferous monocultures is undoubtedly another contributing factor as is the known relatively high deposition of heavy metals and increasing presence of photochemical smog. Of course, we have no way to untangle individual contributions of this combined stress but we can be fairly certain that degradation is not a simple consequence of high acid deposition: even the German report acknowledges as much.

Is it not, then, that the common insistence of holding acid deposition to be *the* cause of forest degradation is merely a result of our desire to do something about the damage? "Whichever cause is labelled primary, the fact remains that drought, frost and other acts of nature will continue. Pollution is the only factor that can readily be controlled" (Roberts, 1983). Hence, the new West German goal is to cut SO_2 emissions from large power plants by half within 10 years. How effective this move will be nobody can predict; moreover, as long as the Czechs and East Germans do not cooperate, some of the most heavily affected border areas will get little relief.

Clearly, acid deposition and forest health will remain controversial for a long time to come and there are many reasons for the persistence of fundamental uncertainties. The most obvious is the detection of a relatively low-level stress of acid deposition amid the natural fluctuations of forest ecosystems. To unequivocally detect acid precipitation-induced changes in forest ecosystems above the wide variations of climate and pest attacks would require acidity inputs greatly surpassing any current and conceivable future levels.

Moreover, the inherent complexities of forest ecosystems make it difficult to appraise the importance and consequences of some clearly demonstrable changes and degradations affecting certain segments of the intricate whole. Much of the concern is with the growth trends of boreal forests but these ecosystems are well adapted to grow on highly acid and nutrient-poor soils and it is obvious that acid deposition could make much difference in soils whose pH is already around 3.5–4.5 but only after a considerable period of time.

Experimental studies will provide few meaningful conclusions. While neither the Scandinavian nor American experiments with forest plots exposed to simulated acid rain uncovered any growth inhibition—even when the treatments included applications of solutions far more acid than commonly encountered in affected ecosystems—the relatively short duration of the tests (between 1 and 6 years) precludes any reliable long-term appraisals. Obviously, studies spanning decades are needed to chart many subtle, cumulative changes in such ecosystems as boreal forests where the trees may take nearly a century to mature.

Still, there are some communalities pointing to workings of a syndrome in which acid deposition may have an important part—though we cannot say if it is one of primary cause or secondary but crippling aggravation or just as a contribution to a complex synergistic mixture.

Johnson and Siccama (1983) list the following five similarities between the recent North American and European coniferous forest declines: tree mortality extends over large areas with a wide variety of soils; drought or unusually warm, dry summers appear to be related to the decline; analyses of the roots of declining spruces and firs suggest calcium deficiency; declining trees have 10% more sulfur in their needles than healthy ones; and the severity of the damage increases with elevation.

Setting aside the still so uncertain conclusions regarding these degradative processes, I will now try to summarize the well-established ecosystemic changes caused by acid deposition and to review the evidence concerning effects on microbiota whose health is, after all, the irreplaceable foundation of plant and animal survival.

4.4.3.5. Ecosystemic Degradation Acid deposition does not work as tropical deforestation or subtropical desertification. Except for the areas near huge point sources of SO_2 where plant poisoning and subsequent erosion may create desolute moonscapes and where heavy acidification may turn small lakes into lifeless reservoirs, the effects are subtle, cumulative—and not even necessarily injurious. Fertilizing effects of nitrates in atmospheric deposition were already noted (Sections 4.3.1.5 and 4.4.3.3) as was the importance of airborne sulfur for crops increasingly grown with sulfur-free fertilizers.

The most frequently cited hallmarks of acidification's negative effects on ecosystems are fish kills, especially the acute episodes associated with spring thaws when the pH of runoff waters can drop as low as 3.0, interference with normal reproductive processes among various aquatic species, leaching of alkaline nutrients from soils, and release of heavy metal cations into lake or underground drinking waters (Cowling and Linthurst, 1981; Cowling, 1982b).

Unlike with aquatic ecosystems, there is, to use Hutchinson's (1980) words, "little solid evidence" concerning not only visible but even detectable damage caused by acid precipitation in terrestrial ecosystems (establishment of a clear link between German forest decline and acid deposition would, of course, be such an evidence). This is not surprising as naturally acidic soils are quite widespread and yet able to support a variety of far from impoverished plant and heterotroph communities: extensive forest areas on granitic bedrock function well with soil pH at 3.5–4.5. Moreover, most soils have enough carbonates, organic compounds, and clay minerals to buffer the acid inputs.

As for the agroecosystems, their long-standing fertilization and liming make any conceivable H^+ ion additions through acid precipitation quite negligible. Not surprisingly, Jones and Suarez (1980) conclude that "at the present rural levels of air SO_2 and subsoil SO_4, the possibility that plant health of eastern U.S. agroecosystems is being influenced by too much or too little atmospheric S is slight."

But should we not put aside our macrobiotic bias and concentrate more on what any environmental changes will do to microorganisms? After all, no flowering plants, no animals, and no people could survive, even in drastically altered forms, without microfloras and microfaunas—while in the converse case, most microbiotic processes could go on without profound alterations and, in Delwiche's (1964) words, "only the superficial features of forests and grass and concrete and steel would be absent."

Stunning quantitative examples of the importance played by microorganisms and also by rarely glimpsed soil invertebrates can be found in Alexander (1977), Edwards and Lofty (1977), and Kitchell et al. (1979) and, fortunately, a relatively large amount of information is available on the effects of soil acidification on microorganisms and microbial processes owing to the earlier interest in acidifying effects of inorganic fertilizers.

Perhaps the single most important shift in microflora accompanying acidification is from bacteria to fungi. Comparisons of mull (pH around 6) and mor (pH 4) soils also show that higher acidity lowers the rate of nitrification, mineralization of nitrogen, and enzymatic activity, inhibits nitrogen fixation by legume symbionts, and reduces diversity of microfauna. However, any substantial effects on microbial processes appear to be limited to extremely acid levels (i.e., below pH 3).

Moreover, many microorganisms have shown adaptability to declining pH and process inhibitions noted in one soil at a given pH cannot be, for totally unexplained reasons, seen in a different soil at the same pH (Alexander, 1980). And as most microbial transformations in soil are carried out by several populations, the disapperance of one is not automatically detrimental as another group will, before too long, fill the vacated niche. Or, as Wainwright (1980) believes, microorganisms will, given the time, develop sufficient resistance so that essential soil biochemical processes will continue unhindered.

And some microorganisms need to do very little adjusting to acid environments. Field experimentation in southern Norway brought many proofs of considerable resistance, even outright preference of certain decomposers for more acid soils (Abrahamsen et al., 1980). Some basidiomycetes grew almost as well as pH 3.0 as at pH 4.5, abundance of collembolans went up significantly as the soil acidity dropped from pH 6.0 to pH 2.5, and total numbers of mites (*Acarina*) also increased. Moreover, lime application led to large declines of density for both collembolans and mites, clearly indicating their preference for more acid conditions.

The increase, however, was species-specific: some decomposers were little affected, some even declined a bit, but other species increased in abundance to such an extent that the *total* density went up. This is a perfect example of a complex ecosystem of soil invertebrates adjusting to a stress, one which the authors find difficult to explain: increase in food availability owing to higher fungal productivity or the reduced numbers of predators are possible partial explanations.

There are, of course, specialized microbial populations—such as nitrifying or nitrogen-fixing bacteria—whose decline or elimination would affect soil processes rather significantly. Nitrification may proceed at pH below 5.0 but its rate declines with acidity and the process basically stops at pH 4.5. Similarly, nodulation by *Rhizobium* spp. will not occur at low acidities—in soybeans at least pH 4.6, in red clover at pH 5.2, in alfalfa no less than pH 6.5. Nonsymbiotic nitrogen fixers are affected in the same way as are many decomposers.

However, these undesirable effects may not be very important—for example, in the case of coniferous forests, inhibition of nitrification would not be destabilizing as the trees are not obligate nitrate plants (Tamm, 1976)—or may be easily countered as in the case of inhibited nitrogen fixation whose higher rates can be restituted by simply adding molybdenum which is less available at lower pH. For pathogens the consequences of acid deposition are equivocal: some bacteria and fungi may thrive, others, including fungal pathogens such as *Microsphaera, Melampsora,* and *Phragmidium,* will be eliminated, and polluted areas may have a lower incidence of some plant diseases.

The practical consequences of acidification on the critical turnover of organic debris may then be minimal. For example, experiments by Abrahamsen et al. (1980) showed that the initial decomposition of plant debris was influenced only negligibly by acidity: lodgepole pine needles decomposed faster during the first 3 months with pH 3.0 "rain" than with pH 5.6, then the rate slowed down and after 9 months there were no significant differences between the treatments. These findings, of course, do not exclude the possibility of a long-term slowdown of organic debris decomposition as a result of acid deposition, a transformation suspected of contributing to the West German *Waldsterben*.

Simple conclusions must be avoided. There is enough evidence to be quite worried about some rather speedy and extensive local or regional degradation of aquatic ecosystems, especially where the affected lakes have been an essential feature of the lucrative recreation industry (Ontario's Muskoka–Haliburton region is perhaps the best North American example) or are a part of natural parks (the Adirondacks). On the other hand, any threats to industrialized nations' agroecosystems seem negligible in view of fertilization-induced acidity and its correction by liming.

The greatest potential concern must be with the health of boreal and mixed forests owing to both their economic value and the inestimable ecosystemic and recreation services they provide to the populations of North American and European nations. Fortunately, neither Scandinavian nor North American evidence points to sweeping irreversible ecosystemic damage on a large scale. The best discernible trend—whatever the actual role of acid deposition in forest decline syndrome—is toward a change of species mix in the heavily affected or sensitive sites, with conifers declining and broadleaves filling the vacated niches. Altogether this is an unwelcome change (especially from an economic point of view) but not an ecosystemic debacle. New evidence will bring us closer to assessing the long-range nature of terrestrial ecosystemic damages but uncertainties may persist as they had for several decades with regard to effects on human health.

4.4.4. Sulfur Compounds and Human Health

The effects of polluted air on human health, apparent for many generations of Western industrial expansion, were not widely investigated in a rigorous and interdisciplinary manner until the 1950s but then the research expanded rapidly, ranging from *in vitro* studies of animal tissues bathed in pollutants to experimental inhalation of gases and aerosols by volunteers, from perfectly controlled exposure of whole menageries of laboratory animals to computerized analyses of complex, collinear data selected to uncover the links between pollutants and excessive death rates.

And, hardly surprisingly, much of this research effort was focused on sulfur compounds, above all on sulfur dioxide, the ubiquitous ingredient of atmospheres polluted by fossil fuel combustion. Rich accumluation of this diverse information makes it possible to judge the acute and chronic tolerances, responses, and effects of SO_2 with a relatively high degree of reliability. More information is needed on various sulfates' effect on human health but in general it appears that in most of the rich countries the air pollution control efforts of the past two decades have virtually eliminated sulfur compounds as a worrisome public health risk.

Several decades of fairly extensive experiments with various animal species (ranging from guinea pigs to monkeys and from grasshoppers to beagles) and human volunteers have established quite well the toxicology of sulfur compounds. Both sulfur dioxide and sulfates irritate the respiratory system but there is much more detailed evidence to assess the effects of SO_2. The latter causes bronchoconstriction, usually measured as an increase in airway resistance. H_2SO_4 is much more irritating than SO_2 but its effects are highly dependent on particle size with tiny aerosols gaining easier ingress into the lungs.

In comparison with ambient air pollution levels now existing even in relatively "dirty" urban and industrial areas, very high SO_2 concentrations are required to elicit measurable response in healthy men. Dalbey (1980), who reviewed all major experimental chronic exposures of laboratory animals to SO_2 and H_2SO_4 (undertaken between 1963 and 1978), concluded that the concentrations of these compounds currently encountered in the cities of industrialized Western nations are too low to cause changes of pulmonary function and structure. Sensitive persons will have detectable, though hardly severe, broncho-constriction at around 3 mg/m^3, but for most of the tested individuals the response does not start until about 14 mg SO_2/m^3, i.e., 175 times the United States animal standard and more than 300 times the typical urban SO_2 concentrations.

Nonetheless, scores of statistical exercises seeking the relationship between air pollution and mortality have been published in Europe, North America, Russia, and Japan, all of them handicapped by less than faultless designs, intractable interferences of unaccounted variables, and, the basic difficulty, impossibility of clearly distinguishing the effects of sulfur pollutants from general particulate matter.

But no mortality analyses, no matter how carefully designed, can indicate specific effects of sulfur compounds on general population or specific group death rates. Ferris's conclusions still stand: general mortality data appear to be too insensitive to estimate effects of chronic low-level pollution with accuracy and confidence (Ferris, 1969) and "it does not seem reasonable to conclude . . . that SO_x or particulates are having an effect on mortality at the levels usually measured at present. It is not possible to conclude unequivocally that there may not be some effect, but this is probably unlikely" (Ferris, 1979).

As for the morbidity effects, two recent controversies will provide a good illustration of the continuing polemics. The first one concerns the conclusions of Croker and colleagues' (1979) USEPA-sponsored study of air pollution control benefits. Pearce et al. (1982) questioned the basic approach (as it "uses naive models of the real world"), pointed out the consistent linearity of the models "to extrapolate results in which no real confidence can be placed," criticized the use of certain data sets, and noted great inadequacies of air quality monitoring in the areas covered by the study. Their conclusion is that "the results, even though some of them suggest a consistent link between air pollution and acute illness, should not be regarded as reliable in any way. Their value can be doubted even as orders of magnitude. . . ."

In a reply to this criticism, Atkinson and Crocker (1982) argue that the estimator risks in studies of this kind can never be eliminated, just realistically minimized, and buttress their conclusion and interpretations by numerous references to the theory of methods and implications of advanced statistics and econometrics. The debate lapses into splitting fine methodological points and at its end, so remote from the real question of morbidity–air pollution links, it is clear that even with the greatest allowances made, such studies cannot serve as guides for legislative action or strategic planning.

And yet estimates of air pollution-induced health costs have been used repeatedly and perhaps nowhere with such largesse as in the OECD's (1981) methodological study on the cost and benefits of SO_2 controls. The study attempted to quantify control costs of three scenarios for all European members of the OECD (i.e., 18 nations of western and southern Europe and Turkey) and the monetary benefits accruing from improvements in human health, lower material and crop losses, and less damage to aquatic ecosystems.

The three control scenarios were: (I) a reference level including planned control technologies to be operative by 1985 and reducing western European SO_2 emissions by 0.8 million t; (II) an effort in which every country spends 0.05% of its GDP on emission controls; and (III) every country reduces its 1985 SO_2 emissions 37% below the 1974 level. In terms of the 1974 emissions, 1985 relative values would then be 125% in the base case, 121% in case I, 97% in case II, and 63% in case III.

Control costs of the other three scenarios were put as \$(1980) 1.17–2.7 billion a year while financial benefits totals ranged from as little as \$(1980) 860 million to as much as \$19.3 billion. This huge range results from uncertainties concerning benefits of lowered mortality and morbidity. While the benefits from lower corrosion range only from about \$500 million to nearly \$1 billion in the three scenarios and the gains for higher crop yields were appraised to be, for different countries and for three crop years in the mid-1970s, between roughly \$100 and 280 million, the difference for health benefits between scenario I and III was only 2.8-fold, but the uncertainty about the health effects was expressed by using a 25-fold range in each case.

Consequently, for each scenario, total benefits would be appreciably below the control costs if the lowest assumptions of health benefits were the best reflection of the actual situation—or the cost–benefit ratio may be extremely favorable (at least 1 : 7) with the highest assumed costs of mortality–morbidity gains. Clearly, the study's conclusions are critically dependent on its treatment of health effects but a close reading will reveal an astonishing admission of inapplicability for any real valuation.

To quantify health benefits, the study assumed linear dose–response relationship (without a low dose threshold) between mortality and SO_4^{2-} levels and morbidity costs were calculated in the same manner. But the authors readily admitted that this approach has no foundation in epidemiological data, especially at low sulfate concentrations. As a result, "there is a considerable range of uncertainty, and indeed there are arguments among the scientific community that the methodology used here may not be a sound basis for cost-benefit analysis in the health field" (OECD, 1981).

Such forthrightness is welcome: many reports of this kind do not make such a critical *caveat*—but one must read the relevant section carefully to find the admission of fundamental uncertainty as the final cost summations and benefit totals based on these theoretical assumptions are presented *without* any warning. Naturally, the conclusions are seen first and disseminated rapidly and one more item of, this time self-admitted, misinformation enters the realm of environmental scares. Barnes et al. (1983) offer a good extended critique of the whole OECD study but nothing can surpass its self-incrimination.

These controversies have become more complicated owing to a gradual shift of the blame: as the innocuous nature of SO_2 at the ambient concentrations now prevailing even in the urban areas of industrialized nations (i.e., concentrations below 50 $\mu g/m^3$) came to be increasingly accepted, sulfates emerged as the prime suspects of health damage formerly attributed to SO_2 (USEPA, 1975b). However, it has long been appreciated that various sulfate species differ significantly in their irritant potency. Guinea pig broncho-constriction testing established H_2SO_4 as the most irritating sulfate, followed by metallic cations (zinc ammonium sulfate, ferric sulfate) and ending with $(NH_4)_2SO_4$ whose potency is less than one-tenth that of H_2SO_4 (Amdur et al., 1978). Yet ammonium sulfate usually accounts for some nine-tenths of all sulfate species in ambient air, and short-term exposures to its very high levels (ten times the concentration in polluted cities) did not cause any demonstrable effects (McCarroll, 1980). Moreover, Sackner et al. (1978) exposed volunteers, including some asthmatic children, to submicrometer H_2SO_4 mist in concentrations up to 1 mg/m^3 for 10 min without any demonstrable pulmonary effects.

Consequently, it might seem that sulfates are not really worrisome—but there is some important evidence, although not directly applicable to people breathing low concentrations of the acid cations, which clearly demands further study: some *in vitro* as well as *in vivo* experiments with animal lung tissues have demonstrated a histamine-mediated broncho-constriction associated with exposure to sulfur compounds, with ammonium sulfate being the most potent compound in triggering the process (Coffin and Knelson, 1976). Much more work in disaggregating the effects of individual sulfate species, and also different aerosol sizes, is required before promulgating rational ambient standards for atmospheric sulfates, an action anticipated in the United States since 1977 (Rowe et al., 1978).

The best epidemiological and toxicological evidence can be summarized in the following ways. First and foremost, no persuasive data can be presented to demonstrate higher mortalities as a result of chronic exposures to ambient SO_2 levels *now* encountered in most urban and industrial areas of North America and Europe. Second, while there is no lack of statistically significant findings linking sulfur compounds (SO_2 alone, SO_2 and SO_4^{2-}, or SO_2 and all particulate matter) to increase in morbidity (especially respiratory diseases), advancing a hypothesis "that bronchitis caused by simple chemical irritation by some substance such as sulfur dioxide in the concentrations found in urban air is even less tenable than the idea that this gas alone is responsible for the observed effects of so-called smog" (Lawther and Bonnell, 1970). As in so many other cases, statistically significant correlation is no proof of a causative relationship.

And evidence from an urban area (London) where the relationship between morbidity and air pollution has been studied more extensively than anywhere else strongly indicates that with the air quality improvements achieved since the 1960s, the link has been virtually broken—if indeed there ever was one. The considerable lowering of peak SO_2 concentrations and the less dramatic reduction of mean SO_2 values in London since the enactment of the Clean Air Act in 1956 refuted the SO_2–morbidity link so that by 1970 there appeared to be little more than random variations. But as particulate levels during that period declined more spectacularly than did SO_2 concentrations (from around 300 $\mu g/m^3$ to well below 100 $\mu g/m^3$), severing of the air pollution–morbidity link most likely resulted as much from this latter change as from the former improvement (Lawther and Bonnell, 1970; Lawther et al., 1970).

Third, the current low ambient SO_2 levels do not appear to cause, alone or in combination with particulate matter (including acid aerosols) normally present in the air, airway resistance or to change any other measure of lung function among healthy adults—this is the only conclusion to be made from hundreds of toxicological studies and experiments done with animals and people. Greater sensitivity among more susceptible population groups after chronic exposure even at fairly low concentrations should always be seen as nontrivial, however.

Consequently, it is difficult to disagree with McCarroll (1980) when he argues that we might want to impose further SO_2 controls aimed at additional lowering of ambient SO_2 and SO_4^{2-} levels in Western nations to reduce acid loading of ecosystems or to improve visibility but that "unless adverse effects on human health can be clearly demonstrated by pollutants at present ambient levels, human health should not be cited as the rationale for their control."

4.5. MANAGEMENT OPTIONS

Controlling sulfur emissions is far from being a straightforward technical challenge—finding the best methods for desulfurizing fuels or flux gases, bringing the costs of such technologies down through large-scale diffusion and continuing innovations, trying to introduce new techniques which would control both sulfur and nitrogen oxides, working with recyclable reagents or yielding salable products at competitive prices. But the broader, and fundamental, questions of actual damages caused by gaseous SO_2 and by acid deposition whose reliable appraisal would be the best guide for determining the extent and the intensity of needed controls have not moved much closer to practical resolution. To an uninitiated observer, this may seem surprising in view of the unmistakably intense interest devoted to acid deposition by public media and governmental bureaucracies—but those readers should have no doubts about the huge uncertainties permeating our understanding of acid deposition sydromes.

4.5.1. Underlying Uncertainties

When the first NAS review devoted to a large extent to acid deposition was published in 1981, it noted that precipitation in sensitive freshwater ecosystems should not have a pH lower than 4.7–4.6, rather than the pH prevalent in the most seriously affected areas (i.e., 4.2–4.1)—and that such an improvement would require a 50% reduction in deposited hydrogen ions (NAS, 1981a). Yet this correct statement was completely misinterpreted (by the media and by numerous politicans eager to project themselves as guardians of the public good) as a recommendation for a 50% reduction in emissions of sulfur and nitrogen oxides.

As a result, a 50% reduction of SO_2 emissions remains the cornerstone of official Canadian plans to control North American acid deposition and a rallying target for many environmental lobbies. But such an interpretation is indefensible: while it is very likely that reduced H^+ ion deposition would follow a reduction in emissions throughout the northern part of the North American continent, the 1981 report did not—and could not—specify how much additional control of SO_2 emissions would bring the desired 50% H^+ deposition decline in some sensitive areas.

A new NAS committee was set up to address this link (NAS, 1983a) but it, too, found the goal elusive as it had to conclude that implications of its findings for choosing among possible emission control strategies ("should they be deemed necessary") are limited: "We do not believe it is practical at this time to rely upon currently available models to distinguish among alternative strategies. . . . If we assume that all other factors, including meteorology, remain unchanged, the annual average concentration of sulfate in precipitation at a given site should be reduced in proportion to a reduction in SO_2 and sulfate transported to that site from a source or region of sources."

But this conclusion has a critical qualifier: "Because we cannot rely on current models or analyses of air-mass trajectories, we cannot objectively predict the consequences for deposition in ecologically sensitive areas of changing the spatial pattern of emissions in Eastern North America, such as by reducing emissions in one area by a larger percentage than in other areas" (NAS, 1983a).

In practical terms, this means that the best scientific evidence does not permit us to order Ohio Valley power plants to cut their emissions by 70% so as to lower acid deposition in the Adirondacks by 50%—or to press for any similar combination of source–impact emission–deposition combination. Still, assorted bureaucracies have been trying to come up with precise quantitative orders for reducing the emissions and I know of no better example of quandaries characterizing the process than the activities accompanying the United States–Canadian efforts to define the extent of the acid deposition problem, to formulate jointly acceptable positions, and to come up with an effective long-term plan of remedial actions.

To have chronicled the major developments in some detail—at least since August 5, 1980 when the two countries signed a Memorandum of Intent to establish a number of working groups charged to gather data forming the basis for negotiations toward a transboundary control agreement—could easily have taken a large section of this book or have been a book in itself; indeed, two small books have been written just about those bilateral dealings (Gold, 1981; Carroll, 1982), and several extensive reports.

Acid rain has become the most important, durable rallying point for Canadian environmentalists, as well as a high-priority item of the government policy, and the high share of United States emissions falling on Canada has served as a convenient outlet for the ever-present anti-Americanism rife in nearly all strata of Canadian society. Yet it is a simple fact that in per capita terms, as well as when prorated for each $1000 of GNP, Canada emits more sulfur than the United States and that the continent's three largest point sources are all in Canada. In the early 1980s, Canadian SO_2 emissions prorated to 100 kg S per capita and a bit over 8 kg per U.S. $(1980) 1000 of GNP while the United States values were, respectively, just under 60 kg and about 4.5 kg.

Similar tensions are much alive in Europe where the current conversion of the West German government to advocacy of strict SO_2 emission controls has put much greater pressure on other large polluters than the situation when the emission cuts were favored only by the two main "target" countries, Norway and Sweden. In spite of this West German change of position, the 1982 Stockholm Conference on Acidification of the Environment did not reach agreement on the levels of acceptable sulfur loading by failing to endorse the Swedish call for a maximum deposition of 0.3–0.5 g total sulfur/m^2 annually, or an equivalent of 9–15 kg of dry and wet sulfate per ha per year.

And with the huge central European emitters—Czechoslovakia, East Germany, and Poland—standing aloof of the western European efforts, any contemplated rounded-value emission cuts (i.e., 25 or 50% reduction) in northwestern Europe are clearly arbitrary and, to no small degree, politically motivated (in the West German case an all too patent maneuver by the ruling party to undercut the environmentalist Greens).

The huge cost-benefit uncertainties, so well illustrated by the OECD (1981) report whose principal results were quoted in the section on epidemiological evidence, and the high expense of even limited local and regional emission reductions—for example, in downtown Los Angeles a 50% improvement in sulfate air quality could be achieved at a cost of no less than $(1980) 100 million annually (Cass, 1981), while Environment Canada estimates the cost of SO_2 controls at the Sudbury smelter at Can. $(1980) 430 million—make any long-term control commitments even more difficult. Nor are many available emission control methods fully commercially established and operationally perfected.

4.5.2. Emission Controls

Virtually all industrialized nations now have emission standards or regulations applicable to releases of sulfur dioxide from a variety of stationary sources. These regulations are specified in a variety of ways: for example, for SO_2 emissions the USEPA prescribes the maximum emissions per heat unit of fuel burned, in Japan there are stack height criteria to maximize the dispersion, and in England and France specified stack heights apply only to coal-fired sources while emissions from liquid fuels are regulated by limiting the sulfur content of fuel oils (Rubin, 1981).

The USEPA's New Source Performance standards for sulfur dioxide, promulgated on June 11, 1979, specify that the emissions of all steam generators with a heat input of more than 250×10^6 Btu/hr (73 MW) must be below 1.2 lb/10^6 Btu (520 mg/J) for coal-fired sources and less than 0.8 lb/10^6 Btu (340 mg/J) for oil- and gas-fired plants; for coal a 90% reduction in potential SO_2 emissions is required at all times except when the emissions are below 0.6 lb/10^6 Btu when 70% reduction is acceptable. For liquid and gaseous fuels, 90% reduction in potential emissions is in force for all fluxes above 0.2 lb/10^6 Btu (80 mg/J).

Compliance with these standards, and even more so with any future stricter limits, is impossible without relying on a single control method—unless one is lucky to have huge, secure supplies of exceedingly pure natural gas to burn—and choices of sulfur emission control strategies can be logically divided by their timing. First, there are precombustion arrangements, selecting fuels with naturally low S content or desulfurizing high-S fuels by selective cleaning or through a conversion to synthetic fuels. Intracombustion cleanup cannot eliminate the formation of SO_2 as the present sulfur is immediately and irreversibly oxidized, but fluidized bed combustion (and use of sorbents inside the conventional boilers) can substantially reduce the sulfur-loading of the escaping gas. Postcombustion options are just two: let the gas go but arrange for its maximum dilution, and hence lower ground concentrations through tall-stack release—or capture it in a variety of wet or dry, throwaway or regenerable flue gas desulfurization devices.

Table 4-5 summarizes these options and lists the typical SO_2 reduction ranges as well as appraisals of commercial availability. Before looking at the possibilities

TABLE 4-5
CONTROL STRATEGIES FOR REDUCING SO_2 EMISSIONS

Control tracing	Control method	Average SO_2 reduction[a]	Availability of controls	Potential for widespread application before 2000
Precombustion	Coal rank switching	30–80	Now	Low
	Switching to low-S oil	90–92	Now	Low
	Switching to natural gas	95–97	Now	Low
	Mechanical coal cleaning	20–80	Now	High
	Chemical coal cleaning	40–90	1990s	Medium
	Coal-derived liquid fuels	90–95	1990s?	Low
	Coal-derived gases	90–95	1990s?	Low
In-flame	Atmospheric fluid bed combustion	80–90	Late 1980s	High
	Pressurized fluid bed combustion	80–95	1990s?	Low
Postcombustion	Dilution with tall stacks	0	Now	Low
	Wet flue gas desulfurization	80–90	Now	High
	Dry flue gas desulfurization	60–80	Now	High
	Combined SO–NO removed	60–90	1990s	Medium

[a]Partially after Reuther (1982).

of clean combustion (desulfurized fuels and fluidized bed boilers) and, above all, the techniques, problems, costs, and innovations of flue gas desulfurization, now the leading choice of sulfur emission controls, and then assessing the prospects of the only control method used in practice to counteract acidification after the pollutants have reached soils and waters—large-scale liming of croplands and lakes—I will deal with the use of now-discredited tall stacks which were considered an essential control tool until the early 1970s.

Any description of tall stacks in the early 1980s as effective sulfur pollution control tools seems quite inappropriate. Obviously, they do not affect the total SO_2 emissions at all and by releasing the hot gases at higher altitudes they only assure that a greater portion of that flux will reside longer in the atmosphere where it will be transported over longer distances and converted to a higher degree to sulfates—thus clearly contributing to acid deposition far away from the combustion sources. But not so long ago the arguments were different and claims made on behalf of tall stacks were all but timid.

How fundamentally the perception has changed is perhaps best illustrated by summarizing a 1970 review of the effectiveness of tall stacks, prepared for the Second International Clean Air Conference by Clarke et al. (1971). These authors regarded tall stacks as a cheap, reliable, and indispensable air pollution control tool whose criticism was largely misguided. Above all, they felt, that "a chimney can in most cases do all that a [sulfur] removal process could do at a fraction of the cost to the community. . . . Sulphur dioxide extraction processes are not promising . . . justification for [their] further development is questionable."

These conclusions arose from a rapid decline of SO_2 ground concentrations throughout England during the 1960s (Ross, 1971a,b). While at the beginning of that decade SO_2 levels in the country's 15 largest cities surpassed 0.5 mg/m^3 nearly half of the time, by the end of the 1960s such levels were reached on only about 10% of all days—in spite of the fact that the total SO_2 emission had been rising with expansion of coal-fired power plants (from 4.54 to 5.69 million t between 1950 and 1970). This lack of correlation between falling ground concentrations and rising emissions led to the conclusion that "emission is not pollution" as long as the increased emissions come from tall stacks, which released 0.91 million t SO_2 in 1950 but 3 million t by 1970.

The whole argument can be reduced to two simple recommendations: use fuels with the lowest practicable sulfur content in all small and medium urban installations where SO_2 is released near the ground—but go up to a sulfur content of 2.5% in coal or 3.5% in residual fuel oil in large sources with tall stacks. As Ross (1971b) put it dramatically: "If a large city allows sulphur from its heating plants to attack the lungs and library books of its citizens while at the same time refusing to allow its electric power stations to supply life-giving sulfur to the surrounding cornfields, the results will be a reduced standard of living."

But by the time these British solutions were being recommended to the American policymakers to "set in motion a train of events that will lead to a fresh evaluation of tall stacks on their merit in reducing ground level concentrations"

(Frankenberg, 1971), the short era of rapidly rising stack heights in the United States was already reaching its end. While in 1960 the tallest power plant stack in the United States was 180 m, by 1970 this size doubled with construction of the Mitchell plant's (Moundsville, West Virginia) stack which surpassed any previously built chimney by about 60 m.

This tall stack, based on a 3.75-m-thick foundation (diameter of nearly 50 m), needed almost 19,000 m^3 of concrete and 3000 t of reinforcing steel (both totals including foundation requirements) to rise to 360 m where its diameter was 11.1 m and its wall thickness a mere 22.5 cm. Rapid construction of this tall stack was made possible by the continuous pour method introduced to the United States from West Germany and at that time it seemed that even taller power plant stacks would soon follow.

The average size of new fossil-fueled power plant stacks also doubled during the 1960s, from 60 m to 120 m, and the increase was even faster for coal-fired power plants, from 73 m in 1960 to 183 m for the plants completed in 1969. But neither the British enthusiasm nor the vigorous commitment of American utilities turned the 1970s into a decade of even taller stacks. Federal standards of performance for new stationary sources forced attention toward flue gas desulfurization as the only long-term solution for sulfur emissions from large thermal stations. And when the awareness of acid rain as a potentially serious environmental threat came shortly afterwards, something the British arguments on behalf of tall stacks either disregarded completely or dismissed as an improbable exaggeration as late as 1971, it could not be dismissed in the environmentally sensitized America of the mid-1970s.

True, the average height of new fossil-fueled power plant stacks continued to rise to its present 200 m but no new stacks topped the record of 360 m, and offering tall stacks as a solution to the SO_2 problem now appears hopelessly improper. In fact, tall stacks must be seen as one of the key contributing factors to acidification of areas remote from emission sources. Reduction of local effects may be quite impressive as perhaps best demonstrable by an analysis of precipitation chemistry from 31 rains in the summer and fall of 1978 and 1979 collected within a 50-km radius of the INCO smelter at Sudbury, Ontario (Chan et al., 1982).

The smelter's 381-m-high stack at that time emitted about 1700 t SO_2/day (ranging from 658 to 2320 t) but the relative contribution of these emissions to the total wet deposition of acids and sulfur around Sudbury appeared small and depended more on the weather system influencing the area. During warm fronts bringing polluted air from the south, only 9% of all free H^+ ions and 12% of total sulfur could be attributed to the smelter, and even for cold fronts and their northerly flow these shares rose only to 21 and 31%.

But while the Sudbury region benefited, the stack's plumes could be detected hundreds of kilometers downwind (Section 4.2.3.2) and the smelter's operation was reduced to limit the daily emissions. Internationally, once the Swedish claims of long-range exports of British sulfur to Scandinavian waters and forests had discredited the overall desirability of tall stacks, there was probably no chance for a reversal: rather than being a nearly perfect solution (if not the solution), they were now seen to be a great part of the problem—and the efforts turned to flue gas desulfurization. But before dealing with this technique at least a brief discussion must be made of clean combustion and the problems which preclude easy cleaning of sulfurous coals so that flue gas controls become necessary.

4.5.3. Clean Combustion

Unless one has access to fuels with very low sulfur content, the fuels must be desulfurized, partially or fully, before combustion, or burned in such a way that sulfur is entrapped during the combustion. As coal combustion is by far the largest source of anthropogenic sulfur (see Section 4.1.3.1), these options reduce in an overwhelming number of cases to three possibilities of coal substitution, coal clean-up, and fluidized bed combustion.

The first choice hardly deserves much discussion unless one wants to get immersed in hypothetical scenarios of nuclear generation, a frustrating subject with a very dubious record of success. Substitution of coal in thermal generation, now by far the leading use of the fuel, by less sulfurous natural gases and oils is clearly not a sensible option on a planet which stores in coals at least five or six times more recoverable energy than in recoverable crude oils and where the two largest industrialized nations (as well as China) have the resources, technology, needs, and plans for further substantial increase of their already high coal extraction. Even long before the first of OPEC's big crude oil price interventions in 1973, coals' use in power generation was climbing in every country with sizeable coal reserves (from the United States to China, from Czechoslovakia to India)—and even great crude oil price moderation or declines will not change that trend.

4.5.3.1. Coal Cleaning Theoretically, the proposition is straightforward and the fuel's makeup should actually help it: as coals are complexes of various constituents, organic and inorganic, the cleaning should first separate them and then the undesirable ingredients should be removed. In reality, complete separation of organic and inorganic constituents is impossible but even should this be achieved sulfur would still contaminate both parts. As noted in the opening of this chapter, the element is an essential ingredient of organic matter and removing this fraction of a coal's sulfur—as little as 10% but as much as 50% of the total S content—is still practically infeasible.

Research and development interest has thus been focused on the removal of inorganic sulfur, most of which is found in coals associated with iron as pyrite (FeS_2). This is an obvious advantage as pyrite has a relatively high specific gravity and hence can be separated in water slurries—but it first must be liberated from the fuel and this is rather difficult. Nor can it be done by obtaining easy-to-process uniform pieces, and when all of this more or less succeeds one is left with another difficult, energy-intensive task of dewatering fine-sized slurries before combustion.

A wide variety of coal processing methods for sulfur removal, ranging from old established techniques to new pilot plant projects, utilize gravity separation.

Reviews edited by Wheelock (1977), Eliot (1978), and Khoury (1982) provide detailed technical descriptions. Principal ways of coarse cleaning are in jigs, heavy media vessels, and heavy-media cyclones, all low-cost, high-capacity devices. In jigs, repeated upward water pulses (30–50 times/min) stratify coal particles before the top-floating clean coal is mechanically separated. In heavy media vessels, finely ground magnetite is added to water to produce a suspension with a specific gravity of about 1.5 on which the lighter coal will float and through which the heavier refuse will sink; for somewhat finer, though still rather large fractions (around 3 cm), the water–magnetite–coal slurry is propelled into a cyclonic flow with centrifugal force aiding the gravity by moving the heavier particles toward the outer wall.

Both jigging and heavy-media cycloning are also used for intermediate and fine cleaning and other methods to process these coal fractions include tabling (separation in a water film on decks of shaking tables), single water cyclones (no magnetite added, with light coal removed in central overflow and heavy refuse in outer wall underflow), and froth flotation. This is an especially effective method in which coal particles in aqueous suspension are removed to the surface by attaching themselves to bubbles generated by special reagents and air introduced into the mixing tanks; skimmers pick the clean coal, leaving ash and pyrite in the suspension.

More than a score of chemical cleaning methods are now in various stages of research and development, some with claims of sulfur removal efficiencies surpassing 90% for pyritic sulfur and 60% for organic sulfur, but no significant commercial applications can be foreseen during the 1980s.

As for the efficiency of various cleaning processes, the common commerical practice of crushing coal to a maximum size of 3.81 cm and separating at a specific gravity of 1.6 g/cm^3 will remove 28–43% of the pyrite from low-sulfur coals, 37–55% from high-sulfur ones; this translates to an average reduction of SO_2 emissions of 11–40% (USEPA, 1981). The best current technology, consisting of crushing to below 0.95 cm and separation at 1.3 g/cm^3, can remove up to 63–80% of all pyrites and thus reduce SO_2 emissions by as much as 20–55%. The best potential technologies would about double the pyrite removal rate for low-sulfur coals but would raise it only about 10% in the case of high-sulfur fuels.

Much more costly chemical desulfurization processes such as Meyers's pyrite leaching process or Batelle's hydrothermal treatment can remove 95% of all pyrites, the latter one also about 25% of organic sulfur so that SO_2 emissions can be reduced by a maximum of 62–73%. But while physical desulfurization processes cost just around \$(1982) 7/t to operate and less than \$(1982) 13 million for a 500 t/hr plant in capital expenditure, the chemical treatments will cost anywhere from two to six times more to run per tonne and about 15 times more to install.

Actual costs of coal cleaning vary greatly with the equipment used, volume flow, and operation arrangements. Rittenhouse (1982) quotes a Hoffman–Munter Corporation study of eight modern facilities whose total operation costs ranged from close to \$(1981) 5/t to just over \$(1980) 9/t, about a 1.9-fold difference. Total annual costs per tonne of sulfur removed range wider, from just \$(1981) 200 to \$(1981) 1900; the high capital charges per tonne arise largely from the fact that the plants averaged just about 30% of actual operating time.

On contract for the Electric Power Research Institute, Bechtel National (1981) prepared a detailed review of the impacts that coal cleaning has on the cost of new coal-fired power generation by comparing three alternatives—no, partial, and intensive cleanup—at seven sites, five in various parts of the eastern United States and one each in Texas and Utah. While the intensive coal cleanup (splitting of the fuel into two or three size fractions each of which is then cleaned separately) requires extra capital investment of up to \$(1978) 37 million per plant, or 3–4% of the plant's total capital cost, in six of seven studied cases this investment was responsible for a decrease in overall capital costs of the plants ranging from 0.1 to 3.0% and in five cases the cleaned fuel reduced levelized revenue requirements by the same margin.

Clearly, even intensive coal cleanup appears to be cost effective in most large coal-fired thermal stations and with much research interest given to the improvement of existing removal processes, coal cleaning will remain an essential control method for decades to come—unless fluidized bed combustion makes unforeseen rapid advances.

4.5.3.2. FLUIDIZED BED COMBUSTION Air pollution control advantages of this method are best seen in comparison with conventional coal-fired combustion. In standard power plant boilers, pulverized coal is burned with excess air at up to 1600°C, producing large quantities of SO_2 and nitrogen oxides which are carried to the stack with hot fire gases; heavy metals are also liberated in the process, clinker ash is difficult to handle, and abrasive ash particles wear down the contact surfaces. In contrast, fluidized bed combustion (FBC) largely eliminates all of these problems by burning coal in contact with crushed limestone (which captures virtually all of the sulfur) and by operating with temperatures of 760–840°C, i.e., hundreds of degrees below the temperature at which NO_x formation becomes objectionable. Dry, powdery ash produced during FBC does not slag and hence it is easier to handle and discharges of heavy metals and alkali salts in flue gases are reduced.

This all clearly adds up to a nearly perfect solution and so it is hardly surprising that the interest in FBC for large-scale steam generation has intensified and that the number of published reviews, technical analyses, and assessments ranks with the most frequent topics of modern power engineering during the 1970s; excellent shorter state-of-the-art reviews are Berman (1979) and Highley (1980); comprehensive technical treatments can be had, among many others, in Babcock and Wilcox Corporation (1976) and USEPA (1980c).

In spite of this enormous amount of attention in the recent past, FBC remains just one of many conversion technologies whose promise is so much better than their actual performance. The technique is not new at all and even a quick retracing of its fortunes will show how the predictions of common commercial applications had to be repeatedly pushed further ahead. Although commercial- and industrial-scale atmospheric FBC units are now available, the whole technology is still largely in the R&D phase.

In existence since the 1920s, FBC was used for relatively small-scale coal gasification, catalytic crude oil cracking and production of various combustible gases and liquids (especially in the chemical industry) but the intensive development of small FBC steam generators began only in the mid-1970s. Major advances have been achieved in several test and pilot-plant boilers and the steady commercialization, which so often appeared so close only to recede again, is now finally getting within practical research. This slow, to some people even disappointing, progress is an expected characteristic of a new technology which, although not forbiddingly complex, requires some intricate design and usual accumulation of operating experience to become successful.

The principle is simple enough: a layer of small, noncombustible particles (usually limestone or dolomite) is forced by air flow through a perforated plate into a fluid motion and combustion of any solid fuels (coals, also char, refuse) proceeds in suspension within this liquidlike bed. Careful design must balance the gas velocity, particle size, and bed depth for the most efficient operation. Heat release per unit area of the bed increases linearly with the velocity of existing gases but as the velocity goes up, so must the size of the bed particles to prevent their entrainment in the gas; yet, obviously, the rate of heat transfer will decline as the particle sizes increase.

When perfected for large steam boilers, the performance of FBC should be roughly comparable to that of conventional coal-fired generation with SO_x emission controls except for the mass and composition of the spent bed material compared to effluent from common flue gas desulfurization (FGD) systems: in atmospheric FBC, almost three times as much limestone is required as for FGD operating with the same SO_2 removal and, moreover, the spent material contains a good deal of aggressive CaO that may cause handling and disposal problems.

Although atmospheric FBC can be used for electricity generation, its efficiency is better suited to process or space heating; pressurized FBC (PFBC) is more suited for electricity generation but two critical problems—achievement of dependable cost-effective operation of boiler tubes embedded in the fluidized bed to reduce boiler size and weight and the ability to clean adequately and efficiently the gas for turbine feed—will have to be solved before any substantial diffusion of the technology can take place.

Testing of small FBC units is now proceeding successfully above all in the United States and United Kingdom. Foster Wheeler was the first American company to build a commercial power plant FBC unit, a 135,000 kg/hr of steam addition to the Monongahela Power Company's Rivesville (West Virginia) station (began 1968, operational since 1976). British Babcock Contractors is offering commercially units with steam capacities up to 225,000 kg/hr, i.e., an equivalent of about 60 MWe. Combustion Engineering is looking, after its 22,500 kg/hr demonstration plant, at a conceptual design of a 200-MWe plant for the late 1980s or early 1990s and Curtiss–Wright is developing larger PFBC units.

The International Energy Agency oversees a large PFBC (85 MWe) test in England, where Stal-Laval is working on another relatively large (50 MWe) unit. Many steam units with capacities equivalent to 10–50 MWe have recently been ordered and are scheduled to start operation in the United States and in several European countries in the years ahead. FBC boilers with capacities of 500–1000 MW, i.e., the most frequent unit range of new large power plants, are hardly around the corner but the technology appears to be definitely entering the real world and the industrial companies seem to be ready to adopt it on a progressively larger scale.

Although the availability of commercial FBC would seem to preclude the need for SO_2 scrubbers, FGD is not going to fade away. Berman (1979) put it well when he wrote that "the power industry, in its quest to use more coal, is now faced with a difficult choice: to use new scrubber technology, or to use new FBC technology." There is no long-term operation and reliability experience with either of the choices but FGD units can be added to the existing plants while the adoption of FBC necessitates rebuilding the whole boiler section; besides, the scrubbers are already here in capacities sufficient to handle flue gases from some very large units while FBC continues its measured climb to commercial importance.

By the end of the 1980s, FBC's fate will be much better discernible as the units of 100–200 MW will have been tested and as further advances should prove the appeal of PFBC–turbine cogeneration technology. Afterwards a real takeoff in FBC orders should come fast but at least until then it will be neither coal cleaning nor fluidized beds but a variety of FGD techniques which will dominate the efforts to control sulfur emissions.

4.5.4. Flue Gas Desulfurization

If commercial FBC has been slow in coming, the prospects for FGD appeared for a long time at least equally discouraging. Translation of apparently simple chemical reactions easily demonstrable in laboratories to commercially available installations able to treat hundreds of thousands to over 1 million m^3 per hr of hot flue gases laden with varying volumes of SO_2 in an efficient, reliable, and economically acceptable manner has been surprisingly difficult.

For several decades FGD was following the fate of magnetohydrodynamics, mass-produced electric cars, and tidal generation: no shortage of encouraging breakthroughs reported in the engineering journals, significant application foreseen not far away—but in the end no practical advances. Decades of tinkering and promises were finally ended in the mid-1960s when commercialization of an FGD system finally got seriously under way and I will first review the rapid advances in the diffusion of various SO_2 control techniques since that time. In the subsequent section I will focus on the costs of FGD methods and their problems as well as upcoming technical innovations.

4.5.4.1. Basic Techniques and Problems Research in FGD started during the 1930s in England, Germany, and the Soviet Union but, as just noted, it was not until the mid-1960s when the serious development of large-scale systems for power plant uses began in earnest. The first two real applications of FGD in commercial coal-fired plants came when Combustion Engineering (Windsor, Connecticut) installed its limestone scrubber system on a 140-MW boiler of the Meramec plant (Union Electric Company, St. Louis) in September 1968 and on a slightly smaller boiler (125 MW) at a station of the Kansas Power and Light Company (Jonakin and McLaughlin, 1969). Interestingly, both pioneering utilities were outside the main concentration of coal-fired generating capacity in the Ohio River Valley.

By 1972 there were 15 FGD processes offered commercially, for demonstration or studies in full-scale power plant settings (Spaite, 1972), and six additional FGD units were installed on utility boilers ranging from 37 to 175 MW (total capacity 722 MW). Five of these six commercial units as well as 11 of the 15 available FGD systems were based on the principle of wet alkali scrubbing which has clearly emerged as the best option.

General uncertainties introduced by OPEC's sharp hikes in crude oil prices and the intensive but much confused search for new energy supplies in 1974 and 1975 were reflected by a temporary slowdown in new FGD installations: only a handful of new installations were begun and the total operating capacity remained nearly stationary. In hindsight this was the last pause before real takeoff as the J-bend of exponential growth followed immediately afterwards. By early 1977, a PEDCo survey found 24 operational systems handling a total capacity of 5997 MW and 33 FGD projects under construction with a capacity of 13,918 MW (Papamarcos, 1977).

The 1977 amendments to the Clean Air Act and USEPA's revised 1979 New Source Performance standard allowed no slackening of SO_2 controls and the exponential trend established in 1976 and 1977 held. In December 1979, FGD systems were handling a total of 21,450 MW (Jahnig and Shaw, 1981) and by late 1981 there were 88 operational systems removing SO_2 from power plants (33,300 MW) and another 40 FGD projects (16,160 MW) were under construction (Orem, 1982). By the mid-1980s, the total FGD capacity is thus almost exactly 50,000 MW and SO_2 removal has been installed at about 14% of all United States coal-fired power plants (versus only 7% in 1979). As shown in Figure 4-16, this rapid rise of FGD orders closely paralleled, with a lag of about 10 years, the diffusion of electrostatic precipitators and since 1978 FGD units have accounted for considerably larger capital outlays than the control of particulates.

This brief overview makes clear the rapid maturity of FGD technology: to move from scratch in the early 1970s when the first commercial installations were still very much pilot plants surrounded by doubts and largely pessimistic predictions to a total investment on the order of \$(1983) 5 billion in a new industry operating with increasing reliability and serving one-seventh of the country's coal-fired generating capacity in just 15 years is a record not shared by many similar technologies.

Limestone or lime slurry scrubbing emerged, as noted in the previous section, as by far the most important FGD systems in power plants, accounting for about 70% of all operational capacity by 1975 and 79% by 1977; by 1980 their share slipped back to 70% and in 1985 it should be up slightly to about 72% (Jahnig and Shaw, 1981). Limestone flyash or lime-flyash scrubbing are a distant second (14% of installed capacity in 1980, 12% by 1985) while among the regenerable processes

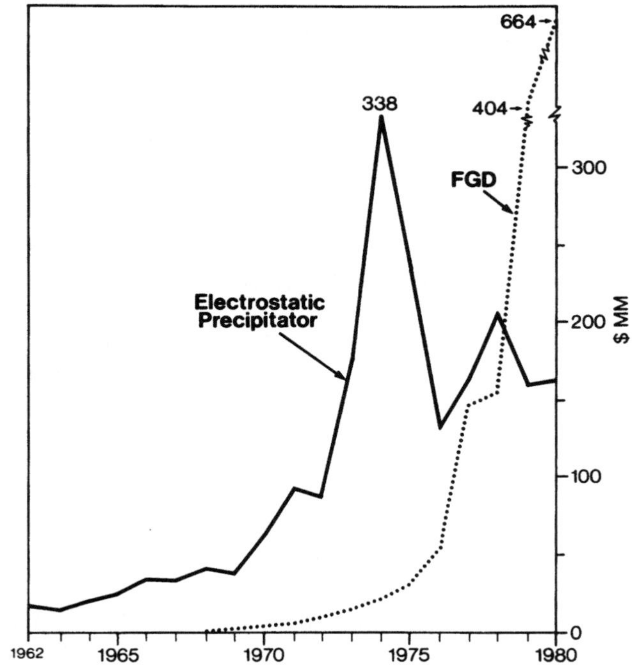

FIGURE 4-16. New orders for electrostatic precipitators and FGD systems in the United States between 1962 and 1980 (Orem, 1982).

Wellman–Lord is the only one of importance (6% in 1980, 4% by 1985). For detailed technical descriptions and comparative assessments of these leading FGD processes, as well as numerous less successful counterparts, I would most recommend, among many other similar publications, Slack and Hollinden (1975), Jahnig and Shaw (1981), and, above all, Hudson and Rochelle (1982).

Lime/limestone scrubbing owes it dominance to the relative simplicity of the process: water slurry of lime or limestone is used to capture SO_2 from hot flue gas (which may be prescribed to remove flyash in a cyclone or in an ESP) in scrubbers using spray, grid, or plate towers, venturis, marble beds, or turbulent-contact absorbers (Figure 4-17 is a typical flow sheet of such a system). The principal reactions in lime/limestone scrubbing are simple, with sulfur dioxide converted to sulfur (water or carbon dioxide as residues) and part of the sulfite oxidized by the flue gas oxygen to sulfate; for lime

$$SO_2 + Ca(OH)_2 \rightarrow CaSO_3 + H_2O$$

for limestone

$$SO_2 + CaCO_3 \rightarrow CaSO_3 + CO_2$$

for oxidation

$$CaSO_3 + \tfrac{1}{2}O_2 \rightarrow CaSO_4$$

However, the actual chemistry during the conversion is extremely complex: Lowell (1970) identified 41 neutral and ionic compounds in solution and seven solid components participating in the reactions, which means that the reaction rates and the influence of design and process variables in this involved three-phase system are nearly impossible to determine (Karlsson and Rosenberg, 1980). Operating experience rather than theoretical knowledge has helped in controlling the process chemistry to raise the SO_2 removal efficiency, eliminate plugging and scaling, achieve maximum reagent utilization, and minimize corrosion.

Scaling has been by far the main problem with lime/limestone scrubbers, especially at the two wet–dry interfaces (when the gas contacts the slurry and when it leaves the scrubber) and with lime FGD. In general, high liquor rate, high content of sulfite and sulfate crystals in the slurry to provide surfaces for crystallizing, and high slurry velocity help to reduce the problem (Slack, 1973). Carbonate and sulfite

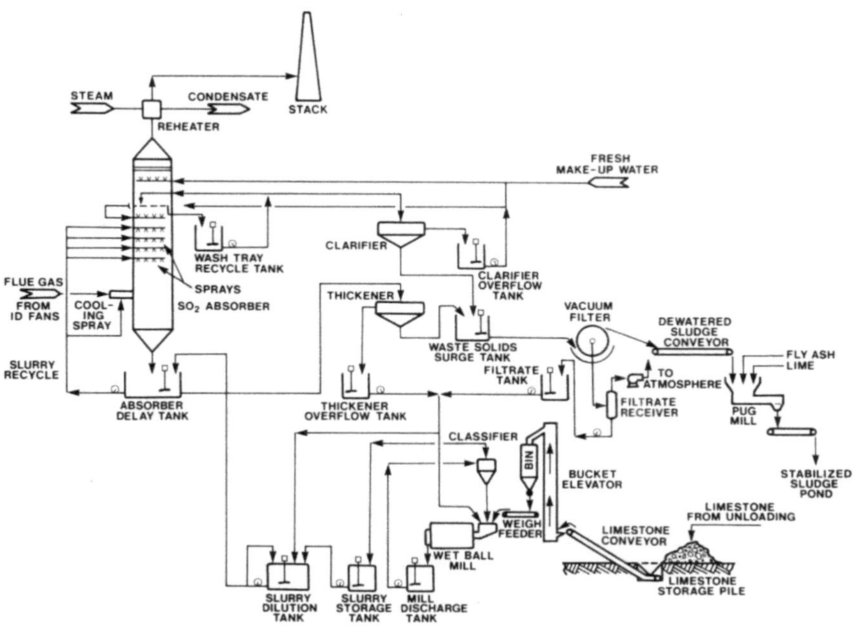

FIGURE 4-17. Flow diagram of limestone flue gas desulfurization (Bechtel National, 1981).

scaling can be mostly controlled by maintaining pH well below 9 (sulfite scaling thus rarely occurs with limestone FGD where pH is lower).

Sulfate scaling is more intractable: the compound's concentration should be kept below saturation level or large amounts of gypsum should be recirculated to initiate sulfate deposition on the crystals rather than on scrubber surfaces (Karlsson and Rosenberg, 1980). Fly ash presence also offers sites for deposition but the slurry's total solid content should not surpass 15% to maintain smooth pumping (the range of 7–15% is now common).

Scaling and plugging were the main reasons for abandoning crushed limestone injection into the boilers, a method used in several early FGD installations: all new systems have tail scrubbing only. Simple scrubber designs also help reduce scaling and hence the earlier choices of marble beds or turbulent-contact absorbers were also abandoned for plain spray towers without packing; the Weir cross-flow absorber is a new suitable simple design.

Corrosion is another inevitable complication and it is worst in closed-loop scrubbers (water pollution regulations usually forbid dumping the scrubbing liquor and hence the recycling) where the recirculation can lead to high buildup of soluble chlorides. Make-up water will always have some chlorides and coal also contains about 0.01% of chlorine and consequently a portion of closed-loop systems may be bathed with solutions of up to 5% of chloride, an environment putting extreme demands on materials.

Fortunately, expensive stainless carbon steels are not the only solution. Thorough washing of scrubber modules once a week nearly eliminates the corrosion (Kopecki and McDaniel, 1976), and appropriate design changes can substantially reduce the overall amount of corrosion-resistant materials by confining high chloride concentrations to small portions of the system (Paul, 1978).

In properly run systems, 75–90% of all input reagents are utilized, with the highest reported value at 98%. In general, SO_2 reactions are slower and less efficient with limestone than with lime but the choice between the two reagents is basically an economic one. Lime, a processed chemical made from limestone, is obviously costlier to produce and also costlier to transport as limestone is merely quarried, moved in open trucks, and crushed. On a molar basis, the cost ratio of lime to limestone is between 2 and 4, depending on the transportation cost (Karlsson and Rosenberg, 1980; see also detailed cost comparisons later in this section). As a result, the mid-1980s should have twice as much capacity in limestone FGD than in lime systems. However, for scrubbing efficiencies in excess of 90% limestone FGD will not do; final cleanup with lime is needed to boost the efficiency, or a lime system (up to 95% SO_2 removal) or a dual alkali FGD (also 95% efficient) must be installed.

Problems do not end with removing sulfur dioxide from the flue gas: large quantities of wastes are generated continuously and must be disposed of in an economically and environmentally acceptable manner. This entails, first, securing a leakproof disposal site in the vicinity of the plant and large enough to accommodate these wastes for several decades (20–50 years). Generated amounts vary with the

FGD method and coal's sulfur content but a typical lime/limestone unit serving a coal-fired boiler burning 3% sulfur coal with 10% ash will produce 3–5 t of wet sludge per MW per day (Taylor, 1973; Cheremisinoff and Fellman, 1974); or in terms of dry sludge (solids are 35–50% of the total) between 1 and 2.5 t/MW per day. A 1000-MWe plant with lime/limestone FGD would thus require during its lifetime (30 years) a sludge disposal area of at least 200–300 ha if the wastes are laid up to 10 m thick.

Nationwide, the mass of FGD wastes to be disposed of annually in the mid-1980s will be about 15 million t compared to about 70 million t of coal ash (of which almost 50 million t is fly ash); however, if all coal-fired capacity is desulfurized by the mid-1980s, the dry sludge would equal, or even surpass, the ash total. FGD waste disposal is thus less space demanding than the required fly ash ponding but is beset with many problems which until recently no utility managers had yet encountered.

Chemically, the wastes with pH 10.5–11 consist of calcium sulfite (about 75% of the sludge for lime, 60% for limestone FGD), calcium sulfate (around 10% in both systems), calcium carbonate (about one-third in limestone, just 5% in lime FGD), and there are also varying amounts of fly ash removed by the scrubber. Shares of individual constituents depend on the additive use, coals' sulfur content, and a scrubber's design. That the waste liquors exceed drinking water standards for total dissolved solids with high concentrations of Ca, SO_4^{2-}, Cl, Mg, and Na is hardly surprising; several trace metals from fly ash are also usually present and potential contamination of groundwaters by the scrubber liquids must always be considered.

Physically, the waste properties vary widely and are determined mainly by the degree of dewatering. Naturally, insufficient dewatering, or inadvertent rewetting, makes the sulfite crystals physically unstable, with little or no compressive strength. Improvements in currently used dewatering equipment offer one solution or the wastes can be oxidized to produce gypsum with crystals larger and thicker than sulfite and hence settling quicker and absorbing less water; however, the additional oxidation outside of the scrubber increases the cost so that perfection of in-the-loop oxidation techniques is needed. Another common aid is chemical fixation to stabilize the wastes with additional benefits of reducing major solubles in the leachate and a substantial (an order of magnitude at least) decline in permeability.

An Electric Power Research Institute study found that for a hypothetical 1000-MW unit, the total (30-year levelized) disposal cost per dry tonne of waste solids (sludge and fly ash) is \$(1978) 10.90 for wet disposal without and \$14.60 with fixation (using the Dravo Calcilox process), while for dry disposal the cost without fixation would be \$11, with fixation \$(1978) 12.40 (Rittenhouse, 1978).

Jones (1979) cites as typical ponding costs \$(1977) 5–8/t dry solids, and about \$(1980) 9/t; for chemical treatment and landfill, the cost rises to about \$(1980) 14/t (the latter figure includes the cost of clarifier and thickener). These costs could be reduced by selling part of the generated waste but although the potential for useful applications certainly exists, the experience with fly ash utilization is hardly encouraging: since 1963 when it was first used as an admixture to concrete, only slow and small inroads were made and by 1980 a mere 13% of the 45 million t of fly ash were not ponded (Jones, 1979) in spite of a special organization's (National Ash Association) promotion efforts. In the United States, only a very limited demand exists for gypsum as a farming liming material and so practically all FGD wastes will be ponded or used as landfill.

Among the regenerable techniques, which will account for a mere 6.2% of United States installed FGD capacity in 1985, the Wellman–Lord process is dominant (67% of the total). In this sulfate scrubbing process, a solution of Na_2SO_3 absorbs SO_2 from precleaned flue gases in a scrubber or absorber, producing bisulfite:

$$Na_2SO_3 + SO_2 + H_2O \rightarrow 2NaHSO_3$$

which is thermally decomposed in an evaporative crystallizer and the regenerated sulfite is reused as the absorbent. Wet SO_2-enriched gas flow can then be converted to H_2SO_4 or to elemental sulfur (for a detailed material balance see Jahnig and Shaw, 1981). The process is, as other regenerable options (magnesia, aqueous carbonate and citrate scrubbing), both more expensive to install and to operate than the throwaway lime/limestone FGD (see the next section). However, the future share of regenerable FGD is almost certainly expected to grow.

The main advantage is, of course, the greatly reduced amount of waste requiring disposal. Sale of the by-products to offset the system's cost appears naturally attractive but will be hampered by difficult marketing and transportation problems which may make any large-scale deals impractical. For example, if by-product sulfuric acid could be sold to a nearby facility, it would be better to produce it instead of elemental sulfur but this course would be foreclosed for any large power plant farther away from H_2SO_4-consuming industries.

4.5.4.2. Costs and Innovations In spite of the fact that the lime/limestone processes are not able to take any by-product credit, they remain the cheapest choice as far as both the capital investment and operating expenses are concerned—although they are a good deal more expensive than anticipated at the time of their introduction. The earliest cost estimates for the limestone process based on a retrofit to a 200-MW coal-fired unit put the capital investment at \$(1970) 13.83/kW, while for a new 1000-MW unit only \$(1970) 6.40 was thought sufficient (Slack et al., 1971). Yet the pilot plant work forced changes in these estimates so that 3 months after their publication the authors mailed out corrections putting the cost at over \$(1971) 30/kW.

Since then a large number of cost esimates have been published but as they refer to different types of coal, different efficiencies of SO_2 removal, various plant sizes and times of construction, comparisons are not meaingful. Only with the accumulated experience has it been possible to prepare comparative assessments by adjustments to a common basis and between 1975 and 1978 nine such reviews were prepared by various United States organizations and companies, the most detailed ones being an evaluation by the Radian Corporation (1977) for the USEPA and a PEDCo (1978) survey. These studies estimated FGD capital costs for a new 500-MW coal-fired unit to be as low as $(1977) 65 for the sodium carbonate process and as high as $(1978) 164 for double alkali cleanup with revenue requirements ranging from 3.5 to 6.4 mills/kWh.

Two comprehensive studies published in the early 1980s provide perhaps the most reliable guides to FGD expenditure and I shall review them in some detail. The first one was prepared by Bechtel National (1980) for model FGD systems installed at a new 2 × 500-MW coal-fired power plant in the Midwest which generates electricity with 70% load factor and complies with the USEPA's New Source Performance standards: the study shows very small investment and operation differences among the leading commercialized processes.

The second study, by Jahnig and Shaw (1981), first provided a convenient graphic comparison of investment costs for limestone scrubbing based largely on previous cost reviews by the USEPA, PEDCo, Bechtel, and Batelle and reproduced

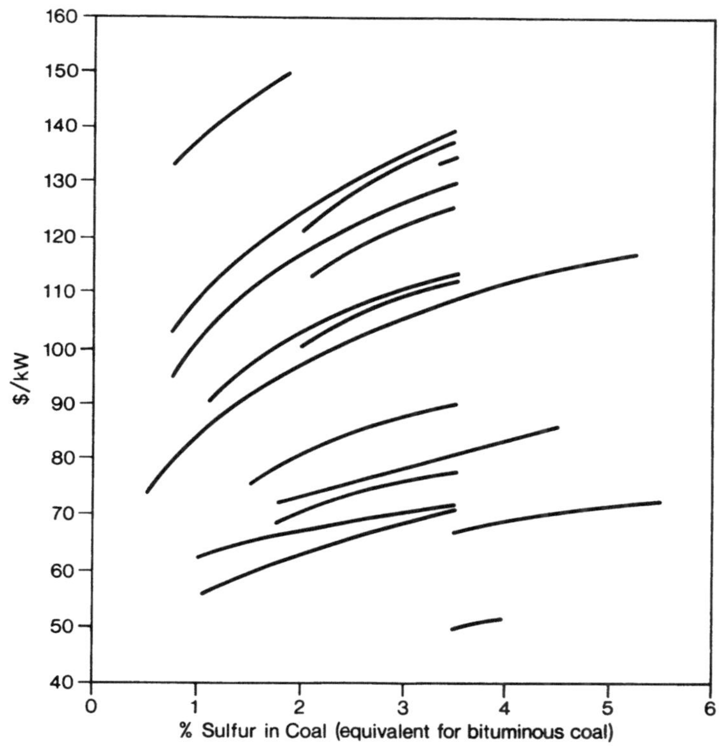

FIGURE 4-18. Comparisons of investment costs for limestone scrubbing; the curves chart the results of 13 different studies or plant experiences during the years 1976–1979 with all costs given in mid-1979 dollars (Jahnig and Shaw, 1981).

in Figure 4-18. With two exceptions, all cost curves intersect or touch the band for coal with a sulfur content of 2–3% and display a wide range of costs from about \$(1979) 60/kW (for a Northern States Power system with no sludge pond) to over \$(1979) 140/kW for a Bechtel study estimate for Western coal with landfill.

As their model for detailed cost comparisons, Jahnig and Shaw (1981) chose a power plant with installed capacity of 800 MW, burning bituminous coal with a sulfur content of 3.9%, and having 90% SO_2 removal for limestone FGD and 95% in other cases with all costs for the middle of 1979.

Their most important assumptions in the process of conversion to a common base were to prorate investment versus plant size at the 0.8 power of size and at the 0.3 power of the amount of sulfur removed. As FGD systems consume both electricity and steam from the power plant, a larger output of both is needed in a station with SO_2 controls to maintain a designed net output and Jahnig and Shaw (1981) correctly included the resulting increase in investment and operating cost as part of the overall FGD expenditure.

Other studies have shown that energy cost of FGD remains virtually constant over a 20-fold variation in emission levels: for liquid sodium systems, about 2% of primary energy input is needed for fans and pumps; for limestone FGD, energy penalty increases nonlinearly with increasing SO_2 removal rate and coal sulfur content, representing at least 3% and as much as 5% of the initial thermal inputs (Rubin, 1981; an interesting detailed study is Schiff, 1979). These shares are much lower than energy costs of coal cleaning: the operation itself is not energy intensive but the penalty is in the coal lost as refuse during the process. Typical cleaning losses translate into 12–16% of the power plant's thermal inputs, the highest value being for deep cleaning of high-sulfur Eastern coal.

Cost estimates for limestone scrubbing are undoubtedly most reliable owing to the relatively large number of installations and the length of construction and operating experience but comparison of 12 published studies shows a surprisingly wide range of costs: for an 800-MW plant with 90% SO_2 removal and 3% S coal, the total investment costs were as low as \$(1979) 65 and as high as \$130. Of this range, Jahnig and Shaw (1981) chose a medium value to represent limestone scrubbing in comparisons with other FGD methods.

Replacement energy investment is nearly identical for all nonregenerable systems—\$(1979) 19–21/kW—and even the process costs are not much different [all values are in \$(1979)/kW]—109 for limestone, 104 for purchased lime, 117 for lime with on-site calciner, 118 for lime–limestone—so that the total costs for the family of scrubbers are between \$(1979) 124 and 140/kW. In comparison, regenerable FGD costs considerably more, \$(1969) 167–190/kW, with the Wellman–Lord process at \$177. Nonregenerable FGD has also a clear operating cost advantage even after the by-product credit of the regenerable process is taken into account. While limestone scrubbing can operate with a mere 5.9 mills (1979/kWh and other throwaway systems cost 6.1–6.3 mills/kWh, regenerable FGD requires 7.4–9.1 mills (1979)/kWh (Wellman–Lord rates 7.6).

Taking \$(1979) 130 as the average per kilowatt investment for nonregenerable systems and \$(1979) 180 as the mean of regenerable processes to weigh the respective mid-1980s' FGD capacities (46,134 and 3078 MW) results in a total investment of \$(1979) 6.55 billion for the country's FGD equipment installed at 15% of all coal-fired power plants. An obvious question concerns what additional burden this represents in building the plants whose basic costs have been rapidly escalating anyway.

Detailed power plant cost studies by Budwani (1978, 1982) offer the best comparative base with the FGD data of Jahnig and Shaw (1981). New coal-fired units built and ordered between 1975 and 1985 range between 250 and 1300 MW, with most at 500–600 MW, and during the late 1970s their cost ranged from \$(1979) 300 to 600/kW, a twofold difference identical with the investment spread for limestone FGD. Taking \$(1979) 450/kW as the average, installation of lime/limestone FGD at \$(1979) 130 raises the base price by 29% to \$580/kW.

This, rounded to 30%, is the best single-figure estimate I can offer. Budwani (1978) lists the costs of a coal-fired plant with FGD at \$(1979) 530–900, estimates that desulfurization adds 25–30% to the total capital cost, and notes that a further 3–4% penalty has to be included owing to the reduction of generating output going for the scrubbers' energy requirements. One can thus safely conclude that FGD raises the capital cost of power plants by at least 25%, more likely by a third, and such a conclusion belies any arguments about marginally higher prices for cleaner air: the cost is clearly high. And operating costs are of the same magnitude. At 6–8 mills/kWh, FGD systems add 25–40% to the basic cost of 20–25 (1979) mills/kWh.

However, long-term costs of meeting clean air requirements are not only dependent on the prescribed controls: as far as SO_2 from power plants is concerned, they will be determined above all by the growth rates of electricity consumption with lower future increments translating into less new capacity and fewer control systems. As our long-term generating capacity and energy consumption forecasts have been nearly uniformly incorrect, many past attempts to estimate the FGD costs can be seen, just a few years after their release, as highly exaggerated.

Efforts to cut the considerable costs of FGD, or at least to keep them from rising substantially, will certaintly bring many modifications and some fundamental innovations in the years ahead. Two such major trends appear to be currently discernible: moves toward dry scrubbing systems, and the increased use of fly ash in desulfurization.

Dry scrubbing systems have several obvious advantages: they need less equipment (no thickeners or vacuum filters, less complex slurry preparation) and hence lower investment, their energy requirements are much lower as there is no reheating of gases, and their wastes are easier to handle. For low-sulfur coals they already appear more cost effective than the combination of conventional particulate-only and SO_2-only controls (Orem, 1982). At the beginning of the 1980s, only one dry FGD process was operating at a 100-MW unit but by 1985 there was a dozen with a total capacity of 3953 MW, with 95% of installations on units burning coal with a sulfur content of less than 1%.

Two key ingredients of dry FGD are spray dryer absorber (a vessel containing flue gas dispenser and atomizer unit, with the hopper bottom) followed by efficient precipitator or baghouse. Very small reagent particles (50–100 μm in diameter) must be generated by the atomizer and intimately mixed with the dirty gas to achieve high rates of SO_2 absorption as well as good reagent drying (Davis et al., 1979). Expected diffusion of dry FGD processes will also mean greater R & D activity in the field of fabric filters for baghouses. So far only a few fibers are available to withstand the demanding thermal (82–120°C) and chemical (alkali sorbents and the potential presence of moisture) conditions encountered in dry FGD.

Use of fly ash as a source of alkali is the most obvious modification to supplant, partially or completely, lime or limestone. Advantages of such an arrangement are numerous: addition of ground alkaline compounds is minimized or eliminated; reaction of SO_2 with fly ash is more stable than with lime or limestone; total volume of waste solids is reduced; the resulting solids may be easier to handle; and liquid/gas ratio, and hence also the pumping power, is reduced owing to more efficient removal (C. Johnson, 1979). Fly ash may either be collected in the usual manner—dry with electrostatic precipitator—and then reintroduced into the wet scrubber's gas stream or a venturi scrubber can be integrated into the SO_2 scrubber for joint removal.

The latter choice is greatly preferable as a significant reduction of capital costs results from the lower cost of venturi scrubbers (an electrostatic precipitator of the same capacity may cost up to 20 times more), from their collection of fly ash regardless of the sulfur content (precipitators are well known to have lower efficiencies with low-sulfur flue gases), and from joint handling and disposal of fly ash and sludge. Three factors will determine the extent to which fly ash can substitute for the outside alkali: quantity of SO_2 to be removed; the amount of fly ash generated; and the fly ash alkali content and its availability.

The first factor is likely to be given by air quality requirements and the second one is an unchangeable natural variable given by a coal's ash content. However, the third factor can be varied as the extractability of fly ash alkali is a function of pH: for the best alkali extraction from the fly ash, the scrubber recycle slurry should be maintained at low pH (around 4).

And while in conventional alkali systems the liquid/gas ratio goes up sharply with lower slurry pH, this relationship is fortunately not true when fly ash is used. As other fly ash cations in the liquid phase of the recycle slurry (Mg^{2+}, Na^+, K^+, Al^{3+}) are extracted with varying degrees, the scrubbing efficiency is enhanced and up to 20% of pumping energy may be saved in comparison with the lime or limestone removal. Finally, scaling problems caused by calcium sulfate supersaturation in conventional FGD systems are eliminated, thus raising the overall process reliability.

Alkaline fly ash scrubbing is especially suited for FGD of low-sulfur coals from the western United States: although their sulfur content is most often well below 1%, their low heat value means that coal with a sulfur content of no more than about 0.4% could be burned without controls to comply with the USEPA's performance standards of 520 mg/J (Maxwell and Shapiro, 1979). As some of these coals contain up to 20% ash, which in turn has up to 40% alkaline constituents, fly ash FGD may be sufficient to bring the emissions below the needed limit.

While fly ashes contain a mixture of alkaline ions which improve the efficiency of the SO_2 removal process, the same effect may be achieved by addition of just one suitable element—and magnesium is perhaps the best choice. Compared to calcium, it is 100–1000 times more soluble in the scrubbing liquor (the difference is determined by pH and ionic effects), yielding more alkali per unit of slurry and lowering pumping rates; magnesium will also eliminate the effects of coal chlorides and will help to prevent calcium sulfate supersaturation and scaling (Josephs, 1980).

The best method of magnesium enrichment is, of course, to use a limestone contaminated with enough $MgCO_3$: theoretically, 2–3% would be sufficient but the carbonate's reactivity must be ascertained first. With very pure limestones, dolomitic lime (the cheapest choice), magnesium oxide, or magnesium hydroxide can be used. Successful operation of magnesium-enriched FGD systems at the Phillips station (Duquesne Light Company) and at the Paddy's Run and Cane Run scrubbers (Louisville Gas & Electric Company) dates to the mid-1970s and the applications are spreading.

A major recent innovation is the removal of SO_2 in an aqueous slurry of melamine to produce solid melamine sulfite (Gautney et al., 1982). Subsequent thermal decomposition at 100–200°C regenerates the melamine completely and yields SO_2 and water vapor. Laboratory scrubbing tests showed high SO_2 removal rates (95.5%), low energy requirements for regeneration, and no scrubbing efficiency reduction from other flue gas constituents but, obviously, extensive bench-scale and pilot-plant testing will have to precede any successful full-scale operation.

An innovation with a prospect of immediate application is the recent identification of adipic acid as a highly effective additive to improve the performance of wet limestone FGD systems (Mobley and Chang, 1981); considerable mass transfer enhancement is achieved by the buffering effect of less than 3000 ppm of the acid without any significant environmental impact and with clear economic advantages owing to the savings of energy, reduction of solid waste, and possibility of utilizing less expensive coal and limestone. Additions of adipic acid can also reduce the electricity demand for 90% SO_2 removal by 30% at a concentration of 1400 ppm (Borgwardt, 1982).

Several Japanese companies have been developing control processes for simultaneous desulfurization and denitrification based either on oxidation–absorption–reduction technique [using O_3 and $CaCO_3$ to produce $CaSO_4$, $Ca(NH_2SO_3)_2$, N_2, and N_2O] or absorption–reduction method working with Na_2SO_3 or $(NH_4)_2SO_3$ as the main reagents and water-soluble ferrous-chelating compounds as catalysts to aid in the absorption of the insoluble NO (Chang et al., 1982). As of 1985, none of these processes has approached the commercial stage.

I believe that in retrospect the 1980s will be seen as the decade of transition in

FGD: wet alkali scrubbing, although now certainly a mature technology, will continually adjust and improve but its dominance will gradually start to slip as a greater variety of processes are installed in North America and abroad. New substantial orders in the EEC countries, above all in West Germany, will be an important stimulus for perfecting the existing method and intensifying research in innovative processes where the Japanese contributions will also be important.

These developments will, almost certainly, bring some notable capital and operation cost reductions but FGD at large thermal plant and major nonferrous smelters will remain expensive and will continue to represent a significant portion of the whole plant investment. Perhaps the least encouraging outlook is for practical commercial uses of recovered wastes or potentially salable products. But availability and reliability of the technology are most definitely no obstacles for controls on almost any scale; uncertainties surrounding the optimum control levels and the relatively high cost of FGD are the key problems to consider. Even so, future expansion of FGD controls is assured as is the diffusion of large-scale liming of acidifying waters.

4.5.5. Liming

Addition of $CaCO_3$ or $Ca(OH)_2$ to acid farm soil, especially to those receiving large quantities of acidyfing ammoniacal fertilizers, is a widespread practice whose detailed review can be found in Pearson and Adams (1967). Liming of lakes and ponds has also been a long-established part of management for higher productivity and since the mid-1970s the practice has been extended to acid lakes in Scandinavia and North America, with Sweden being by far in the forefront of this remedial intervention: by the summer of 1982 some 1500 lakes had been limed at a cost of 100 million Swedish crowns (Swedish Ministry of Agriculture, 1982).

Compounds used most frequently in this extensive Swedish liming have been lye, soda, and quicklime, all much more soluble than limestone and dolomite: about 5 g of the former agents is needed per m^3 to raise the pH from 4.5 to 6.5, half of the requirement of the latter. For lakes with long-retention lime, the least-cost method is to put the lime directly into the water with one application of 10–20 g/m^3 roughly sufficient to neutralize the water for up to 10, even 15, years. The Swedes have also been liming thousands of square kilometers of watersheds (at a rate of about 200 kg $CaCO_3/ha$) but this treatment is recommended only on low-lying areas with a high groundwater table or on waterlogged ground.

The lakes most difficult to lime are those with waters very high in iron—as iron humates precipitated onto the undissolved lime prevent the compound's further dissolution—and streams, not surprisingly, present an even greater challenge. Continuous dispensers (a variety of simple but reliable continuous doser and well arrangements) are much more practical than huge one-shot applications.

The results have been largely satisfactory: significantly higher diversity and biomass of both phytoplankton and zooplankton, and the recovery of young fish and crayfish to preacidification levels are the most notable improvements. Without any doubt, liming of lakes, and even watercourses, is a viable method to restore natural pH in areas sensitive to high acid deposition, especially as the practice is devoid of any serious unwelcome side effects. The only concern arises from transient oxygen deficiency in stationary bottom waters where liming-induced higher rates of decomposition cause increased oxygen demand but this temporary effect is unimportant in comparison with pronounced reduction of toxic aluminum and mercury loading brought by liming. Consequently, the greatest practical complication is to keep the pH high during sudden acid releases associated with the first spring snowmelts.

Needless to say, liming treats the symptoms and not causes of acidification of aquatic or soil ecosystems but it does so at a relatively modest cost and in an environmentally benign fashion: for these reasons it should be practiced as widely as possible in all vulnerable regions of Europe and North America—until the time when acid deposition rates have shrunken to a barely noticeable nuisance.

5 LESSONS, OUTLOOKS, CONSEQUENCES

"There's more evidence to come yet, please your Majesty," said White Rabbit, jumping up in a great hurry.

—Lewis Carroll
Alice's Adventures in Wonderland (1865)

Before closing this book, succinct summaries of the best available evidence on the extent and effects of anthropogenic interference in the three cycles are in order.

The judgment must be equivocal as our understanding of the three key biogeochemical cycles is a mixture of exhaustingly traced minutiae, satisfactory grasp of basics, major gaps in essential segments, and not a little ignorance. About 1000 references cited in the preceding chapters, a fraction of the available material dating mostly since the early 1970s, attest well to the scope and accelerated intensity of scientific interest in studying complex natural cycling and its anthropogenic interferences.

Yet it is not only some marginal flows of obscure, unimportant sections of major cycles that remain only qualitatively known. To name just one example in each of the three cycles, our understanding of carbon releases owing to vegetation changes, denitrification flux, or volcanic contributions to atmospheric sulfur is so uncertain that the totals put on these flows can differ by up to an order of magnitude. And in scores of instances the quantification cannot be done better than within two- to fivefold ranges.

Summarizing major findings and conclusions presented in the three topical chapters will involve unavoidable repetitions but the reviews have the advantage of stressing, for the last time and in a condensed form, the two key impressions which should have emerged from the detailed coverage: uncertainties in our understanding of the three cycles are neither negligible nor removable soon by intensified research; and while human interferences are of such scales and extents that concerns and cautions are imperative, they do not justify any catastrophic conjectures and doomsday scenarios. These messages will unify the following separate reviews.

5.1. CLIMATIC CHANGE AND THE BIOSPHERE

Photosynthetic reduction of CO_2 is the foundation of all complex life on this planet but it was only during the late 1950s and the early 1960s that we came to understand in detail its biochemical paths and only since the mid-1960s that we started to acquire a thorough, systematic understanding of primary productivity in the biosphere's rich variety of ecosystems. In spite of these efforts, global estimates of autotrophic production and standing phytomass storage retain a large margin of uncertainty, a situation largely ascribable to rapid changes in areas of major ecosystems and to the impossibility of offering representative large-scale averages based on sparse localized detailed assessments.

The best available information now places global annual terrestrial primary production at around 100 billion t C and the total storage at about 1 trillion t C; ocean phytomass is negligible compared to this huge land storage but marine autotrophs probably fix no less than 60 billion t C annually. Carbon stored in heterotrophs, a reservoir dominated by microorganisms, is a very small fraction of all biomass.

Human interference in the fixation and storage of living carbon comes from changes in the areal extent of ecosystems (above all tropical deforestation and conversion of forests and grasslands to farmlands) and from biomass combustion. The former process now results in a net annual loss of between 1 and 2 billion t C, i.e., no more than about 0.2% of standing terrestrial phytomass. Household and industrial combustion of fuelwood and crop residues releases about 1 billion t C, and about 2 billion t C is oxidized during regular burning of grasslands and in fires set by shifting cultivators. Total anthropogenic liberation of biomass carbon thus appears to be approaching 5 billion t annually but only the first 1–2 billion t C is produced in excess of the normal photosynthesis–respiration cycle as most of the combustion is just an accelerated respiration.

Calculation of CO_2 generation from burning of coals, liquid fuels, and natural gases can be accounted for with much higher accuracy than biomass carbon releases but numerous weaknesses of global energy statistics combine to cause overestimates on the order of 10% in recent estimates and higher errors in the historic reconstructions commencing in the mid-19th century. Evaluation of coal-derived CO_2 emissions is certainly least accurate, that for natural gas (excluding flaring) most reliable.

Total annual releases of carbon from combustion of fossil fuels are now very close to 5 billion t and the aggregate anthropogenic generation of excess carbon is at least 6, possibly as much as 7 billion t/year. Past releases are, obviously, much more uncertain: combustion of fossil fuels since 1850 has most likely generated between 150 and 180 billion t C; the range for CO_2 from ecosystemic changes (pioneer agriculture and other deforestation) is much more uncertain but totals around 130 billion t C are most plausible.

The grand total of all anthropogenic releases for the period 1850–1980 is thus about 300 billion t C, or 2–3 billion t C/year. Unfortunately, neither of the two naturally changing isotopic ratios—$^{14}C/^{12}C$ and $^{13}C/^{12}C$—is unequivocally helpful in verifying the historic CO_2 releases and hence the relative contributions of fossil fuel combustion and biomass removal shall remain a matter of speculation.

So will the past increases of atmospheric CO_2 concentrations as it is only since the late 1950s that we have reliable, continuous measurements from Hawaii (Mauna Loa) and from the South Pole, more recently from Alaska and American Samoa. These isolated sites are excellent locations for charting the background CO_2 concentrations and although minor complications and errors in calibrating the monitoring devices and comparing the records among stations do not permit attainment of values with accuracies of less than 1 ppm, there is no doubt whatsoever about the global trend: atmospheric CO_2 content has steadily increased.

Regular diurnal (early afternoon minima, late morning peaks) and seasonal (late winter maxima, late summer minima in the Northern Hemisphere) fluctuations caused by photosynthesis–respiration cycles overlap the rise which has averaged about 1.2 ppm/year at Mauna Loa and 1.4 ppm/year for the entire Northern Hemisphere and which has brought atmospheric CO_2 levels to more than 340 ppm in the early 1980s.

Although we do not lack tropospheric CO_2 measurements after 1850, all values before 1870 must be dismissed as unreliable and the later values, too, are beset by biases and errors so that we cannot estimate the preindustrial CO_2 concentrations closer than between 270 and 290 ppm. If all CO_2 from fossil fuel combustion and biomass clearing had remained in the atmosphere, concentrations would have increased by about 140 ppm rather than by the maximum of 70 ppm, and thus about half of the flux has been sequestered, most likely in the ocean.

With recent more accurate data, it is easy to calculate that no more than about 56% of fossil fuel-derived CO_2 remained in the atmosphere between 1960 and 1980 and when the excess CO_2 from ecosystem shifts is added, the retention share drops to as low as 40%. However, annual fluctuations of anthropogenic CO_2 retention in the atmosphere have been very large and remain unexplained.

While any accurate predictions of CO_2 increases are impossible, fixing the uppermost limit is rather easy: combustion of all ultimately recoverable fossil fuels would, with 50% retention, push atmospheric concentrations to nearly 1000 ppm. Although the eventual burning of all this fuel is unlikely, doubling of current CO_2 concentrations with continuous reliance on fossil fuels as the key sources of primary energy is an obvious possibility.

When assuming, as did most of the models offered during the 1970s, that global energy consumption will grow at 4% annually and 60% of generated CO_2 will be retained in the atmosphere, the doubling would come in the third decade of the next century; on the other hand, the slower consumption growth prevalent during the first half of the 1980s and retention rates below 50% would postpone doubling of preindustrial concentrations until the closing decades of the 21st century.

Regardless of the actual doubling time, trophospheric warming should accompany this CO_2 rise as the well-known opacity of the gas in the wavelengths around

15 μm would increasingly block terrestrial infrared radiation. Effects of this warming have often been described in fairly catastrophic terms but they will be, obviously, critically dependent on the magnitude of the eventual temperature rise.

Numerous quantitative models have been presented since the early 1960s, ranging from simple one-dimensional calculations to complex three-dimensional computerized exercises involving modeling of global climate and consuming hours of processing time. The latter efforts have been instrumental in establishing, against some infrequent but strong dissent, a consensus that doubling of preindustrial CO_2 levels will raise trophospheric temperatures by about 3°C.

The other generally accepted conclusions are: trophospheric warming will be about two to three times more pronounced in higher latitudes than in the tropics and greater in the Arctic than in the Antarctic; it will be accompanied by stratospheric cooling; it should lead to higher circumpolar runoffs, later snowfalls, earlier snowmelts and to general intensification of the global water cycle; it will cause some important but difficult to pinpoint shifts in frequency and amount of precipitation; and it could significantly influence planetary primary productivity.

More specific descriptions of changes are abundant but largely speculative. On land the possible shifting of precipitation zones has been studied mainly with the help of climatic analogues (ranging from the early medieval warming all the way down to the Pliocene) and the predictions are for Russia, Europe, Mediterranean Africa, Middle East, and parts of India and China being much wetter and, most notably, the heartland of the North American continent being drier. Informed consensus does not see, however, any drastic changes in global food production capacity.

The extent and rapidity of polar ice melting would determine the rate of ocean level rise and hence the damages and responses accompanying this change. Melting of Arctic ice would not have any great coastal effects but it would shift existing midlatitude climatic zones at least 200 km northward. More unlikely disintegration of the western Antaractic sheet would take hundreds of years and it would raise ocean levels by about 5 m, causing the greatest population displacements in Asia and the highest economic loss in the United States and western Europe. Still, even a 5-m rise would flood only about 1.5% of the United States and displace less than 6% of the country's population.

Uncertainties surrounding all assessments of CO_2-induced warming are not surprising. Revelle (1982) concludes: "Indeed, about the only facts available are the actual measurements of atmospheric carbon dioxide . . . and some fairly reliable data from the UN on the annual consumption of fossil fuels in industrial countries."

Studying changes in a system embracing the whole biosphere and unfolding on a time scale of centuries cannot lead to anything but a succession of very frail and changing conclusions. A closer look at models predicting the extent of atmospheric warming shows a long array of fundamental weaknesses. Even the most complex three-dimensional simulations have limited computational domain, treat the ocean in a variety of simplistic ways, and cannot account reliably for complex cloud behavior, snow and ice cover changes, and water vapor feedbacks.

Predictions of tropospheric temperature increase brought by CO_2 doubling average no more than 2.13°C when all published three-dimensional models are taken into account, while the means for two- and one-dimensional exercises are, respectively, 1.8 and 2°C. Considering these similarities, is there justification to exclude automatically all simpler models and to rely for "consensus" only on select three-dimensional studies predicting increases around 3°C? A more fundamental question is: how much sense does it make to model future atmospheric temperature increases just as a function of CO_2 concentrations when climate is a complex response to a host of terrestrial and extraterrestrial variables?

Climatic variation is an enormous realm of study, an exceedingly difficult topic replete with complex feedbacks where subtle changes are often simply untraceable to initial deviations of a single variable. Besides natural changes caused by variability of solar flux, planetary orbital fluctuations, tidal periodicity, volcanic activity, cryosphere and ocean behavior, there are important anthropogenic factors which must not be neglected. These include, above all, other "greenhouse" gases, ranging from nitrous oxide to methane and halocarbons (freons) as well as particulate matter generated by many human activities.

Discerning an unmistakable signal of CO_2-induced warming in the midst of all of this natural and man-made clutter is, in the least, an extraordinary challenge, more likely an impossibility. The best conclusion is that in the mid-1980s, with CO_2 some 50–70 ppm above the preindustrial level, no trophospheric warming attributable to CO_2 can be confirmed by careful analyses. And even relatively simpler responses are difficult to pin down, foremost among them the effect of CO_2 doubling on plants.

We can say with certainty that higher CO_2 levels will not harm them: for nine-tenths of the long time autotrophs have been around, their environment was warmer and tropospheric CO_2 concentrations higher than today so the future change may be seen as a return to long-term normal. But how significant will plants' positive response be? A wealth of experimental data confirm higher photosynthetic rates, higher dry weight per unit leaf area, better germination, earlier flowering, faster maturity, higher tolerance to some pollutants, higher N fixation, and, perhaps most notably, lower transpiration losses and hence more efficient use of water.

This list of very desirable changes translates into dry mass yields 20–45% higher and as much as doubled efficiency of water use with doubled CO_2 levels. Moreover, the effect is more marked in C_3 species, which include most major food and feed crops, and less pronounced in C_4 plants, a group to which most of the world's principal field weeds belong. Altogether, these are most desirable, even providential changes—but actual performance may be far less glorious.

While one should not overemphasize limiting roles played by water and nutrients in field conditions, shortages of timely moisture and of the three principal as well as numerous micronutrients will restrict the yield response to considerably less than an average gain of 30% measurable in experimental settings. In terms of storing additional biomass, only the response of forest ecosystems matters and although we have no empirical evidence, we cannot doubt highly species-specific reaction to doubled CO_2 and considerations of complex interaction in natural communities suggest that after an initial growth spurt, the later effect will be increasingly buffered and over the long term there may be no significant storage difference, just an interspecies shift. Simply, we should not expect the biosphere to become a huge sink for CO_2.

As for the response of crops to warmer climate, we can draw considerable comfort from the recent past: better farming methods and new plant varieties did succeed in raising the yields in periods of temporary warming (and cooling) as well as in introduction of major crops into new regions previously considered unsuitable for such field farming.

In fact, this adaptation may be seen as a perfect example of perhaps the most rational response to any eventual warming as controlling CO_2 at the source is, in contrast with preventing formation of sulfur and nitrogen oxides or stripping them from combustion gases, decidedly impractical. There are alternatives to adaptation but a closer look reveals their profound limitations. Cutting down growth rates of fossil fuel consumption is a practicable course in the rich world but not among the poor nations with obviously huge needs to expand their fuel use.

Coal burning is the largest anthropogenic source of atmospheric CO_2 but the Soviet Union, United States, and China, the three top producers possessing about three-quarters of all global reserves, all plan to expand its production and the notion of these three nations coming to an agreement on reduced coal use is obviously naive thinking: why should the Soviets consent, when their top climatic experts are convinced that the country will experience a warmer *and* wetter era with higher CO_2? Various evaluations of taxing fossil fuel use also illustrated the relative ineffectiveness of such a measure.

Then there is a large group of sci-fi-type proposals embracing pumping of CO_2 into abandoned mines or giant greenhouses, dumping it into gargantuan oceanic "gigamixers," fertilizing the ocean with phosphates to enhance primary productivity, or manipulating terrestrial albedo. These ideas can be easily dismissed as impractical but arguments about accelerated transition to nonfossil energies deserve a serious response.

Claims of those who argue that any more delays in embarking on such a course are dangerous and that we should, at worst, err conservatively by moving away from fossil fuels, must be countered by pointing out that the risks of an eventual 2–3°C warming are unquantifiable, that the very degree, timing, and consequences of trophospheric temperature rise are unpredictable, and that we simply must know much more before we act.

I am pleased that this attitude is shared by the latest CO_2 assessment by the National Academy of Sciences: *Changing Climate* (NAS, 1983b) stresses the ubiquitous uncertainties, calls for iterative assessments rather than clear-cut proclamations, and recognizes the inevitability of some climatic change regardless of the future makeup and level of energy use. "Overall, we find in the CO_2 issue reason for concern, but not panic . . . we do not believe, however, that the evidence at hand about CO_2-induced climate change would support steps to change current fuel-use patterns away from fossil fuels."

Seeing the CO_2 "problem" in the only sensible perspective—as an ingredient, albeit perhaps a very important one, of a complex multifactorial process of climatic change—allows us to face eventual changes with confidence rather than fears. Climatic change has been an inseparable part of the evolution of the biosphere as well as of the complexification of civilization. Our adaptive capacities, so outstandingly illustrated by voluntary long-distance migrations and by extension of productive farming to various extreme environments, will enable us to meet the challenge.

Obviously, as with all changes, impacts and benefits will not fall evenly and some responses will be much more successful than others; this is no particular attribute of the CO_2 "problem," just an inevitable complication in facing a change—rather than contemplating a planetary catastrophe.

5.2. NITROGEN'S ESSENTIALITY AND LOSSES

The element's protein-building criticality and relative scarcity in compounds digestible by man and assimilable by plants on the one hand and its huge inert atmospheric reservoir and intricate fluxes throughout the biosphere on the other make the nitrogen cycle a fascinating study in opposites. Of the three plant macronutrients, nitrogen received undoubtedly the greatest attention but complexities of its cycling are responsible for still numerous uncertainties in our understanding.

Primary production of natural ecosystems is fundamentally dependent on biofixation carried by free-living and symbiotic microorganisms. The latter ones, above all *Rhizobium*–legume associations, are of the greatest practical importance in managed agroecosystems but surprising weaknesses mark our record in utilizing this precious potential. Incredible as it sounds, there is not a single leguminous crop for which we have very reliable fixation figures in field conditions, the ecology of the genus *Rhizobium* is relatively poorly known, and the agronomy of legumes has been neglected in comparison with efforts put into grain and oil crops.

Although legumes are globally still the second largest source of vegetal protein, worldwide production of edible pulses has not increased during the past generation which saw a 50% rise in human population and rapid expansion of cereal and sugar crops. Leaving soybeans, overwhelmingly destined for feed, aside, per capita consumption of legumes has fallen all around the world, crops are grown largely without inoculation with appropriate rhizobia, there is little agronomic optimization and only some tentative efforts to commercialize many of underused species.

Grand plans for early manipulation of biofixation, offsprings of post-1973 concerns about hydrocarbon scarcity cutting into production of synthetic nitrogen, have been much reduced in scope and much extended in payoff times: these efforts, requiring a great deal of basic research, will face some profound practical obstacles and will take decades before significant advances are reached. Improving the performance of the well-known existing arrangements—legume symbioses, nonlegume associations, and algal systems—is thus certainly the best approach to follow. Yet it would be wishful thinking to believe that any wider integration of food legumes in modern farming will be greatly welcome; they shall remain proteins of the poor.

Similarly, recycling of organic wastes, theoretically such a desirable source of nitrogen owing to the slow release of the nutrient as well as to associated antierosion benefits, is not headed for expansion. In poor countries, where crop residue recycling is most desirable, competing uses of straws and stalks for animal feed, fuel and manufacture greatly limit the available quantities; in rich countries with high average yields, even complete recycling of all crop residues could not supply more than a fraction of nutritional needs.

Analogically, even the most extensive spreading of animal manures would not cover current high nutritional requirements and the practice is limited to utilization of wastes produced in confinement and made less appealing owing to its high cost and often huge volatilization and leaching losses. Composting of plant, animal, and human wastes is highly labor intensive and now on the decline even in China, the only remaining stronghold of the practice where the conversion of wastes into biogas has also shown signs of far from promising advance. In general, there can be no doubt about worldwide shift from organic sources of nitrogen to more efficient and more flexible synthetic fertilizers which, in spite of their recent cost increases, are also surprisingly cheap and readily available in all but the poorest nations.

The Haber–Bosch synthesis of ammonia in the second decade of this century ushered the new era in crop farming and its many improvements led to impressive declines of energy cost of ammonia and to wide affordability of nitrogenous fertilizers: in spite of the rapid price rises of the 1970s, United States ammonia in the early 1980s cost less than it did two decades earlier when expressed in constant monies and did not account for a larger share of farming expenditures although the average application rates rose substantially. Even in poor countries, where costs are much higher and availability much lower, both domestic production and application rates increased greatly during the last generation. The best example is China, which now applies as much synthetic nitrogen per hectare as does the United States! The long-term outlook is for no substantial change in the basic Haber–Bosch process based on natural gas or crude oil; coal feedstocks have perhaps the best prospect as an alternative while various proposed unconventional approaches to provide nitrogen are mostly decades from practical application.

With synthetic nitrogen assuming such a prominent place in modern farming, studies of nutritional requirements of plants became commonplace but our understanding is still full of serious gaps, in no small part because of the impossibility of precise measurement of nitrogen deficiency in crops. Yield response studies performed repeatedly for major grain crops show clear leveling off with increasing nitrogen applications but in optimum conditions the harvests may rise much beyond these common limits.

Clearly, without extensive nitrogen fertilization it would be impossible to support the high-protein diets of the rich world, and the nutritional standards of the poor countries, now ranging from adequate to grossly insufficient, would greatly deteriorate. There can be no doubt that nitrogen fertilization is the least replaceable major anthropogenic interference in a key biogeochemical cycle: we have little choice but to continue, and to expand, the practice if we are to live healthy, active lives and if the prospects of the poor world are to rise above the pains and uncertainties of subsistence survival.

Nitrogen in fertilizers, as well as in gases released during combustion of fossil fuels and biomass (the second most important route of human interference in the cycle, but one much smaller in global mass terms than fertilization), enters, moves, and resides in a variety of compounds in soils, waters, biota, and atmosphere and its fluxes and presence lead to many potentially undesirable or outright harmful effects. Unfortunately, we have had, in spite of a large research effort, only a mediocre success in quantifying major environmental losses of the element and in preparing reliable balance sheets.

Volatilization losses range typically from 5 to 20% of fertilizer when it is incorporated into soil and may easily surpass half of the applied nutrient in organic wastes spread on the surface. Among inorganic fertilizers, urea volatilization may be very high as may the losses from ammonia fertilizers spread onto high-pH waters in wet fields; little-appreciated losses from plant tops are also significant.

Denitrification, a commission to microflora sustaining the nitrogen cycle, increases with higher fertilization and more extensive legume planting but its quantification is exceedingly difficult. Experiments have measured maximum losses of many tens of kilograms of nitrogen per day after application of manure on water-saturated soils and the recorded daily extremes differ by several orders of magnitude. Selecting large-scale averages—such as 1 kg N/ha per year or 1% of applied fertilizer—is thus bound with huge uncertainties.

Releases of N_2O during denitrification became a focus of intensive concern during the 1970s when some simple modeling led to conclusions of worrisome threats this flux poses for stratospheric ozone. After about a decade of inquiries we must view the whole affair in a much calmer way. Seeing field fertilization as *the* culprit is certainly wrong as combustion of fossil fuels and biomass may be as large, or even a much larger source of N_2O; the fate of the gas is uncertain owing to our still evolving understanding of N_2O in oceans and soils (are they a source or a sink?) and to the role played by vegetation.

Modeling of N_2O doubling and its effects brought disparate results, ranging from huge O_3 losses (about 20%) early in the next century to negligible (a few percent) declines many decades later, and all such calculations must be seen as merely suggestive warnings and not as foundations for major modification of our farming practices. Actual rates of N_2O increase are now accurately known from the GMCC monitoring and they indicate doubling no sooner than two centuries from now. Undoubtedly, fertilizer-derived N_2O is only part of this flux and, moreover, whatever the eventual effects on O_3, they cannot be considered in isolation from other stratospheric processes. In sum, what was perceived in the mid-1970s as one of the major future environmental threats must be seen a decade later as a minor aggravation.

By all means, leaching of nitrate nitrogen into waters remains a greater worry, although its urgency also weakened considerably since the scares of the late 1960s and early 1970s when the process was seen by some scientists as a critically dangerous disturbance of the nitrogen cycle. Many environmental and agronomical variables determine the leaching losses which most frequently range from 10 to 20% of applied nitrogen and may reflect not only recent fertilization but also gradual soil nitrogen releases ushered in by the onset of cropping decades ago. As with volatilization, plant-top leaching losses, especially just before maturity, are also significant.

Of many nitrogen forms in waters, nitrates are by far the most important as their high concentration can cause life-threatening methemoglobinemia in infants and can aggravate eutrophication of lakes and reservoirs. Excessive nitrate levels are most commonly found in shallow, unlined wells near animal barns or feedlots or in watersheds where crops receive high rates of fertilization. Deeper wells and water bodies in such areas may also have elevated nitrate levels but only seasonally do these concentrations surpass even conservatively set health standards.

Even in such regions of continuous heavy fertilization as the United States Corn Belt, nitrate levels in waters have been found to be much less hazardous than postulated during the initial period of rising environmental consciousness when this phenomenon occupied a front place among degradative anthropogenic trends. Similarly, eutrophication aided by nitrogen in shallow eutrophic lakes has not become even remotely as worrisome a process as foreseen around 1970 when it was viewed as another proof of the nitrogen cycle being seriously out of balance.

When so many uncertainties surround the quantification of major nitrogen fluxes, it is hardly surprising that preparation of nitrogen budgets in agroecosystems remains in the realm of approximations where only qualitative assurances can be offered. Perhaps the most important of these are that concerns about exhaustion of organic nitrogen reserves in the intensively farmed soils of the rich world are not warranted, that proper field management can establish new soil nitrogen equilibria, and that the overall efficiency of utilizing the applied nitrogen is typically in excess of 50%.

The best possible quantification of nitrogen balances in the United States agroecosystem shows that chemical fertilizers are either the single largest nitrogen input or just a fraction lower than biofixation and that all managed inputs (fertilizers and leguminous crops, including improved pastures) account for about two-thirds of all additions: clearly, human intervention is the critical factor. The country's preference for foods rich in animal protein leads to relatively large transfers as roughly ten times as much nitrogen is removed from fields and pastures than is actually consumed in all foodstuffs.

With actual utilization efficiency not much above 50%, roughly 20 times the mass of nitrogen contained in food must be cycled through the agroecosystems.

This high cycling intensity has inevitable undesirable effects but as these are overwhelmingly local in scope and far from catastrophic in impact, there is no need for any drastic, emergency changes in current farming practices.

In comparison, managing the nitrogen cycle in the tropics is much more difficult. Rapid mineralization, low litter accumulation, and low nutrient efficiency in widespread oligotrophic tropical rain forests lead to irreversible nitrogen losses after deforestation, and subsequent rapid leaching and erosion make productive farming unsustainable. Fertilization and careful management can succeed only at a high cost in special circumstances and the paradigm of intensive temperate zone farming will thus remain inapplicable to most of the equatorial zone.

Health concerns associated with releases of nitrogenous compounds into the environment fall into two broad categories: air pollution caused by nitrogen oxides and nitrates and presence of high nitrate concentrations in waters and vegetables. The first worry is not convincingly supported by any large body of experimental or epidemiological studies. At the ambient levels currently prevailing even in large cities with heavy traffic and many stationary combustion sources, nitrogen oxide concentrations cannot be seen as worrisome contributors to higher incidence of upper respiratory diseases and to increased premature mortality. Surprisingly high indoor concentrations associated with unvented gas stoves and burners should certainly be of higher concern as they affect sensitive infants and children.

In any case, nitrates in waters and vegetables are the dominant concern, not because of their direct toxicity but owing to their enzymatic conversion to nitrites implicated not only in infant methemoglobinemia but also in adult cancer. Acute nitrite toxicity in babies is life-threatening but readily treated with full recovery and avoidable with simple precautions: in rich countries it is now an infrequent phenomenon as public water supplies contain safe nitrate levels and as people in areas with seasonally elevated nitrate concentrations in private shallow wells are aware of the potential risk for infants younger than 3 months.

In contrast, the risks of chronic exposure to potentially carcinogenic substances are virtually impossible to quantify and no "safe" intake levels can be specified. Most of our nitrates—at least nine-tenths—enter our bodies in vegetables and fruits with water and meats accounting for the rest. No simple generalizations regarding the historic changes in nitrate content of vegetables are possible as there is a huge variability even among cultivars of the same species. But there is little doubt that heavier fertilization raises nitrate concentration and that harvesting the crops in sunny weather lowers the nitrate level as radiation activates nitrate reductase.

As the highest nitrate concentrations are in petioles, stems, and outer leaves, removal of these parts in nitrate-accumulating species such as beets, broccoli, celery, radishes, and spinach will lower nitrate ingestion quite considerably. Part of this ingested nitrate is converted to nitrite but the formation of nitrosamines is, most fortunately, slowed by coingested ascorbic acid. Nitrites are also taken directly, above all in cured meats where they prevent lethal botulism and maintain appealing color and flavor.

In the 1970s, concern about nitrite consumption led to major reduction of the compounds in processed meats and, as with vegetables, simple precautions such as moderate intake of bacon and addition of citrus fruits to inhibit nitrosation can further lower the ingested amount. When the average smoker inhales daily about 100 times more nitrosamines than could be swallowed in food, it is clear that less smoking would be a much more sensible line of attack than an economically and epidemiologically (botulism!) costly ban on nitrate and nitrites in foods or even more taxing restrictions on crop fertilization. Put in a wider perspective, human nitrate–nitrite exposure has come to be seen, in contrast to the early scares, as a concern to be followed in careful scientific investigation rather than as a threat necessitating fundamental changes in our food system.

Controls of all the outlined releases of nitrogenous compounds fall in two large categories: lowering the formation and preventing the escape of nitrogen oxides during combustion—and managing fertilization in more efficient ways. Automotive NO_x controls, either by a variety of combustion modification techniques or by catalytic conversion of gases, can meet very strict standards for all internal combustion engines while controls of diesel emissions remain more difficult. Stationary NO_x are also controlled by modifications of the combustion processes and recent improvements such as premix, distributed mixing, and dual register burners can reduce the emissions by as much as 85%; however, controls beyond that level appear to require much more costly flue gas treatment.

Many adjustments and improvements can be combined to minimize nitrogen losses in fertilization. Soil testing and plant analysis are not still fully tapped even in many rich farming areas—but even with their help fertilizing does not become a straightforward science. Unexceptional recommendations include: choice of fertilizer by the least cost of nitrogen rather than by a specific compound as there is no significant difference between the efficiency of properly applied ammoniacal or nitrate compounds; delay of major applications until some time after planting; and incorporation of fertilizers into the soil.

Other measures should include prevention of soil erosion, proper crop rotations, choice of the best varieties, optimum irrigaion, control of weeds and pests, timely harvests, and appropriate postharvest measures. Slow-release fertilizers and nitrification inhibitors are the latest means to boost efficiency of nitrogen fertilizer. However, sulfur-coated urea, the leading slow-release compound, is not suitable for high-yielding field crops, and applications of commercially available nitrification inhibitors whose clear benefits—higher yield, higher protein content, lower nitrate content in plant tissues, lessened susceptibility to diseases—have been experimentally well demonstrated have had mixed field results. Still, this is an approach worthy of further attention.

In sum, prospects for controlling nitrogen releases to the environment both from combustion of fossil fuels and from fertilization of crops are excellent. Most of the methods are readily available, well-tested, and not exceptionally expensive. We have the know-how and technology both to expand the availability of nitrogen where needed and to minimize its undesirable fluxes.

5.3. CAUSES AND COSTS OF ACIDIFICATION

Sulfur, the critical bonding element in living matter, is much less abundant in the biosphere than carbon and nitrogen, although sulfates are dominant in evaporites and are the second most important anion in seawater. But its natural atmospheric fluxes, none of them known with satisfactory accuracy, are numerous, involving volcanic emissions, sea spray, and various biogenic releases. These fluxes, occurring largely as hydrogen sulfide and dimethyl sulfide, have been measured only recently both on land and during research cruises and such records are still far from sufficient to eliminate an uncomfortably huge range of possible flow deduced from simple balancing of existing sulfur cycle models.

Anthropogenic sulfur emissions come above all from coal combustion, but ever-present variation in coals' sulfur content, lack of reliable data on sulfur content of the fuel as actually burned, and the impossibility of representative generalizations on the share of SO_2 retained by alkaline ashes make it impossible to estimate global coal-derived SO_2 flux with accuracy of less than several million tonnes annually. Compared to this flux of at least 60 million t S/year (a flow comparable to terrestrial biogenic flux), SO_2 from coking and cement production (around 500,000 t s/year for each process) appears negligible. Emissions from refining of crude oil and combustion of liquid fuels are currently only about a third of coal-derived flux, while those from combustion of natural gas are insignificantly small. Nonferrous metallurgy adds about 10 million t S/year and while H_2SO_4 synthesis and woodpulping are negligible sources in the global account, their uncontrolled releases may be locally highly objectionable.

Large uncertainties in quantification of natural sulfur fluxes—their aggregate may be as low as or just over 100 and as high as more than 250 million t S/year (i.e., mean close to 200 million t S)—and the smaller, although still notable range for anthropogenic contributions—anywhere between 80 and 110 million t S by 1980—require an interval estimate of human interference in the global sulfur cycle. Anthropogenic "enrichment" may now be as large as all natural fluxes and as small as roughly one-third of their total but most likely equals one-half, i.e., about 30% of all sulfur entering the atmosphere.

Human contributions derive disproportionately from the prosperous countries of the northern temperate belt (almost 95% of the total originates in the Northern Hemisphere) and in the most industrialized regions of these countries anthropogenic sulfur emissions may not only surpass natural flows but could overwhelm them by up to four orders of magnitude! In small industrialized European nations and in many states of the eastern United States, anthropogenic sulfur fluxes now prorate to 5–10 t S/km^2 per year and the single largest point sources (nonferrous smelters and coal-fired power plants) can emit between 0.5 and 2 million t SO_2 annually.

Sulfur dioxide, the dominant sulfurous pollutant, has been extensively monitored since the 1950s in urban regions of Europe and North America and a variety of analytical methods allow for sufficiently accurate assessment of secular trends. The key conclusion is that both average and peak seasonal concentrations declined almost everywhere as cleaner hydrocarbons replaced coal in urban combustion and as coal-fired power stations were located in the mining areas and raised their effective dispersion levels by using tall stacks. While typical urban SO_2 levels used to average several hundred micrograms per cubic meter (compared to natural background of just 5–10 μ/m^3), by the mid-1970s the annual means were commonly down to below 50 $\mu g/m^3$ and only a few percent of urban stations were recording SO_2 in excess of adopted standards. Only in countries heavily dependent on coal for household and industrial use—such as China, Poland, and Czechoslovakia—are the annual means, and even more so winter peaks, much above national standards.

But even in the most polluted areas, SO_2 levels only rarely build up to acutely dangerous levels as the compound is rather rapidly removed by a variety of dry and wet processes which have been much studied lately. Photochemical gas-phase oxidation, proceeding at rates of a few percent per hour, is a very important removal route but aqueous reactions are much faster (complete removal in just 1 hr) and are the dominant global source of sulfates. Dry deposition, much less studied and much less understood, may remove about half of all atmospheric SO_2; another important removal is absorption by water surfaces, vegetation, and alkaline soils.

Sulfates formed by oxidation of SO_2 (primary sulfates, except for $MgSO_4$ in sea spray, are relatively scarce) are dominant aerosols of polluted atmospheres, especially in the fine size category where they account for up to four-fifths of particulate matter. Monitoring abilities have not become sophisticated enough to distinguish among neutral, acidic, and metallic sulfates and various measurement interferences do not make the older total sulfate records very reliable. While terrestrial background levels are below 5 $\mu g/m^3$, current urban concentrations average typically between 10 and 15 $\mu g/m^3$ with clear summer maxima up to and even above 20 $\mu g/m^3$.

The relationship between atmospheric SO_2 and sulfates is surprisingly complex: in polluted areas sulfate levels often closely correlate with SO_2 concentrations but many interesting case studies show a definitely nonlinear relationship. Clearly, generalizations in this case are perilous and multiple uncertainties will require considerable research effort to clarify. Ubiquitous, tiny, hygroscopic sulfate aerosols are the principal source of cloud condensation nuclei and other essential forms of removal are, naturally, gravitational settling of larger particles, washout (high in the initial rain phase), and impaction and adsorption. In dry weather and with sources high above the ground, complete removal may take weeks; in wet weather and with low sources, atmospheric cleansing may be accomplished in a matter of hours. Typical sulfate residence time is, however, 2–4 days and all anthropogenic sulfur species stay in the atmosphere approximately 40 hr before being removed, long enough to permit long-range transport.

Long-range transfer of sulfurous pollution became an avidly studied process, first (in the early 1970s) in Europe where the Swedish charges that England and central Europe were the principal sources of the country's immissions helped to launch the issue of "acid rain," and later in North America. Extensive European studies show the following: sulfur pollution travels usually in the lowest 1–2 km of the troposphere; central European sulfur can reach virtually the entire continent in just 2½ days; dry removal is more important than wet deposition, except in areas of high orographic precipitation; southern Scandinavia is actually not the largest "importer" as about half of all emitted sulfur is deposited outside the nation of origin; weather determines the changing immission patterns and often makes Czechoslovakia and East Germany the largest sulfur "exporters" and their neighbors the most affected recipients.

In North America six of eight major cyclones tracking across the continent converge on New England, four of them after passing over the region of the largest SO_2 emissions in the Ohio Valley dotted with huge coal-burning power stations—and this region's pollution is also carried northeast with the clockwise flow on the backside of frequent summer anticyclones. However, atmospheric mixing and precipitation variations bring a surprisingly uniform deposition pattern with a relatively gentle spatial gradient between the Ohio Valley and upstate New York. These long-range transfers have been modeled frequently but they share unrealistic simplifications as their fundamental undoing. Hence, the best current consensus is that the available atmospheric transport models are not adequate to be applied reliably in predicting response to alternative control strategies.

Long-range transport spreads acid deposition over relatively large areas but misinterpretations, errors, and uncertainties encountered in describing, measuring, and analyzing this complex phenomenon narrow confident appraisals to just a few unassailable generalizations. Errors contributed by alkalinity from glass containers which were generally used before the mid-1960s to collect and to store precipitation samples make it impossible to reconstruct the precipitation acidity changes in any reliable way. Large uncertainties influence recalculations of acidities from ionic measurements and even the current standard methods of rainwater acidity determination can show huge inconsistencies if not employed rigorously by competent personnel. Construction of isopleth maps for large areas based on very thin sampling networks necessitates questionable interpolations and produces misleading patterns.

Consequently, for the United States, where reliable and extensive sampling began only in the late 1970s, historical data are inadequate to establish an indisputable secular trend toward higher precipitation acidity—while two available, spatially limited but accurate and consistent sets show no significant long-term change between the mid-1960s and the early 1980s. In Europe, where large-scale consistent monitoring dates to the mid-1950s, only about a quarter of the stations show a statistically significant trend of rising precipitation acidity and, even more surprisingly, adjacent sites often have a mere 20–30% variance in common!

Complexities of the acid deposition process are no less impressively illustrated by the lack of a simple quantitative yardstick in fixing the very acidity of precipitation. Rain or fog acidified just by the natural presence of CO_2 in the atmosphere would have pH 5.6—but it turns out that not only areas downwind from large sources of SO_2 but also some of the world's most remote locations (in the mid-Pacific, southern Indian Ocean) have precipitation with average pH well below 5, while others have quite alkaline rains in spite of heavy regional pollution (northern China is perhaps the best case).

This is not so surprising once it is realized that precipitation pH is not simply a consequence of emitted acidifying compounds but a result of a complex interplay of many cations and anions. Where alkaline matter is abundant in the atmosphere (as it is in most drier temperate and subtropical regions), it provides a significant buffer and influx of acidifying substances results simply in lowering of alkalinity as the rains continue to have pH between 6 and 7.5. In the plains–prairie heartland of North America, in Mexico, northern India, and northern China, terrigenic alkaline cations can neutralize most or all acid anions—while on Mauna Loa even a very low acid anion loading results in lowered pH in the near absence of alkaline matter high above the ocean, and in the humid, vegetation-covered eastern North America about half of emitted acid anions are not neutralized and must be balanced by H^+ and hence raise the precipitation acidity.

An interesting speculation arising from these acid–base considerations is then the extent to which eastern North America's acid deposition problems have been caused by almost perfect stripping of alkaline fly ashes from combustion gases, by regrowth of forests and extensive grass and tree planting and paving, all processes diminishing the presence of terrigenic alkaline matter: approximate calculations confirm that the effect has been definitely significant.

Yet another major complication concerns the relative importance and roles of sulfates and nitrates in the acid deposition. Higher NO_x emissions have brought an increased share of nitrates in precipitation and the compounds are now the dominant acidifying aerosols during eastern North America's winters. No_x emissions are expected to grow faster than SO_2 releases and the former are also more difficult to control as they originate much more so from a vast number of medium and small sources than does SO_2. However, nitrates have in many cases a clear fertilizing effect and cannot be considered in the same category as sulfates.

Finally, it must be appreciated that dry acid deposition, much less studied and much less understood, is no less important than wet deposition and that both must be considered in assessing effects on biota. Total rates of wet and dry deposition in Europe and North America now surpass 50 kg S/ha per year over areas of several hundred thousand square kilometers and range between 10 and 30 kg S/ha per year over large parts (several million square kilometers) of both continents.

Environmental effects of SO_2 and sulfates include many diverse changes, degradations, and deteriorations, some of them well known and reliably predictable and others insidious and impossible to attribute to any single disturbance. There is no doubt about visibility reductions caused by atmospheric sulfates: haze and closed horizons are now a common summer occurrence over large parts of western Europe and eastern North America. Winter urban visibilities were much improved by substitution of coal in household and small industrial combustion but summer haziness, caused largely by airborne sulfates, persists and may even have worsened although establishment of secular trends is a difficult task.

As the midlatitude pollution is carried northward, it ages, becomes distributed through the whole troposphere, and, in the absence of precipitation removal, markedly thickens the Arctic haze, especially during late winter months. This should obviously influence the region's radiation balance but effects of sulfates on planetary radiation fluxes remain speculative.

Weathering and corrosion of building materials and metals are among the best understood effects of acid deposition. Limestones and marbles are, obviously, greatly susceptible, especially in the presence of high moisture and in locations with freezing–thawing cycles. Some of the world's most admired architectural treasures (Acropolis, Taj Mahal) are now affected and their preservation remains an engineering challenge. Metal corrosion losses run globally into many billions of dollars each year as unprotected carbon steel, galvanized steel, nickel, and zinc corrode much faster in the presence of acid compounds. Damage to buried pipes and structures is also large, as is discoloration and deterioration of paints.

Among biota, effects on aquatic ecosystems are usually noticeable first and have been studied extensively, first in Scandinavia, where many thousands of lakes have been acidified, and later in North America, where hundreds are showing the stress. Marked differences in the way the lakes respond to acid deposition are a complex function of precipitation frequency and distribution, ground characteristics, watershed size, and terrain and its vegetation cover. The commonly observed deteriorative trend accompanying sinking pH are higher sulfate loading and rising releases of Al^{3+} ions and heavy metals into water.

The most dramatic biotic change is, of course, decline and eventual disappearance of sensitive fish species preceded by reproductive failures and higher morbidity. High-acid shocks accompanying spring meltdown are also highly injurious to amphibians while total phytoplankton biomass may change little although its structure shifts considerably as sensitive species are replaced by the more resistant ones. Groundwater acidification raises hardness, sulfate and metal loading of drinking water, with consequences as obvious as corrosion of copper pipes and as speculative as the role of higher Al^{3+} levels in Alzheimer's disease.

Soils respond to acidification in a more complex manner. General shifts are predictable: loss of alkaline cations, above all Ca^{2+} and mobilization of Al^{3+} and heavy metals. However, closer studies show that natural soil formation is often more important than acid deposition in determining soil acidity as soils tend to get more acid regardless of their original material. Moreover, changing land use, such as recovery of forests or large-scale planting of coniferous monocultures, acidifies soils as much and often more than acid deposition.

Fortunately, only relatively small areas of Europe and North America have soils susceptible to acidification—although the commonly reproduced small-scale maps carry an erroneously simplified message of large-scale susceptibilities; besides, wherever the soils of rich nations are now farmed, intensive fertilization by ammonia and its compounds, regularly corrected by liming, dwarfs any conceivable effects of acid deposition. In sum, irreversible acidification of soils appears to be a minor, highly localized worry.

Crops usually require between 10 and 50 kg S/ha per year for healthy growth, a need often partially filled by atmospheric deposition. Excessive SO_2 concentrations cause well-known acute or chronic response (leaf spotting, chlorosis, necrosis) but SO_2 levels currently encountered even in major source areas (such as the Ohio Valley) have only a minor chronic effect on major crops which suffer mostly owing to ozone attack. Open chamber testing of wet acid deposition effects has showed growth inhibition among some dicots but grain crops appear to be rather insensitive and some, especially leguminous, species are actually stimulated: in general, a picture of good resistance.

Severe destruction of forests near large smelters is well documented but chronic effects of relatively low-level acidification are the most controversial point of ecosystemic injuries induced by acid deposition. Scandinavian research shows no decline in long-term forest growth rate as the downward trend in Sweden in the 1950s and early 1960s has been succeeded by renewed good increments and as Norwegian conclusions stress no demonstrable effect.

In contrast, evidence from central Europe and eastern North America is more disquieting. Dieback of red spruces and firs in high elevations in New England appears to be accelerating but the search for clear-cut causes has been unrewarding. Besides acid deposition, drought and infection are certainly implicated but the decline may be a result of a highly complex multiple-stress syndrome. This is also the most likely case for the recent accelerated *Waldsterben* in West Germany and Czechoslovakia where the forest damage is extensive especially among coniferous species.

Ecosystemic degradation on a microorganismic scale is marked by species-specific response as populations of bacteria, fungi, and soil invertebrates shift but overall biomass changes little. Undoubtedly, acid deposition appears to pose the greatest threat (regardless of whether it is of merely contributory or a causative nature) to coniferous forests, while degradation of aquatic communities is a more localized phenomenon and changes in soils and crops are rarely worrisome.

Health effects of high SO_2 concentrations were of much concern in the 1950s and 1960s but current urban SO_2 levels in Europe, Japan, and North America (usually below 50 $\mu g/m^3$) are too low to cause changes in pulmonary structure and function which set in with SO_2 at 3–10 mg/m^3. The effect of low SO_2 levels on general mortality and on higher incidence of respiratory diseases is much more difficult to anser as numerous epidemiological studies are open to conflicting interpretations and offer little solid ground for establishing the dose–response relationship.

However, SO_2 concentrations do not explain excessive death rates even during the well-studied episodes of high air pollution (particulate matter levels do) and problems with intercity comparison (interferences by impossible-to-control variables) negate any useful conclusions. If general mortality data are too insensitive to reveal possible SO_2 effects, studies of respiratory morbidity show much clearer or undeniable links, above all among susceptible people (children, elderly, chronically sick). Still, these findings offer no reliable guide for legislation or long-term planning as there is no justification for assuming a linear, or any other, dose–response relationship.

Shifting the blame to sulfates brings similar uncertainties because they vary widely in their irritant potency with the commonest one (ammonia sulfate) being, in spite of some counterarguments, generally considered the least offensive. Overall, there are hardly any vigorously demonstrated links between atmospheric concentrations of sulfur compounds and general mortality and few clear causative links between those concentrations as currently encountered even in urban air and respiratory morbidity. Clearly, health effects of airborne sulfurous gases and aerosols can no longer be used as the rationale for controls.

Fundamental uncertainty besetting current control strategies concerns the fact that although we are certain that annual acid deposition over a large region will decline in proportion to reduced SO_2 emissions, we are unable to predict the changes of acid deposition in sensitive downwind areas in response to reducing SO_2 emissions in one source region by a larger share than in another. This inability leads to continuing disagreements betwen the United States and Canada and the United Kingdom and continental western Europe.

Once the commitments are made, rapidly maturing control technologies can be used to lower the offensive emissions. They no longer include tall stacks—which have gone from being the method of choice in the 1960s and early 1970s to being seen as one of the major contributors to regional acid deposition (as they promote longer atmospheric residence time and hence more oxidation to sulfates and longer transport)—and rely largely on cleaner combustion and flue gas desulfurization.

Removal of sulfur from coal, the most offensive fuel, is accomplished by a variety of mechanical separation techniques which take advantage of different specific densities of the fuel and its inorganic, largely pyritic, sulfur. These methods commonly remove about one-third of the pyritic sulfur in low-sulfur coals and up to one-half in high-sulfur fuels, the best methods removing up to 80% and thus reducing overall SO_2 emissions by about half as the organic sulfur remains virtually intact. Emerging chemical cleaning methods remove practically all inorganic and up to a quarter of organic sulfur but their capital and operation costs are still prohibitively high. On the other hand, intensive mechanical coal cleaning may increase a typical power plant's capital cost by 3–4% but it is cost-effective owing to its subsequent operation savings.

Fluidized bed combustion, during which the burning of various fuels proceeds in contact with a floating bed of crushed limestone, has long been recognized as an excellent SO_2 control method but its development for large steam generation in power plants has been frustratingly slow. The first commercial units are now in regular operation but they are still an order of magnitude smaller than common units in coal-fired power plants. Consequently, flue gas desulfurization (FGD) has become *the* control method although it, too, was slow in reaching the commercial stage. However, it finally took off in the United States during the 1970s when it moved from a handful of semiexperimental units to installations serving more than 20,000 MW of generating capacity.

Wet alkali scrubbing with limestone or lime slurry had emerged as the leading technique, now accounting for about 70% of all operating FGD units. The technique has its share of problems—corrosion, scaling, and plugging during operation and the necessity to store large volumes of waste which is a potential water contaminant are the main ones—but in spite of its throwaway nature it has a financial edge in comparison with regenerable techniques (sulfite scrubbing being the most promising). Still, the investment costs are multiples of the earlier estimates and in new coal-fired generating units installation of efficient wet alkali scrubbers (removing in excess of 90% of all SO_2) will increase the plant's capital cost by 25–30% and operating expenditures by 20–40%.

Innovations in FGD technology include dry scrubbing with fly ash, suitable for low-sulfur coals with abundant alkaline ashes, addition of magnesium and adipic acid in wet scrubbing to boost the removal efficiency, and attempts at simultaneous removal of sulfur and nitrogen oxides. Undoubtedly, the currently dominant wet alkali scrubbing will eventually decline in importance as some of the innovations become commercial. In any case, SO_2 controls on almost any scale are now technically possible.

The last control option—the one which treats symptoms rather than the cause—is liming of susceptible ecosystems, a practice long regularized in crop farming and recently extended, above all in Sweden, to lakes and their watersheds. Highly soluble compounds (lye, soda, quicklime) are used in preference to limestones and the results have been largely satisfactory, with acidification reversed and biota diversity increasing. Until the acid deposition rates are cut, this inexpensive, benign method should be widely practiced to protect select sensitive ecosystems.

5.4. THE CHOICES

This book has sought to demonstrate that while the anthropogenic interference in the three key biogeochemical cycles now proceeds on scales comparable with leading natural fluxes, there is hardly any evidence that these interventions pose highly unacceptable, irreversible global threats to the integrity of the planet's life-supporting environment. Yet this principal general conclusion must not be interpreted as an endorsement of benign neglect—just as advocacy of measured, thoughtful response.

Thus, it is a call for sustained research and better understanding, rather than a nod to any emergency adjustments and a commitment to accelerated work on alternative ways of securing food, heat, and motive power. Unglamorous as it sounds, this appears to me the only intellectually honest course to follow.

Four fundamental concerns must also be kept in mind: the first one involves the perception and ranking of grand-scale worries; the second, the ways these formidable problems are approached and treated; the third, the almost inexplicably irrational ways in which modern societies go about setting their corrective priorities; and the last, the very availability and practicality of less offensive alternatives.

Shifting perceptions of our critical environmental problems and ever-evolving scientific understanding make it very unlikely that a ranking can be given whose accuracy would last for decades. Fundamental as anthropogenic interferences in the three key cycles undoubtedly are, their long-term results may be measured much more in terms of changes and adjustments (brought by higher mean temperatures, increased nutrient flux or acid deposition burden) rather than representing absolute, irretrievable losses.

In contrast, there are two degradative processes whose consequences are irreparable losses affecting the very survival of ecosystems. Soil erosion, a perennial worldwide problem, has been intensified by extension of crop cultivation in erosion-prone environments, by expansion of row-crop farming in many nations, by accelerating tropical and subtropical deforestation, and by overgrazing and desertification. Brown (1984) estimates that globally we are now losing 23 billion t of cropland soil in excess of new soil formation; with global farmland topsoil reserves at roughly 3.5 trillion t, this loss rate would result in complete depletion in just 150 years.

Other estimates do not indicate such rapid decline. For example, Buringh's (1981) forecast of 1975–2000 erosion losses is 50 million ha for croplands, 15 million ha for grasslands, and 10 million ha for forests, a total of 75 million ha. Although a loss of 50 million ha of farmland would be equivalent to just 3% of the total cropland area in 1975, this disappearance can be more impressively illustrated as a decline of 5 ha every minute or half of all farmland currently feeding 1 billion Chinese. Moreover, observations from various parts of the world document considerable yield decreases for long periods before progressing soil erosion makes cropping impossible (Wolman, 1983).

Considering the fact that only half a dozen of the world's 150 nations are regular foodgrain exporters, and that soil erosion is now advancing not only in China, which has the world's most erodible region in its Loess Plateau and has suffered huge forest losses (Smil, 1984a), India, Brazil, and Nigeria but also in the USSR, the country with the world's largest area under crops, and in the United States, the premier food exporter where more than one-third of cropland has topsoil losses above sustainable rates, almost any erosion loss must be seen with great apprehension.

The other concern with irreplaceable loss is extinction of species and reduction of genetic diversity. This worry has gradually evolved into one of the leading topics of applied biology and it has also reached policy-making bodies such as the Council on Environmental Quality (1980) and the National Academy of Sciences (1980b); reviews by Myers (1979) and Ehrlich and Ehrlich (1981) detail the numerous dimensions of these losses whose consequences may range from hardly noticeable disappearance of a rare, endemic wildflower to catastrophic reduction or cessation of essential ecosystemic services on a large regional scale (e.g., deforestation-caused loss of water retention with subsequent worsening of flood–drought cycles and erosion).

Undoubtedly, changing climate, heavy eutrophication, or high acid deposition may be among the factors promoting extinction of sensitive species so that human interferences in the three cycles must be seen as part of the extinction process—but certainly as only a marginal part compared with such onslaughts as extensive tropical deforestation and farmland degradation (salinization, alkalinization), erosion and desertification brought by improper soil management.

I do not think it is an evasive attempt to avoid the consequences of human interference with major biogeochemical cycles, most importantly the concerns about planetary warming and acid deposition, to ask these questions: Should we worry more about a possible slight tropospheric temperature rise (which may be largely CO_2-induced) which would not at all be unprecedented in the long history of the biosphere and which alone would lead to nothing more destructive than shifts of ecosystem boundaries and adaptation or migration of species—or about the complete loss of whole populations, habitats, and ecosystems as a result of widespread tropical and subtropical deforestation, erosion, and desertification? Should we be concerned more about several additional kilograms of sulfate deposited on acid-prone farmland where we can counteract its effect by liming—or by the fact that our intensive row cropping of that land caused soil erosion losses incompatible with permanent farming?

The difference is between changes—unwelcome, undesirable, often injurious, costly to counteract or remedy but changes nonetheless—and irreparable losses, and the comparisons must be made not to argue that fossil fuel combustion and crop fertilization are hardly worth mentioning as environmental problems but to keep these anthropogenic interferences in a broader perspective, contrasted with other human activities which are, most definitely, more worrisome.

The importance of the second concern—the ways we approach and treat changes whose consequences may have such a worrisome reach—is most profitably illustrated with the CO_2 "problem." This, as stressed before, is the only truly global anthropogenic interference in a major biogeochemical cycle and, as Schelling (1983) aptly remarked, "how the issue is named can affect its apparent character." If we are certain of the global enormity of the threat and believe that only rapid reductions of CO_2 generation will suffice to avert a planetary catastrophe, then we must also insist on a global solution because every molecule of CO_2, no matter where and how generated, brings us all closer to doomsday.

This would call for an unprecedented course of action, a true planetary cooperation, something that can be termed, without being melodramatic about it, a continuous sacrifice—yet any equitable apportioning of such a tax on economic growth and personal well-being is impossible to imagine in our world of national rivalries, expressed so wastefully in prodigious amassing of arms and so repeatedly in failures to conclude workable international agreements demanding incomparably lesser sacrifices. Besides, even those who forecast rapid and high "warming" concede that some nations will benefit—would the latter join in any global rescue?

Let us consider the prospect of a 2–3°C *cooling!* With the world's principal grain-growing regions located between 30° and 50°N, such a cooling would have an incomparably harsher impact on our well-being as it would shorten vegetation periods and subdue the planetary water cycle. Clearly, longer frost-free spans and intensified evaporation–precipitation cycle (although this effect would obviously not be uniformly distributed!) are much preferable. And what of the possibility that anthropogenic warming (if in fact it does occur) may even counteract a coming natural cooling trend and thus prevent much more unpleasant consequences? Only an arrogant computer modeler would have a resounding answer to this question!

And then there are those who see no way out even with fossil fuels gone and civilization in an inextricable corner. For example, Lemon (1983), in introducing a book on CO_2 and plants, stated that we face "a choice between nuclear energy or fossil fuel to meet the energy needs of an increasing population" and that the first option "poses risks of a sudden catastrophe" while the other route means "risks of slow catastrophe—environmental decay, climate change, possible crop failures, and coastal flooding." He is, I will argue, wrong on all counts—and joining him in error are all those who cannot see the future in any way other than as a difficult choice between not only unappealing but even "catastrophic" strategies.

Leaving the nuclear worries aside, this quote is a perfect illustration of our insistence on dealing with the cause of the perceived CO_2 "problem" and not with its symptoms, an approach we do not advocate so eagerly in countless other, even patently life-threatening, situations. Once more let us have Schelling's (1983) sensible look: "But it would be wrong to commit ourselves to the principle that if fossil fuels and carbon dioxide are where the problem arises, that must be also where the solution lies."

This takes us to the third concern, the one about the strange ways in which we choose our priorities, and I will leave the CO_2 "problem" to focus on complexities surrounding acid rain and a widely based clamor to control the offensive emissions as an outstanding illustration of this point. Scores of pages in this book were devoted to demonstrating the impossibility of simple generalizations regarding the levels at which acid deposition could be considered "harmful" to soils, waters, biota, and human health. The impacts and injuries are highly time-, site-, species-, and individual-specific.

In agroecosystems, any negative effects will be dwarfed by acidifying nitrogenous (ammoniacal) fertilization and corrected by liming; in natural terrestrial ecosystems, the impact may range from easily recognizable acute injury in the vicinity of large SO_2 sources to gradual deterioration in which acid deposition is seen as one of several plausible contributing causes; in aquatic ecosystems, variations of watershed characteristics will be mirrored in wide variations of specific morbidity and mortality.

And linking the levels at which SO_2 and SO_4^{2-} are now encountered in the eastern half of North America with clear-cut health effects is plainly impossible as any such low-level impact cannot be satisfactorily separated from that of other environmental contaminants and, whatever their combined long-term result, it will be irretrievably hidden by the much more pronounced influence of varying lifestyles, diets, and smoking. On top of all this is the simple but almost always overlooked fact that dry acid deposition is at best accounted for only very unsatisfactorily, most often not at all, and hence we do not know with any assurance what the total burden is.

Clearly, any regional- or continental-scale generalizations about cause–effect relationships and any setting of quantitative limits—how much can be "safely" released or "safely" deposited in "sensitive" areas (and how loose the areal definition of sensitivity is was abundantly demonstrated earlier with soil maps!)—are not meaningful given the state of our knowledge, and may well never be as the problem is not one open to useful, practical generalizations.

And yet such generalizations have been made repeatedly, the two of the simplest nature and greatest notoriety being the demand to limit deposition to less than 20 kg SO_4^{2-}/ha per year and to cut the current SO_2 emissions by 50%. In this continent these demands are the official positions of the Canadian government which has repeatedly expressed its dismay at America's unwillingness to go along with these goals.

Uncertainties characterizing our understanding of effects preclude any sensible cost/benefit analysis which is a fundamental precondition for establishing the extent of controls: the OECD (1981) exercise, whose loose numbers were cited previously, is certainly an excellent example of an analysis which gives no guidance, just huge ranges of conjectures. Unfortunate and frustrating as it may be, acid deposition is not the only environmental problem whose uncertainties and clear specificities do not make it quantifiable even on the basis of the best scientific evidence.

So why should there be an insistence that every hectare receive less than x kilograms of sulfates and that every power plant cut emissions by a certain percentage? This is not how we regulate elsewhere: not every intersection has traffic lights, our cars still lack air bags—and even the use of seat belts is not mandatory in all states of the United States and in all Canadian provinces. Quantifying the costs and benefits of these car-related regulations is so much easier and so much more meaningful than any attempts at pricing acid deposition costs. About the benefits of these changes, preventing accidents and saving human lives, there can be no doubt—and yet the progress in introducing these patently life-saving regulations has been astonishingly sluggish.

In the realm of environmental regulations, we have not moved aggressively against much more clear-cut offenders such as lead in gasoline or a multitude of toxic pesticide residues in food. When there is no general sense of urgency to eliminate many other obviously or potentially harmful environmental pollutants or to reduce losses incurred in car accidents, one cannot consistently argue for crash programs to reduce SO_2 emissions whose diffuse effects are undoubted but so hard to quantitatively pin down.

The fourth concern, the question of real alternatives, is best illustrated with nitrogen cycle interventions. Unlike SO_2-related acid rain, which could be eliminated with universal and costly installation of FGD technology (hence the problem is "just" the determination of the elusive appropriate level that emissions must not exceed), it is impossible to eliminate volatilization, denitrification, and leaching of nitrogenous compounds as long as we fertilize our fields. Can we do without?

As far as plants are concerned, their energy costs are virtually identical for acquiring the nutrient from fertilizers or from biofixation. Typical *Rhizobium*–plant symbiosis will spend about 6 mg of carbon in the nodule to fix 1 mg of nitrogen—and almost exactly the same amount to reduce 1 mg of nitrate-nitrogen in the root (Minchin and Pate, 1973). Consequently, there is no loss of yield when the fixing is done by living organisms—but we *do* mind because the *Rhizobium* symbioses go along with only a very few of those crops that we prefer to grow for food or feed.

Hence, we greatly favor freeing the fields of legumes and fertilizing the cereal, root, and oil crops with nitrogen we have been able to fix since the beginning of this century. For this choice we have to pay a premium of high-energy cost because the way we fix nitrogen is critically dependent on a reliable availability of natural gas and electricity, two of the most concentrated, versatile, high-quality forms of energy fueling our civilization.

However, we should not see this high-energy obligation as unduly taxing: it claims only a tiny part of our total fossil fuel consumption, there is no better way to convert that energy into edible matter (in spite of often large inefficiencies in fertilizer use, producing bacterial protein from hydrocarbon substrate is even less efficient), the high yields achievable with fertilizers enable cultivation of less land and hence help to preserve natural genetic diversity as well as to conserve landscapes for other natural or recreational services. Most fundamentally, we really do not have any other choice.

While in rich nations the practice of "eating closer to the sun" (i.e., reduction of meat intake) could free large areas of land for food crops which could then be grown in a less intensive manner with lower yields, such a shift cannot work for the three-quarters of mankind who derive less than one-fifth of energy and protein from animal foods. Further sizable population increases on the three poor continents must be accommodated largely through rising yields which are impossible without higher fertilization. And when China or Egypt or El Salvador already use at least as much or more nitrogenous fertilizer per hectare of cropland as does the United States, the future needs will unmistakably be enormous.

Of course, we should strive to increase the efficiency with which the applied nitrogen is used by plants but in spite of many approaches toward that aim, reductions of nitrogen losses can go only so far. And the case of carbon dioxide presents an even more obvious no-choice outlook as none of the noncarbon energy sources—nuclear electricity and heat or solar radiation captured directly by hot water collectors, central power stations, or photovoltaic cells, and indirectly in wind, waves, or falling water—can, singly or in combination, supply economically a substantial share of the global energy demand for several generations to come. Higher combustion efficiencies and vigorous conservation efforts can keep the growth rates of future CO_2 releases down but, as with nitrogenous fertilizers, further large anthropogenic transfers will have to continue.

So will, then, many of the uncertainties: as frustrating as it may be in an era when science is so often treated as carpentry—provide raw materials, tools, and a plan of action and expect a lasting structure at a specified time—we shall not be finished with the study of biogeochemical cycles soon. Most importantly, the prodigious challenge of assessing long-term global effects of many interferences in multivariate unison, rather than in traditional one-dimensionality, has hardly begun. Yet this is the only road to meaningful understanding of the actual processes and the only opening to realistic forecasts.

But, of course, ours is a civilization uncomfortable with feelings of lasting uncertainty. As the spiritual certainties of religious faith weakened during the past three centuries, science came to occupy an even more prominent place in our search for new permanent beliefs. Not only do we want to know now but we want to extend our control far into the future. Utilitarian prescience rather than just exciting intellectual challenge became the leading justification of our scientific endeavors.

Doubtless, one can point to many successes in this quest—but failures abound. And the relationship is frustratingly simple: the more complex, the more vital the concern, the greater the uncertainty about its outcome. Why then do we persist in forecasting tropospheric temperatures a century from now and portraying nitrate-laden waters and acidifying ecosystems in the decades ahead? Is it not that, owing to our intolerance of uncertainty, we insist on fixing the outcomes, unpleasant as they may be, rather than admitting that while our fundamental choices may be more restricted than we would wish, the eventual consequences will only in rare cases coincide with our forecasts?

In the early 1330s, Japan, passing from more than a century of Hōjō's rule to Go-Daigo's hands, was hardly a place to cherish uncertainty. Yet when Yoshida Kenkō took up a brush to write a collection of penetrating essays there was no doubt in his mind: "The most precious thing in life is its uncertainty."

This attitude is far from the *raison d'être* of Western science but its absence in most of our efforts has not given superiority to our thoughts and writings. I shall be most pleased if this book is seen as a thorough affirmation of transience.

POSTSCRIPT

Manuscript of this book was finished in January 1984 and shipped to Plenum in May. A year later, after revision and copyediting, the typescript goes to the compositor and I have an opportunity for a brief postscript. A year in these intensively researched topics brings plenty of news, but this will not be a short systematic update, just a few pages on what caught most of my attention during the first year of the manuscript's inexorable slide to eventual oblivion.

Judging by mass media interest, the CO_2 story, after reaching new high points following the publication of EPA and NAS reports in late 1983, has clearly receded. There are fewer and fewer advocates of rapid (in just several decades) West Antarctic ice sheet melting (and hence of the consequent 5-m sea level rise), and without the prospects of a big deluge the fable loses much of its attraction. There is also a solidifying consensus among plant scientists that further atmospheric CO_2 increases will most likely improve crop harvests and that yields will continue to rise (Waggoner, 1984).

But as a look at a recent issue of the biennial *Carbon Dioxide Information Center Communications* (published at the Oak Ridge National Laboratory) will show, research in this multidisciplinary field is not slackening. According to the Center's survey nearly 2000 persons in some 50 nations are now professing interest in CO_2 research. If only every fourth one writes a paper each year. . . . But of all carbon-cycle-related news the most interesting story to me was about CO: Khalil and Rasmussen (1984) found that the gas's background concentration is increasing. Such a rise could deplete hydroxyl radicals, interfere with chains of removal reactions of many trace gases, and possibly affect stratospheric ozone—but the measurement series was only 3.5 years long and source and sink fluctuations may have affected the observed trend as the gas's short atmospheric lifetime (2–6 months) makes it unlikely that a substantial long-term rise could be sustained. In any case, a new twist to carbon cycle concerns.

Nitrogen-related worries came to the fore in Britain with publication of the Royal Society's report on the country's nitrogen cycle (Royal Society Study Group Report Staff, 1984). The major findings: a mere one-tenth of nitrogen in fertilizers ends up in food and in some localities drinking water contains nitrate levels exceeding allowable limits. This contamination was especially serious during the record 1976 drought—and yet no methemoglobinemia cases were recorded in the country. But it was the growing realization of nitrate contribution to acid deposition which attracted the most attention. The British report concluded that one-third of the acidity in rain is caused by nitrates, while in West Germany nitrogen oxides, especially in uncontrolled emissions from passenger cars, were repeatedly labeled as key contributors to a multifactorial syndrome damaging the country's forests. This NO_x-derived damage was ascribed more to the oxides' involvement in photochemical reactions generating ozone than to formation of hydrogen ions and intensified acidification (Schütt et al., 1983). *The Economist* editorial (October 20, 1984) headlined this link bluntly: "Dirty cars, dying trees," but, of course, the connection is not that simple.

Coniferous stands are heavily damaged in neighboring Czechoslovakia where both the numbers of cars and total multilane highways are fractions of the corresponding numbers for West Germany (hence car emission levels are lower and NO_x generation is considerably lower owing to reduced speeds). In Germany the possibility of limiting speeds on the Autobahn caused a national anguish. But a closer look at Germany's *Waldsterben* shows it to be not as drastic as figures cited earlier in this book might indicate. New work by Karl-Eugen Rehfuess of Munich's Ludwig-Maximilian University shows that only some 3% of affected trees are heavily damaged, that most of the loss is limited to less than 10% of all needles, that most of these lost needles are old ones that had been contributing little to overall photosynthesis, and that many lightly damaged trees do recover.

Yet what if neither the original acid deposition hypothesis, nor the ozone one, nor the syndrome explanation (acids and ozone and hydrocarbons and heavy metals and drought) are correct and if it is simply the overfertilizing effect of nitrogenous compounds which causes the forest damage? Such is indeed a hypothesis advanced by Bengt Nihlgaård (1985). Increased atmospheric ammonium deposition, a consequence of volatilization from farmlands fertilized with NH_3, and higher emissions of NO_x eventually converted to NO_3^-, overload many forests, which have evolved to live with low nutrient levels, with nitrogen. The trees must then intensify photosynthesis and with it the production of wastes that must be released from roots and leaf cells. Nitrogen surplus and high-volume production lead to relative deficiencies of other nutrients (K, P, Mg, Mo) and water, as well as to depression of mycorrhizal conversions (too much NO_3^-) which, in turn, will cut phosphorus intake and further increase accumulation of nitrogen in leaf and needle tissues. When the leaves are exposed to acid deposition or when transpiration is lowered during droughts, nonprotein wastes cannot be eliminated in normal ways and the tree, already more susceptible to frost and pest attack, might shed whole leaves or needles in order to survive. A cogent, sensible explanation—and yet one more reason to avoid any rush into "solving" the acid deposition problem.

Plenty of other evidence points in the same cautious direction. The number of field studies and quantitative models of acid deposition chemistry and meteorology has been increasing rapidly (Bhumralkar, 1984; Calvert, 1984; Durham, 1984; Hicks, 1984), but, as George M. Hidy, whose work was cited with much approval, reminds us, we still cannot prescribe specific remedies for specific receptor regions with any degree of confidence (cited in Samson, 1984).

Jacobson and Showman (1984) published results of their 12-year study of vegetation in the vicinity of four large power stations in the Ohio Valley, which was undertaken to find out if higher SO_2 emissions will cause foliar symptoms on vegetation and disappearance of sensitive lichens. Their conclusions: during the 12 years there was only one episode of acute vegetation injury, no instance of damage, and no change in distribution of lichens. The latter finding is most important, since chronic SO_2 effects on higher plants are unlikely when the distribution of SO_2-sensitive lichens is unaffected.

Pierce (1985) made a detailed survey of the acid tolerance of amphibians and, although he noted obvious cases of high sensitivity (mentioned earlier in this book), he found the whole group relatively acid tolerant, with increasing mortality affecting many species only below ph 4. Substantial intraspecific variation in acid sensitivity among amphibians is yet another warning against unwarranted generalizations about susceptibility, be they about soils or biota.

Huge gaps in our knowledge and fascination with the complexities of the three cycles remain, but I sense that the year that has elapsed since the finishing of this book has been marked by a most welcome shift toward a deeper realization of the nonlinear nature of the challenges and a greater humility in prescribing the "solutions." Often speedy action is essential—but in the realm of understanding the working of the three cycles and correcting our harmful interferences an old Latin proverb fits perfectly: *festina lente*.

REFERENCES

Aalund, L. R. 1976. Wide variety of world crudes gives refiners range of charge stocks. *The Oil and Gas Journal 74*:87–122.

Aalund, L. R. 1983. Assay series begins with Mideast crudes. *The Oil and Gas Journal 81*:71–77.

Abeles, F. B., L. E. Cracker, L. E. Forrence, and G. R. Leather. 1971. Fate of air pollutants: Removal of ethylene, sulfur dioxide and nitrogen dioxide by soil. *Science 173*:914–916.

Abrahamsen, G. 1980. Impact of atmospheric sulphur deposition on forest ecosystems. In: *Atmospheric Sulfur Deposition,* D. S. Shriner, C. R. Richmond, and S. E. Lindberg, eds. Ann Arbor Science, Ann Arbor, Mich., pp. 397–414.

Abrahamsen, G., J. Hovland, and S. Hagvar. 1980. Effects of artificial acid rain and liming on soil organisms and the decomposition of organic matter. In: *Effects of Acid Precipitation on Terrestrial Ecosystems,* T. C. Hutchinson and M. Havas, eds. Plenum Press, New York, pp. 341–362.

Ackerman, T. P. 1979. On the effect of CO_2 on atmospheric heating rates. *Tellus 31*:115–123.

Adams, D. F., and S. O. Farwell. 1981. Sulfur gas emissions from stored flue gas desulfurization sludge. *Journal of the Air Pollution Control Association 31*:557–564.

Adams, D. F., S. O. Farwell, E. Robinson, M. R. Pack, and W. L. Bamesberger. 1981. Biogenic sulfur source strengths. *Environmental Science and Technology 15*:1493–1498.

Adams, J. A. S., M. S. M. Mantovani, and L. L. Lundell. 1977. Wood versus fossil fuel as a source of excess carbon dioxide in the atmosphere: A preliminary report. *Science 196*:54–56.

Agnew, A. F. 1977. Coal reserves, resources and production. In: *Project Interdependence: U.S. and World Energy Outlook Through 1980.* U.S. Government Printing Office, Washington, D.C., pp. 208–262.

Ahmed, S., and H. P. M. Gunasena. 1979. N utilization and economics of some intercropped systems in tropical countries. *Tropical Agriculture 56*:115–123.

Ajtay, G. L., P. Ketner, and P. Duvigneaud. 1979. Terrestrial primary production and phytomass. In: *The Global Carbon Cycle,* B. Bolin, E. T. Degens, S. Kempe, and P. Ketner, eds. Wiley, New York, pp. 129–181.

Albanese, A., and M. Steinberg. 1978. *Environmental Control Technology for Atmospheric Carbon Dioxide.* Brookhaven National Laboratory, Upton, N.Y.

Aldrich, S. R. 1980. *Nitrogen in Relation to Food, Environment, and Energy.* University of Illinois Press, Urbana–Champaign.

Alexander, M. 1977. *An Introduction to Soil Microbiology.* Wiley, New York.

Alexander, M. 1980. Effects of acidity on microorganisms and microbial processes in soil. In: *Effects of Acid Precipitation on Terrestrial Ecosystems,* T. C. Hutchinson and M. Havas, eds. Plenum Press, New York, pp. 363–374.

Alkezweeny, A. J., and D. C. Powell. 1977. Estimation of transformation rate of SO_2 to SO_4 from atmospheric concentration data. *Atmospheric Environment 11*:179–182.

Allison, F. E. 1955. The enigma of soil nitrogen balance sheets. *Advances in Agronomy 7*:213–250.

Allison, F. E. 1966. The fate of nitrogen applied to soils. *Advances in Agronomy 18*:219–258.

Allison, F. E. 1973. *Soil Organic Matter and Its Role in Crop Production.* Elsevier, Amsterdam.

Allison, L. J., and S. S. Talmage. 1982. The carbon dioxide question: An introduction to the literature. In: *Carbon Dioxide Review: 1982,* W. C. Clark, ed. Oxford University Press (Clarendon), London, pp. 415–427.

Alpert, D. J., and P. K. Hopke. 1981. A determination of the sources of airborne particles collected during the regional air pollution study. *Atmospheric Environment 15*:675–687.

Altshuller, A. P. 1973. Atmospheric sulfur dioxide and sulfate. *Environmental Science and Technology 7*:709–712.

Altshuller, A. P. 1976. Regional transport and transformation of sulfur dioxide to sulfate in the United States. *Journal of the Air Pollution Control Association 26*:318–324.

Altshuller, A. P. 1979. Model predictions of the rates of homogeneous oxidation of sulfur dioxide to sulfate in the troposphere. *Atmospheric Environment 13*:1653–1661.

Altshuller, A. P. 1982. Relationships involving particle mass and sulfur content at sites in and around St. Louis, Missouri. *Atmospheric Environment 16*:837–843.

Amarasiri, S. L. 1978. Sri Lanka. In: *Organic Recycling in Asia*. FAO, Rome, pp. 119–133.

Amdur, M. O., J. Bayles, V. Ugro, and D. W. Underhill. 1978. Comparative potency of sulfate salts. *Environmental Research 16*:1–8.

Andersen, N. R., and A. Malahoff, eds. 1977. *The Fate of Fossil Fuel CO_2 in the Oceans*. Plenum Press, New York.

Andreae, M. O., W. R. Barnard, and J. M. Ammons. 1983. The biological production of dimethylsulfide in the ocean and its role in the global atmospheric sulfur budget. *Ecological Bulletin 35*:167–177.

Aneja, V. P., J. H. Overton, Jr., L. T. Cupitt, J. L. Durham, and W. E. Wilson. 1979. Direct measurements of emission rates of some atmospheric biogenic compounds. *Tellus 31*:174–178.

Aneja, V. P., A. P. Aneja, and D. F. Adams. 1982. Biogenic sulfur compounds and the global sulfur cycle. *Journal of the Air Pollution Control Association 32*:803–807.

Armentano, T. V. 1980. Drainage of organic soils as a factor in the world carbon cycle. *BioScience 30*:825–830.

Arnold, J. E. M. 1979. Wood energy and rural communities. *Natural Resources Forum 3*:229–252.

Arrhenius, S. 1896. The influence of the carbonic acid in the air upon the temperature of the ground. *Philosophical Magazine,* Series 5, 41:237–276.

Atkins, D. H. F., R. A. Cox, and A. E. J. Eggleton. 1972. Photochemical ozone and sulphuric acid aerosol formation in the atmosphere over southern England. *Nature 235*:372–376.

Atkinson, S. E., and T. D. Crocker. 1982. On scientific inquiry into the human health effects of air pollution: A reply to Pearce *et al. Journal of the Air Pollution Control Association 32*:1206–1209.

Augustsson, T., and V. Ramanathan. 1977. A radiative–convective model study of the CO_2 climate problem. *Journal of the Atmospheric Sciences 34*:448–451.

Averitt, P. 1975. *Coal Resources of the United States*. U.S. Department of Interior, Washington, D.C.

Ayanaba, A. 1980. The potential contribution of nitrogen from rhizobia—A review. In: *Organic Recycling in Asia*. FAO, Rome, pp. 211–229.

Ayanaba, A., and P. J. Dart, eds. 1977. *Biological Nitrogen Fixation in Farming Systems of the Tropics*. Wiley, New York.

Ayers, R. S., and R. L. Branson. 1973. *Nitrates in the Upper Santa Ana River Basin in Relation to Groundwater Pollution*. University of California Press, Riverside.

Aykroyd, W. R., and J. Doughty. 1964. *Legumes in Human Nutrition*. FAO, Rome.

Aykroyd, W. R., J. Doughty, and A. Walker. 1982. *Legumes in Human Nutrition*. FAO, Rome.

Babcock and Wilcox Corporation. 1976. *Summary Evaluation of Atmospheric Pressure Fluidized Bed Combustion Applied to Electric Utility Large Steam Generators*. EPRI, Palo Alto, Calif.

Babich, H., and D. L. Davis. 1981. Food tolerances and action levels: do they adequately protect children? *BioScience 31*:429–438.

Bacastow, R. B., and C. D. Keeling. 1973. Atmospheric carbon dioxide and radiocarbon in the natural carbon cycle. In: *Carbon and the Biosphere,* G. M. Woodwell and E. V. Pecan, eds. U.S. Atomic Energy Commission, Washington, D.C., pp. 86–135.

Bach, W., J. Pankrath, and W. Kellogg, eds. 1979. *Man's Impact on Climate*. Elsevier, Amsterdam.

Bach, W., J Pankrath, and S. H. Schneider, eds. 1981. *Food–Climate Interactions*. Reidel, Dordrecht.

Bache, B. W. 1978. The acidification of soils. In: *Effects of Acid Precipitation on Terrestrial Ecosystems,* T. C. Hutchinson and M. Havas, eds. Plenum Press, New York, pp. 183–202.

Baes, C. F. 1982. Effects of ocean chemistry and biology on atmospheric carbon dioxide. In: *Carbon Dioxide Review: 1982,* W. C. Clark, ed. Oxford University Press (Clarendon), London, pp. 187–204.

Baes, C. F., and R. E. Mesmer. 1976. *The Hydrolysis of Cations*. Wiley–Interscience, New York.

Baes, C. F., Jr., H. E. Goeller, J. S. Olson, and R. M. Rotty. 1976. *The Global Carbon Dioxide Problem*. Oak Ridge National Laboratory, Oak Ridge, Tenn.

Baker, D. N., and H. Z. Enoch. 1983. Plant growth and development. In: *CO_2 and Plants,* E. R. Lemon, ed. Westview Press, Boulder, Colo., pp. 107–130.

Baker, M. B., D. Caniparoli, and H. Harrison. 1981. An analysis of the first year of MAP3S rain chemistry measurements. *Atmospheric Environment 15*:43–55.

Balderston, W. L., B. F. Sherr, and W. J. Payne. 1976. Blockage by acetylene of nitrous oxide reduction in *Pseudomonas perfectomarinus. Applied Environmental Microbiology 31*:504–508.

Barber, J., ed. 1976–1979. *Topics in Photosynthesis.* Elsevier, Amsterdam.

Bardach, J. F., and R. M. Santerre. 1981. Climate and the fish in the sea. *BioScience 31*:206–215.

Barnard, W. R., M. O. Andreae, W. E. Watkins, H. Bingemer, and H. W. Georgii. 1982. The flux of dimethylsulfide from the oceans to the atmosphere. *Journal of Geophysical Research 87*:8787–8793.

Barnes, R. A. 1979. The long range transport of air pollution: A review of European experience. *Journal of the Air Pollution Control Association 29*:1219–1235.

Barnes, R. A., G. S. Parkinson, and A. E. Smith. 1983. The costs and benefits of sulphur oxide control. *Journal of Air Pollution Control Association 33*:737–741.

Barrie, L. A., and J. L. Walmsley. 1978. A study of sulphur dioxide deposition velocities to snow in northern Canada. *Atmospheric Environment 12*:2321–2332.

Barry, R. G. 1978. Cryospheric responses to a global temperature increase. In: *Carbon Dioxide, Climate and Society,* J. Williams ed. Pergamon Press, Elmsford, N.Y., pp. 169–180.

Barsin, J. A. 1981. Fossil steam generator NO_x control update. In: *Proceedings of the Joint Symposium on Stationary Combustion NO_x Control.* EPRI, Palo Alto, Calif., pp. 98–122.

Bartholomew, W. V. 1977. Soil nitrogen changes in farming systems in the humid tropics. In: *Biological Nitrogen Fixation in Farming Systems of the Tropics,* A. Ayanaba and P. J. Dart, eds. Wiley, New York, pp. 27–42.

Bartholomew, W. V., and F. E. Clark, eds. 1965. *Soil Nitrogen.* American Society of Agronomy, Madison, Wisc.

Bassham, J. A. 1977. Increasing crop production through more controlled photosynthesis. *Science 197*:630–638.

Bassham, J. A., and M. Calvin. 1957. *The Path of Carbon in Photosynthesis.* Prentice–Hall, Englewood Cliffs, N.J.

Bazilevich, N. I., L. Y. Rodin, and N. N. Rozov. 1971. Geographical aspects of biological productivity. *Soviet Geography 12*:293–317.

Bechtel National, Inc. 1980. *Economic and Design Factors for Flue Gas Desulfurization Technology.* EPRI, Palo Alto, Calif.

Bechtel National, Inc. 1981. *Impact of Coal Cleaning on the Cost of New Coal-Fired Power Generation.* EPRI, Palo Alto, Calif.

Beevers, L. 1976. *Nitrogen Metabolism in Plants.* Arnold, London.

Beilke, S., and H.-W. Georgii. 1968. Investigation on the incorporation of sulfur-dioxide into fog-and-rain droplets. *Tellus 20*:436–441.

Beilke, S., and D. Lamb. 1974. On the absorption of SO_2 in ocean water. *Tellus 26*:268–271.

Bell, P. R. 1982. Methane hydrate and the carbon dioxide question. In: *Carbon Dioxide Review: 1982,* W. C. Clark, ed. Oxford University Press (Clarendon), London, pp. 401–406.

Benemann, J. R. 1978. *Biofuels: A Survey.* EPRI, Palo Alto, Calif.

Bengtson, G. W. 1979. Forest fertilization in the United States: Progress and outlook. *Journal of Forestry 77*:222–229.

Berger, A., ed. 1981. *Climatic Variations and Variability: Facts and Theories.* Reidel, Dordrecht.

Bergersen, F. J. 1977. Factors controlling nitrogen fixation by rhizobia. In: *Biological Nitrogen Fixation in Farming Systems of the Tropics,* A. Ayanaba and P. J. Dart, eds. Wiley, New York, pp. 153–165.

Berman, I. M. 1979. Fluidized bed combustion systems: Progress and outlook. *Power Engineering 83*(11):46–56.

Berry, R. S., T. V. Long, II, and H. Makino. 1975. An international comparison of polymers and their alternatives. *Energy Policy 3*:144–155.

Bhumralkar, C., ed. 1984. *Meteorological Aspects of Acid Rain*. Butterworths, London.

Bhumralkar, C. M., R. L. Mancuso, D. E. Wolf, W. B. Johnson, and J. Pankrath. 1981. Regional air pollution model for calculating short-term (daily) patterns and transfrontier exchanges of airborne sulfur in Europe. *Tellus 33*:142–161.

Biggar, J. W. 1978. Spatial variability of nitrogen in soils. In: *Nitrogen in the Environment*, Volume I, D. R. Nielsen and J. G. MacDonald, eds. Academic Press, New York, pp. 201–211.

Biggins, P. D. E., and R. M. Harrison. 1979. Characterization and classification of atmospheric sulfates. *Journal of the Air Pollution Control Association 29*:838–840.

Bischof, W. 1973. Carbon dioxide concentration in the upper troposphere and lower stratosphere. Part 3. *Tellus 25*:305–308.

Bischof, W. 1974. The measurement of atmospheric CO_2. In: *World Meteorological Operations Manual No. 299*. WMO, Geneva.

Bischof, W. 1977. Comparability of CO_2 measurements. *Tellus 29*:435–444.

Blakeslee, C. F., and H. E. Burbach. 1972. Controlling NO_x emissions from steam generators. Paper presented at the Air Pollution Control Association Meeting, Miami Beach, June 18–22, 1972.

Blouin, G. M. 1974. *Effects of Increased Energy Costs on Fertilizer Production Costs and Technology*. Tennessee Valley Authority, Muscle Shoals, Ala.

Blount, C. W., and F. W. Dickson. 1973. Gypsum–anhydrite equilibria in sysetms $CaSO_4$–H_2O and $CaSO_4$–NaCl–H_2O. *American Mineralogy 58*:323–331.

Bodhaine, B. A., and J. M. Harris, eds. 1982. *Geophysical Monitoring for Climatic Change No. 10*. National Oceanic and Atmospheric Administration, Boulder, Colo.

Bohn, H. L. 1976. Estimate of organic carbon in world soils. *Soil Science Society of America Journal 40*:468–470.

Bolin, B. 1970. The carbon cycle. *Scientific American 223*(3):124–132.

Bolin, B. 1977. Changes of land biota and their importance for the carbon cycle. *Science 196*:613–615.

Bolin, B. 1979. On the role of the atmosphere in biogeochemical cycles. *Quarterly Journal of the Royal Meteorological Society 105*:25–42.

Bolin, B., ed. 1981. *Carbon Cycle Modelling*. SCOPE 16. Wiley, New York.

Bolin, B., and F. Arrhenius. 1977. Nitrogen—An essential life factor and a growing environmental hazard. *Ambio 6*:96–105.

Bolin, B., and C. Persson. 1975. Regional dispersion and deposition of atmospheric pollutants with particular application to sulfur pollution over western Europe. *Tellus 27*:286–310.

Bolin, B. and R. J. Charlson. 1976. On the role of the troposphere sulfer cycle in the shortwave radiative climate of the earth. *Ambio 5*: 47–54.

Bolin, B., E. J. Degens, S. Kempe, and P. Ketner, eds. 1979. *The Global Carbon Cycle*. SCOPE 13. Wiley, New York.

Bond, G. 1976. The results of the IBP survey of root-nodule formation in nonleguminous angiosperms. In: *Symbiotic Nitrogen Fixation in Plants*, P. S. Nutman, ed. Cambridge University Press, London, pp. 443–474.

Borgwardt, R. H. 1982. Energy requirements for SO_2 absorption in limestone scrubbers. In: *Flue Gas Desulfurization*, J. L. Hudson and G. T. Rochelle, eds. American Chemical Society, Washington, D.C., pp. 307–324.

Botkin, D. B. 1977. Forests, lakes and anthropogenic production of carbon dioxide. *BioScience 27*:325–331.

Botkin, D. B., J. F. Janak, and J. R. Wallis. 1973. Estimating the effects of carbon fertilization on forest composition by ecosystem simulation. In: *Carbon and the Biosphere*, G. M. Woodwell and E. V. Pecan, eds. U.S. Atomic Energy Commission, Washington, D.C., pp. 328–342.

Braekke, F. H., ed. 1976. *Impact of Acid Precipitation on Forest and Freshwater Ecosystems in Norway*. SNSF Project, As, Norway.

Brame, J. S. S., and J. G. King. 1967. *Fuel: Solid, Liquid and Gaseous*. Arnold, London.

Braun, R. C., and M. J. G. Wilson. 1970. The removal of atmospheric sulphur by building stones. *Atmospheric Environment 4*:371–378.

Bremner, J. M., and A. M. Blackmer. 1981. Terrestrial nitrifcation as a source of atmospheric nitrous oxide. In: *Denitrification, Nitrification, and Atmospheric Nitrous Oxide*, C. C. Delwiche, ed. Wiley, New York, pp. 151–170.

Brewer, P. G. 1983. Carbon dioxide and the oceans. In: *Changing Climate*. NAS, Washington, D.C., pp. 188–215.

Brimblecombe, D. 1978. 'Dew' as a sink for sulphur dioxide. *Tellus 30*:151–157.

Brink, R. A., J. W. Densmore, and G. A. Hill. 1977. Soil deterioration and the growing world demand for food. *Science 197*:625–630.

Brinkmann, W. L. F., and U. D. M. Santos. 1974. The emission of biogenic hydrogen sulfide from Amazonian floodplain lakes. *Tellus 26*:261–267.

Broecker, W. S. 1973. Factors controlling CO_2 content in the oceans and atmosphere. In: *Carbon and the Biosphere*, G. M. Woodwell and E. V. Pecan, eds. U.S. Atomic Energy Commission, Washington, D.C., pp. 32–49.

Broecker, W. S. 1975. Climatic change: Are we on the brink of a pronounced global warming? *Science 189*:460–463.

Broecker, W. S., and T. Takahashi. 1977. Neutralization of fossil fuel CO_2 by marine calcium carbonate. In: *The Fate of Fossil Fuel CO_2 in the Oceans*, N. R. Andersen and A. Malahoff, eds. Plenum Press, New York, pp. 213–241.

Brosset, C. 1978. Water-soluble sulphur compounds in aerosols. *Atmospheric Environment 12*:25–38.

Brown, L. R. 1978. *The Worldwide Loss of Cropland*. Worldwatch Institute, Washington, D.C.

Brown, L. R. 1984. Conserving soils. In: *State of the World 1984*, L. Starke, ed. Worldwatch Institute, Washington, D.C., pp. 53–73.

Brykowski, F. J., ed. 1982. *Ammonia and Synthesis Gas: Recent and Energy Saving Processes*. Noyes Data, Park Ridge, N.J.

Bryson, R. A. 1974. A perspective on climatic change. *Science 184*:753–760.

Bryson, R. A., and G. J. Dittberner. 1976. A non-equilibrium model of hemispheric mean surface temperature. *Journal of the Atmospheric Sciences 33*:2094–2106.

Budwani, R. N. 1978. Fossil-fired power plants: What it takes to get them built. *Power Engineering 82*(5):36–42.

Budwani, R. N. 1982. Power plant scheduling construction, and costs: 10-year analysis. *Power Engineering 86*(8):36–49.

Budyko, M. I. 1982. *The Earth's Climate: Past and Future*. Academic Press, New York.

Buresh, R. J., M. E. Casselman, and K. H. Pasnick, Jr. 1980. Nitrogen fixation in flooded soil systems. *Advances in Agronomy 33*:149–192.

Buringh, P. 1981. *An Assessment of Losses and Degradation of Productive Agricultural Land in the World*. FAO, Rome.

Burns, R. C., and R. W. F. Hardy. 1975. *Nitrogen Fixation in Bacteria and Higher Plants*. Springer-Verlag, Berlin.

Burris, R. H. 1980. The global nitrogen budget—Science or seance? In: *Nitrogen Fixation*, Volume I, W. E. Newton and W. H. Orme-Johnson, eds. University Park Press, Baltimore, pp. 7–16.

Burton, T. M., R. M. Stanford, and J. W. Allan. 1982. The effects of acidification on stream ecosystems. In: *Acid Precipitation: Effects on Ecological Systems*, F. M. D'Itri, ed. Ann Arbor Science, Ann Arbor, Mich., pp. 209–236.

Cadle, R. D. 1975. Volcanic emissions of halides and sulphur compounds to the troposphere and stratosphere. *Journal of Geophysical Research 80*:1650–1652.

Cadle, S. H., D. P. Chock, P. R. Monson, and J. M. Heuss. 1977. General Motors sulfate dispersion experiment: Experimental procedures and results. *Journal of the Air Pollution Control Association 27*:33–38.

Callendar, G. S. 1938. The artificial production of carbon dioxide and its influence on temperature. *Quarterly Journal of the Royal Meteorological Society 64*:223–237.

Callendar, G. S. 1958. On the measurement of carbon dioxide in the atmosphere. *Tellus 10*:241–248.

Calvert, J. G., and J. N. Pitts, Jr., eds. 1966. *Photochemistry*. Wiley, New York.

Calvert, J. G., S. Fu, J. W. Bottenheim, and O. P. Strausz. 1978. Mechanism of the homogeneous oxidation of sulfur dioxide in the troposphere. *Atmospheric Environment 12*:197–226.

Calvert, J. G., ed. 1984. *SO_2, NO and NO_2 Oxidation Mechanisms: Atmospheric Considerations*. Butterworths, London.

Carroll, J. E. 1982. *Environmental Diplomacy: An Examination and Prospective of Canadian–U.S. Environmental Relations*. University of Michigan Press, Ann Arbor.

Cass, G. R. 1979. On the relationship between sulfate air quality and visibility with examples in Los Angeles. *Atmospheric Environment 13*:1069–1084.

Cass, G. R. 1981. Sulfate air quality control strategy design. *Atmospheric Environment 15*:1227–1249.

Cassens, R. G., T. Ito, H. Lee, and D. Buege. 1978. The use of nitrite in meat. *BioScience 28*:633–637.

Cavender, J. H., D. S. Kircher, and A. J. Hoffman. 1973. *Nationwide Air Pollutant Emission Trends 1940–1970*. EPA, Research Triangle Park, N.C.

Central Intelligence Agency. 1980. *USSR: Coal Industry Problems and Prospects*. CIA, Washington, D.C.

Chamberlain, J. W., H. M. Foley, G. J. MacDonald, and M. A. Ruderman. 1982. Climate effects of minor atmospheric constituents. In: *Carbon Dioxide Review: 1982*, W. C. Clark, ed. Oxford University Press (Clarendon), London, pp. 253–277.

Chamberlin, T. C. 1897. A group of hypotheses bearing on climatic changes. *Journal of Geology 5*:653–683.

Chan, W. H., R. J. Vet, M. A. Lusis, J. E. Hunt, and R. D. S. Stevens. 1980. Airborne sulfur dioxide to sulfate oxidation studies of the INCO 381 m chimney plume. *Atmospheric Environment 14*:1159–1170.

Chan, W. H., C. U. Ro, M. A. Lusis, and R. J. Vet. 1982. Impact of the INCO nickel smelter emissions on precipitation quality in the Sudbury area. *Atmospheric Environment 16*:801–814.

Chang, S. G., D. Littlejohn, and N. H. Lin. 1982. Kinetics of reactions in a wet flue gas simultaneous desulfurization and denitrification system. In: *Flue Gas Desulfurization*, J. L. Hudson and G. T. Rochelle, eds. American Chemical Society, Washington, D.C., pp. 127–152.

Chang, T. Y., J. M. Norbeck, and B. Weinstock. 1979. An estimate of the NO_x removal rate in an urban atmosphere. *Environmental Science and Technology 13*:1534–1537.

Charlson, R. J., D. S. Covert, T. V. Larson, and A. P. Waggoner. 1978. Chemical properties of tropospheric sulfur aerosols. *Atmospheric Environment 12*:39–53.

Chass, R. L., W. B. Krenz, J. S. Nevitt, and J. A. Danielson. 1972. Los Angeles County acts to control emissions of nitrogen oxides from power plants. *Journal of the Air Pollution Control Association 22*:15–19.

Chatt, J., and G. J. Leigh. 1968. The inactivity and activation of nitrogen. In: *Recent Aspects of Nitrogen Metabolism in Plants*, E. J. Hewitt and C. V. Cutting, eds. Academic Press, New York, pp. 3–12.

Chen, K., R. C. Winter, and M. K. Bergman. 1980. Carbon dioxide from fossil fuels. *Energy Policy 8*:318–330.

Cheremisinoff, P., and W. B. Eaton. 1976. Combustion control methods to limit NO_x emissions. *Power Engineering 80*(9):56–58.

Cheremisinoff, P. N., and R. T. Fellman. 1974. Optimizing SO_2 scrubbing processes. *Power Engineering 78*(10):54–56.

Child, J. J. 1976. New developments in nitrogen fixation research. *BioScience 26*:614–617.

Chou, M., P. Li, and A. Arking. 1982. Climate studies with a multi-layer energy balance model. Part II. The role of feedback mechanisms in the CO_2 problem. *Journal of the Atmospheric Sciences 39*:2657–2666.

Cichanowicz, J. F., and R. E. Hall, cochairmen. 1981. *Proceedings of the Joint Symposium on Stationary Combustion NO_x Control*, two volumes. EPRI, Palo Alto, Calif.

Clark, F. E. 1981. The nitrogen cycle viewed with poetic license. In: *Terrestrial Nitrogen Cycles: Processes, Ecosystem Strategies and Management Impacts*, F. E. Clark and T. Rosswall, eds., *Ecological Bulletin 33*:13–24.

Clark, F. E., and T. Rosswall, eds. 1981. *Terrestrial Nitrogen Cycles: Processes, Ecosystem Strategies and Management Impacts. Ecological Bulletin 33.*

Clark, J. A., and C. S. Lingle. 1977. Future sea–land changes due to west Antarctic ice sheet fluctations. *Nature 269*:206–209.

Clark, T. L. 1980. Annual anthropogenic pollutant emissions in the United States and southern Canada east of the Rocky Mountains. *Atmospheric Environment 14*:961–970.

Clark, W. C., ed. 1982. *Carbon Dioxide Review: 1982*. Oxford University Press (Clarendon), London.

Clark, W. C., K. H. Cook, G. Marland, A. M. Weinberg, R. M. Rotty, P. R. Bell, L. J. Allison, and C. L. Cooper. 1982. The carbon dioxide question: Perspectives for 1982. In: *Carbon Dioxide Review: 1982*, W. C. Clark, ed. Oxford University Press (Clarendon), London, pp. 3–44.

Clarke, A. J., D. H. Lucas, and F. F. Ross. 1971. *Tall Stacks—How Effective are They?* Central Electricity Generating Board, London.

Clarkson, D. T., and J. B. Hanson. 1980. The mineral nutrition of higher plants. *Annual Review of Plant Physiology 31*:239–298.

Cleveland, W. S., and T. E. Graedel. 1979. Photochemical air pollution in the northeastern United States. *Science 204*:1273–1278.

Climatic Impact Assessment Program. 1975. *Impacts of Climatic Change on the Biosphere*. U.S. Department of Transportation, Washington, D.C.

Cloud, P. 1980. Early biogeochemical systems. In: *Biogeochemistry of Ancient and Modern Environments*, P. A. Trudinger, M. R. Walter, and B. J. Ralph, eds. Springer-Verlag, Berlin, pp. 7–27.

Coffer, W. R., D. R. Schryer, and R. S. Rogowski. 1980. The enhanced oxidation of SO_2 to NO_2 on carbon particulates. *Atmospheric Environment 14*:571–575.

Coffer, W. R., D. R. Schryer, and R. S. Rogowski. 1981. The oxidation of SO_2 on carbon particles in the presence of O_3, NO_2 and N_2O. *Atmospheric Environment 15*:1281–1286.

Coffin, D. L., and J. H. Knelson. 1976. Effects of sulfur dioxide and sulfate aerosol particles on human health. *Ambio 5*:239–242.

Cogbill, C. V., and G. E. Likens. 1974. Acid precipitation in the northeastern United States. *Water Resources Research 10*:1133–1137.

Coleman, R. 1966. The importance of sulphur as a plant nutrient in world crop production. *Soil Science 101*:230–239.

Commoner, B. 1968. Threats to integrity of the nitrogen cycle: Nitrogen compounds in soil, water, atmosphere and precipitation. Paper presented at the AAAS Annual Meeting, Dallas.

Commoner, B. 1971. *The Closing Circle*. Knopf, New York.

Commoner, B. 1975. Threats to the integrity of the nitrogen cycle: Nitrogen compounds in soil, water, atmosphere and precipitation. In: *The Changing Global Environment*, J. F. Singer, ed. Reidel, Dordrecht, pp. 341–366.

Commoner, B. 1977. Cost-risk-benefit analysis of nitrogen fertilizers: A case history. *Ambio 6*:157–161.

Conrad, R., and W. Seiler. 1980. Field measurements of the loss of fertilizer nitrogen into the atmosphere as nitrous oxide. *Atmospheric Environment 14*:555–558.

Considine, B. M., ed. 1977. *Energy Technology Handbook*. McGraw–Hill, New York.

Cook, C. W. 1979. Meat production potential on rangelands. *Journal of Soil and Water Conservation 34*:168–171.

Cooke, M. J., and R. A. Wadden. 1981. Atmospheric factors influencing daily sulfate concentrations in Chicago air. *Journal of the Air Pollution Control Association 31*:1197–1199.

Cooper, C. F. 1982. Food and fiber in a world of increasing carbon dioxide. In: *Carbon Dioxide Review: 1982*, W. C. Clark, ed. Oxford University Press (Clarendon), London, pp. 277–320.

Cottenie, A. 1980. *Soil and Plant Testing as a Basis of Fertilizer Recommendations*. FAO, Rome.

Council for Agricultural Science and Technology. 1976. *Effect of Increased Nitrogen Fixation on Stratospheric Ozone*. CAST, Ames, Iowa.

Council on Environmental Quality. 1980. *Environmental Quality 1980*. CEQ, Washington, D.C.

Coupland, R. T., ed. 1979. *Grassland Ecosystems of the World: Analysis of Grasslands and Their Uses*. Cambridge University Press, London.

Coutant, R. W. 1977. Effect of environmental variables on collection of atmospheric sulfate. *Environmental Science and Technology 11*:873–878.

Cowling, F. B. 1982a. Acid precipitation in historical perspective. *Environmental Science and Technology 16*:110A–123A.

Cowling, F. B. 1982b. A status report on acid precipitation and its biological consequences as of April 1982. In: *Acid Precipitation: Effects on Biological Systems,* F. M. D'Itri, ed. Ann Arbor Science, Ann Arbor, Mich., pp. 3–20.

Cowling, F. B., and R. A. Linthurst. 1981. The acid precipitation phenomenon and its ecological consequences. *BioScience 31*:649–654.

Cox, R. A. 1974. Particle formation from homogeneous reactions of sulphur dioxide and nitrogen dioxide. *Tellus 26*:235–240.

Crocker, J. D., W. Schulze, S. Ben-David, and A. V. Kneese. 1979. *Methods Development for Assessing Air Pollution: Control Benefits,* Volume I. EPA, Washington, D.C.

Croft, J. A. 1976. Nighttime images of the earth from space. *Scientific American 239*:86–98.

Cronan, C. S. 1980. Consequences of sulfuric acid inputs to a forest soil. In: *Atmospheric Sulfur Deposition,* D. S. Shriner, C. R. Richmond, and S. E. Lindberg, eds. Ann Arbor Science, Ann Arbor, Mich., pp. 335–343.

Cronn, D. R., and W. Nutmagul. 1982. Volcanic gases in the April 1979 Soufriere eruption. *Science 216*:1121–1123.

Crouse, W. H., and D. L. Anglin. 1977. *Automotive Emission Control.* McGraw–Hill, New York.

Crutzen, P. J. 1970. The influence of nitrogen oxides on the atmosphere ozone content. *Quarterly Journal of the Royal Meteorological Society 96*:320–325.

Crutzen, P. J. 1974. Estimates of possible variations in total ozone due to natural causes and human activities. *Ambio 3*:201–210.

Crutzen, P. J. 1976a. Upper limits on atmospheric ozone reductions following increased application of fixed nitrogen to the soil. *Geophysical Research Letters 3*:169–172.

Crutzen, P. J. 1976b. The possible importance of COS for the sulfate layer of the stratosphere. *Geophysical Research Letters 3*:73.

Crutzen, P. J. 1981. Atmospheric chemical processes of the oxides of nitrogen, including nitrous oxide. In: *Denitrification, Nitrification, and Atmospheric Nitrous Oxide,* C. C. Delwiche, ed. Wiley, New York, pp. 17–44.

Crutzen, P. J., L. E. Heidt, J. P. Krasnec, W. H. Pollock, and W. Seiler. 1979. Biomass burning as a source of atmospheric gases CO, H_2, N_2O, NO, CH_3Cl and COS. *Nature 282*:253–256.

Culbertson, W. C., and J. K. Pitman. 1973. Oil shale. In: *United States Mineral Resources,* D. A. Brobst and W. P. Pratt, eds. USGS, Washington, D.C., pp. 497–503.

Cullis, C. F., and M. M. Hirschler. 1980. Atmospheric sulphur: Natural and man-made sources. *Atmospheric Environment 14*:1263–1278.

Cumberland, J. C., J. R. Hibbs, and I. Hoch, eds. 1982. *The Economics of Managing Chlorofluorocarbons.* Resources for the Future, Washington, D.C.

Cummings, M. B., and C. H. Jones. 1918. *The Aerial Fertilization of Plants with Carbon Dioxide.* Vermont Agricultural Station, Burlington.

Daigger, L. 1974. Scientific analysis compares cost of fertility. *Upbeet 62*:10.

Dalbey, W. E. 1980. Laboratory studies of biological effects of sulfur oxides. In: *Atmospheric Sulfur Deposition,* D. S. Shriner, C. R. Richmond, and S. E. Lindberg, eds. Ann Arbor Science, Ann Arbor, Mich., pp. 69–75.

Daniel, G. H. 1956. The energy requirements of the United Kingdom. In: *Proceedings of the International Conference on the Peaceful Uses of Atomic Energy.* UNO, New York, pp. 239–244.

Darzi, M., and J. W. Winchester. 1982. Aerosol characteristics at Mauna Loa Observatory, Hawaii, after east Asian dust storm episodes. *Journal of Geophysical Research 87*:1251–1258.

Davey, T. R. A. 1978. Anthropogenic balance for Australia, 1976. Paper presented at the Australian Mineral IIndustries Council Environmental Workshop, Hobart, October 1978.

Davies, T. D. 1976. Precipitation scavenging of sulphur dioxide in an industrial area. *Atmospheric Environment 10*:879–890.

Davies, T. D. 1983. Sulphur dioxide precipitation scavenging. *Atmospheric Environment 17*:797–805.

Davis, R. A., J. A. Meyler, and K. E. Gude. 1979. Dry scrubber maintains high efficiency. *Power Engineering 83*(10):83–85.

Davis, W. T., and M. A. Fiedler. 1982. The retention of sulfur in fly ash from coal-fired boilers. *Journal of the Air Pollution Control Association 32*:395–397.

Davison, R. L., D. F. S. Natusch, J. R. Wallace, and C. H. Evans. 1974. Trace elements in fly ash. *Environmental Science and Technology 8*:1107–1113.

Dawson, G. A. 1977. Atmospheric ammonia from undisturbed land. *Journal of Geophysical Research 82*:3125–3133.

Dawson, G. A. 1980. Nitrogen fixation by lightning. *Journal of Atmospheric Sciences 37*:174–178.

de Bary, E., and C. Junge. 1963. Distribution of sulfur and chlorine over Europe. *Tellus 15*:370–381.

DeCarlo, J. A., E. T. Sheridan, and Z. E. Murphy. 1966. *Sulfur Content of United States Coals*. U.S. Bureau of Mines Information Circular 8312. U.S. Bureau of Mines, Washington, D.C.

Deevey, E. S., Jr. 1970. Mineral cycles. *Scientific American 223*:149–158.

Delmas, R., and C. Boutron. 1978. Sulfate in Antarctic snow: Spatio-temporal distribution. *Atmospheric Environment 12*:723–728.

Delmas, R., J. Baudet, and J. Servant. 1978. Mise en evidence des sources naturelles de sulfate en milieu tropical humide. *Tellus 30*:158–168.

Delmas, R., J. Baudet, J. Servant, and Y. Baziard. 1980. Emissions and concentrations of hydrogen sulfide in the air of the tropical forest of the Ivory Coast and of temperate regions in France. *Journal of Geophysical Research 85*:4468–4474.

DeLuisi, J. J., ed. 1981. *Geophysical Monitoring for Climatic Change No. 9*. USDC, NOAA, Boulder, Colo.

Delwiche, C. C. 1964. The cycling of carbon and nitrogen in the biosphere. In: *Microbiology and Soil Fertility*, C. M. Gilmour and O. N. Allen, eds. Oregon State University Press, Corvallis, pp. 29–58.

Delwiche, C. C. 1970. The nitrogen cycle. *Scientific American 223*:136–147.

Delwiche, C. C. 1977. Energy relations in the global nitrogen cycle. *Ambio 6*:106–111.

Delwiche, C. C. 1978. Legumes—Past, present, and future. *BioScience 28*:565–570.

Denmead, O. T., J. R. Freney, and J. R. Simpson. 1979. Nitrous oxide emission during denitrification in a flooded field. *Soil Science Society of America Journal 43*:716–718.

De Vooys, C. G. N. 1979. Primary production in aquatic environments. In: *The Global Carbon Cycle*, B. Bolin, E. T. Degens, S Kempe, and P. Ketner, eds. Wiley, New York, pp. 259–292.

Dickinson, R. E. 1982. Modeling climate changes due to carbon dioxide increases. In: *Carbon Dioxide Review: 1982*, W. C. Clark, ed. Oxford University Press (Clarendon), London, pp. 101–133.

Dickson, W. 1975. *Acidification of Swedish Lakes*. Institute of Freshwater Research, Drottningholm.

Dijkshoorn, W., and A. L. van Wijk. 1967. The sulphur requirements of plants as evidenced by the sulphur–nitrogen ratio in the organic matter. *Plant and Soil 26*:129–157.

D'Itri, F. M., ed. 1982. *Acid Precipitation: Effects of Biological Systems*. Ann Arbor Science, Ann Arbor, Mich.

Döbereiner, J. 1977. Biological nitrogen fixation in tropical grasses—Possibilities for partial replacement of mineral N fertilizers. *Ambio 6*:174–177.

Dochinger, L. S., and T. A. Seliga, eds. 1976. *Proceedings of the First International Symposium on Acid Precipitation and the Forest Ecosystem*. USDA Forest Service, Upper Darby, Pa.

Dockery, D. W., and J. D. Spengler. 1981. Indoor–outdoor relationships of respirable sulfates and particles. *Atmospheric Environment 15*:335–343.

Donahue, J. D. 1983. The political economy of milk. *The Atlantic Monthly 252*(10):59–68.

Doyle, J. J. 1966. *The Response of Rice to Fertilizer*. FAO, Rome.

Drabløs, D., and A. Tollan, eds. 1980. *Ecological Impact of Acid Precipitation*. SNSF Project, As, Norway.

Dumas, M. J., and M. J. B. Boussingault. 1844. *The Chemical and Physiological Balance of Organic Nature*. Saxton & Miles, New York.

Dunigan, E. P., O. B. Sober, J. L. Rabb, and D. J. Boquet. 1980. *Effects of Various Inoculants on Nitrogen Fixation and Yield of Soybeans.* Bulletin 726, Agricultural Experiment Station, Louisiana State University, Baton Rouge.

Durham, J. L., ed. 1984. *Chemistry of Particles, Fogs and Rain.* Butterworths, London.

Economic Commission for Europe. 1981. *The Influence of Sulphur Pollutants on Atmospheric Corrosion of Important Materials.* ECE, Geneva.

Edens, J. J., and S. O. Soldberg. 1977. Nutrient discharge from a 90 km^2 watershed. *Progress in Water Technology* 8(4–5):85–89.

Edwards, C. A., and J. R. Lofty. 1977. *Biology of Earthworms.* Chapman & Hall, London.

Edwards, D. G. 1977. Nutritional factors limiting nitrogen fixed by rhizobia. In: *Biological Nitrogen Fixation in Farming Systems of the Tropics,* A. Ayanaba and P. J. Dart, eds. Wiley, New York, pp. 189–204.

Ehrlich, P. R., and A. H. Ehrlich. 1981. *Extinction: The Causes and Consequences of the Disappearance of Species.* Random House, New York.

el-Din, M. A., A. Talib, and J. Fritz. 1980. Biomass energy potential in Egypt: A realistic appraisal. In: *Proceedings Bio-Energy '80.* The Bio-Energy Council, Washington, D.C., pp. 448–452.

Electric Power Research Institute. 1979. *Ecological Effects of Acid Precipitation.* EPRI, Palo Alto, Calif.

Electric Power Research Institute. 1981. *EPRI Sulfate Regional Experiment: Results and Implications.* EPRI, Palo Alto, Calif.

Eliassen, A., and J. Saltbones. 1975. Decay and transformation rates of SO_2 as estimated from emission data, trajectories and measured air concentrations. *Atmospheric Environment* 7:425–429.

Eliot, R. C., ed. 1978. *Coal Desulfurization Prior to Combustion.* Noyes Data Corporation, Park Ridge, N.J.

Ellsaesser, H. W. 1975. The upward trend in airborne particulates that isn't. In: *The Changing Global Environment,* S. F. Singer, ed. Reidel, Dordrecht, pp. 235–269.

Elshout, A. J., J. W. Viljeer, and H. van Duuren. 1978. Sulphates and sulphuric acid in the atmosphere in the years 1971–1976 in the Netherlands. *Atmospheric Environment 12*:785–790.

Emanuelsson, A., E. Eriksson, and H. Egnér. 1954. Composition of atmospheric precipitation in Sweden. *Tellus 6*:261–267.

Engelstad, O. P., and D. A. Russel. 1975. Fertilizers for use under tropical conditions. *Advances in Agronomy 27*:175–208.

Environment Canada. 1981. *The Clean Air Act Annual Report 1980–1981.* Environment Canada, Ottawa.

Eriksson, E. 1959. Atmospheric chemistry. Svensk Kemisk Tidskrift 71:15–32.

Eriksson, E. 1960. The yearly circulation of chloride and sulphur in nature: Meteorological, geochemical and pedological implications II. *Tellus 12*:63–109.

Eriksson, E. 1963. The yearly circulation of sulfur in nature. *Journal of Geophysical Research 68*:4001–4008.

Evans, H. J., ed. 1975. *Enhancing Biological Nitrogen Fixation.* National Science Foundation, Washington, D.C.

Evans, H. J., and L. F. Barber. 1977. Biological nitrogen fixation for food and fiber production. *Science 197*:332–339.

Evendijk, J. F., and P. A. R. Post van der Burg. 1977. Monitoring 'Rijnmond smog' for alert conditions. *Environmental Science and Technology 11*:450–455.

Fensterstock, J. C., and R. K. Fankhauser. 1968. *Thanksgiving 1966 Air Pollution Episode in the Eastern United States.* National Air Pollution Control Administration, Durham, N.C.

Ferman, M. A., R. S. Eisinger, and P. R. Monson. 1977. Characterization of Denver air quality. In: *Denver Air Pollution Study—1973,* Volume II, P. A. Russell, ed. EPA, Research Triangle Park, N.C.

Ferman, M. A., G. T. Wolff and N. A. Kelly. 1981. The nature and source of haze in the Shenandoah Valley/Blue Ridge Mountains area. *Journal of Air Pollution Control Association 31*:1074–1082.

Ferris, B. G. 1969. Chronic low-level air pollution. *Environmental Research* 2:79–87.

Ferris, B. G. 1979. Health aspects of fossil fuel electric power plants—Air pollution. In: *Symposium on Energy and Human Health: Human Costs of Electric Power Generation*. EPA, Washington, D.C., pp. 178–203.

Fisher, B. E. A. 1975. The long-range transport of sulphur dioxide. *Atmospheric Environment* 9:1063–1070.

Fisher, B. E. A. 1982. The transport and removal of sulphur dixoide in a rain system. *Atmospheric Environment* 16:775–783.

Floate, M. J. S. 1977. Changes in soil pools. *Agro-Ecosystems* 4:292–295.

Flohn, H. 1961. Man's activity as a factor in climatic change. *Annals of the New York Academy of Sciences* 95:271–281.

Flohn, H. 1979. A scenario of possible future climates—natural and manmade. In: *Proceedings of the World Climate Conference*. WMO, Geneva, pp. 243–266.

Flohn, H. 1982. Climate change and an ice-free Arctic Ocean. In: *Carbon Dioxide Review: 1982*, W. C. Clark, ed. Oxford University Press (Clarendon), London, pp. 143–179.

Focht, D. D., and W. Verstraete. 1977. Biochemical ecology of nitrification and denitrification. *Advances in Microbial Ecology* 1:135–214.

Folsom, B. A., L. P. Nelson, and J. Vatsky. 1981. The development of a low NO_x distributed mixing burner for pulverized coal boilers. In: *Proceedings of the Joint Symposium on Stationary Combustion NO_x Control*. EPRI, Palo Alto, Calif., pp. 451–490.

Food and Agriculture Organization. 1955. *World Forest Resources*. FAO, Rome.

Food and Agriculture Organization. 1958. *Efficient Use of Fertilizers*. FAO, Rome.

Food and Agriculture Organization. 1959. *Tabulated Information on Tropical and Subtropical Grain Legumes*. FAO, Rome.

Food and Agriculture Organization. 1960. *World Forest Inventory*. FAO, Rome.

Food and Agriculture Organization. 1963. *World Forest Inventory*. FAO, Rome.

Food and Agriculture Organization. 1966a. *Statistics of Crop Responses to Fertilizers*. FAO, Rome.

Food and Agriculture Organization. 1966b. *Rice Grain of Life*. FAO, Rome.

Food and Agriculture Organization. 1976. *Forest Resources in Asia and Far East Region*. FAO, Rome.

Food and Agriculture Organization. 1978. *Fertilizers and Their Use*. FAO, Rome.

Food and Agriculture Organization. 1980a. *Maximizing the Efficiency of Fertilizer Use by Grain Crops*. FAO, Rome.

Food and Agriculture Organization. 1980b. *Soil and Plant Testing and Analysis*. FAO, Rome.

Food and Agriculture Organization. 1981a. *Blue-green Algae for Rice Production*. FAO, Rome.

Food and Agriculture Organization. 1981b. *Crop Production Levels and Fertilizer Use*. FAO, Rome.

Food and Agriculture Organization and United Nations Development Programme. 1977. *China: Recycling of Organic Wastes in Agriculture*. FAO, Rome.

Forrest, J., and L. Newman. 1973. Ambient air monitoring for sulfur compounds: A critical review. *Journal of the Air Pollution Control Association* 23:761–768.

Forrest, J., and L. Newman. 1977. Further studies on the oxidation of sulfur dioxide in coal-fired power plant plumes. *Atmospheric Environment* 11:465–474.

Forsberg, C. 1977. Nitrogen as a growth factor in fresh water. *Progress in Water Technology* 8:275–290.

Forziati, A. 1982. The chlorofluorocarbon problem. In: *The Economics of Managing Chlorofluorocarbons*, J. C. Cumberland, J. R. Hibbs, and I. Hoch, eds. Resources for the Future, Washington, D.C., pp. 36–63.

Foster, P. M. 1969. The oxidation of sulphur dioxide in power station plumes. *Atmospheric Environment* 3:157–175.

Fowler, D. 1978. Dry deposition of SO_2 on agricultural crops. *Atmospheric Environment* 12:369–373.

Franda, M. 1980. Kettering's nitrogen fixers. *American University Field Staff Reports* 1980(15).

Frankenberg, T. T. 1971. What sulphur dioxide problem? Preamble. *Combustion* 43(2):2–3.

Freney, J. A., O. T. Denmead, and J. R. Simpson. 1979. Nitrous oxide emission from soils at low moisture contents. *Soil Biology and Biochemistry 11*:167–173.

Friend, J. P. 1973. The global sulfur cycle. In: *Chemistry of the Lower Atmosphere,* S. I. Rasool, ed. Plenum Press, New York, pp. 177–201.

Frink, C. R., and Voigt, G. K. 1976. *Potential Effects of Acid Precipitation on Soils of the Humid Temperate Zone.* USDA, Washington, D.C.

Frissel, M. J., ed. 1977a. *Cycling of Mineral Nutrients in Agricultural Ecosystems.* First International Environmental Symposium, Amsterdam, May 31–June 4, 1976. *Agro-Ecosystems 4*:i–viii and 1–354.

Frissel, M. J. 1977b. Application of nitrogen fertilizers: Present trends and projections. *Ambio 6*:152–156.

Frissel, M. J., and G. J. Kolenbrander. 1977. The nutrient balances: Summarizing graphs and tables. *Agro-Ecosystems 4*:277–292.

Fukushima, T., H. Nakamura, and T. Sakai. 1977. Exhaust emission control of S.I. engines by engine modification. Paper presented at the Annual Meeting of the Society of Automotive Engineering, Detroit, February 28–March 4, 1977.

Gagan, E. W. 1974. *Air Pollution Emissions and Control Technology: Cement Industry.* Environment Canada, Ottawa.

Galloway, J. N., and E. B. Cowling. 1978. The effects of precipitation on aquatic and terrestrial ecosystems: A proposed precipitation chemistry network. *Journal of the Air Pollution Control Association 28*:229–235.

Galloway, J. N., and G. E. Likens. 1981. Acid precipitation: The importance of nitric acid. *Atmospheric Environment 15*:1081–1085.

Galloway, J. N., and D. M. Whelpdale. 1980. An atmospheric sulfur budget for eastern North America. *Atmospheric Environment 14*:409–417.

Galloway, J. N., G. E. Likens, W. C. Keene, and J. C. Miller. 1982. The composition of precipitation in remote areas of the world. *Journal of Geophysical Research 87*:8771–8786.

Gandrud, B. W., and A. L. Lazrus. 1981. Filter measurements of stratospheric sulfate and chloride in the eruption plume of Mount St. Helens. *Science 211*:826–827.

Gardner, D. E., J. A. Graham, and D. Menzel. 1979. Health consequences of nitrogen dioxide exposure. In: *Energy/Environment IV.* EPA, Washington, D.C., pp. 261–292.

Garland, J. A. 1977. The dry deposition of sulphur dioxide to land and water surfaces. *Proceedings of the Royal Society of London Series A 354*:245–268.

Garland, J. A. 1978. Dry and wet removal of sulphur from the atmosphere. *Atmospheric Environment 12*:349–362.

Garland, J. A. 1981. Enrichment of sulphate in maritime aerosols. *Atmospheric Environment 15*:787–791.

Garratt, J. R., and G. I. Pearman. 1973. CO_2 concentration in the atmosphere boundary-layer over southeast Australia. *Atmospheric Environment 7*:1257–1266.

Gartrell, F. E., F. W. Thomas, and S. B. Carpenter. 1963. Atmospheric oxidation of sulphur dioxide in coal-burning power plant plumes. *American Industrial Hygiene Association Journal 24*:113–120.

Gates, W. L., and K. H. Cook. 1980. *Preliminary Analysis of Experiments on the Climatic Effects of Increased* CO_2 *with the O.S.U. Atmospheric General Circulation Model.* Oregon State University, Corvallis.

Gauri, K. L., and G. C. Holdren, Jr. 1981. Pollutant effects on stone monuments. *Environmental Science and Technology 15*:386–390.

Gautney, J., Y. K. Kim, and J. D. Hatfield. 1982. Melamine: A regenerative SO_2 absorbent. *Journal of the Air Pollution Control Association 32*:260–265.

Gelinas, C. G., and H. S. L. Fan. 1979. Reducing air pollutant emissions at airports by controlling aircraft ground operations. *Journal of the Air Pollution Control Association 29*:125–132.

Georgii, H.-W. 1978. Large scale spatial and temporal distribution of sulfur compounds. *Atmospheric Environment 12*:681–690.

Gibbs, R. J. 1972. Water chemistry of the Amazon River. *Geochimica et Cosmochimica Acta 36*:1061–1066.

Gilbertson, C. B., D. L. Van Dyne, C. J. Clanton, and R. K. White. 1978. *Nutrients Available in Livestock and Poultry Manure Residues as Reflected by Management Systems.* American Society of Agricultural Engineers Technical Paper TP78-3064.

Gillette, D. G. 1975. Sulfur dioxide and material damage. *Journal of the Air Pollution Control Association 25*:1238–1243.

Gilliland, R. L. 1982. Commentary. In: *Carbon Dioxide Review: 1982,* W. C. Clark, ed. Oxford University Press (Clarendon), London, pp. 242–244.

Glass, N. R., G. E. Glass, and P. J. Rennie, 1979. Effects of acid precipitation. *Environmental Science and Technology 13*:1350–1355.

Glass, N. R., D. E. Arnold, J. N. Galloway, G. R. Hendrey, J. L. Lee, W. W. McFee, S. A. Norton, C. F. Powers, D. L. Rambo, and C. L. Schofield. 1982. Effects of acid precipitation. *Environmental Science and Technology 16*:162A–169A.

Gold, P. S., ed. 1981. *Acid Rain: A Transjurisdictional Problem in Search of Solution.* State University of New York, Buffalo.

Goldhaber, M. B., and I. R. Kaplan. 1974. The sulfur cycle. In: *The Sea,* E. D. Goldberg, ed. Wiley, New York, pp. 569–655.

Goldman, M. A. 1974. Carbon dioxide measurements and local wind patterns at Mauna Loa Observatory, Hawaii. *Journal of Geophysical Research 79*:4550–4554.

Good, N. E., and D. H. Bell. 1980. Photosynthesis, plant productivity, and crop yield. In: *The Biology of Crop Productivity,* P. S. Carlson, ed. Academic Press, New York, pp. 3–51.

Goodland, R. J. A. 1980. Environmental ranking of Amazonian development projects in Brazil. *Environmental Conservation 7*:9–26.

Gorham, E. 1981. Scientific understanding of atmosphere–biosphere interactions: A historical overview. In: *Atmosphere–Biosphere Interaction.* NAS, Washington, D.C., pp. 9–21.

Grahn, O. H. 1980. Fish kills in two moderately acid lakes due to high aluminum concentration. In: *Ecological Impact of Acid Precipitation.* SNSF Project, As, Norway, pp. 310–311.

Grahn, O. H., H. Hultberg, and L. Landner. 1974. Oligotrophication—A self-accelerating process in lakes subjected to excessive supply of acid substances. *Ambio 3*:93–94.

Granat, L. 1972. On the relation between pH and the chemical composition in atmospheric precipitation. *Tellus 24*:550–560.

Granat, L. 1976. A global atmospheric sulphur budget. *Ecological Bulletin 22*:102–122.

Granat, L. 1978. Sulfate in precipitation as observed by the European atmospheric chemistry network. *Atmospheric Environment 12*:413–424.

Greenberg, G. E., C. T. Hill, and D. J. Newburger. 1977. *Regulation, Market Prices, and Process Innovation: The Case of Ammonia Industry.* Westview Press, Boulder, Colo.

Greenwood, E. A. N. 1976. Nitrogen stress in plants. *Advances in Agronomy 28*:1–25.

Gregory, D. P., and J. B. Pangborn. 1976. Hydrogen energy. *Annual Review of Energy 1*:279–310.

Grey, D. C., and M. L. Jensen. 1972. Bacteriogenic sulfur in air pollution. *Science 177*:1099–1100.

Gribben, J., ed. 1978. *Climatic Change.* Cambridge University Press, London.

Griffing, G. W. 1977. Ozone and oxides of nitrogen production during thunderstorms. *Journal of Geophysical Research 82*:943–950.

Gupta, R. K., and R. S. Ambasht. 1979. Use and management. In: *Grassland Ecosystems of the World,* R. T. Coupland, ed. Cambridge University Press, London, pp. 241–244.

Gutschick, V. P. 1977. *Long-Term Strategies for Supplying Nitrogen to Crops.* Los Alamos Scientific Laboratory, Los Alamos, N.M.

Gutschick, V. P. 1980. Energy flows in the nitrogen cycle, especially in fixation. In: *Nitrogen Fixation,* Volume I, W. E. Newton and W. H. Orme-Johnson, eds. University Park Press, Baltimore, pp. 17–27.

Gutschick, V. P. 1981. Evolved strategies in nitrogen acquisition by plants. *The American Naturalist 118*:607–637.

Haagen-Smit, A. J., and L. G. Wayne. 1968. Atmospheric reactions and scavenging processes. In: *Air Pollution,* Volume 1, A. C. Stern, ed. Academic Press, New York, pp. 149–186.

Hageman, R. H. 1980. Effect of form of nitrogen on plant growth. In: *Nitrification Inhibitors—*

Potentials and Limitations, M. Stelly, ed. American Society of Agronomy, Madison, Wisc., pp. 47–61.

Hahn, J. 1974. The North Atlantic Ocean as a source of atmospheric N_2O. *Tellus 26*:160–168.

Hahn, J. 1981. Nitrous oxide in the oceans. In: *Denitrification, Nitrification, and Atmospheric Nitrous Oxide,* C. C. Delwiche, ed. Wiley, New York, pp. 191–240.

Hahn, J., and C. Junge. 1977. Atmospheric nitrous oxide: A critical review. *Zeitschrift fuer Naturforschung 32a*:190–214.

Hales, J. M., and M. T. Dana. 1979. Regional-scale deposition of sulfur dioxide by precipitation and scavenging. *Atmospheric Environment 13*:1121–1132.

Hall, M. C. G., D. G. Cacuci, and M. E. Schlesinger. 1982. Sensitivity analysis of a radiative–convective model by adjoint method. *Journal of the Atmospheric Sciences 39*:2038–2050.

Hallberg, R. O. 1976. A global sulphur cycle based on a pre-industrial steady-state of the pedosphere. *Ecological Bulletin 22*:93–101.

Hallberg, R., ed. 1983. *Environmental Biogeochemistry. Ecological Bulletin* No. 35, Stockholm.

Hamilton, P. M., R. H. Varey, and M. M. Millan. 1978. Remote sensing of sulphur dioxide. *Atmospheric Environment 12*:127–133.

Hansen, D. A., and G. M. Hidy. 1982. Review of questions regarding rain acidity data. *Atmospheric Environment 16*:2107–2126.

Hansen, J., D. Johnson, A. Lacis, S. Lebedeff, P. Lee, D. Rind, and G. Russell. 1981. Climate impact of increasing atmospheric carbon dioxide. *Science 213*:957–966.

Hansen, M. H., K. Ingvorsen, and B. B. Jorgensen. 1978. Mechanisms of hydrogen sulfide release from coastal marine sediments to the atmosphere. *Limnology and Oceanography 23*:68–76.

Hanson, H., N. E. Borlaug, and R. G. Anderson. 1982. *Wheat in the Third World.* Westview Press, Boulder, Colo.

Harcombe, P. A. 1977. Nutrient accumulation by vegetation during the first year of recovery of a tropical forest ecosystem. In: *Recovery and Restoration of Damaged Ecosystems,* J. Cairns, K. Dickson, and E. Herricks, eds. University Press of Virginia, Charlottesville, pp. 347–378.

Hardy, R. W. F., and A. H. Gibson, eds. 1977. *A Treatise on Dinitrogen Fixation.* Wiley, New York.

Hargett, N. L., and J. T. Berry. 1981. *1980 Fertilizer Summary Data.* National Fertilizer Development Center, Muscle Shoals, Ala.

Harkov, R. 1981. Ecosystem acidity. *Journal of the Air Pollution Control Association 32*:344.

Harremoës, P., chairman. 1975. *Proceedings of the Conference on Nitrogen as a Water Pollutant.* International Association on Water Pollution Research, Copenhagen, August 18–20, 1975.

Harrison, R. M., and H. A. McCartney. 1979. Some measurements of ambient air pollution arising from the manufacture of nitric acid and ammonium nitrate fertilizer. *Atmospheric Environment 13*:1105–1120.

Hart, M. H. 1978. The evolution of the atmosphere of the earth. *Icarus 33*:23–39.

Hauck, R. D. 1980. Mode of action of nitrification inhibitors. In: *Nitrification Inhibitors—Potentials and Limitations,* M. Stelly, ed. American Society of Agronomy, Madison, Wisc., pp. 19–32.

Havelka, U. D., M. G. Boyle, and R. W. F. Hardy. 1982. Biological nitrogen fixation. In: *Nitrogen in Agricultural Soils,* F. J. Stevenson, ed. American Society of Agronomy, Madison, Wisc., pp. 365–422.

Hayes, E. T. 1976. Energy implications of materials processing. *Science 191*:661–665.

Haynes, R. J., and K. M. Goh. 1978. Ammonium and nitrate nutrition of plants. *Biological Reviews of the Cambridge Philosophical Society 53*:465–510.

Hays, J. D., J. Imbrie, and N. J. Shackleton. 1976. Variations in the earth's orbit: Pacemaker of the ice ages. *Science 194*:1121–1132.

Hecht, N. L., and D. S. Duvall. 1975. *Characterization and Utilization of Municipal and Utility Sludges and Ashes,* Volume II. EPA, Cincinnati, Ohio.

Hegg, D. A., and P. V. Hobbs. 1978. Oxidation of sulfur dioxide in aqueous systems with particular reference to the atmosphere. *Atmospheric Environment 12*:241–253.

Hegsted, D. M. 1978. Protein-caloric malnutrition. *American Scientist 60*:61–65.

Heinsdijk, D. 1975. *Forest Assessment.* Center for Agricultural Publishing and Documentation, Wageningen, The Netherlands.

Heitschmidt, R. K., W. K. Lauenroth, and J. L. Dodd. 1978. Effects of controlled levels of sulphur dioxide on western wheatgrass in a southeastern Montana grassland. *Journal of Applied Ecology* *15*:859–868.

Henderson, G. S., W. T. Swank, and J. W. Hornbeck. 1980. Impact of sulfur deposition on the quality of water from forested watersheds. In: *Atmospheric Sulfur Deposition,* D. S. Shriner, C. R. Richmond, and S. E. Lindberg, eds. Ann Arbor Science, Ann Arbor, Mich., pp. 431–442.

Hendrey, G. R., K. Baalsrud, T. S. Traaen, M. Laake, and G. Raddum. 1976. Some hydrobiological changes. *Ambio 5*:224–227.

Henry, R. C., and G. M. Hidy. 1979. Multivariate analysis of particulate sulfate and other air quality variables by principal components. Part I. Annual data from Los Angeles and New York. *Atmospheric Environment 13*:1581–1596.

Hergert, G. W., and R. A. Wiese. 1980. Performance of nitrification inhibitors in the Midwest (west). In: *Nitrification Inhibitors—Potentials and Limitations,* M. Stelly, ed. American Society of Agronomy, Madison, Wisc., pp. 89–106.

Hibbard, W. R., Jr. 1979. Policies and constraints for major expansion of U.S. coal production and utilization. *Annual Review of Energy 4*:147–174.

Hicks, B. B., ed. 1984. *Deposition Both Wet and Dry.* Butterworths, London.

Hidy, G. M., P. K. Mueller, and E. Y. Tong. 1978. Spatial and temporal distributions of airborne sulfate in parts of the United States. *Atmospheric Environment 12*:735–752.

Highley, J. 1980. The development of fluidized bed combustion. *Environmental Science and Technology 14*:270–275.

Hileman, B. 1983. Arctic haze. *Environmental Science and Technology 17*:232A–236A.

Hills, F. J., R. Sailsbery, and A. Ulrich. 1978. *Sugarbeet Fertilization.* Division of Agricultural Sciences Bulletin 1891, University of California, Davis.

Hirschler, M. M. 1981. Man's emission of carbon dioxide into the atmosphere. *Atmospheric Environment 15*:719–727.

Hitchock, D. R. 1976. Atmospheric sulfates from biological sources. *Journal of the Air Pollution Control Association 26*:210–215.

Hobbs, P. V., and D. A. Hegg. 1982. Sulfate and nitrate mass distributions in the near fields of some coal-fired power plants. *Atmospheric Environment 16*:2657–2662.

Hobbs, P. V., L. F. Radke, M. W. Eltgroth, and D. A. Hegg. 1981. Airborne studies of the emissions from the volcanic eruptions of Mount St. Helens. *Science 211*:816–818.

Hobson, P. N., and A. H. Robertson. 1977. *Waste Treatment in Agriculture.* Applied Science Publishers, London.

Hodgeson, J. A., W. A. McClenny, and P. L. Hanst. 1973. Air pollution monitoring by advanced spectroscopic techniques. *Science 182*:248–258.

Hoffer, J., J. Kliwer, and J. Moyer. 1979. Sulfate concentrations in the southwestern desert of the United States. *Atmospheric Environment 13*:619–627.

Högberg, P., and M. Kvarnström. 1982. Nitrogen fixation by the woody legume *Leucaena leucocephala* in Tanzania. *Plant and Soil 66*:21–28.

Holland, H. D. 1978. *The Chemistry of the Atmosphere and Oceans.* Wiley, New York.

Hollin, J. T. 1972. Interglacial climates and Antarctic ice surges. *Quaternary Research (New York) 2*:401–408.

Holser, W. T., and I. R. Kaplan. 1966. Isotope geochemistry of sedimentary sulfates. *Chemical Geology 1*:93–135.

Horie, Y., and V. Mirabella. 1982. Estimating future NO_2 levels in the greater Los Angeles area by source-type contribution. *Journal of the Air Pollution Control Association 32*:266–273.

Hoveland, C. S., ed. 1976. *Biological N Fixation in Forage-Livestock Systems.* American Society of Agronomy, Madison, Wisc.

Huber, D. M., H. L. Warren, D. W. Nelson, and C. Y. Tsai. 1977. Nitrification inhibitors—New tools for food production. *BioScience 27*:523–529.

Hudson, J. L., and G. T Rochelle, eds. 1982. *Flue Gas Desulfurization.* American Chemical Society, Washington, D.C.

Hultberg, H., and O. Grahn. 1975. Effects of acid precipitation on macrophytes in oligotrophic Swedish lakes. In: *First Special Symposium on Atmospheric Contribution to the Chemistry of Lake Waters.* International Association of Great Lakes Research, Ann Arbor, Michigan, pp. 208–217.

Hung, R. J., and G. S. Liaw. 1981. Advection fog formation in a polluted atmosphere. *Journal of the Air Pollution Control Association 31*:55–61.

Hunt, W. F., ed. 1976. *National Air Quality and Emissions Trends Report, 1975*. EPA, Research Triangle Park, N.C.

Husar, R. B., J. P. Lodge, Jr., and D. J. Moore, eds. 1978. *Sulfur in the Atmosphere*. Proceedings of the International Symposium on Sulfur in the Atmosphere, Dubrovnik, Yugoslavia, September 7–14, 1977.

Husar, R. B., J. M. Holloway, D. E. Patterson, and W. E. Wilson. 1981. Spatial and temporal pattern of eastern U.S. haziness: A summary. *Atmospheric Environment 15*:1919–1928.

Hutchinson, G. E. 1944. A century of atmospheric biogeochemistry. *American Scientist 32*:129–132.

Hutchinson, T. C. 1980. Conclusions and recommendations. In: *Effects of Acid Precipitation on Terrestrial Ecosystems,* T. C. Hutchinson and M. Havas, eds. Plenum Press, New York, pp. 617–627.

Hutchinson, T. C., and M. Havas, eds. 1980. *Effects of Acid Precipitation on Terrestrial Ecosystems.* Proceedings of the NATO Conference on Effects of Acid Precipitation on Vegetation and Soils, Toronto, May 21–27, 1978. Plenum Press, New York.

Hutzinger, O., ed. 1981. *The Natural Environment and Biogeochemical Cycles.* Springer-Verlag, Berlin.

Idso, S. B. 1980. The climatological significance of a doubling of earth's atmospheric carbon dioxide concentration. *Science 207*:1462–1463.

Idso, S. B. 1982. *Carbon Dioxide: Friend or Foe?* IBR Press, Tempe, Ariz.

Innes, W. B. 1979. *Effect of Nitric Oxide Emissions on Photochemical Smog.* Purad Publication, Upland, Calif.

Innes, W. B. 1981. Effect of nitrogen oxide emissions on ozone levels in metropolitan regions. *Environmental Science and Technology 15*:904–912.

International Energy Agency. 1982. *World Energy Outlook.* OECD, Paris.

Ivanov, M. V. 1981. The global biogeochemical sulphur cycle. In: *Some Perspectives of the Major Biogeochemical Cycles,* G. E. Likens, ed. Wiley, New York, pp. 61–78.

Jacobson, J. S. 1981. Acid rain and environmental policy. *Journal of the Air Pollution Control Association 31*:1071–1073.

Jacobson, J. S., and R. E. Showman. 1984. Field surveys of vegetation during a period of rising electric power generation in the Ohio Valley. *Journal of the Air Pollution Control Association 34*:48–51.

Jahnig, C. F., and H. Shaw. 1981. A comparative assessment of flue gas treatment processes. Part I—Status and design basis. Part II—Environmental and cost comparison. *Journal of the Air Pollution Control Association 31*:421–428, 596–604.

Jaiswal, P. L., ed. 1971. *Handbook of Manures and Fertilizers.* Indian Center for Agricultural Research, New Delhi.

Jansson, S. L., and J. Persson. 1982. Mineralization and immobilization of soil nitrogen. In: *Nitrogen in Agricultural Soils,* F. J. Stevenson, ed. American Society of Agronomy, Madison, Wisc., pp. 229–252.

Jenny, H. 1941. *Factors of Soil Formation.* McGraw-Hill, New York.

Johnson, A. H., and T. G. Siccama. 1983. Acid deposition and forest decline. *Environmental Science and Technology 17*:294A–305A.

Johnson, C. 1979. Fly ash removes SO_2 effectively from boiler flue gases. *Power Engineering 83*(6):61–63.

Johnson, J. W., L. F. Welch, and L. T Kurtz. 1975. Environmental implications of N fixation by soybeans. *Journal of Environmental Quality 4*:303–306.

Jonakin, J., and J. F. McLaughlin. 1969. *Operating Experience with the First Full Scale System for Removal of SO_2 and Dust from Stack Gases.* Paper presented at the American Power Conference, Chicago.

Jones, J. W. 1979. Disposal of wastes from coal-fired power plants. In: *Energy/Environment IV.* EPA, Washington, D.C., pp. 117–130.

Jones, P. D., and T. M. L. Wigley. 1980. Northern Hemisphere temperatures, 1881–1979. *Climate Monitor* 9:43–47.

Jones, U. S., and E. L. Suarez. 1980. Impact of atmospheric deposition on agroecosystems. In: *Atmospheric Sulfur Deposition,* D. S. Shriner, C. R. Richmond, and S. E. Lindberg, eds. Ann Arbor Science, Ann Arbor, Mich., pp. 377–396.

Jonsson, B., and L. Svensson. 1982. *Studie över luft föroreningarnas inverkan på skogsproduktionen.* Instituet för biometri och skogsindelning, Umea, Sweden.

Jordan, B. C., and A. J. Broderick. 1979. Emissions of oxides of nitrogen from aircraft. *Journal of the Air Pollution Control Association* 29:119–124.

Jordan, C. F., and R. Herrera. 1981. Tropical rain forests: Are nutrients really critical? *The American Naturalist* 117:167–180.

Josephs, D. X. 1980. Magnesium enrichment improves flue gas scrubbing. *Power Engineering* 84(9):71–72.

Judeikis, H. S., and T. B. Stewart. 1976. Laboratory measurements of SO_2 deposition velocities on selected building materials and soils. *Atmospheric Environment* 10:769–776.

Junge, C. E. 1956. Recent investigations in air chemistry. *Tellus* 8:127–139.

Junge, C. E. 1958. The distribution of ammonia and nitrate in rain water over the United States. *Transactions of American Geophysical Union* 39:241–248.

Junge, C. E. 1960. Sulfur in the atmosphere. *Journal of Geophysical Research* 65:227–237.

Junge, C. E. 1963. *Air Chemistry and Radioactivity.* Academic Press, New York.

Junge, C. E., and R. T. Werby. 1958. The concentration of chloride, sodium, potassium, calcium and sulfate in rain water over the United States. *Journal of Meteorology* 5:417–425.

Kallend, A. S., A. R. W. Marsh, J. H. Pickles, and M. V. Proctor. 1983. Acidity of rain in Europe. *Atmospheric Environment* 17:127–137.

Kandel, R. S. 1981. Surface temperature sensitivity to increased CO_2. *Nature* 293:634–636.

Kaplan, I. R. 1972. Sulfur cycle. In: *The Encyclopedia of Geochemistry and Environmental Sciences,* R. W. Fairbridge, ed. Van Nostrand–Reinhold, Princeton, N.J., pp. 1148–1152.

Kaplan, L. D. 1960. The influence of carbon dioxide variations on the atmospheric heat balance. *Tellus* 12:204–208.

Karlovsky, J. 1981. Cycling of nutrients and their utilisation by plants in agricultural ecosystems. *Agro-Ecosystems* 7:127–144.

Karlsson, H. T., and H. S. Rosenberg. 1980. Technical aspects of lime/limestone scrubbers for coal-fired power plants. Part I: Process chemistry and scrubber systems. Part II: Instrumentation and technology. *Journal of the Air Pollution Control Association* 30:710–714, 822–826.

Kawamura, T., and Frey, D. J. 1981. Current developments in low NO_x firing systems. In: *Proceedings of the Joint Symposium on Stationary Combustion NO_x Control.* EPRI, Palo Alto, Calif., pp. 123–159.

Keeling, C. D. 1973a. Industrial production of carbon dioxide from fossil fuels and limestone. *Tellus* 25:174–198.

Keeling, C. D. 1973b. The carbon dioxide cycle. In: *Chemistry of the Lower Atmosphere,* S. Rasool, ed. Plenum Press, New York, pp. 251–329.

Keeling, C. D., and R. B. Bacastow. 1977. Impact of industrial gases on climate. In: *Energy and Climate.* NAS, Washington, D.C., pp. 72–95.

Keeling, C. D., R. B. Bacastow, A. E. Bainbridge, C. A. Ekdahl, Jr., P. R. Guenther, and L. S. Waterman. 1976a. Atmospheric carbon dioxide variations at Mauna Loa Observatory, Hawaii. *Tellus* 28:538–551.

Keeling, C. D., J. A. Adams, Jr., C. A. Ekdahl, Jr., and P. R. Guenther. 1976b. Atmospheric carbon dioxide variations at the South Pole. *Tellus* 28:552–564.

Keeling, C. D., R. B. Bacastow, and T. P. Whorf. 1982. Measurements of the concentration of carbon dioxide at Mauna Loa Observatory, Hawaii. In: *Carbon Dioxide Review: 1982,* W. C. Clark, ed. Oxford University Press (Clarendon), London, pp. 377–385.

Keeney, D. R. 1972. *The Fate of Nitrogen in Aquatic Ecosystems.* University of Wisconsin Resources Center, Madison.

Keeney, D. R. 1980a. Prediction of soil nitrogen availability in forest ecosystems: A literature review. *Forest Science 26*:159–171.

Keeney, D. R. 1980b. Factors affecting the persistence and bioactivity of nitrification inhibitors. In: *Nitrification Inhibitors—Potentials and Limitations,* M. Stelly, ed. American Society of Agronomy, Madison, Wisc., pp. 33–46.

Keeney, D. R. 1982. Nitrogen management for maximum efficiency and minimum pollution. In: *Nitrogen in Agricultural Soils,* F. J. Stevenson, ed. American Society of Agronomy, Madison, Wisc., pp. 605–650.

Keeney, D. R., and W. R. Gardner. 1975. The dynamics of nitrogen transformations in the soil. In: *The Changing Global Environment,* S. F. Singer, ed. Reidel, Dordrecht, pp. 367–375.

Kellogg, W. W. 1978. Is mankind warming the earth? *The Bulletin of the Atomic Scientists 34*(2):11–19.

Kellogg, W. W., R. D. Cadle, E. R. Allen, A. L. Lazrus, and E. A. Martell. 1972. The sulfur cycle. *Science 175*:587–596.

Kempe, S. 1979. Carbon in the rock cycle. In: *The Global Carbon Cycle,* B. Bolin, E. T. Degens, S. Kempe, and P. Ketner, eds. Wiley, New York, pp. 343–377.

Khalil, M. A. K. and R. A. Rasmussen. 1984. Carbon monoxide in the Earth's atmosphere: increasing trend. *Science 224*:54–56.

Khoury, D. L., ed. 1982. *Coal Cleaning Technology.* Noyes Data Corporation, Park Ridge, N.J.

Killough, G. G., and W. R. Emanuel. 1981. A comparison of several models of carbon turnover in the ocean with respect to their distributions of transit time and age, and responses to atmospheric CO_2 and ^{14}C. *Tellus 33*:280–290.

Kimball, B. A. 1982. *Carbon Dioxide and Agricultural Yield.* U.S. Water Conservation Laboratory, WCL Report 11, Phoenix, Ariz.

King, P. J., F. Morton, and A. Sagarra. 1973. Chemistry and physics of petroleum. In: *Modern Petroleum Technology,* G. D. Hobson, ed. Wiley, New York, pp. 186–219.

Kinzelbach, W. K. H. 1983. China: Energy and environment. *Environmental Management 7*:303–310.

Kirkby, E. A., ed. 1970. *Nitrogen Nutrition of the Plant.* University of Leeds, Leeds.

Kitchell, J. F., R. V. O'Neill, D. Webb, G. W. Gallepp, S. M. Bartell, J. F. Koonce, and B. S. Ausmus. 1979. Consumer regulation of nutrient cycling. *BioScience 29*:28–34.

Kneip, T. J., and P. J. Lioy, eds. 1980. *Aerosols: Anthropogenic and Natural Sources of Transport. Annals of the New York Academy of Sciences* Vol. 338.

Knelson, J. H., and R. E. Lee. 1977. Oxides of nitrogen in the atmosphere: Origin, fate and public health implications. *Ambio 6*:126–130.

Knowles, R. 1981. Denitrification. In: *Terrestrial Nitrogen Cycles,* F. E. Clark and T. Rosswall, eds. *Ecological Bulletin* No. 33, Stockholm, pp. 315–329.

Koblents-Mishke, O. J., V. V. Volkovinskiy, and Y. G. Kabanova. 1970. Plankton primary production of the world ocean. In: *Scientific Exploration of the South Pacific,* W. S. Wooster, ed. NAS, Washington, D.C., pp. 183–193.

Kohl, D. H., G. B. Shearer, and B. Commoner. 1971. Fertilizer nitrogen: Contribution to nitrate in surface water in a corn belt watershed. *Science 174*:1331–1334.

Kohl, D. H., G. Shearer, and F. Vithayanthil. 1978. Nitrogen mass balance studies. In: *Nitrogen in the Environment,* Volume I, D. R. Nielsen and J. G. MacDonald, eds. Academic Press, New York, pp. 183–200.

Kolenbrander, G. J. 1972. Does leaching of fertilizers affect the quality of ground water at the waterworks? *Stikstof 15*:8–15.

Kolenbrander, G. J. 1977. Nitrogen in organic matter and fertilizer as a source of pollution. *Progress in Water Technology 8*:67–84.

Konishi, M., N. Nakamura, E. Oono, T. Baika, and S. Sanda. 1979. Prechamber decreases NO_x emissions. *Atutmotive Engineering 87*(5):72–77.

Kopecki, E. S., and C. F. McDaniel. 1976. Corrosion minimized, efficiency enhanced in wet limestone scrubbers. *Power Engineering 80*(4):86–89.

Kozlowski, T. T. 1980. Impacts of air pollution on forest ecosystems. *BioScience 30*:88–93.

Kramer, J. R. 1978. Acid precipitation. In: *Sulfur in the Environment*, Part I, J. O. Nriagu, ed. Wiley, New York, pp. 325–372.

Kramer, J., and A. Tessier. 1982. Acidification of aquatic ecosystems: A critique of chemical approaches. *Environmental Science and Technology 16*:606A–615A.

Kramer, P. J. 1981. Carbon dioxide concentration, photosynthesis, and dry matter production. *BioScience 31*:29–33.

Kreijger, P. C. 1976. Energy analysis of materials and structures in the building industry. In: *The Energy Accounting of Materials, Products, Processes and Services*, 9th International TNO Conference, Rotterdam, pp. 141–160.

Kritz, M. A. 1982. Exchange of sulfur between the free troposphere, marine boundary layer, and the sea surface. *Journal of Geophysical Research 87*:8795–8803.

Krueger, A. J. 1983. Sighting of El Chichón sulfur dioxide clouds with the Nimbus 7 total ozone mapping spectrometer. *Science 220*:1377–1379.

Krug, E. C., and C. R. Frink. 1983. Acid rain on acid soil: A new perspective. *Science 221*:520–525.

Kucera, V. 1976. Effects of sulphur dioxide and acid precipitation on metals and anti-rust painted steel. *Ambio 5*:243–248.

Kuroda, H., Y. Nakajima, K. Sugihara, Y. Takagi, and S. Muranaka. 1978. Fast burn-heavy EGR improves economy, reduces NO_x. *Automotive Engineering 86*(8):56–62.

La Bastille, A. 1981. Acid rain: How great a menace? *National Geographic 162*:652–681.

Ladd, J. N., and R. B. Jackson. 1982. Biochemistry of ammonification. In: *Nitrogen in Agricultural Soils*, F. J. Stevenson, ed. American Society of Agronomy, Madison, Wisc., pp. 173–228.

Lalas, D. P., V. R. Veirs, G. Karras, and G. Kallos. 1982. An analysis of the SO_2 concentration levels in Athens, Greece. *Atmospheric Environment 16*:531–544.

Lamb, H. H. 1970. Volcanic dust in the atmosphere. *Philosophical Transactions of the Royal Society of London Series A 266*:425–533.

Lamborg, M. R., and R. W. F. Hardy. 1983. Microbial effects. In: *CO_2 and Plants*, E. R. Lemon, ed. Westview Press, Boulder, Colo., pp. 131–176.

Lanly, J. P., and J. Clement. 1979. Present and future natural forest and plantation areas in the tropics. *Unasylva 31*(123):12–20.

Lanly, J. P., and M. Gillis. 1980. *Provisional Results of the FAO/UNEP Tropical Forest Resources Assessment Project*. FAO, Rome.

La Rivière, J. W. M. 1979. Foreword. In: *The Global Carbon Cycle*, B. Bolin, E. T. Degens, S. Kempe, and P. Ketner, eds. Wiley, New York, pp. v–vi.

Larsen, R. I. 1971. *A Mathematical Model for Relating Air Quality Measurements to Air Quality Standards*. EPA, Research Triangle Park, N.C.

Larson, T. V. 1980. Secondary aerosol: Production mechanisms of sulfate compounds in the atmosphere. *Annals of the New York Academy of Sciences 338*:26–38.

Larson, W. E., F. J. Pierce, and R. H. Dowdy. 1983. The threat of soil erosion to long-term crop production. *Science 219*:458–465.

La Rue, T. A., and T. G. Patterson. 1981. How much nitrogen do legumes fix? *Advances in Agronomy 34*:15–38.

Lasaga, A. C. 1981. The sulfur cycle. *Earth and Mineral Sciences 51*(1):7–11.

Lasaga, A. C., and H. D. Holland. 1976. Mathematical aspects of non-steady-state diagenesis. *Geochimica et Cosmochimica Acta 40*:257–266.

Lauer, D. A. 1975. Limitations of mineral waste replace ment for inorganic fertilizers. In: *Energy, Agriculture and Waste Management*, W. J. Jewell, ed. Ann Arbor Science, Ann Arbor, Mich., pp. 409–432.

Lawther, P. J., and J. A. Bonnell. 1970. *Some Recent Trends in Pollution and Health in London and Some Current Thoughts*. Central Electricity Generating Board, London.

Lawther, P. J., R. E. Waller, and M. Henderson. 1970. Air pollution and exacerbations of bronchitis. *Thorax 25*:525–539.

Leach, G., and M. Slesser. 1973. *Energy Equivalents of Network Inputs to Food Producing Processes.* University of Strathclyde, Glasgow, Scotland.

Leaderer, B. P., and J. A. Stolwijk. 1981. Seasonal visibility and pollutant sources in the northeastern United States. *Environmental Science and Technology 15*:305–309.

Lee, D. H. K. 1970. Nitrates, nitrites and methemoglobinemia. *Environmental Research 3*:484–511.

Lee, J. J. 1982. The effects of acid precipitation on crops. In: *Acid Precipitation: Effects on Ecological Systems,* F. M. D'Itri, ed. Ann Arbor Science, Ann Arbor, Mich., pp. 453–468.

Lee, J. J., and D. E. Weber. 1980. *Effects of Sulfuric Acid Rain on Two Model Hardwood Forests: Throughfall, Litter Leachate, and Soil Solution.* EPA, Corvallis, Oreg.

Lee, S. D., ed. 1980 *Nitrogen Oxides and Their Effects on Health.* Ann Arbor Science Publishers, Ann Arbor, Michigan.

Lee, S. K. 1978. Korea. In: *Organic Recycling in Asia.* FAO, Rome, pp. 88–92.

Legg, J. O., and J. J. Meisinger. 1982. Soil nitrogen budgets. In: *Nitrogen in Agricultural Soils,* F. J. Stevenson, ed. American Society of Agronomy, Madison, Wisc., pp. 503–566.

Leighton, P. A. 1961. *Photochemistry of Air Pollution.* Academic Press, New York.

Lemon, E. R., ed. 1983. *CO_2 and Plants.* Westview Press, Boulder, Colo.

Lenhard, U., and G. Gravenhorst. 1980. Evaluation of ammonia fluxes into the free atmosphere over western Germany. *Tellus 32*:48–55.

Liberti, A., D. Brocco, and M. Possanzini. 1978. Adsorption and oxidation of sulfur dioxide on particulates. *Atmospheric Environment 12*:255–261.

Liebsch, E. J., and R. G. de Pena. 1982. Sulfate aerosol production in coal-fired power plant plumes. *Atmospheric Environment 16*:1323–1331.

Lieth, H. 1975. Measurement of caloric values. In: *Primary Productivity of the Biosphere,* H. Lieth and R. H. Whittaker, eds. Springer-Verlag, Berlin, pp. 119–129.

Lieth, H., and R. H. Whittaker, eds. 1975. *Primary Productivity of the Biosphere.* Springer-Verlag, Berlin.

Likens, G., ed. 1981. *Some Perspectives of the Major Biogeochemical Cycles.* SCOPE 17. Wiley, New York.

Likens, G. E., and T. J. Butler. 1981. Recent acidification of precipitation in North America. *Atmospheric Environment 15*:1103–1109.

Likens, G. E., F. H. Bormann, N. M. Johnson, D. W. Fisher, and R. S. Pierce. 1970. Effects of forest cutting and herbicide treatment on nutrient budgets in the Hubbard Brook watershed-ecosystem. *Ecological Monographs 40*(1):23–47.

Likens, G. E., R. F. Wright, J. N. Galloway, and T. J. Butler. 1979. Acid rain. *Scientific American 241*:43–51.

Liljestrand, H. M., and J. J. Morgan. 1979. Error analysis applied to indirect methods for precipitation acidity. *Tellus 31*:421–431.

Linzon, S. N., P. J. Temple, and R. G. Pearson. 1979. Sulfur concentrations in plant foliage and related effects. *Journal of the Air Pollution Control Association 29*:520–525.

Liu, S. C., R. J. Cicerone, T. M. Donahue, and W. L. Chameides. 1977. Sources and sinks of atmospheric N_2O and possible ozone reduction due to industrial fixed nitrogen fertilizers. *Tellus 29*:251–263.

Livingstone, D. A. 1963. Chemical composition of rivers and lakes. In: *Data of Geochemistry* (6th ed.), M. Fleischer, ge. USGS, Washington, D.C.

Lodge, J. P. 1980. Book reviews. *Atmospheric Environment 14*:281–284.

Lodge, J. P., Jr., and J. B. Pate. 1966. Atmospheric gases and particulates in Panama. *Science 153*:408–410.

Lodge, J. P., Jr., P. A. Machado, J. B. Pate, D. C. Sheesley, and A. F. Wartburg. 1974. Atmospheric trace chemistry in the American humid tropics. *Tellus 26*:250–253.

Lorenz, O. A. 1978. Potential nitrate levels in edible plant parts. In: *Nitrogen in the Environment,* Volume II, D. R. Nielsen and J. G. MacDonald, eds. Academic Press, New York, pp. 201–219.

Loucks, O. L., and T. V. Armentano. 1982. Estimating crop yield effects from ambient air pollutants in the Ohio River valley. *Journal of the Air Pollution Control Association 32*:146–150.

Lovelock, J. E. 1979. *Gaia: A New Look at Life on Earth.* Oxford University Press, London.
Lovelock, J. E., R. J. Maggs, and R. A. Rasmussen. 1972. Atmospheric dimethyl sulphide and the natural sulphur cycle. *Nature 237*:452–453.
Lowell, P. S. 1970. *A Theoretical Description of the Limestone Injection–Wet Scrubbing Process.* NTIS, Springfield, Va.
Lowenstam, H. 1981. Minerals formed by organisms. *Science 211*:1126–1131.
Lundegardh, H. 1924. *Der Kreislauf der Kohlensäure in der Natur.* Fischer Verlag, Jena.
Lusis, M. A., and H. A. Wiebe. 1976. The rate of oxidation of sulfur dioxide in the plume of a nickel smelter stack. *Atmospheric Environment 10*: 793–798.
Lynch, J. M. 1979. Straw residues as substrates for growth and product formation by soil micro-organisms. In: *Straw Decay and Its Effect on Disposal and Utilization,* E. Grossbard, ed. Wiley, New York., pp. 47–56.
MacDonald, G. A. 1972. *Volcanoes.* Prentice–Hall, Englewood Cliffs, N.J.
Machta, L. 1983. Effects of non-CO_2 greenhouse gases. In: *Changing Climate.* NAS, Washington, D.C., pp. 285–291.
MacNeil, D., T. MacNeil, and W. J. Brill. 1978. Genetic modifications of N_2-fixing systems. *Bio-Science 28*:576–579.
Madden, R. A., and V. Ramanathan. 1980. Detecting climate change due to increasing carbon dioxide. *Science 209*:763–767.
Madsen, B. C. 1981. Acid rain at Kennedy Space Center, Florida: Recent observations. *Atmospheric Environment 15*:853–862.
Malmer, N. 1976. Chemical changes in the soil. *Ambio 5*:231–234.
Manabe, S. 1971. Estimates of future change of climate due to the increase of carbon dioxide concentration in air. In: *Man's Impact on Climate.* MIT Press, Cambridge, Mass., pp. 249–264.
Manabe, S., and R. T. Wetherald. 1967. Thermal equilibrium of the atmosphere with a given distribution of relative humidity. *Journal of the Atmospheric Sciences 24*:241–259.
Manabe, S., and R. T. Wetherald. 1975. The effects of doubling the CO_2 concentration on the climate of a general circulation model. *Journal of the Atmospheric Sciences 32*:3–15.
Manabe, S., and R. T. Wetherald. 1980. On the distribution of climatic change resulting from an increase in CO_2 content of the atmosphere. *Journal of the Atmospheric Sciences 37*:99–118.
Marchetti, C. 1977. On geoengineering and the CO_2 problem. *Climatic Change 1*:59–68.
Marchetti, C., and N. Nakicenovic. 1979. *The Dynamics of Energy Systems and the Logistic Substitution Model.* IIASA, Laxenburg.
Marion, G. M. 1979. Biomass and nutrient removal in long rotation stands. In: *Impact of Intensive Harvesting on Forest Nutrient Cycling.* State University of New York, Syracuse, pp. 98–110.
Markham, J. W. 1958. *The Fertilizer Industry.* Vanderbilt Press, Nashville, Tenn.
Marland, G., and R. M. Rotty. 1979. Carbon dioxide and climate. *Reviews of Geophysics and Space Physics 17*:1813–1824.
Marsh, A. R. W. 1978. Sulphur and nitrogen contributions to the acidity of rain. *Atmospheric Environment 12*:401–406.
Martin, D. M., and D. R. Goff. 1973. *The Role of Nitrogen in Aquatic Environment.* Department of Limnology, Academy of Natural Sciences of Philadelphia, Philadelphia.
Martin, G. B., and J. S. Bowen. 1979. Nitrogen oxides control. In: *Energy/Environment IV,* Proceedings of the Fourth National Conference on the Interagency Energy/Environment R & D Program, June 7–8, 1979.
Martin, W., and A. C. Stern. 1974. *The Collection, Tabulation, Codification and Analysis of the World's Air Quality Management Standards.* EPA, Washington, D.C.
Matsuda, S., M. Takeichi, T. Hishinuma, F. Nakajima, T. Narita, Y. Watanabe, and M. Imanari. 1978. Selective reduction of nitrogen gases in combustion flue gases. *Journal of the Air Pollution Control Association 28*:350–353.
Matthews, W. H., W. W. Kellogg, and G. D. Robinson, eds. 1971. *Man's Impact on the Climate.* MIT Press, Cambridge, Mass.

Matthias, A. D., A. M. Blackmer, and J. M. Bremner. 1979. Diurnal variability in the concentration of nitrous oxide in surface air. *Geophysical Research Letters 6*:441–443.

Maxwell, M. A., and M. D. Shapiro. 1979. Sulfur oxides control: Flue gas desulfurization. In: *Energy/Environment IV*. EPA, Washington, D.C., pp. 49–68.

Mayer, R., and B. Ulrich. 1978. Input of atmospheric sulfur by dry and wet deposition to two central European forest ecosystems. *Atmospheric Environment 12*:375–377.

Maynard, D. N., A. V. Barker, P. L. Minotti, and N. H. Peck. 1976. Nitrate accumulation in vegetables. *Advances in Agronomy 28*:71–118.

McCalla, T. M. 1974. Use of animal wastes as a soil amendment. *Journal of Water and Soil Conservation 29*:213–216.

McCarroll, J. 1980. Health effects associated with increased use of coal. *Journal of the Air Pollution Control Association 30*:652–656.

McConnell, J. C. 1973. Atmospheric ammonia. *Journal of Geophysical Research 78*:7812–7821.

McDermott, D. L., K. D. Reiszner, and P. W. West. 1979. Development of long-term sulfur dioxide monitor using permeation sampling. *Environmental Science and Technology 13*:1087–1090.

McElroy, M. B. 1975. Testimony before U.S. House of Representatives, Committee on Science and Technology. U.S. Government Printing Office, Washington, D.C.

McElroy, M. B., J. W. Elkins, S. C. Wofsy, and Y. L. Yung. 1976. Sources and sinks for atmospheric N_2O. *Review of Geophysics and Space Physics 14*:143–150.

McElroy, M. B., S. C. Wofsy, and Y. L. Yung. 1977. The nitrogen cycle: Perturbations due to man and their impact on atmospheric N_2O and O_3. *Royal Society of London Transactions 277B*:159–181.

McFee, W. W. 1980. *Sensitivity of Soil Regions to Acid Precipitation*. EPA, Corvallis, Oreg.

McKenney, D. J., D. L. Wade, and W. I. Findlay. 1978. Rates of N_2O evolution from N-fertilized soil. *Geophysical Research Letters 5*:530–532.

McLean, R. A. N. 1981. The relative contributions of sulfuric and nitric acid in acid rain to the acidification of the ecosystem. *Journal of the Air Pollution Control Association 31*:1184–1187.

McMahon, T. A., and P. J. Denison. 1979. Empirical atmospheric deposition parameters—A survey. *Atmospheric Environment 13*:571–585.

Meagher, J. F., L. Stockburger, E. M. Bailey, and O. Huff. 1978. The oxidation of sulfur dioxide to sulfate aerosols in the plume of a coal-fired power plant. *Atmospheric Environment 12*:2197–2203.

Meagher, J. F., E. M. Bailey, and M. Luria. 1982. The impact of mixing tower and power plant plumes on sulfate aerosol formation. *Journal of the Air Pollution Control Association 32*:389–391.

Meetham, A. R. 1952. *Atmospheric Pollution: Its Origin and Prevention*. Pergamon Press, Elmsford, N.Y.

Menees, G. P., and C. Park. 1976. Nitric oxide formation by meteoroids in the upper atmosphere. *Atmospheric Environment 10*:535–545.

Mengel, K., and E. A. Kirkby. 1978. *Principles of Plant Nutrition*. International Potash Institute, Bern.

Mercer, J. H. 1978. West Antarctic ice sheet and CO_2 greenhouse effect: A threat of disaster. *Nature 271*:321–325.

Messer, J., and P. L. Brezonik. 1983. Agricultural nitrogen model: A tool for regional environmental management. *Environmental Management 7*:177–187.

Mészáros, E. 1978. Concentration of sulfur compounds in remote continental and oceanic areas. *Atmospheric Environment 12*:699–705.

Mészáros, E., G. Várhelyi, and L. Haszpra. 1978. On the atmospheric sulfur budget over Europe. *Atmospheric Environment 12*:2273–2277.

Mikkelsen, D. S., S. K. De Datta, and W. N. Obcemea. 1978. Ammonia volatilization losses from flooded rice soils. *Soil Science Society of America Journal 42*:725–730.

Millan, M. M., and Y. S. Chung. 1977. Detection of a plume 400 km from the source. *Atmospheric Environment 14*:939–944.

Miller, J. M., ed. 1974. *Geophysical Monitoring for Climatic Change No. 1*. National Oceanic and Atmospheric Administration, Boulder, Colo.

Miller, J. R., ed. 1978. *Prospects for Man: Climate Change*. York University Press, Toronto.

Miller, P. R., and J. R. McBride. 1975. Effects of air pollution on forests. In: *Responses of Plants to Air Pollution*, J. B. Mudd and T. T. Kozlowski, eds. Academic Press, New York, pp. 196–235.

Miller, R. J., and D. W. Wolfe. 1978. Nitrogen inputs and outputs: A valley basin study. In: *Nitrogen in the Environment,* Volume I, D. R. Nielsen and J. G. MacDonald, eds. Academic Press, New York, pp. 163–172.

Milne, J. W., D. B. Roberts, and D. J. Williams. 1979. The dry deposition of sulphur dioxide—Field measurements with a stirred chamber. *Atmospheric Environment 13*:373–379.

Minchin, F. R., and J. S. Pate. 1973. The carbon balance of a legume and the functional economy of its root nodules. *Journal of Experimental Botany 24*:259–271.

Mitchell, J. M., Jr. 1975. A reassessment of atmospheric pollution as a cause of long-term changes of global temperature. In: *The Changing Global Environment,* S. F. Singer, ed. Reidel, Dordrecht, pp. 149–173.

Mobley, J. D., and J. C. S. Chang. 1981. The adipic acid enhanced limestone flue gas desulfurization process. *Journal of the Air Pollution Control Association 31*:1249–1253.

Möhr, P. J., and E. B. Dickinson. 1979. Mineral nutrition in maize. In: *Maize,* Ciba-Geigy, Basel, Switzerland, pp. 26–32.

Möller, D. 1980. Kinetic model of atmospheric SO_2 oxidation based on published data. *Atmospheric Environment 14*:1067–1076.

Möller, F. 1963. On the influence of changes in the CO_2 concentration in air on the radiation balance of the earth's surface and on the climate. *Journal of Geophysical Research 68*:3877–3886.

Morgan, G. B., G. Ozolins, and E. C. Tabor. 1970. Air pollution surveillance systems. *Science 170*:289–296.

Morisot, A. 1981. Erosion and nitrogen losses. In: *Terrestrial Nitrogen Cycles,* F. E. Clark and T. Rosswall, eds. *Ecological Bulletin* No. 33, Stockholm, pp. 353–361.

Morse, J. W., and R. A. Berner. 1972. Dissolution kinetics of calcium carbonate in sea water. *American Journal of Science 272*:840–851.

Moss, D. N., and R. B. Musgrave. 1971. Photosynthesis and crop production. *Advances in Agronomy 23*:317–336.

Motor Vehicle Manufacturers Association. 1983. *Facts and Figures '83.* MVMA, Detroit, Mich.

Mudahar, M. S., and T. P. Hignett. 1981. *Energy and Fertilizer.* International Fertilizer Development Center, Muscle Shoals, Ala.

Mudd, J. B., and T. T. Kozlowski, eds. 1975. *Responses of Plants to Air Pollution.* Academic Press, New York.

Mullins, P. V. 1977. Nitrogen removal from natural gas. In: *Gas Engineers Handbook.* Industrial Press, New York, pp. 4/96–4/99.

Munger, J. W. 1982. Chemistry of atmospheric precipitation in the north-central United States: Influence of sulfur, nitrate, ammonia and calcareous soil particulates. *Atmospheric Environment 16*:1633–1645.

Munger, J. W., and S. J. Eisenreich. 1983. Continental-scale variations in precipitation chemistry. *Environmental Science and Technology 17*:32A–42A.

Muniz, I. P. 1981. Acidification—Effects on aquatic organisms. In: *Beyond the Energy Crisis,* Volume IV, R. A. Fazzolare and C. B. Smith, eds. Pergamon Press, Elmsford, N.Y., pp. A101–A123.

Mustacchi, C., P. Armenante and V. Cena. 1978. Carbon dioxide in the ocean. In: *Carbon Dioxide, Climate and Society,* J. Williams, ed. Pergamon Press, Elmsford, N.Y., pp. 283–289.

Myers, N. 1979. *The Sinking Ark.* Pergamon Press, Elmsford, N.Y.

Myers, N. 1980. The present status and future prospects of tropical moist forests. *Environmental Conservation 7*:101–114.

National Academy of Sciences. 1972. *Accumulation of Nitrate.* NAS, Washington, D.C.

National Academy of Sciences. 1975. *Understanding Climatic Change: A Program for Action.* NAS, Washington, D.C.

National Academy of Sciences. 1977a. *Nitrogen Oxides.* NAS, Washington, D.C.

National Academy of Sciences. 1977b. *Leucaena: Promising Forage and Tree Crop for the Tropics.* NAS, Washington, D.C.

National Academy of Sciences. 1977c. *Energy and Climate.* NAS, Washington, D.C.

National Academy of Sciences. 1978a. *Nitrates: An Environmental Assessment.* NAS, Washington, D.C.

National Academy of Sciences. 1978b. *Sulfur Oxides*. NAS, Washington, D.C.
National Academy of Sciences. 1979a. *Microbial Processes: Promising Technologies for Developing Countries*. NAS, Washington, D.C.
National Academy of Sciences. 1979b. *Tropical Legumes: Resources for the Future*. NAS, Washington, D.C.
National Academy of Sciences. 1979c. *Carbon Dioxide and Climate: A Scientific Assessment*. NAS, Washington, D.C.
National Academy of Sciences. 1980. *Research Priorities in Tropical Biology*. NAS, Washington, D.C.
National Academy of Sciences. 1981a. *Atmosphere–Biosphere Interactions: Toward a Better Understanding of the Ecological Consequences of Fossil Fuel Combustion*. NAS, Washington, D.C.
National Academy of Sciences. 1981b. *The Health Effects of Nitrate, Nitrite, and N-Nitroso Compounds*. NAS, Washington, D.C.
National Academy of Sciences. 1982a. *Carbon Dioxide and Climate: A Second Assessment*. NAS, Washington, D.C.
National Academy of Sciences. 1982b. *United States–Canadian Tables of Feed Composition, Third Revision*. NAS, Washington, D.C.
National Academy of Sciences. 1982c. *Causes and Effects of Stratospheric Ozone Reduction: An Update*. NAS, Washington, D.C.
National Academy of Sciences. 1983a. *Acid Deposition: Atmospheric Processes in Eastern North America*. NAS, Washington, D.C.
National Academy of Sciences. 1983b. *Changing Climate*. NAS, Washington, D.C.
National Defense University. 1978. *Climatic Change to the Year 2000: A Survey of Expert Opinion*. U.S. Department of Defense, Washington, D.C.
National Defense University. 1980. *Crop Yields and Climate Change to the Year 2000*. National Defense University, Fort Lesley J. McNair, Washington, D.C.
National Research Council of Canada. 1977. *Sulphur and Its Inorganic Derivatives in the Canadian Environment*. NRC, Ottawa.
National Research Council of Canada. 1981. *Acidification in the Canadian Aquatic Environment*. NRC, Ottawa.
Naugle, D. F., and D. L. Fox. 1981. Aircraft and air pollution. *Environmental Science and Technology 15*:391–395.
Needham, A. E. 1965. *The Uniqueness of Biological Materials*. Pergamon Press, Elmsford, N.Y.
Nelson, D. W. 1982. Gaseous losses of nitrogen other than through denitrification. In: *Nitrogen in Agricultural Soils*, F. J. Stevenson, ed. American Society of Agronomy, Madison, Wisc., pp. 327–364.
Nelson, D. W., and D. M. Huber. 1980. Performance of nitrification inhibitors in the Midwest (east). In: *Nitrification Inhibitors—Potentials and Limitations*, M. Stelly, ed. American Society of Agronomy, Madison, Wisc., pp. 75–88.
Nelson, W. L. 1972. What's the average sulfur content vs. gravity? *The Oil and Gas Journal 70*:59.
Newell, R. E., and T. G. Dopplick. 1979. Questions concerning the possible influence of anthropogenic CO_2 on atmospheric temperature. *Journal of Applied Meteorology 18*:822–825.
Newhall, C. G., and S. Self. 1982. The volcanic explosivity index (VEI): An estimate of explosive magnitude for historical volcanism. *Journal of Geophysical Research 87*:1231–1238.
Newman, L. 1981. Atmospheric oxidation of sulfur dioxide: A review as viewed from power plant and smelter plume studies. *Atmospheric Environment 15*:2231–2239.
Newton, W. E., and C. J. Nyman, eds. 1976. *Proceedings of the First International Symposium on Nitrogen Fixation*. Washington State University Press, Pullman.
Newton, W. E., and W. H. Orme-Johnson, eds. 1980. *Nitrogen Fixation*. University Park Press, Baltimore.
Neyra, C. A., and J. Döbereiner. 1977. Nitrogen fixation in grasses. *Advances in Agronomy 30*:1–38.
Nguyen, B. C., B. Bonsang, and G. Lambert. 1974. The atmospheric concentration of sulfur dioxide and sulfate aerosols over antarctic, subantarctic areas and oceans. *Tellus 26*:241–249.

Nguyen, B. C., A. Gaudry, B. Bonsang, and G. Lambert. 1978. Reevaluation of the role of dimethylsulphide in the sulphur budget. *Nature 275*:637–639.

Nichols, D. E., P. C. Williamson, and D. R. Waggoner. 1980. Assessment of alternatives to present-day ammonia technology with emphasis on coal gasification. In: *Nitrogen Fixation,* Volume I, W. E. Newton and W. H. Orme-Johnson, eds. University Park Press, Baltimore, pp. 43–60.

Nickell, L. G. 1977. Sugar cane. In: *Ecophysiology of Tropical Crops,* P. de T. Alvin and T. T. Kozlowski, eds. Academic Press, New York, pp. 89–111.

Nielsen, D. R., and J. G. MacDonald, eds. 1978. *Nitrogen in the Environment,* two volumes. Academic Press, New York.

Nihlgård, B. 1985. The ammonium hypothesis—an additional explanation to the forest dieback in Europe. *Ambio 14*:2–8.

Noggle, J. C. 1980. Sulfur accumulation by plants; the role of gaseous sulfur in crop nutrition. In: *Atmospheric Sulfur Deposition,* D. S. Shriner, C. R. Richmond, and S. E. Lindberg, eds. Ann Arbor Science, Ann Arbor, Mich., pp. 289–296.

Nommik, H., and K. Vahtras. 1982. Retention and fixation of ammonium and ammonia in soils. In: *Nitrogen in Agricultural Soils,* F. J. Stevenson, ed. American Society of Agronomy, Madison, Wisc., pp. 123–172.

Nordhaus, W. D. 1977. Economic growth and climate: The carbon dioxide problem. *American Economic Review 67*:341–346.

Nordhaus, W. D. 1979. *Efficient Use of Energy Resources.* Yale University Press, New Haven, Conn.

Nordhaus, W. D., and G. W. Yohe. 1983. Future carbon dioxide emissions from fossil fuels. In: *Changing Climate.* NAS, Washington, D.C., pp. 87–153.

Noxon, J. F. 1976. Atmospheric nitrogen fixation by lightning. *Geophysical Research Letters 3*:463–465.

Nriagu, J. O., ed. 1976. *Environmental Biogeochemistry,* Volume 1. Ann Arbor Science, Ann Arbor, Mich.

Nriagu, J. O., ed. 1978a. *Sulfur in the Environment.* Wiley, New York.

Nriagu, J. O. 1978b. Deteriorative effects of sulfur pollution on materials. In: *Sulfur in the Environment,* Part II, J. O. Nriagu, ed. Wiley, New York, pp. 1–35.

Nutman, P. S., ed. 1976. *Symbiotic Nitrogen Fixation in Plants.* Cambridge University Press, London.

Nye, P. H., and D. J. Greenland. 1960. *The Soil Under Shifting Cultivation.* Commonwealth Agricultural Bureaux, Farnham Royal, England.

Odum, E. P. 1969. The strategy of ecosystem development. *Science 164*:262–270.

Oeschger, H., U. Siegenthaler, V. Schotterer, and A. Guegelmann. 1975. A box diffusion model to study the carbon dioxide exchange in nature. *Tellus 27*:168–192.

Ohring, G., and S. Adler. 1978. Some experiments with a zonally averaged climatic model. *Journal of the Atmospheric Sciences 35*:186–205.

Olson, J. S. 1970. Carbon cycles and temperate woodlands. In: *Analysis of Temperate Forest Ecosystems,* D. E. Reichle, ed. Springer-Verlag, Berlin, pp. 226–241.

Olson, J. S. 1982. Earth's vegetation and atmospheric carbon dioxide. In: *Carbon Dioxide Review: 1982,* W. C. Clark, ed. Oxford University Press (Clarendon), London, pp. 388–398.

Olson, J. S., H. A. Pfuderer, and Y.-H. Chan. 1978. *Changes in the Global Carbon Cycle and the Biosphere.* Oak Ridge National Laboratory, Oak Ridge, Tenn.

Olson, J. S., C. J. Allison, and D.-N. Collier. 1980. *Carbon Cycles and Climate: A Selected Bibliography.* Oak Ridge National Laboratory, Oak Ridge, Tenn.

Onwueme, I. C. 1978. *The Tropical Tuber Crops.* Wiley, New York.

Openshaw, K. 1978. Woodfuel—A time for re-assessment. *Natural Resources Forum 3*:35–51.

Orel, A. E., and J. H. Seinfeld. 1977. Nitrate formation in atmospheric aerosols. *Environmental Science and Technology 11*:1000–1006.

Orem, S. R. 1982. Air pollution control technology: An overview. *Journal of the Air Pollution Control Association 32*:246–249.

Organization for Economic Cooperation and Development. 1977. The OECD *Programme on Long Range Transport of Air Pollutants.* OECD, Paris.

Organization for Economic Cooperation and Development. 1981. *The Costs and Benefits of Sulphur Oxide Control*. OECD, Paris.
Organization for Economic Cooperation and Development. 1982. *The Use of Coal in Industry*. OECD, Paris.
Ottar, B. 1976. Monitoring long-range transport of air pollutants: The OECD Study. *Ambio* *5*:203–206.
Ottar, B. 1978. An assessment of the OECD study on long-range transport of air pollutants (LRTAP). *Atmospheric Environment* *12*:445–454.
Overrein, L. N., H. M. Seip, and A. Tollan. 1980. *Acid Precipitation—Effects on Forests and Fish*. Final Report of the SNSF Project 1972–1980. SNSF Project, Oslo.
Owers, M. J., and A. W. Powell. 1974. Deposition velocity of sulphur dioxide on land and water surfaces using a ^{35}S tracer method. *Atmospheric Environment* *8*:63–67.
Pack, D. H., G. J. Ferber, J. C. Heffler, K. Telegadas, J. W. Angell, W. H. Hoecker, and L. Machta. 1978. Meteorology of long-range transport. *Atmospheric Environment* *12*:425–444.
Pahl, D. 1983. EPA's program for establishing standards of performance for new stationary sources of air pollution. *Journal of the Air Pollution Control Association* *33*:468–482.
Papamarcos, J. 1977. Stack gas cleanup. *Power Engineering* *81*(6):56–64.
Parker, G. G., S. E. Lindberg, and J. M. Kelly. 1980. Atmospheric–canopy interactions of sulfur in the southeastern United States. In: *Atmospheric Sulfur Deposition*, D. S. Shriner, C. R. Richmond, and S. E. Lindberg, eds. Ann Arbor Science, Ann Arbor, Mich., pp. 477–491.
Parker, R. H. 1979. Is 0.4 g/mi of NO_x achievable? *Automotive Engineering* *87*(2):42–45.
Parungo, F., C. Nagamoto, I. Nolt, M. Dias, and E. Nickerson. 1982. Chemical analysis of cloud water collected over Hawaii. *Journal of Geophysical Research* *87*:8805–8810.
Patrick, W. H., and I. C. Mahapatra. 1968. Transformation and availability to rice of nitrogen and phosphorus in waterlogged soils. *Advances in Agronomy* *22*:323–359.
Paul, D. A., and R. L. Kilmer. 1977. *The Manufacturing and Marketing of Nitrogen Fertilizers in the United States*. USDA, Washington, D.C.
Paul, D. A., R. L. Kilmer, M. A. Altobello, and D. N. Harrington. 1977. *The Changing U.S. Fertilizer Industry*. USDA, Washington, D.C.
Paul, G. 1978. Corrosion resistant materials for SO_2 scrubbers. *Power Engineering* *82*(5):54–57.
Payne, A. J., and J. A. Canner. 1969. Urea. *Chemical and Process Engineering* *50*:81–88.
Payne, W. J. 1983. Bacterial denitrification: Asset or deficit? *BioScience* *33*:319–325.
Payne, W. J., J. J. Rowe, and B. F. Sherr. 1980. Denitrification: A plea for attention. In: *Nitrogen Fixation*, Volume I, W. E. Newton and W. H. Orme-Johnson, eds. University Park Press, Baltimore, pp. 29–42.
Pearce, D. W., A. Harris, G. Mooney, R. L. Alchurst, and P. W. West. 1982. Air pollution control and morbidity. *Journal of the Air Pollution Control Association* *32*:1201–1206.
Pearcy, R. W., and O. Björkman. 1983. Physiological effects. In: *CO_2 and Plants*, E. R. Lemon, ed. Westview Press, Boulder, Colo., pp. 65–105.
Pearman, G. I. 1977. Further studies of the comparability of baseline atmospheric carbon dioxide measurements. *Tellus* *29*:171–181.
Pearman, G. I., and J. R. Garratt. 1975. Errors in atmospheric CO_2 concentration measurements arising from the use of reference gas mixtures different in composition to the sample air. *Tellus* *27*:62–66.
Pearson, R. W., and F. Adams, eds. 1967. *Soil Acidity and Liming*. American Society of Agronomy, Madison, Wisc.
PEDCo. 1978. *Particulate and Sulfur Dioxide Emission Control Costs for Large Coal-Fired Boilers*. EPA, Washington, D.C.
Persson, R. 1974. *World Forest Resources*. Skogshögskolan, Stockholm.
Persson, R. 1978. The need for a continuous assessment of the forest resources of the world. Paper presented at the Eighth World Forestry Congress, October 16–28, 1978, Jakarta, Indonesia.
Peters, G. A. 1978. Blue-green algae and algal associations. *BioScience* *28*:580–585.
Petersen, L. 1978. Podzolization: Mechanism and possible effects of acid precipitation. In: *Effects of Acid Precipitation on Terrestrial Ecosystems*, T. C. Hutchinson and M. Havas, eds. Plenum Press, New York, pp. 223–237.

Philander, S. G. H. 1983. Anomalous El Niño of 1982–83. *Nature 305*:16.

Phillips, D. A. 1980. Efficiency of symbiotic nitrogen fixation in legumes. *Annual Review of Plant Physiology 31*:29–49.

Pierce, B. A. 1985. Acid tolerance in amphibians. *BioScience 35*:239–243.

Pierotti, D., and R. A. Rasmussen. 1978. Continuous measurements of nitrous oxide in the troposphere. *Nature 274*:574–576.

Pierson, W. R., W. W. Brachaczek, T. J. Truex, J. W. Butler, and T. J. Korniski. 1980. Ambient sulfate measurements on Alleghany Mountain and the question of atmospheric sulfate in the northeastern United States. *Annals of the New York Academy of Sciences 338*:145–173.

Plass, G. N. 1956. The carbon dioxide theory of climatic change. *Tellus 8*:140–154.

Poincelot, R. P. 1975. *The Biochemistry and Methodology of Composting*. Connecticut Agriculture Experiment Station, New Haven.

Pollack, R. I. 1975. *Studies of Pollutant Concentration Frequency Distributions*. EPA, Research Triangle Park, N.C.

Postgate, J. R. 1977. Consequences of the transfer of nitrogen fixation genes to new hosts. *Ambio 6*:178–180.

Power, J. F. 1981. Nitrogen in the cultivate ecosystem. In: *Terrestrial Nitrogen Cycles*, F. E. Clark and T. Rosswall, eds. *Ecological Bulletin* No. 33, Stockholm, pp. 529–546.

Powers, W. L., G. W. Wallingford, and L. S. Murphy. 1975. Formulas for applying organic wastes to land. *Journal of Soil and Water Conservation 30*:286–289.

Prasad, R., G. B. Rajale, and B. A. Lakhdive. 1971. Nitrification retarders and slow-release nitrogen fertilizers. *Advances in Agronomy 23*:337–383.

Prentice, K. C., and J. C. Coiner. 1980. Agriculturally induced vegetation change, 1950–1975. *Human Ecology 8*:105–116.

Preussmann, R., D. Schmahl, G. Eisenbrand, and R. Port. 1977. In: *Proceedings of 2nd International Symposium on Nitrite in Meat Products*, B. J. Tinbergen and B. Krol, eds. Pudoc, Wageningen, the Netherlands.

Priestley, J. 1790. *Experiments and Observations on Different Kinds of Air*. Reprinted in H. Boynton, *The Beginnings of Modern Science*, Black, Roslyn, N.Y.

Prince, R., and F. F. Ross. 1972. Sulphur in air and soil. *Water, Air and Soil Pollution 1*:286–302.

Quartulli, O. J. 1974. *Developments in Ammonia Production Technology*. M. W. Kellogg, Houston, Tex.

Quispel, A. M., ed. 1974. *The Biology of Nitrogen Fixation*. North-Holland, Amsterdam.

Rachie, K. O. 1977. The nutritional role of grain legumes in the lowland humid tropics. In: *Biological Nitrogen Fixation in Farming Systems of the Tropics*, A. Ayanaba and P. J. Dart, eds. Wiley, New York, pp. 45–60.

Rachie, K. O., and L. M. Roberts. 1974. Grain legumes of the lowland tropics. *Advances in Agronomy 26*:1–132.

Radian Corporation. 1977. *Interagency Flue Gas Desulfurization Evaluation*, two volumes. EPA, Washington, D.C.

Radmer, R., and B. Kok. 1977. Photosynthesis: Limited yields, unlimited dreams. *BioScience 27*:599–605.

Rahn, K., ed. 1981. *Arctic Air Chemistry. Atmospheric Environment 15*:1345–1516.

Rajagopal, R., and R. L. Talcott. 1983. Patterns in groundwater quality: Selected observations in Iowa. *Environmental Management 7*:465–474.

Rao, K. K. P. N., ed. 1976. *Food Consumption and Planning*. Pergamon Press, Elmsford, N.Y.

Rao, S. T., and G. Sistla. 1982. Relationship between urban and rural sulfate levels. *Journal of the Air Pollution Control Association 32*:645–648.

Rasmussen, R. A. 1974. Emission of biogenic hydrogen sulfide. *Tellus 26*:254–260.

Rasool, S. I., and S. H. Schneider. 1971. Atmospheric carbon dioxide and aerosols: Effects of large increases on global climate. *Science 173*:138–141.

Reck, R. A., and V. R. Hummel, eds. 1982. *Interpretation of Climate and Photochemical Models, Ozone and Temperature Measurements*. American Institute of Physics, New York.

Reiners, W. A. 1973. The carbon cycle. In: *Carbon and the Biosphere,* G. M. Woodwell and E. V. Pecan, eds. AEC, Washington, D.C., pp. 368–382.

Reisinger, L. M. and T. L. Crawford. 1980. Sulfate flux through the Tennessee Valley Region. *Journal of the Air Pollution Control Association 30*:1230–1231.

Reisinger, L. M., and T. L. Crawford. 1982. Interregional transport: Case studies of measurements versus model predictions. *Journal of the Air Pollution Control Association 32*:629–633.

Rennie, P. J., and R. L. Halstead. 1977. The effects of sulphur on plants in Canada. In: *Sulphur and Its Inorganic Derivatives in the Canadian Environment.* NRC, Ottawa, pp. 67–179.

Reuther, J. J. 1982. The fate of sulphur in coal combustion. *Earth and Mineral Sciences 51*:50–53.

Revelle, R. 1976. Energy use in rural India. *Science 192*:969–975.

Revelle, R. 1982. Carbon dioxide and world climate. *Scientific American 247*:35–43.

Revelle, R. R. 1983. Methane hydrates in continental slope sediments and increasing atmospheric carbon dioxide. In: *Changing Climate.* NAS, Washington, D.C., pp. 252–261.

Revelle, R., and W. Munk. 1977. The carbon cycle and the biosphere. In: *Energy and Climate.* NAS, Washington, D.C., pp. 140–158.

Revelle, R., and H. E. Suess. 1957. Carbon dioxide exchange between atmosphere and ocean and the question of an increase of atmospheric CO_2 during the past decades. *Tellus 9*:18–27.

Revelle, R., and D. E. Waggoner. 1983. Effects of a carbon dioxide-induced climatic change on water supplies in the western United States. In: *Changing Climate.* NAS, Washington, D.C., pp. 419–432.

Revelle, R., W. Broecker, H. Craig, C. D. Keeling, and J. Smagorinsky. 1965. Atmospheric carbon dioxide. In: *Restoring the Quality of Our Environment.* President's Science Advisory Committee, Washington, D.C., pp. 111–133.

Richards, J. F., J. S. Olson, and R. M. Rotty. 1983. *Development of a Data Base for Carbon Dioxide Releases Resulting from Conversion of Land to Agricultural Uses.* Oak Ridge National Laboratory, Oak Ridge, Tenn.

Ripperton, L. A., L. Kornreich, and J. J. B. Worth. 1970. Nitrogen dioxide and nitric oxide in non-urban air. *Journal of the Air Pollution Control Association 20*:589–592.

Rittenhouse, R. C. 1978. Coping with pollution control requirements for power plants. *Power Engineering 82*(7):42–48.

Rittenhouse, R. C. 1981. Air pollution control for power plants. *Power Engineering 85*(9):62–70.

Rittenhouse, R. C. 1982. Coal supplies and the economics of treatment. *Power Engineering 86*(6):44–52.

Roberts, E. 1982. Potential therapies in aging and senile dementias. *Annals of the New York Academy of Sciences 396*:165–178.

Roberts, L. 1983. Is acid deposition killing west German forests? *BioScience 33*:302–305.

Robertson, L. S., D. D. Warncke, and J. D. Baker. 1980. *Chemical Test Variability Within Soil Types.* Michigan State University Agricultural Experiment Station, East Lansing.

Robinson, E., and R. C. Robbins. 1970. Gaseous nitrogen compound pollutants from urban and natural sources. *Journal of the Air Pollution Control Association 20*:303–306.

Robinson, E., and R. C. Robbins. 1972. Emissions, concentrations and fate of gaseous atmospheric pollutants. In: *Air Pollution Control,* Part 2, W. Strauss, ed. Wiley, New York, pp. 1–93.

Rodhe, H. 1970. On the residence time of anthropogenic sulfur in atmosphere. *Tellus 22*:137–139.

Rodhe, H. 1976. An atmospheric sulphur budget for N. W. Europe in nitrogen, phosphorus and sulphur-global cycles. *Ecological Bulletin 22*:123–134.

Rodhe, H. 1978. Budgets and turn-over times of atmospheric sulfur compounds. *Atmospheric Environment 12*:671–680.

Rodhe, H., P. Crutzen, and A. Vanderpol. 1981. Formation of sulfuric and nitric acid in the atmosphere during long-range transport. *Tellus 33*:132–141.

Rohrman, F. A., and J. H. Ludwig. 1965. Sources of sulphur dioxide pollution. *Chemical Engineering Progress 61*(9):59–63.

Rolston, D. E. 1981. Nitrous oxide and nitrogen gas production in fertilizer loss. In: *Denitrification, Nitrification, and Atmospheric Nitrous Oxide,* C. C. Delwiche, ed. Wiley, New York, pp. 127–149.

Rolston, D. E., D. L. Hoffman, and D. T. Toy. 1978. Field measurement of denitrification. I. Flux of N_2 and N_2O. *Soil Science Society of America Journal 42*:863–869.

Roosen, R. G., R. S. Harrington, J. Giles, and I. Browning. 1976. Earth tides, volcanoes and climatic change. *Nature 261*:680–682.

Root, B. D. 1976. An estimate of annual global atmospheric pollutant emissions from grassland fires and agricultural burning in the tropics. *The Professional Geographer 28*:349–352.

Rose, A. W., E. G. Williams, and R. B. Parizek. 1983. Predicting potential for acid drainage from coal mines. *Earth and Mineral Sciences 52*:37–41.

Rosenberg, N. J. 1982. Commentary. In: *Carbon Dioxide Review: 1982,* W. C. Clark, ed. Oxford University Press (Clarendon), London, pp. 324–328.

Ross, F. F. 1971a. Sulphur dioxide over Britain and beyond. *New Scientist 50*:373–375.

Ross, F. F. 1971b. What sulphur dioxide problem? *Combustion 43*(2):3–7.

Rosswall, T. 1977. Exchange of nutrients between atmosphere and vegetation-soil. *Agro-Ecosystems 4*:296–302.

Rosswall, T. 1981. The biogeochemical nitrogen cycle. In: *Some Perspectives of the Major Biogeochemical Cycles,* G. E. Likens, ed. Wiley, New York, pp. 25–49.

Rotty, R. M. 1974. First estimates of global flaring of natural gas. *Atmospheric Environment 8*:681–686.

Rotty, R. M. 1982. Fossil fuel and cement production, 1860–1980. In: *Carbon Dioxide Review: 1982,* W. C. Clark, ed. Oxford University Press (Clarendon), London, pp. 456–460.

Rotty, R. M., and G. Marland. 1980. *Constraints on Carbon Dioxide Production from Fossil Fuel Use.* Institute for Energy Analysis, Oak Ridge, Tenn.

Rowe, M. D., S. C. Morris, and L. D. Hamilton. 1978. Potential ambient standards for atmospheric sulfates: An account of a workshop. *Journal of the Air Pollution Control Association 28*:772–775.

Royal Ministry for Foreign Affairs. 1971. *Air Pollution Across National Boundaries: The Impact on the Environment of Sulfur in Air and Precipitation.* Royal Ministry for Foreign Affairs, Stockholm.

Royal Society Study Group Report Staff. 1984. *Nitrogen Cycle of the United Kingdom: A Study Group Report.* Royal Society, London.

Rubin, E. S. 1981. Air pollution constraints on increased coal use by industry. *Journal of the Air Pollution Control Association 31*:349–360.

Rutger, J. N., and D. M. Brandon. 1981. California rice cultivation. *Scientific American 244*(2):42–51.

Ryden, J. C., L. J. Lund, J. Lesy, and D. D. Focht. 1979. Direct measurement of denitrification from soils. II. Development and application of field methods. *Soil Science Society of America Journal 43*:110–118.

Sackner, M. A., D. Ford, R. Fernandez, J. Cipley, D. Perez, M. Kwoka, M. Reinhart, E. D. Michaelson, R. Schreck, and A. Wanner. 1978. Effects of sulfuric acid aerosol on cardiopulmonary function of dogs, sheep and humans. *American Review of Respiratory Diseases 118*:497–510.

Sadler, J. C., C. S. Ramage, and A. M. Hori. 1982. Carbon dioxide variability and atmospheric circulation. *Journal of Applied Meteorology 21*:793–805.

Salmon, R. L. 1970. *Systems Analysis of the Effects of Air Pollution on Materials.* Department of Health, Education and Welfare, Raleigh, N.C.

Salter, R. M., and L. J. Schollenberger. 1938. Farm manure. In: *Soils and Men.* USDA, Washington, D.C., pp. 445–461.

Samson, P. J. 1984. The meteorology of acid deposition. *Journal of the Air Pollution Control Association 34*:20–24.

Sanchez, P. A. 1976. *Properties and Management of Soils in the Tropics.* Wiley, New York.

Sanchez, P. A., D. E. Bandy, J. H. Villachica, and J. J. Nicholaides. 1982. Amazon basin soils: Management for continuous crop production. *Science 216*:821–827.

Schelling, T. C. 1983. Climatic change: Implications for welfare and policy. In: *Changing Climate.* NAS, Washington, D.C., pp. 449–482.

Schiff, D. 1979. Available energy analysis of an FAO system. *Power Engineering 83*(11):68–69.

Schlesinger, M. E. 1983. A review of climate models and their simulation of CO_2-induced warming. *International Journal of Environmental Studies 20*:103–114.

Schlesinger, W. H., and J. M. Melack. 1981. Transport of organic carbon in the world's rivers. *Tellus 33*:172–187.

Schmidt, E. L. 1982. Nitrification in soil. In: *Nitrogen in Agricultural Soils,* F. J. Stevenson, ed. American Society of Agronomy, Madison, Wisc., pp. 253–288.

Schneider, S. H. 1975. On the carbon dioxide–climate confusion. *Journal of the Atmospheric Sciences 32*:2060–2066.

Schneider, S. H., and R. S. Chen. 1980. CO_2 warming and coastline flooding. *Annual Review of Energy 5*:107–140.

Schneider, S. H., and L. E. Mesirow. 1976. *The Genesis Strategy.* Plenum Press, New York.

Schneider, S. H., and S. L. Thompson. 1980. Cosmic conclusions from climatic models: Can they be justified? *Icarus 41*:456–469.

Schneider, S. H., W. W. Kellogg, and V. Ramanathan. 1980. Carbon dioxide and climate. *Science 210*:6.

Schnell, R. C., S.-A. Odh, and L. N. Njau. 1981. Carbon dioxide measurements in tropical east African biomes. *Journal of Geophysical Research 86*:5364–5372.

Schnitzer, M. 1978. Effect of low pH on the chemical structure and reactions of humic substances. In: *Effects of Acid Precipitation on Terrestrial Ecosystems,* T. C. Hutchinson and M. Havas, eds. Plenum Press, New York, pp. 203–222.

Schofield, C. L. 1976. Effects on fish. *Ambio 5*:228–230.

Schroeder, E. D. 1981. Denitrification in wastewater management. In: *Denitrification, Nitrification and Atmospheric Nitrous Oxide,* C. C. Delwiche, ed. Wiley-Interscience, New York, pp. 105–125.

Schütt, P., W. Koch, H. Blaschke, K. J. Lang, H. J. Schuck and H. Summer. *So stirbt der Wald.* Verlagsgesselschaft BLV, München.

Sehmel, G. A. 1980. Model predictions and a summary of dry deposition velocity data. In: *Atmospheric Sulfur Deposition,* D. S. Shriner, C. R. Richmond, and S. E. Lindberg, eds. Ann Arbor Science, Ann Arbor, Mich., pp. 223–234.

Seliber, J. 1981. Comments. *Journal of the Air Pollution Control Association 31*:958–960.

Sellers, W. D. 1974. A reassessment of the effect of CO_2 variations on a simple global climatic model. *Journal of Applied Meteorology 13*:831–833.

Semb, A. 1978. Sulphur emissions in Europe. *Atmospheric Environment 12*:455–460.

Sensenbaugh, J. D. 1966. *Formation and Control of Oxides of Nitrogen in Combustion Processes.* Combustion Engineering, Windsor, Conn.

Sequeira, R. 1982. Acid rain: An assessment based on acid–base considerations. *Journal of the Air Pollution Control Association 32*:241–245.

Sereda, P. J. 1977. Effects of sulphur on building materials. In: *Sulphur and Its Inorganic Derivatives in the Canadian Environment.* NRC, Ottawa, pp. 359–426.

Shaw, R. W. 1979. Acid precipitation in Atlantic Canada. *Environmental Science and Technology 13*:406–411.

Sheih, C. M., M. L. Wesely and B. B. Hicks. 1979. Estimated dry deposition velocities of sulfur over the eastern United States and surrounding regions. *Atmospheric Environment 13*:1361–1368.

Sherff, J. L. 1975. Energy use and economics in the manufacture of fertilizers. In: *Energy, Agriculture and Waste Management,* W. J. Jewell, ed. Ann Arbor Science, Ann Arbor, Mich., pp. 433–441.

Shinn, J. H., and S. Lynn. 1979. Do man-made sources affect the sulfur cycle of northeastern states? *Environmental Science and Technology 13*:1062–1067.

Shoulders, E., and R. F. Wittver. 1979. Fertilizing for high fiber yields in intensively managed plantations. In: *Impact of Intensive Harvesting on Forest Nutrient Cycling.* State University of New York, Syracuse, pp. 343–359.

Shriner, D. S., C. R. Richmond, and S. E. Lindberg, eds. 1980. *Atmospheric Sulfur Deposition: Environmental Impact and Health Effects.* Proceedings of the Second Life Sciences Symposium, Gatlinburg, Tenn., Oct. 14–18, 1979. Ann Arbor Science, Ann Arbor, Mich.

Shull, R. P., project officer. 1979. *Resources and Pollution Control.* EPA, Washington, D.C.

Siddiqi, T. A., and C. Zhang. 1982. Ambient air quality standards in China. *Environmental Management. 8*:473–479.

Siegenthaler, U., and H. Oeschger. 1978. Predicting future atmospheric carbon dioxide levels. *Science 199*:388–395.

Silver, W. S., and R. W. F. Hardy. 1976. *Biological Nitrogen Fixation in Forage and Livestock Systems*. American Society of Agronomy, Madison, Wisc.

Sims, J. A. and L. E. Johnson. 1972. *Animals in the American Economy*. Iowa State University Press, Ames.

Singer, R. 1982. Effects of acidic precipitation on benthos. In: *Acid Precipitation: Effects on Biological Systems*, F. M. D'Itri, ed. Ann Arbor Science, Ann Arbor, Mich., pp. 329–352.

Sioli, H. 1973. Recent human Activities in the Brazilian Amazon region and their biological effects. In: *Tropical Forest Ecosystems in Africa and South America: A Comparative Review*, B. J. Meggers, E. S. Ayensu, and W. D. Duckworth, eds. Random House (Smithsonian Institution Press), New York.

Slack, A. V. 1966. *Chemistry and Technology of Fertilizers*. Wiley, New York.

Slack, A. V. 1973. Removing SO_2 from stack gases. *Environmental Science and Technology* *7*:110–119.

Slack, A. V., and G. A. Hollinden. 1975. *Sulfur Dioxide Removal from Waste Gases*. Noyes Data Corporation, Park Ridge, N.J.

Slack, A. V., and R. James. 1973–1979. *Ammonia*. Parts 1–4. Dekker, New York.

Slack, A. V., G. C. McGlamery, and H. L. Falkenberry. 1971. Economic factors in recovery of sulfur dioxide from power plant stack gas. *Journal of the Air Pollution Control Association 21*:9–15.

Slatt, B. J., D. F. S. Natusch, J. M. Prospero, and D. L. Savoie. 1978. Hydrogen sulfide in the atmosphere of the northern equatorial Atlantic Ocean and its relation to the global sulfur cycle. *Atmospheric Environment 12*:981–990.

Sloane, C. S. 1982. Visibility trends. II. Mideastern United States 1948–1978. *Atmospheric Environment 16*:2309–2321.

Smil, V. 1981. China's agro-ecosystem. *Agro-Ecosystems 7*:27–46.

Smil, V. 1983. *Biomass Energies: Resources, Links, Constraints*. Plenum Press, New York.

Smil, V. 1984a. *The Bad Earth: Environmental Degradation in China*. Sharpe, Armonk, N.Y.

Smil, V. 1984b. *China's Energy: Advances and Limitations*. Paper prepared for the International Development Research Center, Ottawa.

Smil, V., and V. Karfik. 1968. Jenom cast spinaveho vzduchu. *Zivotne prostredie 2*:190–191.

Smil, V., and D. Milton. 1974. Carbon dioxide—Alternative futures. *Atmospheric Environment 8*:1213–1223.

Smil, V., P. Nachman, and T. V. Long, II. 1982. *Energy Analysis and Agriculture*. Westview Press, Boulder, Colo.

Smith, C. M. 1980. Scope and possibilities of soil testing for nitrogen. In: *Soil and Plant Testing and Analysis*. FAO, Rome, pp. 122–133.

Smith, N. J. H. 1981. Colonization lessons from a tropical forest. *Science 214*:755–761.

Smith, R. A. 1872. *Air and Rain: The Beginnings of Chemical Climatology*. Longmans, Green, London.

Smith, S. V., and D. W. Kinsey. 1976. Calcium carbonate production, coral reef growth and sea level change. *Science 194*:937–939.

Smith, W. H. 1981. *Air Pollution and Forests: Interaction Between Air Contaminants and Forest Ecosystems*. Springer-Verlag, Berlin.

Söderlund, R. 1977. NO_x pollutants and ammonia emissions—A mass balance for the atmosphere over NW Europe. *Ambio 6*:118–122.

Söderlund, R. 1981. Dry and wet deposition of nitrogen compounds. In: *Terrestrial Nitrogen Cycles*, F. E. Clark and T. Rosswall, eds. *Ecological Bulletin* No. 33, Stockholm, pp. 123–130.

Söderlund, R., and B. H. Svensson. 1976. The global nitrogen cycle. In: *Nitrogen, Phosphorus and Sulphur—Global Cycles*, B. H. Svensson and R. Söderlund, eds. *Ecological Bulletin 22*:23–73.

Sommer, A. 1976. Attempt at an assessment of the world's tropical moist forests. *Unasylva 28*(112–113):5–24.

Sorenson, J. R. J. 1977. Aluminum in relation to the environment and human health. In: *Environmental Biogeochemistry*, J. O. Nriagu, ed. Ann Arbor Science, Ann Arbor, Mich., pp. 427–450.

Spaite, P. W. 1972. SO_2 control: Status, cost and outlook. *Power Engineering 76*(10):34–37.

Spengler, J. D., B. G. Ferris, Jr., D. W. Dockery, and F. E. Speizer. 1979. Sulfur dioxide and nitrogen dioxide levels inside and outside homes and the implications on health effects research. *Environmental Science and Technology 13*:1276–1280.

Spicer, C. W. 1974. *The Fate of Nitrogen Oxides in the Atmosphere.* Battelle Columbus Laboratories, Columbus, Ohio.

Spicer, C. W. 1977. Photochemical atmospheric pollutants derived from nitrogen oxides. *Atmospheric Environment 11*:1089–1095.

Spittlehouse, D. L., and E. A. Ripley. 1977. Carbon dioxide concentrations over a native grassland in Saskatchewan. *Tellus 29*:54–65.

Sposito, G., A. L. Page, and M. E. Frink. 1980. *Effects of Acid Precipitation on Soil Leachate Quality.* EPA, Corvallis, Oreg.

Stanford, E., C. B. England, and A. W. Taylor. 1970. *Fertilizer and Water Quality.* USDA, Washington, D.C.

Staniforth, A. R. 1979. *Cereal Straw.* Oxford University Press (Clarendon), London.

Stelly, M., ed. 1980. *Nitrification Inhibitors—Potentials and Limitations.* American Society of Agronomy, Madison, Wisc.

Stevenson, F. J. 1972. Nitrogen: Element and geochemistry. In: *The Encyclopedia of Geochemistry and Environmental Sciences,* Volume IV, R. W. Fairbridge, ed. Van Nostrand–Reinhold, Princeton, N.J., pp. 795–801.

Stevenson, F. J. 1982a. *Nitrogen in Agricultural Soils.* American Society of Agronomy, Madison, Wisc.

Stevenson, F. J. 1982b. Origin and distribution of N in soil. In: *Nitrogen in Agricultural Soils,* F. J. Stevenson, ed. American Society of Agronomy, Madison, Wisc., pp. 1–42.

Stewart, W. D. P., ed. 1975. *Nitrogen Fixation by Free-Living Microorganisms.* Cambridge University Press, London.

Stewart, W. D. P. 1977. Present-day nitrogen fixing plants. *Ambio 6*:166–173.

Stoiber, R. E., and A. Jepsen. 1973. Sulfur dioxide contributions to the atmosphere by volcanoes. *Science* 182:577–578.

Strain, B. R., ed. 1978. *Report of the Workshop on Anticipated Plant Responses to Global Carbon Dioxide Enrichment.* Duke University Press, Durham, N.C.

Strain, B. R., and F. A. Bazzaz. 1983. Terrestrial plant communities. In: *CO_2 and Plants,* E. R. Lemon, ed. Westview Press, Boulder, Colo., pp. 177–222.

Study of Critical Environmental Problems. 1970. *Man's Impact on the Global Environment: Assessment and Recommendations for Action.* MIT Press, Cambridge, Mass.

Study of Man's Impact on Climate. 1971. *Inadvertent Climate Modification.* MIT Press, Cambridge, Mass.

Stumm, W., ed. 1977. *Global Chemical Cycles and Their Alterations by Man.* Dahlem Konferenzen, Abakon Verlagsgesellschaft, Berlin.

Stumm, W., and J. J. Morgan. 1970. *Aquatic Chemistry.* Wiley–Interscience, New York.

Sub-committee on Acid Rain of the Standing Committee on Fisheries and Forestry. 1981. *Still Waters.* House of Commons, Ottawa.

Sukhatme, P. V. 1977. Nitrogen in malnutrition. *Ambio 6*:137–140.

Svensson, B. H., and R. Söderlund, eds. 1976. *Nitrogen, Phosphorus and Sulphur—Global Cycles.* Ecological Bulletin No. 22, Stockholm.

Swedish Ministry of Agriculture. 1982. *Acidification Today and Tomorrow.* Swedish Ministry of Agriculture, Stockholm.

Sweeney, R. E., K. K. Liu, and I. R. Kaplan. 1978. Oceanic nitrogen isotopes and their uses in determining the source of sedimentary nitrogen. In: *Stable Isotopes in the Earth Sciences,* B. W. Robinson, ed. New Zealand Department of Scientific and Industrial Research, Wellington, pp. 9–26.

Sze, N. D., and M. K. W. Ko. 1980. Photochemistry of COS, CS_2, CH_3SCH_3 and H_2S: Implications for the atmospheric sulfur cycle. *Atmospheric Environment 14*:1223–1239.

Takahashi, J. 1978. Role of night soil in Japanese agriculture. In: *Organic Recycling in Asia.* FAO, Rome, pp. 363–364.

Tamm, C. O. 1976. Biological effects in soil and on forest vegetation. *Ambio 5*:235–238.

Tanaka, A. 1973. Methods of handling rice straw in various countries. *International Rice Commission Newsletter* 2(2):1–20.

Tang, I. N., W. T. Wong, and H. R. Munkelwitz. 1981. The relative importance of atmospheric sulfates and nitrates in visibility reduction. *Atmospheric Environment 15*:2463–2471.

Taylor, W. L. 1973. 42 million tons of scrubber sludge, and what do you get? *Power Engineering* 77(9):64–67.

Terman, G. L. 1979. Volatilization losses of nitrogen as ammonia from surface-applied fertilizers, organic amendments, and crop residues. *Advances in Agronomy 31*:189–223.

The Fertilizer Institute. 1976. *The Fertilizer Handbook*. The Fertilizer Institute, Washington, D.D.

The MAP3S/RAINE Research Community. 1982. The MAP3S/RAINE precipitation chemistry network: Statistical overview for the period 1976–1980. *Atmospheric Environment 16*:1603–1631.

Thomas, G. W., and J. W. Gilliam. 1977. Agro-ecosystems in the U.S.A. *Agro-Ecosystems 4*:182–243.

Thomas, R. H., T. J. O. Sanderson, and K. E. Rose. 1979. Effect of climatic warming on the west Antarctic ice sheet. *Nature 277*:355–358.

Thompson, J. R., I. K. Smith, and D. P. Moore. 1970. Sulfur requirement and metabolism in plants. In: *Symposium: Sulfur in Nutrition*, O. H. Muth and J. E. Oldfield, eds. Avi Publishing, Westport, Conn., pp. 80–96.

Tisdale, S. L., and W. L. Nelson. 1975. *Soil Fertility and Fertilizers*. Macmillan Co., New York.

Tolbert, N. F., and I. Zelitch. 1983. Carbon metabolism. In: *CO_2 and Plants*, E. R. Lemon, ed. Westview Press, Boulder, Colo., pp. 21–64.

Tomlinson, G. H. 1983. Air pollutants and forest decline. *Environmental Science and Technology 17*:246A–256A.

Tong, D., and T. Bao. 1978. On the policy for the construction of the Northwest Plateau. *Renmin Ribao (People's Daily)*, Nov. 16, 1978, p. 2.

Torrey, J. G. 1978. Nitrogen fixation by actinomycete-nodulated angiosperms. *BioScience 28*:586–592.

Trijonis, J. 1982. Visibility in California. *Journal of the Air Pollution Control Association 32*:165–167.

Trijonis, J., and K. Yuan. 1978. *Visibility in the Northeast: Long-Term Visibility Trends and Visibility/Pollutant Relationships*. EPA, Washington, D.C.

Trudinger, P. A., M. R. Walter, and B. J. Ralph, eds. 1980. *Biogeochemistry of Ancient and Modern Environments*. Springer-Verlag, Berlin.

Tsentralnoye Statisticheskoye Upravleniye. 1982. *Narodnoe Khozyaystvo SSSR*. Statistika, Moscow.

Tuesday, C. S., ed. 1971. *Chemical Reactions in the Urban Atmosphere*. American Elsevier, New York.

Turman, B. N., and B. C. Edgar. 1982. Global lightning distributions at dawn and dusk. *Journal of Geophysical Research 87*:1191–1206.

Turner, W. 1974. *Ten Years of Single Train Ammonia Plants*. Kellogg, Houston, Tex.

Tyndall, J. 1861. On the absorption and radiation of heat by gases and vapours, and on the physical connection of radiation, absorption and conduction. *Philosophical Magazine*, Series 4, *22*:169–194, 273–285.

Ulrich, B. 1978. Production and consumption of hydrogen ions in the ecosphere. In: *Effects of Acid Precipitation on Terrestrial Ecosystems*, T. C. Hutchinson and M. Havas, eds. Plenum Press, New York, pp. 255–282.

United Nations Organization. 1956. World energy requirements in 1975 and 2000. In: *Proceedings of the International Conference on the Peaceful Uses of Atomic Energy*, Volume 1. UNO, New York, pp. 3–33.

United Nations Organization. 1977. *Desertification: Its Causes and Consequences*. Pergamon Press, Elmsford, N.Y.

United Nations Organization. 1980. *Yearbook of World Energy Statistics 1970–1979*. UNO, New York.

United States Bureau of Mines. 1954. *Technology of Lignitic Coals*. U.S. Bureau of Mines Information Circular No. 769, Washington, D.C.

United States Bureau of the Census. 1975. *Historical Statistics of the United States: Colonial Times to 1970*. USBC, Washington, D.C.

United States Department of Agriculture. 1971. *Consumption of Commercial Fertilizers, Primary Plant Nutrients, and Micronutrients*. USDA, Washington, D.C.

United States Department of Agriculture. 1978. *Improving Soils with Organic Wastes*. USDA, Washington, D.C.

United States Department of Agriculture. 1980. *Report and Recommendations on Organic Farming*. USDA, Washington, D.C.

United States Environmental Protection Agency. 1971a. *Air Quality Criteria for Nitrogen Oxides*. EPA, Washington, D.C.

United States Environmental Protection Agency. 1971b. *Air Pollution Aspects of Emission Sources: Sulfuric Acid Manufacturing*. EPA, Research Triangle Park, N.C.

United States Environmental Protection Agency. 1973a. *Compilation of Air Pollutant Emission Factors*. EPA, Research Triangle Park, N.C.

United States Environmental Protection Agency. 1973b. *Air Pollution Aspects of Emission Sources: Pulp and Paper Industry*. EPA, Research Triangle Park, N.C.

United States Environmental Protection Agency. 1975a. *Supplement No. 4 for Compilation of Air Pollutant Emission Factors*. EPA, Washington, D.C.

United States Environmental Protection Agency. 1975b. *Position Paper on Regulation of Atmospheric Sulfates*. EPA, Washington, D.C.

United States Environmental Protection Agency. 1978. *Mobile Source Emission Factors*. EPA, Washington, D.C.

United States Environmental Protection Agency. 1980a. *Environmental Outlook 1980*. EPA, Washington, D.C.

United States Environmental Protection Agency. 1980b. *Acid Rain*. EPA, Washington, D.C.

United States Environmental Protection Agency. 1980c. *Environmental, Operational and Economic Aspects of Thirteen Selected Energy Technologies*. EPA, Washington, D.C.

United States Environmental Protection Agency. 1981. *Control Techniques for Sulfur Oxide Emissions from Stationary Sources*. EPA, Research Triangle Park, N.C.

United States Environmental Protection Agency. 1982. *National Air Pollutant Emission Estimates, 1970–1981*. EPA, Washington, D.C.

Urone, P., and W. H. Schroeder. 1969. SO_2 in the atmosphere: A wealth of monitoring data, but few reaction rate studies. *Environmental Science and Technology 3*:436–445.

U.S./Canada Work Group. 1982. *United States–Canada Memorandum of Intent on Transboundary Air Pollution*. Environment Canada, Downsview, Ontario.

van Dop, H., and S. Kruizinga. 1976. The decrease of sulphur dioxide concentrations near Rotterdam and their relation to some meteorological parameters during thirteen consecutive winters (1961–1974). *Atmospheric Environment 10*:1–4.

van Dop, H., T. B. Ridder, J. F. den Tonkelaar, and N. D. van Egmond. 1980. Sulphur dioxide measurements on the 213 metre tower at Cabauw, The Netherlands. *Atmospheric Environment 14*:933–945.

Van Dyne, D. L., and C. B. Gilbertson. 1978. *Estimated Inventory of Livestock and Poultry Manure Resources in the United States*. St. Joseph, Michigan. American Society of Agricultural Engineers Technical Paper TP78–2057.

van Egmond, N. D., O. Tissing, and H. Kesseboom. 1979. Estimating contributions of source areas to the measured yearly average SO_2-concentration field in the Netherlands by dispersion-model parameter-optimization. *Atmospheric Environment 13*:1551–1557.

Van Veen, J. A., W. B. McGill, H. W. Hunt, M. J. Frissel, and C. V. Cole. 1981. Simulation models of the terrestrial nitrogen cycle. In: *Terrestrial Nitrogen Cycles*, F. E. Clark and T. Rosswall, eds. *Ecological Bulletin* No. 33, Stockholm, pp. 25–48.

Vatsky, J. 1981. Development and field operation of the controlled flow/split-flame burner. In: *Proceedings of the Joint Symposium on Stationary Combustion NO_x Control*. EPRI, Palo Alto, Calif., pp. 54–95.

Venkatasubramanian, K., and B. Bowonder. 1980. Forest resources in India. *Futures 12*:317–324.

Viets, F. G. 1978. Mass balance and flue of nitrogen as aids in control and prevention of water pollution. In: *Nitrogen in the Environment*, Volume I, D. R. Nielsen and J. G. MacDonald, eds. Academic Press, New York, pp. 173–182.

Viets, F. G., Jr. 1980. Present status of soil and plant analysis for fertilizer recommendations and improvement of soil fertility. In: *Soil and Plant Testing and Analysis*. FAO, Rome, pp. 9–20.
Vincent, J. M., ed. 1982. *Nitrogen Fixation in Legumes*. Academic Press, New York.
Vitousek, P. M. 1982. Nutrient cycling and nutrient efficiency. *The American Naturalist 119*:553–572.
Vitousek, P. M., and W. A. Reiners. 1975. Ecosystem succession and nutrient retention: A hypothesis. *BioScience 25*:376–381.
Vitousek, P. M., J. R. Gosz, C. C. Grier, J. M. Melillo, W. A. Reiners, and R. L. Todd. 1979. Nitrate losses from disturbed ecosystems. *Sciences 204*:469–474.
Vollenweider, R. A. 1968. *Scientific Fundamentals of the Eutrophication of Lakes and Flowing Waters with Particular Reference to Nitrogen and Phosphorus as Factors in Eutrophication*. OECD, Paris.
Vömel, A. 1965. Nutrient balance in various lysimeter soils. *Zeitschrift für Acker- und Pflanzenbau 123*:155–188.
Voorburg, J. H. 1974. Some waste problems in pig production. *Agriculture and Environment 1*:175–190.
Waddell, T. E. 1974. *The Economic Damages of Air Pollution*. EPA, Washington, D.C.
Wagener, K. 1979. The carbonate system of the ocean. In: *The Global Carbon Cycle*, B. Bolin, E. T. Degens, S. Kempe, and P. Ketner, eds. Wiley, New York, pp. 251–258.
Waggoner, P. E. 1983. Agriculture and a climate changed by more carbon dioxide. In: *Changing Climate*. NAS, Washington, D.C., pp. 383–418.
Waggoner, P. E. 1984. Agriculture and carbon dioxide. *American Scientist 72*:179–184.
Wainwright, M. 1980. Man-made emissions of sulphur and the soil. *International Journal of Environmental Studies 14*:279–288.
Waldman, J. M., J. W. Munger, D. J. Jacob, R. C. Flanagan, J. J. Morgan, and M. R. Hoffman. 1982. Chemical composition of acid fog. *Science 218*:677–680.
Walker, A. B., and R. F. Brown. 1971. Statistics on utilization and economics of electrostatic precipitation for air pollution control. In: *Proceedings of the Second International Clean Air Congress*, H. M. Englund and W. T. Beery, eds. Academic Press, New York, pp. 724–730.
Walsh, L. M., and J. D. Beaton, eds. 1973. *Soil Testing and Plant Analysis*. Soil Science Society of America, Madison, Wisc.
Wang, W. C., Y. L. Yung, A. A. Lacis, T. Mo, and J. E. Hansen. 1976. Greenhouse effects due to man-made perturbations of trace gases. *Science 194*:685–690.
Watt, B. K., and A. L. Merrill. 1963. *Handbook of the Nutritional Contents of Foods*. USDA, Washington, D.C.
Weare, B. C., and F. M. Snell. 1974. A diffuse thin cloud atmospheric structure as a feedback mechanism in global climatic modeling. *Journal of the Atmospheric Sciences 31*:1725–1733.
Wedding, J. B., M. Ligotke, and F. D. Hess. 1979. Effects of sulfuric acid mist on plant canopies. *Environmental Science and Technology 13*:875–885.
Weiss, R. F., and H. Craig. 1976. Production of atmospheric nitrous oxide by combustion. *Geophysical Research Letters 3*:751–753.
Weller, G., D. J. Baker, Jr., W. L. Gates, M. C. MacCracken, S. Manabe, and T. H. V. Haar. 1983. Detection and monitoring of CO_2-induced climate changes. In: *Changing Climate*. NAS, Washington, D.C., pp. 292–382.
Wetselaar, R., and G. D. Farquhar. 1980. Nitrogen losses from tops of plants. *Advances in Agronomy 33*:263–302.
Wheelock, T. D., ed. 1977. *Coal Desulfurization: Chemical and Physical Methods*. American Chemical Society, Washington, D.C.
Whitaker, F. D., H. G. Heinemann, and R. E. Burwell. 1978. Fertilizing corn adequately with less nitrogen. *Journal of Soil and Water Conservation 32*(1):28–32.
Whitby, K. T. 1978. The physical characteristics of sulfur aerosols. *Atmospheric Environment 12*:135–159.
White-Stevens, R. 1977. Perspectives on fertilizer use, residue utilization and food production. In: *Food, Fertilizer and Agricultural Residues*, R. C. Loehr, ed. Ann Arbor Science, Ann Arbor, Mich., pp. 5–26.

Whittaker, R. H., and G. E. Likens. 1975. The biosphere and man. In: *Primary Productivity of the Biosphere*, H. Lieth and R. H. Whittaker, eds. Springer-Verlag, Berlin, pp. 305–328.

Whittaker, R. H., and G. M. Woodwell. 1969. Structure, production and diversity of the oak-pine forest at Brookhaven, New York. *Journal of Ecology 57*:157–174.

Wigley, T. M. L., P. D. Jones, and P. M. Kelly. 1980. Scenario for a warm, high-CO_2 world. *Nature 283*:17–21.

Wiklander, L. 1978. Interaction between cations and anions influencing adsorption and leaching. In: *Effects of Acid Precipitation in Terrestrial Ecosystems*, T. C. Hutchinson and M. Havas, eds. Plenum Press, New York, pp. 239–254.

Wilkins, E. T. 1954. Air pollution aspects of the London Fog of December 1952. *Quarterly Journal of the Royal Meteorological Society 80*:267–279.

Williams, C. N. 1975. *The Agronomy of the Major Tropical Crops*. Oxford University Press, London.

Williams, D. J., J. N. Carras, J. W. Milne, and A. C. Heggie. 1981. The oxidation and long-range transport of sulphur dioxide in a remote region. *Atmospheric Environment 15*:2255–2262.

Williams, J., ed. 1978a. *Carbon Dioxide and Society*. Pergamon Press, Elmsford, N.Y.

Williams, J. 1978b. Introduction to the climate/environment aspect of CO_2 (a pessimistic view). In: *Carbon Dioxide, Climate and Society*, J. Williams, ed. Pergamon Press, Elmsford, N.Y., pp. 131–139.

Wilson, J. W., V. A. Mohnen, and J. A. Kadlecek. 1982. Wet deposition variability as observed by MAP3S. *Atmospheric Environment 16*:1667–1676.

Winchester, J. W. 1980. Sulfate formation in urban plumes. *Annals of the New York Academy of Sciences 338*:297–309.

Windhorst, H. W. 1974. *Studien zur Waldwirtschaftsgeographie: Das Ertragspotential der Wälder der Erde*. Franz Steiner Verlag, Wiesbaden.

Wittwer, S. H. 1980. Carbon dioxide and climatic change: An agricultural perspective. *Journal of Soil and Water Conservation 35*:116–120.

Wolff, G. T., P. J. Lioy, H. Golub, and J. S. Hawkins. 1979. Acid precipitation in the New York metropolitan area: Its relationship to meteorological factors. *Environmental Science and Technology 13*:209–212.

Wolman, G. 1983. Soil erosion and crop productivity: a worldwide perspective. Erosion and Crop Productivity Symposium, Denver, Colo., March 1–3, 1983.

Wong, C. S. 1978. Atmospheric input of carbon dioxide from burning wood. *Science 200*:197–199.

Woodmansee, R. G. 1978. Additions and losses of nitrogen in grassland ecosystems. *BioScience 28*:448–453.

Woodwell, G. M. 1983. Biotic effects on the concentration of atmospheric carbon dioxide: A review and projection. In: *Changing Climate*. NAS, Washington, D.C., pp. 216–241.

Woodwell, G. M., and E. V. Pecan, eds. 1973. *Carbon and the Biosphere*. United States Atomic Energy Commission, Washington, D.C.

Woodwell, G. M., B. A. Houghton, and N. R. Tempel. 1973. Atmospheric CO_2 at Brookhaven, Long Island, New York: Patterns of variation up to 125 meters. *Journal of Geophysical Research 78*:932–940.

Woodwell, G. M., R. M. Whittaker, W. A. Reiners, G. E. Likens, C. C. Delwiche, and D. B. Botkin. 1978. The biota and the world carbon budget. *Science 199*:141–146.

World Energy Conference. 1974. *Survey of Energy Resources*. WEC, London.

World Energy Conference. 1978. *World Energy Resources 1985–2020*. IPC Science and Technology Press, Guildford, England.

World Health Organization. 1973. *Energy and Protein Requirements*. WHO, Geneva.

World Health Organization. 1976. *Selected Methods of Measuring Air Pollutants*. WHO, Geneva.

World Meteorological Organization. 1977. *Report of the Scientific Workshop on Atmospheric Carbon Dioxide*. WMO, Geneva.

World Meteorological Organization. 1979. *World Climate Conference*. WMO, Geneva.

World Power Conference. 1929. *Power Resources of the World (Potential and Developed)*. WPC, London.

Wright, R. F., and E. T. Gjessing. 1976. Changes in the chemical composition of lakes. *Ambio 5*:219–223.

Wu, Z. 1980. Solving the energy crisis from the viewpoint of energy science and technology. *Hongqi (Red Flag)* No. 17, pp. 32–43.

Yaverbaum, L. H. 1979. *Nitrogen Oxides Control and Removal.* Noyes Data Corporation, Park Ridge, N.J.

Yocom, J. E. 1979. Air pollution damage to buildings on the Acropolis. *Journal of the Air Pollution Control Association 29*:333–338.

Yocom, J. E. 1982. Indoor–outdoor air quality relationships. *Journal of the Air Pollution Control Association 32*:506–520.

Yoshinari, T. 1976. Nitrous oxide in the sea. *Marine Chemistry 4*: 189–202.

Young, R. A., and R. F. Holt. 1977. Winter-applied manure: Effects on annual runoff, erosion, and nutrient movement. *Journal of Soil and Water Conservation 32*:219–222.

Zelitch, I. 1971. *Photosynthesis, Photorespiration and Plant Productivity.* Academic Press, New York.

Zettwoog, P., and R. Haulet. 1978. Experimental results on the SO_2 transfer in the Mediterranean obtained with remote sensing devices. *Atmospheric Environment 12*:795–796.

Zhao, D., S. Mou, L. Chen, and K. Liu. 1981. Acid rain investigation in Beijing. *Huanjing Kexue (Environmental Science)* 2(2):50–54.

Zimmermeyer, G. 1978. Will hypotheses about properties of CO_2 affect energy conceptions? In: *Carbon Dioxide, Climate and Society,* J. Williams, ed. Pergamon Press, Elmsford, N.Y., pp. 275–278.

INDEX

Vaclav Smil is a professor in the Department of Geography at the University of Manitoba, Canada. He studied at the Carolinum University in Prague, Czechoslovakia, and received his doctorate in 1972 from The Pennsylvania State University. Before coming to the US to study, Dr. Smil served as a consultant for energy and environmental problems in his native Czecholovakia, concentrating on air pollution and power generation. In addition to writing over 100 papers, Dr. Smil has authored or edited six books, including *Biomass Energies* (Plenum, 1983). Currently, he is on the editorial boards of *Environmental Management* and *Current History*. His research interests include the global potential of biomass energies and their environmental impact, energy flows in the developing world, energy in food production, international energy developments and politics, and human interference in major biospheric cycles.

MIX
Papier aus verantwortungsvollen Quellen
Paper from responsible sources
FSC® C105338

FSC
www.fsc.org

If you have any concerns about our products,
you can contact us on
ProductSafety@springernature.com

In case Publisher is established outside the EU,
the EU authorized representative is:
Springer Nature Customer Service Center GmbH
Europaplatz 3, 69115 Heidelberg, Germany

Printed by Libri Plureos GmbH
in Hamburg, Germany